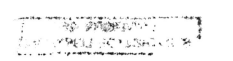

With complecte

Ian Barr

Additional tables for the hydraulic design of pipes, sewers and channels

D. I. H. Barr and HR Wallingford

 Thomas Telford, London

Published by Thomas Telford Services Ltd, Thomas Telford House, 1 Heron Quay, London E14 4JD, UK

Distributors for Thomas Telford books are
USA: American Society of Civil Engineers, Publications Sales Department, 345 East 47th Street, New York, NY 10017-2398
Japan: Maruzen Co Ltd, Book Department, 3-10 Nihonbashi 2-chome, Chuo-ku, Tokyo 103
Australia: DA Books and Journals, 11 Station Street, Mitcham 2131, Victoria

First published 1993

A catalogue record for this book is available from the British Library

ISBN: 0 7277 1667 0

© D. I. H. Barr and HR Wallingford, 1992

While all reasonable efforts have been made to ensure the accuracy of the information given in these Additional Tables, no warranty, express or implied, is given by the publishers or by the authors

Preface

These Additional Tables are intended to be used in conjunction with the existing HR Wallingford *Tables for the hydraulic design of pipes and sewers* (3rd, 4th or 5th editions). It is intended that the two sets of tables will be amalgamated to form the 6th edition of the Wallingford Tables.

The Additional Tables increase the range of diameters covered in the 5th Edition of the Wallingford Tables.

There is also a new, unified, table-based approach to the solution of uniform flow problems with non-circular flow cross-section. A wide range of conduit and channel shapes is covered by tables of properties based on unit size. This solution procedure is introduced on the basis of the use of the Colebrook-White equation, but it can also be used with the Manning equation. Additional tables and information to cover the use of the Manning equation are also provided.

The adoption of the unit size concept for tabulation also allows provision of a simple direct solution for critical depth. Illustration is given of the use of these Additional Tables in profile assessments of gradually varied flows.

D. I. H. Barr is greatly indebted to the Court, the Principal, the Department of Civil Engineering and the Research and Development Services of the University of Strathclyde for facilities and support granted during the period of his appointment as Honorary Professor. He would also like to express his gratitude to colleagues and to others for various helpful comments, for advice and for much invaluable assistance. In particular, the programming of Ronald Baron has been vital in the production of the various forms of table.

Users of these Additional Tables are invited to provide comments or corrections, particularly on conduit or channel shapes which are in common use but which are not covered.

Contents

References

Nomenclature

Tables within text
Table 1: Overall solution paths for uniform flow problems
Table 2: Values of multiplying factor for SU Colebrook-White equations
Table 3: Predictions of proportional depth in Form 1 egg-shape with range of extreme combinations of conditions
Table 4: Computation of M1 flow profile in trapezoidal channel

Figures within text
Fig. 1: Colebrook-White equation and direct solution approximations
Fig. 2: Solution of Colebrook-White equation in simplified usage mode (SU)
Fig. 3: Solution routes for uniform flow in non-circular cross-sections

Appendix 1: Recommended roughness values

Appendix 2: Typical values of Manning roughness coefficient n

Appendix 3: Velocity correction for variation in temperature

Appendix 4: Multiplying factors for discharges in pipes and lined tunnels

Table sequence A
Tables of Colebrook-White solutions
Diameters 2.400 m to 20.00 m

Table A34: k_s = 0.015 mm
Table A35: k_s = 0.030 mm
Table A36: k_s = 0.060 mm
Table A37: k_s = 0.150 mm
Table A38: k_s = 0.30 mm
Table A39: k_s = 0.60 mm
Table A40: k_s = 1.50 mm
Table A41: k_s = 3.00 mm
Table A42: k_s = 6.00 mm
Table A43: k_s = 15.0 mm
Table A44: k_s = 30.0 mm
Table A45: k_s = 60.0 mm
Table A46: k_s = 150 mm
Table A47: k_s = 300 mm
Table A48: k_s = 600 mm

Table sequence B
Values of proportioning exponents in equations (8), (9) and (10)

Table B1: Values of exponent x
Table B2: Values of exponent y

Table sequence C

Tables of properties of unit sections (and proportional flow details for circular pipes only)

Table sequence D

Values of nV and nQ for solutions using Manning equation

Table D1: Diameters 0.350 m to 0.825 m

Table D2: Diameters 0.900 m to 2.100 m

Table D3: Diameters 2.400 m to 20.00 m

Introduction

The Wallingford Charts and the Wallingford Tables
The first editions of the Wallingford Charts[1] and the Wallingford Tables[2] were published in 1958 and 1963 respectively. Corresponding 6th and 5th editions were published in 1990. The continuing availability of these Charts and Tables has greatly facilitated the general adoption of the Colebrook-White resistance equation[3] in the UK and elsewhere. This equation relates specifically to steady turbulent flow in circular pipes flowing full, but more and more it is also adopted for flow calculation within other shapes of pipes or channels. However, the Tables have concentrated on circular pipe flow, with coverage of part-full flow in such pipes and of flow in two forms of egg-shape.

The Additional Tables
This volume is a companion to Tables 1 - 33 of the 3rd, 4th or 5th editions of the Wallingford Tables, these being the tables of Colebrook-White solutions. For such tables, the coverage of diameters is now increased to be as follows: WT before table numbers indicating the Wallingford Tables; and A indicating in this volume

0.050 m (50 mm) - 0.300 m (875 mm) (WT1 - WT11)
0.350 m (350 mm) - 0.825 m (875 mm) (WT12 - WT22)
0.900 m (900 mm) - 2.100 m (2100 mm) (WT23 - WT33)
2.400 m (2400 mm) - 20.00 m (20,000 mm) (A34 - A48)

The increase to the existing cover is required for wider application of a new approach to flows in non-circular conduits and channels which is introduced here. For this, the tables of sequence C give geometrical properties of channel shapes and hydraulic properties for a wide range of conduit and channel shapes. They are based on a conduit or channel of unit size. A linear multiplying factor, M, can then be used to give values for any size of conduit or channel of the given geometry.

Table 1 provides a summary of the main capacities of this extended tabular approach. In Table 1 there is reference to table sequences B and D. The first of these aids interpolation between entries in the Colebrook-White tables. The second permits the use of the Manning[4] equation in conjunction with tables C should the user so prefer. The latter equation does continue to be much used, especially for open channels.

Review of hydraulic resistance

The Colebrook-White equation
In 1939 Colebrook[3] published this equation for turbulent flow in circular pipes flowing full. It followed from the smooth and rough turbulent logarithmic resistance laws for circular tubes. These had been evaluated experimentally by Nikuradse[5,6], after work by Prandtl

1

TABLE 1 : Overall solution paths for uniform flow problems

Applications	Resistance equations	Table sequences				Numerical (Fig. 2; Table 2)
		A	B	C	D	
Water or sewage at normal temperature in circular pipe, full bore flow.	C-W equation or Manning eq.	◎	○		◆	
As above but any fluid.	C-W equation					△
Water or sewage in circular pipe flowing part-full. Solution for discharge or for proportional depth.	C-W equation	◎	○	C1(a)		
Water or sewage in non-circular flow sections, including part-full circular pipes. All solutions covered.	C-W equation or Manning eq.	◎	○	■ ■	◆	
As above but any fluid.	C-W equation			■		△

◎ Tables A of Colebrook-White solutions including Tables WT 1-33 from the 3rd, 4th or 5th Editions of the existing Wallingford Tables. Use for water at normal temperature (15°C), and with enhancement of range of water temperature using Appendix 3.

○ Tables B give a full range of proportioning exponents for use in interpolations of Tables A (including tables WT 1-33), if and as required.

■ Tables C give details of geometric and hydraulic properties of cross-sections on a unit size basis. Specific illustration of the combination of these tables with Tables A or D, or with numerical solution of the Colebrook-White equation, is give in Fig. 3.

◆ Tables D allow of tabular solution of the Manning equation in a parallel mode to that of the Colebrook-White equation with tables A and B. Use for water or sewage at normal temperature: the Manning equation is less appropriate for smoother surfaces.

△ Figure 2 with Table 2 provides for non-tabular solution of the Colebrook-White equation for full bore flow of fluid of any viscosity. The method given is the simplest possible once basic SI units are adopted. Note that roughness size, k_s, is entered into the equations in metres. Thereby, the general method for non-circular flow sections, using Tables C, can be widely applied. This includes Colebrook-White solutions for the of water in smaller containments should tables WT 1-33 not be available. Alternatively, the Wallingford Charts also can be so used.

Note: The use of these tables for gradually varied flow assessments and in treatment of different roughnesses around the wetted perimeter is covered in the text.

and von Karman. The equations describing smooth and rough turbulent flow are

$$\frac{1}{\sqrt{\lambda}} = 2 \log \left\{ \frac{R\sqrt{\lambda}}{2.51} \right\} \qquad \text{(Smooth turbulent flow)} \qquad (1)$$

$$\frac{1}{\sqrt{\lambda}} = 2 \log \left\{ \frac{3.7D}{k_s} \right\} \qquad \text{(Rough turbulent flow)} \qquad (2)$$

The following combination provides a transition between the individual laws. It is known as the Colebrook-White equation to indicate the influence of C. M. White, Colebrook's collaborator and former research supervisor.

$$\frac{1}{\sqrt{\lambda}} = -2 \log \left\{ \frac{k_s}{3.7D} + \frac{2.51}{R\sqrt{\lambda}} \right\} \qquad (3)$$

Here :

Friction factor $\lambda = 2(Sg)D/V^2 = 1.2337(Sg)D^5/Q^2$ (4)

Reynolds number $R = VD/\nu = 4Q/\pi\nu D$ (5)

Mean velocity $V = Q/(\pi/4)D^2$ (6)

where D, k_s, Q (Sg), and ν are diameter, roughness size, discharge, product of piezometric gradient and acceleration due to gravity and kinematic viscosity respectively.

If Eqs (4), (5) and (6) are substituted in Eq. (3), one obtains

$$\frac{V}{\sqrt{(2SgD)}} = \frac{0.9003 Q}{\sqrt{(Sg)D^{2.5}}} = -2 \log \left\{ \frac{k_s}{3.7D} + \frac{1.775 \nu}{\sqrt{(Sg)D^{1.5}}} \right\} \qquad (7)$$

Simplified forms of the Colebrook-White equation
The SU (simplified usage) equations of Fig. 2 derive from the non-dimensional equations[3,7,8] of Fig. 1 and apply to fresh water at 15°C and hence of kinematic viscosity 1.141×10^{-6} m^{3}s^{-1}. Where the kinematic viscosity differs significantly from this value, a simple correction detailed in Table 2 may be applied.

Tables of Colebrook-White solutions (tables WT1-WT33 and A34-A48)
The original concept of the Wallingford Tables was to provide tabulated solutions to the Colebrook-White equation. Each table relates to a particular roughness value. Within each table, solutions are given for a range of diameters and gradients. For each of the

TABLE 2 : Values of multiplying factor for SU Colebrook-White equations

Fluid	Density, ρ (kgm^{-3})	Kin. visc., ν (m^{2}s^{-1}×10^6)	Factor
Water (0°C)	999.9	1.792	1.57
Water (15°C)	999.1	1.141	**1.00**
Water (30°C)	995.7	0.804	0.705
Water (60°C)	983.2	0.477	0.418
Water (100°C)	958.4	0.296	0.259
Paraffin oil (20°C)	800	2.375	2.08
Mercury (20°C)	13546	0.114	0.10
Crude oil (-10°C)	925	2000	1753
Crude oil (20°C)	855	74	64.9
Petrol (0°C)	716	0.8	0.70
Petrol (20°C)	716	0.59	0.52
Petrol (60°C)	716	0.4	0.35
Fuel oil (20°C)	940	1200	1052
Glycerin (20°C)	1258	1183	1037
SAE 10 oil (20°C)	918	89.3	78.3
SAE 30 oil (20°C)	918	479	420
Air (Atmos. 20°C)	1.205	14.9	13.1
Hydrogen (Atmos. 20°C)	0.0839	107	93.8

Note: SU Colebrook-White equations are given in Fig. 2. Then the above values are used as multiplier at * in these equations, for fluids other than water at normal temperature (15°C).

Figure 1: Colebrook-White equation and direct solution approximations

Solution for Q (or V) (i.e. Colebrook-White equation)

$$\frac{V}{\sqrt{(2SgD)}} = \frac{0.9003\,Q}{\sqrt{(Sg)}D^{2.5}} = -2\log\left\{\frac{k_s}{3.7D} + \frac{1.775\,\nu}{\sqrt{(Sg)}D^{1.5}}\right\} \tag{7}$$

Solution for S (Q as input variable for flow) (Barr[9] approximation)

$$\frac{0.9003\,Q}{\sqrt{(Sg)}D^{2.5}} = -1.9\log\left\{\left(\frac{k_s}{3.7D}\right)^{1.053} + \left(\frac{4.932\,\nu D}{Q}\right)^{0.937}\right\}$$

Solution for D (Q as input variable for flow) (Pham[10] approximation)

$$\frac{0.9003\,Q}{\sqrt{(Sg)}D^{2.5}} = -1.8844\log\left\{\frac{0.365(Sg)^{0.2}\,k_s}{Q^{0.4}} + \frac{3.55\,\nu}{Q^{0.6}\,(Sg)^{0.2}}\right\}$$

D is diameter of circular pipe flowing full
V is mean velocity
Q is discharge
S is piezometric gradient
g is acceleration due to gravity
ν is kinematic viscosity
k_s is equivalent sand roughness size

The foregoing variables are in any coherent system of units such as pure SI (kg-m-s units)

Figure 2: Colebrook-White equations in simplified usage mode (SU)

For SI units (D and k_s in m, Q in m^3s^{-1}); fluid flowing in the first instance is water at 15°C (kinematic viscosity 1.141×10^{-6} m^2s^{-1}).

Solution for Q

$$\frac{Q}{\sqrt{S}\ D^{2.5}} = -6.957 \log \left\{ \frac{k_s}{3.7D} + \frac{(0.647 \times 10^{-6})^*}{\sqrt{S}\ D^{1.5}} \right\}$$

Solution for S

$$\frac{Q}{\sqrt{S}\ D^{2.5}} = -6.61 \log \left\{ \left(\frac{k_s}{3.7D} \right)^{1.053} + \left(\frac{(5.63 \times 10^{-6})^* D}{Q} \right)^{0.937} \right\}$$

Solution for D (Q given as flow variable)

$$\frac{Q}{\sqrt{S}\ D^{2.5}} = -6.555 \log \left\{ \frac{0.576\ S^{0.2}\ k_s}{Q^{0.4}} + \frac{(2.566 \times 10^{-6})^*}{Q^{0.6}\ S^{0.2}} \right\}$$

For other viscosities, multiply at * by the factor :-

$$\frac{\text{actual kinematic viscosity in } m^2s^{-1}}{1.141 \times 10^{-6} \ m^2s^{-1}}$$

For other than circular pipes flowing full, D and Q are the values found for the equivalent pipe.

combinations of roughness size, diameter and gradient, the mean velocity and the discharge for water at 15°C may be obtained.

In this volume, diameters are given in metres and roughness sizes in millimetres. Ratio values, D/k_s and the like, must be obtained with consistent units.

The Colebrook-White equation applies to turbulent flows. Velocity and discharge values are omitted where Reynolds numbers (Eq. (5)) are less than 2000. Also excluded are solutions corresponding to values of D/k_s less than about 5.

The linear measure of surface roughness

In his experiments, Nikuradse[6] used pipes which were roughened by a uniformly graded sand glued to the surface. The diameter of those grains has provided a standard of comparison against which other surfaces may be evaluated. Thus the roughness of another surface is given as the size of sand which, as a uniform coating, would give the same resistance under rough turbulent flow. Such k_s values are available for most of the surfaces likely to be used in drainage works. The values are listed in Appendix 1, which is based on recommendations made by Colebrook[3], Rouse[9], King[10], Lamont[11], and Perkins and Gardiner[12].

The Colebrook-White equation is most appropriate where the roughness consists of separate protuberances of random height and spacing. Some classes of pipe, not having this form of roughness, do not follow the same resistance function in the transition zone between smooth-turbulence and rough-turbulence. However in such cases flow is often rough turbulent, or close to this, in the circumstances of most concern.

Design of circular section pipelines and sewers

Use of the Tables

For a given roughness, the Tables relate values of pipe diameter, gradient and discharge or velocity. If any two are specified, the Tables provide the appropriate value of the third variable.

> *What is the mean velocity and the discharge in a 4.0 m diameter tunnel with roughness size 0.6 mm flowing full under a piezometric gradient of 0.00020? (Numerical solution gives 1.0860 ms^{-1} and 13.6468 m^{3}s^{-1})*

From Table A39, read the entries for the stated combination of diameter and gradient, i.e. 1.086 ms^{-1} and 13.647 m^{3}s^{-1}

Interpolation between entries

If intermediate values between the entries in the Table are required then interpolation should be used. Simple linear interpolation may be sufficient in many cases.

In the preceding example, the pipe size is increased to 4.5 m and the gradient to 0.00021, roughness size remaining the same. Obtain the new flow values. (Numerical solution gives 1.195 ms^{-1} and 19.012 m^{3}s^{-1})

On Table A39, linear proportioning between the values for 4.4 and 5.0 m diameter and 0.00020 and 0.00022 gradients gives a velocity of 1.193 ms^{-1} and a discharge of 19.495 m^{3}s^{-1}. If linear interpolation is not sufficiently accurate more sophisticated methods of interpolation can be adopted.

Tables of proportioning exponents (Table sequence B)
As a more accurate alternative to linear interpolation, the tables of sequence B give proportioning exponents. The interpolation is based on the values of D/k$_s$ and of DV for the table point which is selected as the base for the interpolation. The value of the product DV is the convenient substitute for Reynolds number where the value of kinematic viscosity is fixed. Reynolds number values are given also in the tables of sequence B.

In the proportioning equations (Eqs (8)-(10)), given values, table values and required values are indicated by subscripts G, T and R respectively.

$$Q_R = Q_T \left[\frac{D_G}{D_T} \right]^x \left[\frac{S_G}{S_T} \right]^{1/y} \left[\frac{(k_s)_T}{(k_s)_G} \right]^u \tag{8}$$

For V_R, Q_T is replaced by V_T and the exponent x is replaced by (x - 2)

$$S_R = S_T \left[\frac{Q_G}{Q_T} \right]^y \left[\frac{D_T}{D_G} \right]^z \left[\frac{(k_s)_G}{(k_s)_T} \right]^v \tag{9}$$

$$D_R = D_T \left[\frac{Q_G}{Q_T} \right]^{1/x} \left[\frac{S_T}{S_G} \right]^{1/z} \left[\frac{(k_s)_G}{(k_s)_T} \right]^w \tag{10}$$

8

Solution using proportioning exponents

For water at normal temperature, determine the discharge (Q_R) in a 0.56 m (560 mm) diameter pipe (D_G) of 1 mm roughness size ($(k_s)_G$) under a piezometric gradient of 0.00525 (S_G) (Numerical solution gives 0.3901 m^3s^{-1}).

From Table WT18, the discharge for a 0.525 m (525 mm) diameter pipe (D_T) with 0.60 mm roughness size ($(K_s)_T$) under 0.005 gradient (S_T) is read as 0.342 m^3s^{-1}(Q_T).

At this table point

$$D/k_s = 0.525/0.0006 \text{ (or } 525/0.6) = 875$$

$$DV = 0.525 \times 1.578 = 0.828$$

Thus from Tables B1, B2 and B4, x, y and u values are read as 2.63, 1.98 and 0.12 respectively. Then substituting in Eq. (8)

$$Q_R = 0.342 \left[\frac{0.560}{0.525} \right]^{2.63} \left[\frac{0.00525}{0.0050} \right]^{1/1.98} \left[\frac{0.6}{1.0} \right]^{0.12}$$

$$= 0.342 \times 1.1850 \times 1.0249 \times 0.9405 = 0.3907 \ m^3s^{-1}$$

Multiplying factors on tabulated discharges for standard but non-tabulated diameters

Estimates are often required of pipe full discharges in a pipe of a standard diameter which is not included in the main tabulations. Close estimates can be made quite simply, using the multiplying factors provided in Appendix 4. These were obtained using 2.62 as a medial value of exponent x.

Estimate the discharge in a 1.275 m pipe with roughness size 0.30 mm, under a gradient of 0.0012. (Numerical solution gives 1.8207 m^3s^{-1})

From table WT28 for k_s of 0.30 mm, read 1.554 m^3s^{-1} for 1.200 m diameter and gradient 0.0012. From Appendix 4, read factor 1.17 for the adjustment to 1.275 m (D_G) from 1.200 m diameter. The estimate is then 1.554 × 1.17 = 1.82 m^3s^{-1}.

Perimeters involving dissimilar roughness

The wetted perimeter of a channel or pipe may be composed of surfaces of dissimilar roughness. This occurs, for example, where sliming has taken place only below the usual water level or where the invert is of concrete and the walls of brick. Then there are two ways[1,2]

9

of computing discharges using these Tables.

The first method is based on the concept of an equivalent grain roughness for the whole perimeter. It utilises the expression

$$k_s = p_1 k_{s1} + p_2 k_{s2} \qquad (11)$$

where p_1, p_2 denote the proportions of the total perimeter occupied by surfaces 1 and 2, and k_{s1}, k_{s2} denote the equivalent sand grain roughness of surfaces 1 and 2.

This method may be used in situations where the difference in roughness values is not excessive and where the two surfaces occupy similar proportions of the total wetted perimeter. It will also give approximate answers outside these ranges. These ranges can be defined (somewhat arbitrarily) as

$$0.05 < k_{s1}/k_{s2} < 20 \text{ and } 0.33 < p_1/p_2 < 3.0$$

This method provides a direct solution to the problem of composite roughness. The tabular system can be used without the need to resort to a successive approximation technique.

The second method is based on the concept of an equivalent friction factor for the whole wetted perimeter. It utilises the expression

$$\lambda_s = P_1\lambda_1 + P_2\lambda_2 \qquad (12)$$

where the suffixes denote surfaces 1 and 2 as before. The method uses the following approximation[13] to the Colebrook-White equation

$$\frac{1}{\sqrt{\lambda}} = -2 \log \left\{ \frac{k_s}{3.7D} + \frac{5.1286}{R^{0.89}} \right\} \qquad (13)$$

Later the use of an equivalent pipe diameter is explained, and on this basis the RHS of Eq. (13) can be written

$$\left\{ \frac{k_s}{3.7D_{ep}} + \frac{5.1286 \, \nu^{0.89}}{(D_{ep}V)^{0.89}} \right\} \qquad (13a)$$

Or, for water at normal temperature

$$\left\{ \frac{k_s}{3.7D_{ep}} + \frac{2.63 \times 10^{-5}}{(D_{ep}V)^{0.89}} \right\} \qquad (13b)$$

This second method is not so restricted as the first method, but errors will occur where there are large differences in the grain roughness of the two surfaces, say $k_{s1}/k_{s2} < 0.01$ or $k_{s1}/k_{s2} > 100$. The method requires a successive approximation approach as follows

(i) Approximate to the composite value of k_s using $k_s = p_1 k_{s1} + p_2 k_{s2}$.

(ii) Use the appropriate tables to determine the flow, Q, and also the mean velocity, V, for the actual pipe flowing full, or for the equivalent pipe.

(iii) Determine values C_1 and C_2 of the expression (13b) by adopting k_{s1} and k_{s2} respectively.

(iv) Determine C from
$$\log C = \frac{\log C_1}{\sqrt{\{ p_1 + (\log C_1 / \log C_2)^2 \, p_2 \}}}$$

(v) Calculate the composite k_s value from
$$k_s = 3.7 \, D_{ep}\left\{ C - \frac{2.63 \times 10^{-5}}{(D_{ep}V)^{0.89}} \right\}$$

(vi) Use the appropriate tables to determine the flow, Q.

(vii) Repeat (iii) to (vi) until Q changes by less than the required tolerance.

Non-circular cross-sections of flow

Calculation of discharge and velocity in part-full circular pipes
Ackers[1] defined the transitioning parameter θ. The value of θ determines the pattern of proportional flow variation in circular tubes of widely varying degrees of roughness. With SI units and water at standard temperature 15°C, θ is evaluated as follows

$$\theta = \left\{ \frac{k_s}{D} + \frac{1}{3600 \, D \, S^{1/3}} \right\}^{-1} \qquad (14)$$

Mid-column values of θ are shown at the bottom of each column of velocity and discharge values on the basic Tables. Table C1(a) shows values of proportional discharge against proportional depth for a range of values of θ, for circular pipes. The following example shows how data from the tables are combined.

Estimate the discharge at 0.66 proportional depth in a 3.40 m diameter conduit at a gradient of 0.0008. Roughness size is 1.5 mm. (Numerical solution 12.574 m^3s^{-1})

The full conduit discharge read from the 3.400 m diameter column at

0.0008 gradient on Table A40 is 16.401 m^3s^{-1}, with θ shown as 760 as medial for the column. On Table C1(a), the proportional discharge interpolates as 0.7665 giving 12.571 m^3s^{-1}. Table C1 gives unit sectional areas, i.e. 0.5499 m^2 at 0.66 proportional depth. If an estimate of mean velocity is also required, the discharge of 12.571 m^3s^{-1} should be divided by (0.5499 x 3.40^2) m^2.

Calculation of depth in part-full circular pipes
The converse problem of estimating depth, given discharge, also can be solved using Table C1(a). Where necessary, interpolation can be used between table entries. However, for direct assessment of gradient for a part-full circular pipe, or of size of circular pipe to run at a stipulated proportional depth, a different approach should be adopted. This utilises the concept of 'unit size' and is now described.

Hydraulic equivalence
For any steady flow in a non-circular section one can find an equivalent circular cross-section where the flow characteristics are the same, in terms of hydraulic gradient and mean velocity, because the hydraulic mean depth (hydraulic radius) is the same. This includes part-full flow in a circular pipe. According to Chezy[14], the hydraulic mean depth is A/P, which is one quarter of the diameter of the equivalent pipe. Following Johnson[15] and Ackers[16], one can make the following comparison between a non-circular flow section and its equivalent pipe. Since

$$V = V_{ep}$$
$$Q/A = Q_{ep} / [(\pi/4)(4A/P)^2]$$

Thus

$$Q_{ep} = Q \times 4\pi A/P^2 \tag{15}$$

Here A and P are cross-sectional flow area and wetted perimeter of the non-circular flow section, V and Q represent mean velocity and discharge, and the subscript ep indicates equivalent pipe in the sense of providing equivalent velocity, though not, of course, equivalent conveyance. The discharge conversion factor $J = 4\pi A/P^2$ must also apply to the relation between flow areas, because discharge is mean velocity times flow area.

Therefore the core of 'solving flow in a non-circular cross-section reduces to finding the equivalent circular pipe and then solving for flow in this equivalent pipe, using the methods already presented.

'Unit size' measures for shapes of conduits and channels
For any geometrically similar series of conduit or channel shapes, a 'unit size' can be defined in terms of a key dimension. Then, a multiplying factor M specifies the size of an actual example of the shape within the geometrically similar series. Also for each proportional or relative depth value, there is a 'unit' case wetted perimeter value P_u and a 'unit' case cross-sectional area of flow A_u. Hence for an actual example, MP_u and M^2A_u are the size of the wetted

12

perimeter and of the area, respectively, at the specified proportional or relative depth. It follows that there is a 'unit' case value of hydraulic mean depth (hydraulic radius), i.e. $R_u = A_u/P_u$ and hence a 'unit' case value of diameter of equivalent pipe, i.e. $D_{ep(u)} = 4R_u$.

In these Tables, the sizes of $D_{ep(u)}$ and P_u, etc., are quoted in metres, while the values of the other variables in the tables of sequence C are quoted in metre-second units as appropriate. Thus with values of $D_{ep(u)}$ made available in table sequence C

$$D_{ep} = MD_{ep(u)} \qquad (16)$$

In using the Tables, Eqs (15) and (16) provide the essential relationships between a given non-circular flow circumstance and the corresponding equivalent pipe, whether a free surface is involved or not. The gradient is the same for both cases, as is the roughness size and the kinematic viscosity. Thus the completing relationship for the calculations is

$$S_{ep} = S \qquad (17)$$

Tables of properties of unit sections (Table sequence C)
Details of the shape are given at the start of each table. The cross-sections fall into groups as follows.

(i) Conduits running full or part-full. Here, overall height is the natural choice for unit size, leading to the conventional definition of proportional depth of flow. All the section shapes most commonly used in the UK and in North America for drainage and sewerage conduits are covered. For the UK, there is included the coverage of shapes by Wallingford Software's WALLRUS suite of programs[17] and by the Water Research Centre's Sewerage Rehabilitation Manual[18]. Also, the effect of relining of egg-shapes is treated. The practices in North American sewerage have been illustrated by Metcalf and Eddy[19] and summarised into key shapes by both Metcalf and Eddy[20] and Babbitt and Baumann[21]. These summarising shapes are included in the Tables, as are a range of standard corrugated metal conduit shapes[22,23].

(ii) Narrow channels and related conduits. For narrow channels, breadth is the natural choice for unit size, and relative depth of flow is the ratio of depth of flow to this chosen breadth. Limitations on space require that conduit shapes such as closed rectangles and, particularly, arch culverts and ovals be grouped with rectangular and U-shaped channels respectively. This allows that one table only need be allocated to each such case, with the table dealing with a range of relative geometries but always for conduit full flow. An example of this device is the relating of Tables C12 and C13 for U-shaped capped and oval respectively, to Table C11 for U-shaped free surface. Then for part-full arch culverts and ovals and for box-culverts, the effect of the soffit arching, or of the upper splays, is not included for nearly full conditions. However, in all cases there is coverage of a wide range of relative proportions for the full condition.

(iii) Wide open channels. To deal with wide, shallower flowing, open channels, it is more expedient to define arbitrarily a unit depth in relation to the specified unit geometry. This applies to canal and river circumstances in Tables C61 to C88 and accommodates sufficient detail within the standard table arrangement for cases of small relative depths in terms of channel breadth. There are three series of cross-sections. Firstly, there are wider trapezoidal sections termed 'Regime' in Tables C61-66. Secondly, there is a series with a semi-circular bed between notional slope change points between bed and banks in Tables C70-78. This concave element of the cross-section always occupies 0.2 of the depth range covered in the table. There is a range of ratios of the depth of the concave section to its width. For example '10% concave bed river' dips one-tenth of the distance between the slope change points. Thirdly, there is a series with tangential meetings of side slopes and a circularly curved element through the centre line in Tables C79-88. Hence '1.0 to 1 tangent river', where the side-slope meeting the curve is the defining factor.

(iv) Triangular open channels. For the case of triangular channels, determination of depth of flow is the same as determination of size. Table C89 is modified to show side-slope instead of proportional depth. In consequence, there is in this case a direct solution for side-slope, given discharge and required depth of flow, as well as the alternative solutions for discharge or for depth of flow.

(v) Flow in ducts. Table C90 links the 'unit size' approach for non-circular flow sections with some standard cases of flow in ducts. There is restriction of application to flows that are clearly turbulent. For refinement of the estimates in cases (a), (c) and (d) of Table C90, material from ESDU International[24] should be consulted, with respect to adjustment of value of S_{ep}.

There may be required direct solution for depth of flow in a part-full conduit, or in a channel, where shape, size, gradient and roughness are known. This involves comparison of the given discharge with that occurring at a defined depth in the same conduit or channel. Tables C include discharge ratios for this purpose. The proportional depth adopted as base is shown at the start of each table. Such ratios are invariant for any given shape when the Manning equation is adopted, but may vary slightly with change of values of Reynolds number and relative roughness when the Colebrook-White equation is used. Thus, for the latter case, there is shown the adopted medial conditions for which the calculation of the tabulated ratios has been made. Selection of these medial conditions has been made taking into account typical usage of the tables. By comparing the values using both the Colebrook-White equation and the Manning equation, it can be seen that, in general, they provide very similar results.

Finding discharge in a rectangular open channel
The steps for the 'unit size' method of finding a discharge are as follows.

FIGURE 3: Solution routes for uniform flow in non-circular cross-sections

For the three explicit problems of finding (i) discharge (Q), (ii) gradient (S) or (iii) size (factor M), the proportional or the relative depth (Y_N) is known. Hence the values of equivalent diameter for the unit case ($D_{ep(u)}$) and of the equivalent discharge factor (J) can be read from the appropriate table of sequence C.

Then (i) *Find Q*:

$$M \times D_{ep(u)} = D_{ep} \searrow \qquad k_s$$
$$\downarrow$$
$$\boxed{\text{C-W}} \rightarrow Q_{ep} = Q \times J$$
$$S = S_{ep} \nearrow \qquad\qquad \text{Hence Q}$$

(ii) *Find S*:

$$M \times D_{ep(u)} = D_{ep} \searrow \qquad k_s$$
$$\downarrow$$
$$\boxed{\text{C-W}} \rightarrow S_{ep} = S$$
$$Q \times J = Q_{ep} \nearrow$$

(iii) *Find size (i.e. factor M)*:

$$Q \times J = Q_{ep} \searrow \qquad k_s$$
$$\downarrow$$
$$\boxed{\text{C-W}} \rightarrow D_{ep} = M \times D_{ep(u)}$$
$$S = S_{ep} \nearrow \qquad\qquad \text{Hence M}$$

For the inherently implicit problem of finding (iv) normal depth (y_N) in a channel of known shape and size, find Q_s by route (i) above, where Q_s corresponds to that proportional, or that relative, depth which is specified for the medial condition on the appropriate table of sequence C, and using the gradient and roughness size stipulated for the problem. Evaluate Q / Q_s where Q is the stipulated discharge.

Then (iv) *Find y_N*: $Q / Q_s \rightarrow$ value of Y_N on the table of sequence C.
Hence $y_N = M \times Y_N$

Note: The Colebrook-White element of a solution, is indicated by the block containing "C-W". This may be aided by Tables, by Charts or be accomplished by solution of an equation. For the Manning equation, the same overall routes apply, with tabular or numerical solution, and with the Manning discharge ratios from the tables of sequence C for problem (iv).

1) From table sequence C determine, for the given shape, $D_{ep(u)}$ and J.
2) Determine the multiplying factor M between 'unit' size used in Table C and real size.
3) Determine the equivalent diameter of circular pipe, D_{ep} using Eq. (16).
4) Use the main Tables to determine the discharge for pipe with diameter D_{ep} and given slope and roughness size, i.e. Q_{ep}.
5) Determine the required discharge $Q = Q_{ep}/J$.

Estimate the uniform flow discharge at 1.20 m depth in a 2.40 m wide rectangular channel where the gradient is 0.0020 and the roughness size is 1.50 mm. (Numerical solution 6.649 m^3s^{-1})

1) The relative depth is 0.50 and from Table C13 $D_{ep(u)}$ and J are 1.00 m and 1.5708 respectively.
2) From the diagram at the head of Table C13 the value of M is 2.40.
3) Then the diameter of the equivalent circular pipe, D_{ep}, is 2.40 x 1.00 = 2.40 m.
4) From Table A40 for k_s of 1.50 mm, the discharge for 2.40 m diameter and 0.002 gradient is 10.445 m^3s^{-1}, i.e. Q_{ep}.
5) The estimated discharge in the channel is Q_{ep}/J, i.e. 10.445/1.5708 = 6.649 m^3s^{-1}, or in practical terms, 6.6 m^3s^{-1}.

This solution route is shown schematically in Fig. 3, together with the solutions for gradient and for the size of conduit or channel (of stipulated proportional depth of flow where there is a free surface).

Solutions for egg-shape sewer

Finding (i) discharge, or (ii) gradient, or (iii) size where proportional depth is stipulated

In the case of the rectangular channel just treated, both the gradient and the diameter of the equivalent pipe corresponded to table values. Typically this is often true for gradient but rarely so when equivalent diameter is the unknown. To illustrate solution procedures more generally, consider a Form 1 egg-shape (see Table C2). The overall height is 1.8 m, so M is 1.8. At a depth of flow of 0.540 m (i.e. proportional depth (Y_N) of 0.30) with roughness size of 15 mm (k_s) and at gradient 0.001 (S), the numerical solution for discharge is 0.2222 m^3s^{-1} (compared with 1.3839 m^3s^{-1} for just full flow).

For 0.30 proportional depth, Table C2 gives the values of $D_{ep(u)}$ and J as 0.5123 m and 1.9042 respectively. From Fig. 3, these values are seen both to be needed for each of the three examples, as follows.

(i) To estimate the discharge Q, given gradient, size of conduit, roughness size and proportional depth

Substituting in Eq. (16)

16

$$D_{ep} = 1.8 \times 0.5123 = 0.9222 \text{ m}$$
$$S_{ep} = 0.001 \text{ (given)}$$

Here, the stipulated gradient corresponds to a table value, on the Colebrook-White solution tables, as does the roughness size. Thus the nearest table value on Table WT33 for k_s of 15 mm is the discharge for 0.900 m (900 mm) diameter and 0.001 gradient. It is 0.396 $m^3 s^{-1}$, and at this table point $D/k_s = 60$ and $DV = 0.560$, giving (on Table B1) exponent x as 2.69.

$$Q_{ep} = 0.396 \times [0.9222 / 0.900]^{2.69} \quad \text{(i.e. from Eq. (8))}$$

$$= 0.4230 \text{ m}^3 s^{-1} = Q \text{ J} \quad \text{(i.e. from Eq. (15))}$$

Then $Q = 0.4230 / 1.9042 = 0.2221 \text{ m}^3 s^{-1}$

(ii) To estimate the gradient S, now given that Q = 0.2222 m³s⁻¹)

$$D_{ep} = 0.9222 \text{ m (as before)}$$
$$Q_{ep} = 0.2222 \times J = 0.4231 \text{ m}^3 s^{-1}$$

The discharge value on Table WT33, for 0.900 m (900 mm) diameter, 15 mm roughness size and gradient 0.0011, is 0.416 $m^3 s^{-1}$. At this point D/k_s remains as 60 and DV is 0.588, giving y and z as 2.00 and 5.37 from Tables B2 and B3 respectively.

Substituting in Eq. (9)

$$S_{ep} = 0.0011 \left[\frac{0.4231}{0.525} \right]^{2.00} \left[\frac{0.900}{0.0050} \right]^{5.37}$$

$$= 0.000998 = S$$

(iii) To find directly the size of the Form 1 egg-shape, k_s = 1.50, which will run at 0.30 proportional depth with Q = 0.2222 m³s⁻¹ and S = 0.001

$$Q_{ep} = 0.4231 \text{ m}^3 s^{-1} \text{ (as before)}$$
$$S_{ep} = S = 0.001$$

The table point is that for 0.900 m (900 mm) diameter (and gradient 0.0010) on Table WT33 for roughness size of 15 mm. This shows a discharge of 0.396 $m^3 s^{-1}$. Then D/k_s and DV are 60 and 0.560 respectively, with the x value remaining as 2.69.

Substituting as relevant in Eq. (8)

$$D_{ep} = 0.900 \times [0.4229 / 0.396]^{1/2.69} = 0.9224 \text{ m}$$

However

$$D_{ep} = M \, D_{ep(u)} = M \times 0.5132$$

This gives scale factor M as 1.797 on the basis of 1.000 m unit height of egg-shape. Thus the required egg-shape is estimated as 1.797 m (1797 mm) in overall height.

Finding depth of flow in a conduit of specified boundary shape and size, with discharge, gradient and roughness size fixed.

Consider again the egg-shape sewer example. We wish to find the depth corresponding to a discharge of 0.2222 m^3s^{-1}. Using the unit size method for estimation of discharge, as already illustrated, the discharge at 0.60 proportional depth in this particular conduit is estimated as 0.8074 m^3s^{-1}. Thus the discharge of 0.2222 m^3s^{-1} as stipulated gives a $Q/Q_{0.60}$ ratio of 0.2752. From Table C2, this corresponds almost exactly to a proportional depth of 0.30 and hence to a depth of flow of 1.80 x 0.30 = 0.540 m.

The values of the Colebrook-White discharge ratio in Table C2 were calculated for the case of a 1.5 m high egg-shape at a gradient of 0.001 and with a roughness size of 3 mm. For changes in conduit size, gradient or roughness size, the corresponding changes in discharge ratio are small. Fig. 3 includes an abstract of the foregoing route for determination of normal depth.

Table C1(b), which applies directly to circular pipes, provides a means of final corrections to estimates of proportional depth when conditions vary from medial as defined by variation of the obtaining value of θ. The pattern of corrections shown serves to emphasise the small changes in proportional flows caused by large changes in conduit size, gradient and roughness size. Indeed, because of its non-dimensional nature, the values given on Table C1(b) can be applied to quite different basic shapes as in Table 3. This demonstrates that the methods for estimation of depth of flow give accuracy of solution which is well within normal design requirements.

Solutions for trapezoidal open channel

Find the size (i.e. bottom width) of a 45° side slopes (1 to 1) open channel to convey 15 m³s⁻¹ at a surface breadth to depth ratio of 3. The gradient is 0.00095 and the roughness size is 1.5 mm. Then find the depth of flow 10 m³s⁻¹ in the same channel. (Numerical solutions are 4.0365 m and 1.060 m respectively.)

The geometric parameters for the appropriate trapezoidal channel can be found in Table C54. For 0.333 relative depth, linear interpolation gives $D_{ep(u)}$ and J values of 0.9143 m and 1.4750 respectively.

$$Q_{ep} = Q \, J = 15 \times 1.4750 = 22.125 \; m^3s^{-1}$$

From Table A40, for k_s = 1.5 mm, the table point D = 4.000 m and S = 0.00095 gives Q = 27.329 m^3s^{-1} and V = 2.175 ms^{-1}.

TABLE 3 : Predictions of proportional depth in Form 1 egg-shape (egg-s.) with range of extreme combinations of conditions.

Height of egg-s. (m)	Gradient S	Roughness size k_s (mm)	First assess't for prop. depth of 0.150	0.850	θ/θ_{med}	Corrected assess't for prop. depth of 0.150	0.850
0.50	0.00015	15.0	0.143	0.859	0.089	0.149	0.852
0.50	0.00015	0.006	0.147	0.854	0.346	0.150	0.851
0.50	0.075	15.0	0.143	0.859	0.116	0.149	0.853
0.50	0.075	0.006	0.151	0.848	2.721	0.150	0.851
4.00	0.00015	15.0	0.149	0.851	0.715	0.150	0.850
4.00	0.00015	0.006	0.151	0.849	2.765	0.149	0.850
4.00	0.075	15.0	0.149	0.841	0.924	0.149	0.851
4.00	0.075	0.006	0.154	0.845	21.77	0.150	0.849

Note: The first assessments were made using discharge ratio values from Table C2; the corrections were made using Table C1(b).

Then DV = 8.69 and D/k_s = 2666, giving, from Table B1, x = 2.615.

$$D_{ep} = 4.0 \times [22.125 / 27.329]^{1/2.615} = 3.6896 = M\, D_{ep(u)}$$

With the $D_{ep(u)}$ value as already determined, M = 4.0354. Thus the bottom width of the channel is estimated as 4.0354 m (cf. 4.0365 m).

Then to determine the depth of flow of a discharge of 10 $m^3 s^{-1}$ in this same channel, one first requires the discharge at the proscribed relative depth on Table C54, i.e. $Q_{0.50}$. At relative depth 0.50 on this table, $D_{ep(u)}$ and J are read as 1.2426 m and 1.6170 respectively.

$$D_{ep(0.50)} = 4.0354 \times 1.2426 = 5.0144 \text{ m}$$

From Table A40, for k_s = 1.5 mm, table point D = 5.000 m and S = 0.00095 gives Q = 48.917 $m^3 s^{-1}$ and V = 2.49 ms^{-1}.

Then DV = 12.45 and D/k_s = 3333, giving x = 2.61.

$$\begin{aligned} Q_{ep(0.50)} &= 48.917 \times [5.0144 / 5.0]^{2.61} \\ &= 49.286 \text{ } m^3 s^{-1} = Q_{0.50}\, J \end{aligned}$$

$$Q_{0.50} = 49.286 / 1.6170 = 30.480 \text{ } m^3 s^{-1}$$

$$Q/Q_{0.50} = 10 / 30.480 = 0.3281$$

With this discharge ratio, linear interpolation on Table C54 gives the relative depth as 0.2633. Then normal depth = 0.2633 x 4.0354 = 1.0625 m (cf. 1.060 m).

Other sources of resistance

When designing a pipeline on the basis of its full-bore capacity, allowances must be made for head losses which will occur at bends, manholes or other appurtenances involving changes in cross-section. Similar allowances should also be made under conditions of free-surface flow. So far as pipelines are concerned, a customary practice is to take these into account as an equivalent additional length of pipe. Where such losses are a small part of the whole resistance, this is a satisfactory procedure. Hydraulic handbooks, manuals and other technical publications give head loss coefficients, ϵ, in the expression $h_1 = \epsilon V^2/2g$. Head losses at straight through open-channel manholes are generally small, the head loss coefficients being of the following approximate magnitudes.

	Part-full	Full-bore
Open-channel manhole	< 0.1	0.05-0.25
Open-channel manhole bend	~ 0.3	~ 1.5
Open-channel manhole with pipe bend beyond manhole	~ 0.3	~ 0.3

If a manhole incorporates a junction, losses are much higher and depend on the relative magnitudes of the branches and the geometry of the junction.

In addition, there is some loss of head at the entry to a pipe, which will depend upon the sharpness of the arris. Furthermore, the kinetic energy of flow ($V^2/2g$) is generally not recoverable at the exit. These factors may well prove important if the pipe or conduit is relatively short. The head loss a sharp-edged re-entrant inlet is approximately $V^2/2g$. a flush headwall the loss coefficient drops to about 0.4, whilst a rounding as little as a sixth or seventh of the pipe diameter will almost eliminate the entrance loss. Much information on losses at features and bends is given by Miller[25].

In short culverts and conduits at hydraulically steep slopes, separation may occur at the inlet if it is square-edged. Then the inlet acts as a controlling section and precludes full-bore operation. For very short culverts and conduits, friction loss may not be significant. For intermediate cases, these Tables may aid assessments of gradually varied flow conditions, as is shown in a following section.

Checks on mean velocity, Reynolds number and Froude number

With the tabular method now introduced (i.e. for solution for discharge, for gradient or for size), mean velocity in a section is the same as that for the flow in the equivalent pipe. Thus mean velocities may be read directly from the Colebrook-White solution tables.

The Reynolds number is DV/ν where the variables D, V and ν are in consistent units. The value is then given by $DV/(1.141 \times 10^{-6})$ for water at normal temperature (15°C) with kinematic viscosity value 1.141×10^{-6} m^2s^{-1}, where D is the diameter for pipe-full flows and is the diameter of the equivalent pipe for cases other than pipe full.

The Froude number is $V/\sqrt{(gy_{mean})}$ where mean depth y_{mean} is the cross-sectional area of flow divided by the free surface width. This can be evaluated using the tables of sequence C.

Viscosities other than that of water at 15°C

In the Wallingford Tables[2] is provided a system to determine factors to be applied to velocity (and hence discharge) for a change in temperature of water over the normal range of variation of temperature in the civil engineering context. The resulting table of velocity corrections is given here as Appendix 3. Alternatively, the equations of Fig. 2 can be used. This means that one can no longer use the tables to determine the solution. However, it does allow that the tables of sequence C be applied to the solution of flow problems involving fluids of any known viscosity, provided only that checks are made that the Reynolds number, in terms of D_{ep}, remains above 2000.

The Manning equation

Arrangement in terms of equivalent diameter

Some engineers prefer to work the Manning[4] equation. Strictly, this is an equation for rough turbulent flow only. For SI (metre-second) units the Manning equation is stated conventionally as

$$V = (1/n) \; R^{2/3} \; S^{1/2} \tag{18}$$

where n is the Manning coefficient.

Alternatively

$$nV = R^{2/3} \; S^{1/2} = 0.397(D_{ep})^{2/3} \; S^{1/2} \tag{19}$$

and then

$$nQ_{ep} = (\pi/4) \; (D_{ep})^2 \; (D_{ep}/4)^{2/3} \; S^{1/2} \tag{20}$$

$$= 0.3117 \; (D_{ep})^{8/3} \; S^{1/2} \tag{20a}$$

Tables of values of nV and nQ_{ep} are provided in table sequence D. These have the same general pattern as those for sequence A. Only one table is needed for each grouping of diameters, and the grouping of the smallest diameters of the Wallingford Tables is not included.

Parallel solutions of key examples

Suppose that the trapezoidal channel example considered earlier is to be solved as a rough turbulent case using the Manning equation. Williamson's[26] equation provides the following approximate relation[27] between n and k_s for metre-second units.

$$n = (k_s)^{1/6}/26.3 \tag{21}$$

Then, for k_s = 0.0015 m, Eq. (21) gives n = 0.0129. The same values of $D_{ep(u)}$ and J apply as previously, i.e. 0.9143 and 1.4750 respectively, giving Q_{ep} = 22.125 m^3s^{-1} and thus nQ_{ep} = 0.2854 m^3s^{-1}

On Table D3, table point D = 3.400 m and S = 0.00095 gives nQ = 0.2511 m^3s^{-1}

$$D_{ep} = 3.4 \; [0.2854 / 0.2511]^{1/2.67} \; m$$

$$= 3.567 = M \; D_{ep(u)}$$

Then M = 3.901 indicating a bottom width of 3.901 m (cf. 4.0365 m using the Colebrook-White equation).

(Eqs (8), (9) and (10) still apply, with the effect of change of k_s ignored, and with x, y and z values constant at 2.67, 2 and 5.333 respectively as resulting from Eq. (20)).

To find the depth with a flow of 10 m^3s^{-1}

$$D_{ep(0.50)} = 3.903 \times 1.2426 = 4.850 \text{ m}$$

(i.e. using the same values as previously for $D_{ep(u)}$ and J at 0.50 relative depth, i.e 1.2426 and 1.6170 respectively).

Reading nQ for D = 5.000 m and S = 0.00095 from Table D3 as 0.7023

$$nQ_{ep(0.50)} = 0.7023 \times [4.850 / 5.000]^{2.67}$$
$$= 0.6474 \text{ m}^3\text{s}^{-1} = nQ_{0.05} \text{ J}$$

$$Q_{0.50} = 0.6474 / (0.129 \times 1.6170) = 31.04 \text{ m}^3\text{s}^{-1}$$

Thus the flow ratio is 10 / 31.04 = 0.3222, and from Table C54 the relative depth is 0.2641. Then depth of flow = 0.2641 × 3.901 = 1.030 m (cf. 1.060 m by the Colebrook-White equation).

These results both show slight underestimation of linear size because the flow conditions are actually in the transition turbulent region, and this is not taken into account. The results from the egg-shape sewer example already treated are less affected by the adoption of the Manning equation, because flow is effectively roughly turbulent in this case. This is confirmed by undertaking parallel solutions to those already given. The Williamson equation gives n as 0.0189, and the results are

Find discharge	: 0.221 m^{3}s^{-1}, cf. 0.2222 m^{3}s^{-1}
Find gradient	: 0.00101, cf. 0.00100
Find size	: 1.804 m, cf. 1.800 m
Find depth of flow	: 0.301 proportional depth, cf. 0.300

Typical values of the Manning coefficient are given in Appendix 2. More comprehensive lists are available, especially in Ven Te Chow's[28] 'Open Channel Hydraulics'.

Critical depth and critical discharge

The tables of sequence C include values of critical discharge for the unit case, Q_{uc}, this being the product of unit area and the critical velocity for unit mean depth. Division of discharge in a conduit or channel by $M^{2.5}$ gives the corresponding value of unit critical discharge. Hence, the corresponding proportional or relative depth is defined on the table, and then multiplication of this by M gives the critical depth.

Alternatively, in a given shape of channel for which the scale factor M is known, division of a selected depth by M gives the proportional or relative depth. Then the corresponding unit critical discharge can be read from the appropriate table and multiplied by $M^{2.5}$ to obtain the critical discharge for the chosen depth.

Gradually varied flow in prismatic channels

For gradually varied flows, the relative values of normal and of critical depths define the circumstances for the occurrence of flow profiles. This is described in appropriate texts. Then profile details can be estimated using the direct step method, adapted for use the tables of sequence C.

This level of application fits well with the use of a brief routine on a programmable hand calculator, but this is not essential. With or without the use of such a routine, the tables of sequence C do allow profile calculations to be undertaken immediately for any of the shapes of conduit or channel covered, using Eq. 22. The same increment of unit proportional or relative depth will hold for a sequence of steps, as indicated in the appropriate table of sequence C. Then for each successive step, the only new information needed is the new end-of-step value of unit cross-sectional area A_u (subscript 2) and the new end-of-step value of unit wetted perimeter. These are operated on by M^2 and M respectively, as is unit Δy. Here it is assumed that the brief routine includes transfer of the end-of-step values for one step to be the start of the step values (subscript 1) for the next.

$$\Delta x = \left[\Delta y + \frac{\alpha}{2g} (V_2^2 - V_1^2) \right] \Big/ (S_o - S_f) \qquad (22)$$

Δx is the resulting distance along the conduit or channel, positive indicating the direction of flow. Δy is the direct step value, M times Δy_u, or M x $(Y_2 - Y_1)$. α is the Coriolis coefficient. g is acceleration due to gravity. V is mean velocity, Q/M^2A_u, with subscripts 1 and 2 corresponding to successive end of step positions. S_o is channel gradient, positive when sustaining flow. S_f is the friction gradient, which for the Manning equation is given by

$$S_f = \left[\frac{nQ_{ep}}{0.3117(D_{ep})^{2.667}} \right]^2 \qquad (20b)$$

Alternatively, the friction gradient can be assessed using the second equation of Fig. 2, where the Colebrook-White equation is preferred.

In comparing the procedure outlined here with existing solutions, the average value of unit area and wetted perimeter at the beginning and end of each step has been used before applying M^2 and M, respectively.

Solution for gradually varied flow in a trapezoidal channel
Consider the application of the system to the following problem[29].

> *A 1:1 side slopes trapezoidal channel of 3 m bed width carries 19 m ^{3}s^{-1} at a bed slope of 0.0015. Trace the water surface*

TABLE 4 : Computation of M1 flow profile in trapezoidal channel

Given : 45° (1 to 1) side-slopes with multiplying factor M = 3; Discharge Q = 19 m³s⁻¹; Gradient S_o = 0.0015; Manning coef. n = 0.017; Coriolis coef. α = 1.1; Normal depth y_N = 1.725 m and Critical depth y_c = 1.364 m.

Y	Δy_u (m)	P_u (m)	A_u (m²)	Δy (m)	ΔK (m)	J (calc.)	D_{ep} (m)	Q_{ep} (m³s⁻¹)	S_f	S_o-S_f	Δx (m)	$\Sigma \Delta x$ (m)	y (m)
1.333	--	4.7618	3.0992	--	--	--	--	--	--	--	--	--	4.00
1.32	− 0.013	4.7335	3.0624	− 0.04	0.00063	1.7175	7.7869	32.634	5.58×10^{-5}	0.00144	− 27.26	− 27.26	3.96
1.28	− 0.04	4.6204	2.9184	− 0.12	0.00270	1.7178	7.6727	32.641	6.04×10^{-5}	0.00144	− 81.49	− 108.75	3.84
1.24	− 0.04	4.5072	2.7776	− 0.12	0.00305	1.7181	7.4885	32.647	6.88×10^{-5}	0.00143	− 81.71	− 190.46	3.72
:													:
0.64	− 0.04	2.8102	1.0496	− 0.12	0.03536	1.6757	4.5878	31.842	8.93×10^{-4}	0.00061	− 139.44	− 1592.40	1.92
0.60	− 0.04	2.6971	0.9600	− 0.12	0.04433	1.6654	4.3788	31.639	0.00113	0.00037	− 204.88	− 1797.28	1.80

Notes:

(a) For assessment of change in kinetic head, ΔK, velocities are $Q/M^2 A_u$.

(b) Assessments of J, D_{ep}, Q_{ep} and S_f are for mid-step averaged values of P_u and A_u. The latter values are read directly from Table C52, except for the starting values. Here linear interpolation values have been assessed from the adjoining table values.

profile back from a control giving 4 m depth, to the point where the depth is 1.8 m. Take Manning's n as 0.017 and the Coriolis coefficient α as 1.1.

Unit critical discharge is $19/3^{2.5} = 1.2188$ m^3s^{-1}. On Table C54, this corresponds to a relative depth of 0.455, giving critical depth as 1.365 m.

To find normal depth directly, $Q_{0.50}$ is required. From Table C54, $D_{ep(u)}$ and J at 0.50 relative depth are 1.2426 and 1.6170 respectively)

$$D_{ep(0.50)} = 3 \times 1.2426 = 3.7278 \text{ m}$$

From Table D3, nQ at D = 3.400 m and S = 0.0015 is 0.3155 m^3s^{-1}. Using Eq.(8), 0.3155 x [3.7278 / 3.4] 2.667 = 0.4033 m^3s^{-1} (i.e. $nQ_{ep(0.50)}$). But

$$nQ_{0.50} = 0.4033 / 1.6170 = 0.2494$$

Then

$$Q_{0.50} = 0.2494 / 0.017 = 14.672 \text{ m}^3\text{s}^{-1}$$

and hence

$$Q/Q_{0.50} = 1.295$$

On Table C54, this value corresponds to a Y value of 0.5750, giving normal depth as 0.5750 x 3 = 1.725 m. Thus the flow is subcritical throughout and the profile is of the M1 form.

The adoption of 19 steps is imposed by the relative depth increments used in Table C54. Elements of a calculator-aided solution are given in Table 4, showing a result of 1797 m upstream. Clearly, it is simple to amplify such a calculator routine to cover this or any other trapezoidal channel shape, without recourse to the values from the appropriate table of sequence C. However, this does not hold for many more complex shapes. A basic routine for use with these Tables, once prepared, equally can be applied to any of the shapes treated in table sequence C.

Review

The first function of this volume is as a companion to Tables 1-33 of the 3rd, 4th or 5th Editions of the Wallingford Tables. This is illustrated in Table 1, where its other capacities are also listed. Thus there are minor duplications with the 5th Edition of the Wallingford Tables so that this volume can be used in isolation, if necessary, not only for Manning equation based work but also where the Colebrook-White equation is adopted. It is intended that the new computations and forms of settings of tables will be adopted in due course for the 6th Edition of the Wallingford Tables with these Additional Tables incorporated.

References

1. HYDRAULICS RESEARCH LTD., WALLINGFORD. *Charts for the hydraulic design of channels and pipes*, 6th edition. Thomas Telford, London, 1990. (Note: the earlier editions of the Wallingford Charts were published by HMSO in 1958, 1963, 1969 and 1978, in the earlier cases under the authorship of P. Ackers, and by Hydraulics Research Ltd. in 1983).

2. HYDRAULICS RESEARCH LTD., WALLINGFORD. *Tables for the hydraulic design of pipes and sewers*, 5th edition. Thomas Telford, London, 1990. (Note: the earlier editions of the Wallingford Tables were published by HMSO in 1963, 1969 and 1977, in the first two cases under the authorship of P. Ackers and by Hydraulics Research Ltd. in 1983).

3. COLEBROOK, C.F. Turbulent flow in pipes, particular reference to the transition region between the smooth and the rough pipe laws. *J. Instn Civ. Engrs*, 1939, Vol. 11, pp 133-156.

4. MANNING, R. On the flow of water in open channels and pipes. *Proc. Instn Civ. Engrs*, Ire., 1891, Vol. 20, p161; 1895, Vol. 24, p 179.

5. NIKURADSE, J. Gesetzmäßigkeit der turbulenten Strömung in glatten Rohren. *Forsch. Arb. Ing.-Wes.* No.356 (1932).

6. NIKURADSE, J. Strömungsgesetze in rauhen Rohren. *Forsch. Arb. Ing.-Wes.* No.361 (1933).

7. BARR, D.I.H. Explicit Colebrook-White solutions. *Civil Engineering*, Sept. 1986, pp 19-31.

8. PHAM, Q.T. Explicit equations for the solution of turbulent pipe-flow problems. *Trans. Instn Chem. Engrs*, 1979, Vol. 57, pp 281-283.

9. ROUSE, H. Evaluation of boundary roughness. *Proc. 2nd Hydraulics Conf.*, University of Iowa, 1943, Bulletin 27.

10. KING, H. W. *Handbook of Hydraulics*, McGraw-Hill, N.York, 1954, Section 6.

11. LAMONT, P. A. A review of pipe friction data and formulae, a proposed set of exponential formulae based on the theory of roughness. *Proc. Instn Civ. Engrs*, Part 3, 1954, 3, p 248.

12. PERKINS, J. A. and GARDINER, I. M. *The effect of sewage slime on the hydraulic roughness of pipes*. Report IT 218, Hydraulics Research Station, Wallingford, 1982.

13. BARR, D. I. H. Two additional methods of direct solution of the Colebrook-White function. TN 128, *Proc. Instn Civ. Engrs*, Part

2, Dec. 1975, 3, p 827.

14. MOURET, G. Antoine Chezy, histoire d'une formule d'hydraulique. Annales des *Ponts et Chaussees*, 1921-II. (See *History of Hydraulics* by Hunter Rouse and Simon Ince, Dover N.Y., 1963).

15. JOHNSON, S.P. *A survey of flow calculation methods*. Pre-printed programme for June 19-21, 1934, meeting of the Amer. Soc. Mech. Engrs., University of California, Berkeley.

16. ACKERS, P. *Resistance of fluids flowing in channels and pipes*. Hydraulics Research Paper No. 1, HMSO, London, 1958.

17. HYDRAULICS RESEARCH LTD., WALLINGFORD. *Wallingford Software User Manual for WALLRUS*, 4th Edition, 1991.

18. WATER RESEARCH CENTRE. *Sewerage Rehabilitation Manual*. 2nd Edition, Swindon, UK, 1986.

19. METCALF, L. and EDDY, H.P. *American Sewerage Practice*, 2nd Edition, Vol. I, Design of sewers. McGraw-Hill, N.Y., 1928.

20. METCALF and EDDY INC. *Wastewater engineering*. McGraw-Hill, N.Y., 1972.

21. BABBITT, H. E. and BAUMANN, E. R. *Sewerage and sewage treatment*, 8th Edition. Wiley, N.Y. 1958.

22. ARMCO INT. CORP. *Handbook of drainage and construction products*. Middleton, Ohio, USA, 1958.

23. ASSET INT. LTD. *Design Manual*, Newport, U.K., 1989.

24. ESDU International, Item No 66027. *Friction losses for fully developed flow in straight pipes*. 1966 and amended 1971.

25. MILLER, D. S. *Internal flow systems* - 2nd Edition. BHRA (Information Services), Bedford, 1990.

26. WILLIAMSON, J. The laws of flow in rough pipes, Strickler, Manning, Nikuradse and drag-velocity. *La Houille Blanche*, 1951, Vol. 6, No. 5, pp 738-57.

27. BARR, D. I. H. and DAS, M. M. Direct solutions for normal depth using the Manning equation. *Proc. Instn Civ. Engrs*, Part 2, 1986, Vol. 81, pp 315-333.

28. CHOW, V.-T. *Open channel hydraulics*. McGraw-Hill, New York, 1959.

29. FEATHERSTONE, R.E. and NALLURI, C. *Civil engineering hydraulics - essential theory worked examples*. 2nd Edn., BSP Prof. Books, Oxford, 1988.

Nomenclature

A   Area of cross-section of flow

B   Width of channel; horizontal size used in specifying relative depth.

D   Diameter. When used in resistance formula there is the implication that flow is pipe-full, or that the equivalent diameter measure is being utilised. The diameter of a full pipe, or that of the equivalent pipe to a specified non-circular section of flow, is used in ratio D/k_s and in product DV. Values are given in metres, and must be used consistently in respect of ratio D/k_s and in metres in product DV.

g   Acceleration due to gravity, 9.80665 ms^{-2} in SI.

h   Head loss in uniform flow.

i   Hydraulic gradient, resistance head loss per unit length of uniform flow h/l. Used in 3rd and 4th editions of Wallingford Tables and identical with S.

J   Discharge conversion factor, $4\pi A/P^2$ where A and P relate to a particular flow circumstance.

l   Length over which head loss is assessed.

k_s Nikuradse equivalent sand roughness size; the linear measure of roughness size. Stated in millimetres but to be used with consistency in evaluating D/k_s and the like.

log   Common logarithm (base 10).

M   Multiplying factor on unit measure of conduit or channel size to specify size of given example.

n   Manning friction coefficient.

p   A proportion of total wetted perimeter.

P   Wetted perimeter of flow section.

Q   Discharge (volume per unit time).

R   Hydraulic mean depth (hydraulic radius) of flow section, A/P, i.e. $(\pi/4)D^2/\pi D$ or $(1/4)D$.

R   Reynolds number, VD/ν or $4Q/\pi\nu D$.

S   Hydraulic (piezometric) gradient; head loss per unit length of uniform flow, h/l. Identical with i in 3rd and 4th editions of the

Wallingford Tables.

V Mean velocity of flow through cross-section.

Δx Distance along channel in gradually varied flow, corresponding to Δy.

Δy a depth of flow; vertical measure of size.

Y a non-dimensional depth of flow in terms of either vertical or horizontal measure of conduit or channel size. For tables for only conduit full flow, $Y = y_f/B$.

Δy Direct step value in gradually varied flow.

α Coriolis coefficient; $\alpha V^2/2g$ is kinetic energy head.

ϵ Head loss coefficient, normally applied to $V^2/2g$.

θ Proportional flow parameter for circular pipes.

λ Darcy-Weisbach friction factor, $2(Sg)D/V^2$.

ν Kinematic viscosity, i.e. dynamic viscosity divided by mass density. This is 1.141×10^{-6} m^{2}s^{-1} for water at 15°C.

Subscripts

ep Relating to equivalent pipe, i.e. circular pipe flowing full at same gradient and with same hydraulic mean depth (hydraulic radius).

ep(u) Relating to equivalent pipe for unit case.

f Relating to vertical measure of conduit size for tables dealing only with conduit full flow. Also, indicating friction gradient in gradually varied flow.

G Given value between table values in Colebrook-White (or Manning) solutions.

m Medial: typical case without necessarily implying specific numerical basis.

N Relating to normal depth.

o Relating to overall height measure in closed conduit, or to depth corresponding to upper limit of tabulated values in certain open channel flows, i.e. y_N/y_o is (normal) proportional depth Y_N for flow which has y_N as normal depth. Also, channel gradient in gradually varied flow.

0.50 (for example) - A proportional or a relative depth so defining a

30

flow condition in a given conduit or channel.

R   Required value between table values in Colebrook-White (or Manning) solutions.

T   Table value in Colebrook-White (or Manning) solutions.

u   Relating to unit section.

uc   Unit section - critical flow.

um   Unit section - mean depth.

Exponents

u,v,w,x,y,z   exponents in proportioning equations (8), (9) and (10), as evaluated in tables of sequence B.

APPENDIX 1 : Recommended roughness values

Classification (assumed clean and new unless otherwise stated)	Suitable values of k_s (mm)		
	Good	Normal	Poor
Smooth materials (pipes)			
Drawn non-ferrous pipes of aluminium, brass, copper, lead etc, and non-metallic pipes of Alkathene, glass, perspex etc	--	0.003	--
Asbestos-cement	0.015	0.03	--
Metal			
Spun bitumen or concrete lined	--	0.03	--
Wrought iron	0.03	0.06	0.15
Rusty wrought iron	0.15	0.6	3.0
Uncoated steel	0.015	0.03	0.06
Coated steel	0.03	0.06	0.15
Galvanised iron, coated cast iron	0.06	0.15	0.3
Uncoated cast iron	0.15	0.3	0.6
Tate relined pipes	0.15	0.3	0.6
Old tuberculated water mains as follows:			
Slight degree of attack	0.6	1.5	3.0
Moderate degree of attack	1.5	3.0	6.0
Appreciable degree of attack	6.0	15	30
Severe degree of attack	15	30	60
(Good: up to 20 years use; Normal: 40 to 50 years use; Poor: 80 to 100 years use)			
Wood			
Wood stave pipes, planed plank conduits	0.3	0.6	1.5
Concrete			
Precast concrete pipes with "O" ring joints	0.06	0.15	0.6
Spun precast concrete pipes with "O" ring joints	0.06	0.15	0.3
Monolithic construction against steel forms	0.3	0.6	1.5
Monolithic construction against rough forms	0.6	1.5	--
Clayware			
Glazed or unglazed pipe:			
With sleeve joints	0.03	0.06	0.15
With spigot and socket joints and "O" ring seals			
-- dia < 0.150 m	--	0.03	--
-- dia > 0.150 m	--	0.06	--
Pitch fibre			
(lower value refers to full bore flow)	0.003	0.03	
Glass fibre	--	0.06	--
uPVC			
With chemically cemented joints	--	0.03	--
With spigot and socket joints, "O" ring seals at 6 to 9 metre intervals	--	0.06	--

Classification (assumed clean and new unless otherwise stated)	Suitable values of k_s (mm)		
	Good	Normal	Poor

Brickwork

	Good	Normal	Poor
Glazed	0.6	1.5	3.0
Well pointed	1.5	3.0	6.0
Old, in need of pointing	--	15	30

Slimed sewers •

Sewers slimed to about half depth; velocity, when flowing half full, approximately 0.75 ms^{-1}:

	Good	Normal	Poor
Concrete, spun or vertically cast	--	3.0	6.0
Asbestos cement	--	3.0	6.0
Clayware	--	1.5	3.0
uPVC	--	0.6	1.5

Sewers slimed to about half depth; velocity, when flowing half full, approximately 1.2 ms^{-1}:

	Good	Normal	Poor
Concrete, spun or vertically cast	--	1.5	3.0
Asbestos cement	--	0.6	1.5
Clayware	--	0.3	0.6
uPVC	--	0.15	0.3

Sewer rising mains

All materials, operating as follows	Good	Normal	Poor
Mean velocity 1 ms^{-1}	0.15	0.3	0.6
Mean velocity 1.5 ms^{-1}	0.06	0.15	0.30
Mean velocity 2 ms^{-1}	0.03	0.06	0.15

Unlined rock tunnels

	Good	Normal	Poor
Granite and other homogeneous rocks	60	150	300
Diagonally bedded slates	--	300	600

(values to be used with *design* diameter)

Earth channels

	Good	Normal	Poor
Straight uniform artificial channels	15	60	150
Straight natural channels, free from shoals, boulders and weeds	150	300	600

• The roughness of a slimed sewer varies considerably during any year. The normal value is that roughness which is exceeded for approximately half of the time. The poor value is that which is exceeded, generally on a continuous basis, for one month of the year. The value of k_s should be interpolated for velocities between 0.75 and 1.2 ms^{-1}.

The lists of roughness values in Appendices 1 and 2 should not be taken as absolving the engineer of the responsibility for checking the actual surface roughness achieved in particular projects by precise hydraulic tests whenever possible. Where such direct evidence is available from comparable projects it should clearly take precedence over the values quoted here, whether in sand roughness size or in coefficient form.

APPENDIX 2 : Values of Manning roughness coefficient n

	Minimum		Normal		Maximum	
	n	k_s(mm)	n	k_s(mm)	n	k_s(mm)
Closed conduits, partially full						
Smooth pipe, brass or glass	0.009	(0.18)	0.010	(0.33)	0.013	(1.60)
Coated cast iron	0.010	(0.33)	0.013	(1.60)	0.014	(2.50)
Galvanised wrought iron	0.011	(0.59)	0.014	(2.50)	0.016	(5.55)
Corrugated metal storm drain	0.021	(28.4)	0.024	(63.2)	0.030	(240)
Straight clean conc. culvert	0.010	(0.33)	0.011	(0.59)	0.013	(1.60)
Finished concrete	0.011	(0.59)	0.012	(0.99)	0.014	(2.49)
Conc. against poor formwork	0.015	(3.77)	0.017	(8.0)	0.020	(21.2)
Clay drainage tile	0.011	(0.59)	0.013	(1.60)	0.017	(8.0)
Vit. pipe sewer with features	0.013	(1.60)	0.015	(3.77)	0.017	(8.0)
Used san. sewers with features	0.012	(0.99)	0.013	(1.60)	0.016	(5.55)
Rubble masonry, cemented	0.018	(11.3)	0.025	(80.8)	0.030	(240)
Open channels						
Corrugated metal lining	0.021	(28.4)	0.025	(80.8)	0.030	(240)
Trowel finish concrete	0.011	(0.59)	0.013	(1.60)	0.015	(3.77)
Unfinished concrete	0.014	(2.50)	0.017	(8.0)	0.20	(21.2)
Gunite, poor	0.018	(11.3)	0.022	(37.5)	0.025	(80.8)
Glazed brick	0.011	(0.59)	0.013	(1.60)	0.015	(3.77)
Cemented rubble masonry	0.017	(8.0)	0.025	(80.8)	0.030	(240)
Rip rap	0.023	(50.0)	0.033	(427)	0.036	(720)
Asphalt (good)	0.013	(1.60)	0.013	(1.60)		
Asphalt (rough)	0.016	(5.55)	0.016	(5.55)		
New earth channel	0.016	(5.55)	0.018	(11.3)	0.020	(21.2)
Cleaned earth channel	0.018	(11.3)	0.022	(37.5)	0.025	(80.8)
Gravel channel, well maintained	0.022	(37.5)	0.025	(80.8)	0.030	(240)
Smooth rock cut	0.025	(80.8)	0.035	(608)	0.040	(1360)

These values derive from Chow[28], and the corresponding values of the Nikuradse equivalent sand roughness size k_s are given by the Williamson[26] formula.

APPENDIX 3 : Velocity correction for variation in temperature

$$V = V_{15} + at_{15} \quad \text{(in ms}^{-1}) \quad \text{where } t_{15} = t - 15 \text{ (Celsius)}$$

D S (m)	Roughness values, k_s in mm							
	0.006	0.015	0.03	0.06	0.15	0.30	0.60	1.5
2.0	0.0327	0.0151	0.0080	0.0041	0.0017	0.0008	0.0004	0.0002
0.80	0.0290	0.0142	0.0077	0.0040	0.0017	0.0008	0.0004	0.0002
0.40	0.0257	0.0134	0.0075	0.0040	0.0016	0.0008	0.0004	0.0002
0.20	0.0222	0.0124	0.0071	0.0039	0.0016	0.0008	0.0004	0.0002
0.080	0.0174	0.0107	0.0065	0.0037	0.0016	0.0008	0.0004	0.0002
0.040	0.0140	0.0093	0.0060	0.0035	0.0016	0.0008	0.0004	0.0002
0.020	0.0110	0.0079	0.0054	0.0033	0.0015	0.0008	0.0004	0.0002
0.0080	0.0077	0.0060	0.0044	0.0029	0.0014	0.0008	0.0004	0.0002
0.0040	0.0057	0.0048	0.0037	0.0026	0.0013	0.0008	0.0004	0.0002
0.0020	0.0042	0.0037	0.0030	0.0022	0.0012	0.0007	0.0004	0.0002
0.00080	0.0028	0.0025	0.0022	0.0017	0.0011	0.0007	0.0004	0.0002
0.00040	0.0020	0.0019	0.0017	0.0014	0.0009	0.0006	0.0004	0.0002
0.00020	0.0014	0.0014	0.0013	0.0011	0.0008	0.0005	0.0003	0.0002
0.000080	0.0009	0.0009	0.0008	0.0008	0.0006	0.0004	0.0003	0.0002
0.000040	0.0007	0.0006	0.0006	0.0006	0.0005	0.0004	0.0003	0.0002

The values of "a" tabulated above show the increase in velocity in ms^{-1} per 1°C increase in temperature. (This table derives from Table 34 of the 5th edition of the Wallingford Tables.)

APPENDIX 4: Multiplying factors for discharges in pipes and lined tunnels

To estimate discharge in a pipe of standard but non-tabulated diameter, D_G, apply the factor to the tabulated value of discharge, for the stipulated gradient and roughness size, as given for pipe of the nearest tabulated diameter, D_T, to that under consideration.

D_G (m)	D_T (m)	Factor	D_G (m)	D_T (m)	Factor
0.040	0.050	0.56	1.830	1.800	1.04
0.060	0.050	1.61	1.875	1.800	1.11
0.065	0.075	0.69	1.900	1.800	1.15
0.090	0.100	0.76	2.000	2.100	0.88
0.110	0.100	1.28	2.130	2.100	1.04
0.160	0.150	1.18	2.200	2.100	1.13
0.315	0.300	1.14	2.250	2.100	1.20
0.355	0.350	1.04	2.290	2.400	0.88
0.560	0.600	0.84	2.300	2.400	0.89
0.630	0.600	1.14	2.440	2.400	1.04
1.125	1.100	1.06	2.500	2.400	1.11
1.275	1.200	1.17	2.550	2.400	1.17
1.300	1.350	0.91	2.590	2.700	0.90
1.400	1.350	1.10	2.600	2.700	0.91
1.425	1.350	1.15	2.740	2.700	1.04
1.520	1.500	1.04	2.850	2.700	1.15
1.575	1.500	1.14	2.900	3.000	0.92
1.600	1.650	0.92	3.050	3.000	1.04
1.680	1.650	1.05	3.200	3.000	1.18
1.700	1.650	1.08	3.600	3.400	1.16
1.725	1.650	1.12	3.800	4.000	0.87

$k_s = 0.015$ mm
$S = 0.00010$ to 0.00048

Water (or sewage) at 15°C;
full bore conditions.

ie hydraulic gradient =
1 in 10000 to 1 in 2083

velocities in ms⁻¹
discharges in m³s⁻¹

Gradient	(Equivalent) Pipe diameters in m											
	2.400	2.700	3.000	3.400	4.000	5.000	6.400	8.000	10.00	12.60	16.00	20.00
0.00010 1/10000	0.648 2.9296	0.698 3.9947	0.746 5.2705	0.807 7.3238	0.893 11.222	1.026 20.147	1.195 38.455	1.371 68.921	1.572 123.43	1.809 225.52	2.090 420.17	2.390 750.83
0.00011 1/ 9091	0.682 3.0848	0.735 4.2060	0.785 5.5490	0.849 7.7100	0.940 11.813	1.080 21.204	1.258 40.466	1.443 72.516	1.653 129.85	1.902 237.22	2.198 441.91	2.513 789.58
0.00012 1/ 8333	0.715 3.2337	0.770 4.4086	0.823 5.8158	0.890 8.0801	0.985 12.378	1.132 22.217	1.318 42.393	1.511 75.960	1.732 136.00	1.992 248.42	2.301 462.72	2.631 826.67
0.00013 1/ 7692	0.746 3.3768	0.804 4.6034	0.859 6.0725	0.929 8.4361	1.028 12.923	1.181 23.191	1.375 44.246	1.577 79.269	1.807 141.91	2.079 259.19	2.401 482.72	2.745 862.30
0.00014 1/ 7143	0.777 3.5149	0.837 4.7914	0.894 6.3200	0.967 8.7794	1.070 13.447	1.229 24.130	1.431 46.032	1.641 82.461	1.879 147.61	2.162 269.57	2.497 501.99	2.854 896.64
0.00015 1/ 6667	0.806 3.6484	0.869 4.9731	0.928 6.5594	1.004 9.1115	1.110 13.955	1.275 25.038	1.485 47.759	1.702 85.545	1.950 153.11	2.242 279.60	2.589 520.61	2.960 929.81
0.00016 1/ 6250	0.835 3.7779	0.899 5.1493	0.961 6.7915	1.039 9.4333	1.150 14.447	1.320 25.918	1.537 49.432	1.761 88.534	2.017 158.45	2.320 289.31	2.679 538.65	3.062 961.94
0.00017 1/ 5882	0.863 3.9037	0.929 5.3204	0.993 7.0169	1.073 9.7458	1.188 14.924	1.364 26.772	1.587 51.056	1.819 91.434	2.083 163.63	2.396 298.74	2.766 556.15	3.161 993.12
0.00018 1/ 5556	0.890 4.0260	0.958 5.4869	1.024 7.2361	1.107 10.050	1.225 15.389	1.406 27.603	1.636 52.635	1.875 94.255	2.147 168.66	2.469 307.90	2.851 573.16	3.258 1023.4
0.00019 1/ 5263	0.916 4.1452	0.987 5.6491	1.054 7.4496	1.140 10.346	1.261 15.841	1.447 28.412	1.684 54.173	1.930 97.002	2.210 173.56	2.541 316.83	2.933 589.73	3.352 1052.9
0.00020 1/ 5000	0.942 4.2615	1.014 5.8073	1.083 7.6580	1.171 10.635	1.296 16.282	1.487 29.202	1.731 55.674	1.983 99.681	2.271 178.34	2.611 325.53	3.013 605.88	3.443 1081.7
0.00022 1/ 4545	0.992 4.4862	1.068 6.1130	1.140 8.0606	1.233 11.193	1.364 17.135	1.565 30.726	1.821 58.571	2.086 104.85	2.388 187.57	2.745 342.33	3.168 637.06	3.620 1137.2
0.00024 1/ 4167	1.039 4.7015	1.119 6.4059	1.195 8.4462	1.292 11.727	1.429 17.952	1.639 32.186	1.907 61.345	2.185 109.81	2.501 196.41	2.874 358.41	3.317 666.90	3.789 1190.3
0.00026 1/ 3846	1.085 4.9086	1.168 6.6876	1.247 8.8171	1.348 12.241	1.491 18.737	1.711 33.590	1.990 64.012	2.279 114.57	2.609 204.90	2.998 373.86	3.459 695.56	3.951 1241.3
0.00028 1/ 3571	1.129 5.1083	1.215 6.9592	1.298 9.1747	1.403 12.737	1.551 19.494	1.780 34.943	2.070 66.582	2.370 119.15	2.713 213.08	3.118 388.74	3.597 723.17	4.108 1290.5
0.00030 1/ 3333	1.172 5.3014	1.261 7.2219	1.347 9.5204	1.456 13.216	1.610 20.226	1.846 36.251	2.147 69.066	2.459 123.59	2.814 220.98	3.233 403.12	3.729 749.85	4.259 1338.0
0.00032 1/ 3125	1.213 5.4885	1.306 7.4764	1.394 9.8555	1.507 13.681	1.666 20.935	1.911 37.518	2.222 71.473	2.544 127.88	2.911 228.64	3.345 417.05	3.858 775.68	4.405 1383.9
0.00034 1/ 2941	1.253 5.6703	1.349 7.7235	1.440 10.181	1.556 14.131	1.721 21.623	1.973 38.748	2.294 73.808	2.627 132.05	3.006 236.07	3.453 430.56	3.983 800.74	4.547 1428.5
0.00036 1/ 2778	1.292 5.8470	1.391 7.9639	1.485 10.497	1.605 14.570	1.774 22.292	2.034 39.944	2.365 76.079	2.708 136.10	3.098 243.29	3.558 443.70	4.104 825.10	4.685 1471.9
0.00038 1/ 2632	1.331 6.0192	1.432 8.1981	1.529 10.805	1.652 14.997	1.826 22.944	2.094 41.108	2.434 78.290	2.786 140.04	3.187 250.32	3.661 456.48	4.222 848.81	4.819 1514.1
0.00040 1/ 2500	1.368 6.1872	1.472 8.4265	1.571 11.106	1.698 15.413	1.876 23.580	2.151 42.244	2.501 80.446	2.863 143.89	3.274 257.18	3.761 468.95	4.337 871.92	4.950 1555.2
0.00042 1/ 2381	1.404 6.3513	1.511 8.6495	1.613 11.400	1.742 15.820	1.926 24.201	2.208 43.353	2.566 82.552	2.937 147.64	3.360 263.87	3.859 481.12	4.449 894.48	5.078 1595.3
0.00044 1/ 2273	1.439 6.5117	1.549 8.8676	1.653 11.687	1.786 16.218	1.974 24.808	2.263 44.437	2.630 84.609	3.010 151.31	3.443 270.41	3.954 493.01	4.558 916.52	5.203 1634.5
0.00046 1/ 2174	1.474 6.6687	1.586 9.0811	1.693 11.967	1.829 16.607	2.021 25.402	2.317 45.498	2.693 86.622	3.082 154.90	3.524 276.81	4.047 504.64	4.666 938.08	5.325 1672.9
0.00048 1/ 2083	1.508 6.8225	1.623 9.2902	1.732 12.243	1.871 16.988	2.068 25.984	2.370 46.537	2.754 88.594	3.152 158.42	3.604 283.07	4.138 516.03	4.771 959.19	5.444 1710.4
	550	600	650	750	900	1100	1400	1800	2200	2800	3600	4500

θ_{medial} for part-full circular pipes.

$k_s = 0.015$ mm $S = 0.00010$ to 0.00048

$k_s = 0.015$ mm
$S = 0.00050$ to 0.00280

Water (or sewage) at 15°C;
full bore conditions.

ie hydraulic gradient =
1 in 2000 to 1 in 357

velocities in ms⁻¹
discharges in m³s⁻¹

Gradient	(Equivalent) Pipe diameters in m											
	2.400	2.700	3.000	3.400	4.000	5.000	6.400	8.000	10.00	12.60	16.00	20.00
0.00050 1/ 2000	1.541 6.9733	1.658 9.4952	1.770 12.512	1.912 17.362	2.113 26.554	2.422 47.555	2.814 90.527	3.220 161.86	3.682 289.21	4.228 527.19	4.874 979.88	5.562 1747.2
0.00055 1/ 1818	1.622 7.3383	1.745 9.9914	1.863 13.165	2.012 18.266	2.223 27.934	2.547 50.019	2.959 95.202	3.386 170.20	3.871 304.06	4.444 554.18	5.122 1029.9	5.845 1836.2
0.00060 1/ 1667	1.699 7.6879	1.828 10.467	1.951 13.790	2.107 19.132	2.328 29.255	2.668 52.378	3.098 99.677	3.545 178.18	4.052 318.27	4.652 580.01	5.360 1077.8	6.116 1921.3
0.00065 1/ 1538	1.774 8.0240	1.908 10.923	2.036 14.391	2.199 19.964	2.429 30.525	2.783 54.645	3.232 103.98	3.697 185.84	4.226 331.92	4.850 604.81	5.589 1123.7	6.376 2002.9
0.00070 1/ 1429	1.845 8.3480	1.985 11.364	2.118 14.971	2.287 20.766	2.526 31.749	2.894 56.829	3.361 108.12	3.844 193.22	4.394 345.07	5.042 628.70	5.809 1168.0	6.626 2081.6
0.00075 1/ 1333	1.915 8.6612	2.059 11.789	2.197 15.531	2.373 21.542	2.621 32.931	3.002 58.939	3.485 112.12	3.986 200.35	4.555 357.77	5.227 651.77	6.022 1210.7	6.868 2157.6
0.00080 1/ 1250	1.982 8.9646	2.131 12.202	2.274 16.073	2.455 22.293	2.712 34.077	3.106 60.983	3.606 116.00	4.123 207.26	4.712 370.06	5.406 674.10	6.227 1252.1	7.102 2231.1
0.00085 1/ 1176	2.047 9.2591	2.201 12.602	2.348 16.599	2.536 23.021	2.800 35.188	3.207 62.966	3.723 119.76	4.257 213.96	4.864 381.99	5.580 695.77	6.427 1292.2	7.329 2302.4
0.00090 1/ 1111	2.110 9.5455	2.269 12.991	2.421 17.111	2.614 23.730	2.886 36.269	3.305 64.894	3.836 123.41	4.386 220.47	5.011 393.58	5.749 716.82	6.621 1331.2	7.549 2371.7
0.00095 1/ 1053	2.172 9.8244	2.335 13.370	2.491 17.609	2.690 24.420	2.970 37.321	3.401 66.771	3.947 126.97	4.512 226.80	5.155 404.86	5.913 737.31	6.809 1369.1	7.764 2439.1
0.00100 1/ 1000	2.232 10.096	2.400 13.739	2.560 18.095	2.764 25.093	3.052 38.347	3.494 68.601	4.055 130.44	4.635 232.98	5.295 415.86	6.073 757.28	6.993 1406.1	7.973 2504.8
0.00110 1/ 909	2.348 10.622	2.524 14.453	2.693 19.034	2.907 26.391	3.209 40.327	3.674 72.133	4.263 137.13	4.872 244.90	5.565 437.08	6.382 795.80	7.348 1477.4	8.377 2631.6
0.00120 1/ 833	2.459 11.124	2.644 15.136	2.820 19.932	3.044 27.635	3.360 42.223	3.846 75.513	4.462 143.53	5.099 256.31	5.823 457.37	6.678 832.65	7.687 1545.6	8.762 2752.8
0.00130 1/ 769	2.566 11.608	2.758 15.792	2.942 20.795	3.175 28.829	3.505 44.044	4.011 78.759	4.653 149.68	5.317 267.26	6.072 476.86	6.961 868.03	8.013 1611.1	9.133 2869.1
0.00140 1/ 714	2.669 12.073	2.869 16.425	3.060 21.627	3.302 29.980	3.644 45.798	4.170 81.887	4.837 155.61	5.527 277.81	6.311 495.63	7.235 902.09	8.327 1674.2	9.489 2981.1
0.00150 1/ 667	2.768 12.523	2.975 17.036	3.173 22.430	3.425 31.092	3.779 47.493	4.324 84.908	5.015 161.33	5.729 287.99	6.541 513.75	7.498 934.99	8.629 1735.0	9.833 3089.2
0.00160 1/ 625	2.865 12.959	3.079 17.628	3.283 23.208	3.543 32.169	3.910 49.134	4.473 87.833	5.187 166.87	5.926 297.85	6.765 531.29	7.754 966.82	8.922 1794.0	10.17 3193.8
0.00170 1/ 588	2.958 13.382	3.179 18.202	3.390 23.964	3.658 33.214	4.037 50.726	4.618 90.671	5.354 172.24	6.116 307.42	6.981 548.31	8.001 997.69	9.207 1851.1	10.49 3295.3
0.00180 1/ 556	3.049 13.794	3.277 18.761	3.494 24.698	3.770 34.229	4.160 52.274	4.758 93.429	5.517 177.47	6.301 316.71	7.192 564.83	8.242 1027.7	9.483 1906.6	10.80 3393.9
0.00190 1/ 526	3.138 14.194	3.372 19.304	3.595 25.412	3.879 35.218	4.280 53.780	4.895 96.113	5.675 182.55	6.481 325.76	7.397 580.92	8.476 1056.9	9.751 1960.6	11.11 3489.7
0.00200 1/ 500	3.224 14.584	3.464 19.834	3.694 26.109	3.985 36.182	4.397 55.249	5.028 98.730	5.828 187.50	6.656 334.57	7.596 596.60	8.704 1085.3	10.01 2013.2	11.41 3583.2
0.00220 1/ 455	3.390 15.338	3.643 20.858	3.884 27.453	4.190 38.042	4.622 58.083	5.285 103.78	6.126 197.06	6.994 351.57	7.981 626.83	9.144 1140.2	10.52 2114.7	11.98 3763.3
0.00240 1/ 417	3.550 16.050	3.814 21.837	4.066 28.740	4.386 39.822	4.838 60.794	5.531 108.61	6.410 206.20	7.318 367.84	8.349 655.75	9.565 1192.6	11.00 2211.7	12.53 3935.6
0.00260 1/ 385	3.703 16.752	3.978 22.777	4.241 29.976	4.574 41.531	5.045 63.398	5.768 113.25	6.683 214.98	7.628 383.45	8.703 683.51	9.968 1243.0	11.46 2304.8	13.05 4100.9
0.00280 1/ 357	3.851 17.419	4.136 23.683	4.409 31.167	4.756 43.178	5.245 65.906	5.995 117.71	6.945 223.43	7.928 398.48	9.043 710.23	10.36 1291.4	11.91 2394.4	13.56 4260.0
	900	1000	1100	1300	1500	1900	2400	3000	3800	4800	6000	7500

θ_{medial} for part-full circular pipes.

$k_s = 0.015$ mm $S = 0.00050$ to 0.00280

$k_s = 0.015$ mm
$S = 0.0030$ to 0.011

Water (or sewage) at 15°C;
full bore conditions.

ie hydraulic gradient =
1 in 333 to 1 in 71

velocities in ms^{-1}
discharges in m^{3}s^{-1}

Gradient	(Equivalent) Pipe diameters in m											
	2.400	2.700	3.000	3.400	4.000	5.000	6.400	8.000	10.00	12.60	16.00	20.00
0.00300 1/ 333	3.993 18.064	4.289 24.558	4.572 32.317	4.931 44.768	5.437 68.328	6.215 122.03	7.199 231.59	8.216 413.00	9.371 736.03	10.73 1338.2	12.34 2481.0	14.05 4413.6
0.00320 1/ 313	4.131 18.689	4.437 25.406	4.729 33.430	5.100 46.308	5.624 70.673	6.427 126.20	7.444 239.49	8.496 427.04	9.689 761.00	11.10 1383.5	12.76 2564.7	14.52 4562.2
0.00340 1/ 294	4.265 19.294	4.581 26.228	4.882 34.510	5.265 47.801	5.805 72.947	6.634 130.25	7.683 247.15	8.767 440.67	9.998 785.21	11.45 1427.4	13.16 2645.9	14.98 4706.3
0.00360 1/ 278	4.395 19.883	4.720 27.027	5.031 35.560	5.425 49.253	5.981 75.157	6.834 134.18	7.914 254.59	9.030 453.90	10.30 808.73	11.79 1470.0	13.55 2724.7	15.43 4846.2
0.00380 1/ 263	4.522 20.456	4.856 27.804	5.175 36.582	5.580 50.665	6.152 77.308	7.029 138.01	8.139 261.83	9.286 466.78	10.59 831.61	12.12 1511.5	13.93 2801.4	15.86 4982.3
0.00400 1/ 250	4.645 21.015	4.989 28.562	5.316 37.577	5.732 52.042	6.319 79.404	7.219 141.75	8.358 268.89	9.536 479.32	10.87 853.91	12.45 1552.0	14.30 2876.1	16.28 5114.9
0.00420 1/ 238	4.766 21.560	5.118 29.302	5.454 38.549	5.880 53.386	6.482 81.450	7.404 145.39	8.572 275.78	9.779 491.56	11.15 875.66	12.76 1591.4	14.67 2949.1	16.69 5244.3
0.00440 1/ 227	4.884 22.093	5.244 30.025	5.588 39.499	6.025 54.699	6.641 83.448	7.586 148.94	8.782 282.50	10.02 503.52	11.42 896.91	13.07 1629.9	15.02 3020.3	17.10 5370.7
0.00460 1/ 217	4.999 22.614	5.368 30.732	5.719 40.428	6.166 55.983	6.796 85.403	7.763 152.42	8.986 289.08	10.25 515.21	11.68 917.69	13.37 1667.6	15.37 3089.9	17.49 5494.2
0.00480 1/ 208	5.112 23.124	5.488 31.424	5.848 41.337	6.305 57.240	6.948 87.317	7.936 155.83	9.186 295.52	10.48 526.66	11.94 938.02	13.67 1704.4	15.71 3158.0	17.87 5615.1
0.00500 1/ 200	5.222 23.624	5.607 32.103	5.974 42.228	6.440 58.472	7.098 89.191	8.106 159.16	9.382 301.83	10.70 537.87	12.20 957.94	13.96 1740.5	16.04 3224.8	18.25 5733.5
0.00550 1/ 182	5.489 24.833	5.893 33.743	6.279 44.382	6.768 61.449	7.458 93.724	8.517 167.23	9.856 317.07	11.24 564.97	12.81 1006.1	14.66 1827.8	16.84 3386.0	19.16 6019.7
0.00600 1/ 167	5.745 25.990	6.167 35.312	6.570 46.443	7.082 64.297	7.803 98.058	8.910 174.94	10.31 331.65	11.76 590.88	13.40 1052.1	15.33 1911.2	17.61 3540.2	20.03 6293.2
0.00650 1/ 154	5.990 27.100	6.430 36.818	6.850 48.421	7.383 67.032	8.134 102.22	9.287 182.34	10.74 345.64	12.25 615.74	13.96 1096.3	15.97 1991.2	18.34 3688.1	20.87 6555.6
0.00700 1/ 143	6.227 28.170	6.684 38.269	7.120 50.326	7.673 69.664	8.453 106.23	9.650 189.47	11.16 359.11	12.73 639.68	14.50 1138.8	16.59 2068.3	19.05 3830.4	21.67 6808.1
0.00750 1/ 133	6.455 29.203	6.929 39.669	7.380 52.166	7.953 72.206	8.761 110.09	10.000 196.35	11.57 372.11	13.19 662.78	15.02 1179.8	17.18 2142.6	19.73 3967.8	22.45 7051.8
0.00800 1/ 125	6.676 30.202	7.165 41.025	7.632 53.946	8.224 74.667	9.059 113.84	10.34 203.01	11.96 384.69	13.63 685.13	15.53 1219.5	17.76 2214.6	20.40 4100.8	23.20 7287.6
0.00850 1/ 118	6.890 31.172	7.395 42.340	7.876 55.673	8.487 77.052	9.348 117.47	10.67 209.46	12.34 396.89	14.06 706.80	16.02 1258.0	18.32 2284.3	21.04 4229.6	23.92 7516.1
0.00900 1/ 111	7.099 32.114	7.618 43.618	8.113 57.350	8.742 79.370	9.628 120.99	10.99 215.73	12.71 408.74	14.48 727.85	16.49 1295.4	18.86 2352.0	21.66 4354.7	24.63 7738.1
0.00950 1/ 105	7.301 33.031	7.835 44.861	8.344 58.982	8.990 81.625	9.901 124.42	11.30 221.83	13.06 420.27	14.89 748.33	16.96 1331.7	19.39 2417.9	22.26 4476.4	25.32 7954.0
0.01000 1/ 100	7.499 33.924	8.047 46.072	8.569 60.573	9.232 83.823	10.17 127.77	11.60 227.78	13.41 431.50	15.28 768.28	17.41 1367.2	19.91 2482.1	22.85 4595.0	25.99 8164.3
0.01100 1/ 91	7.880 35.647	8.455 48.409	9.003 63.641	9.699 88.062	10.68 134.21	12.18 239.24	14.09 453.16	16.05 806.75	18.28 1435.5	20.90 2605.8	23.99 4823.6	27.28 8569.7
0.01200 1/ 83	8.244 37.295	8.845 50.643	9.418 66.574	10.15 92.115	11.17 140.38	12.74 250.20	14.73 473.86	16.78 843.52	19.11 1500.7	21.85 2724.1	25.08 5042.1	28.51 8957.2
0.01300 1/ 77	8.594 38.877	9.220 52.788	9.817 69.390	10.57 96.004	11.64 146.29	13.28 260.72	15.35 493.72	17.48 878.80	19.91 1563.4	22.76 2837.5	26.12 5251.7	29.69 9329.0
0.01400 1/ 71	8.930 40.400	9.580 54.853	10.20 72.100	10.99 99.749	12.09 151.99	13.79 270.84	15.94 512.84	18.16 912.76	20.67 1623.7	23.63 2946.7	27.12 5453.4	30.83 9686.7
	1600	1700	1900	2200	2600	3200	4100	5000	6500	8000	10000	13000

θ_{medial} for part-full circular pipes.

$k_s = 0.015$ mm $S = 0.0030$ to 0.014

Gradient	(Equivalent) Pipe diameters in m											
	2.400	2.700	3.000	3.400	4.000	5.000	6.400	8.000	10.00	12.60	16.00	20.00
0.01500 1/ 67	9.255 41.871	9.929 56.847	10.57 74.717	11.38 103.36	12.53 157.49	14.29 280.61	16.52 531.30	18.81 945.53	21.41 1681.8	24.48 3052.1	28.09 5648.1	31.93 10032
0.01600 1/ 62	9.570 43.294	10.27 58.776	10.93 77.249	11.77 106.86	12.96 162.81	14.77 290.07	17.07 549.15	19.44 977.24	22.13 1738.1	25.30 3154.1	29.03 5836.4	33.00 10366
0.01700 1/ 59	9.875 44.674	10.59 60.646	11.28 79.705	12.14 110.25	13.37 167.96	15.24 299.23	17.61 566.46	20.05 1008.0	22.83 1792.7	26.09 3252.9	29.94 6018.9	34.03 10689
0.01800 1/ 56	10.17 46.014	10.91 62.463	11.61 82.090	12.51 113.55	13.76 172.97	15.69 308.14	18.13 583.27	20.65 1037.8	23.50 1845.7	26.86 3348.8	30.82 6196.1	35.03 11004
0.01900 1/ 53	10.46 47.318	11.22 64.231	11.94 84.410	12.86 116.75	14.15 177.85	16.13 316.80	18.64 599.62	21.22 1066.9	24.16 1897.2	27.61 3442.2	31.67 6368.5	36.00 11309
0.02000 1/ 50	10.74 48.589	11.52 65.953	12.26 86.671	13.20 119.87	14.53 182.59	16.56 325.23	19.13 615.55	21.79 1095.1	24.80 1947.4	28.34 3533.1	32.51 6536.5	36.95 11607
0.02200 1/ 45	11.28 51.040	12.10 69.275	12.88 91.030	13.87 125.89	15.26 191.75	17.39 341.50	20.09 646.27	22.87 1149.7	26.03 2044.2	29.74 3708.4	34.12 6860.2	38.77 12181
0.02400 1/ 42	11.80 53.383	12.65 72.451	13.47 95.198	14.50 131.65	15.96 200.50	18.18 357.05	21.00 675.63	23.91 1201.8	27.21 2136.7	31.08 3875.9	35.66 7169.6	40.52 12730
0.02600 1/ 38	12.30 55.631	13.19 75.498	14.03 99.197	15.11 137.17	16.62 208.89	18.94 371.97	21.88 703.79	24.90 1251.8	28.34 2225.5	32.37 4036.6	37.13 7466.4	42.20 13256
0.02800 1/ 36	12.78 57.796	13.70 78.431	14.58 103.05	15.69 142.49	17.27 216.98	19.68 386.33	22.72 730.89	25.86 1299.9	29.42 2310.9	33.61 4191.3	38.56 7752.0	43.81 13763
0.03000 1/ 33	13.24 59.885	14.19 81.263	15.10 106.76	16.26 147.62	17.89 224.78	20.38 400.19	23.53 757.05	26.79 1346.4	30.47 2393.3	34.81 4340.5	39.93 8027.6	45.36 14251
0.03200 1/ 31	13.68 61.906	14.67 84.002	15.61 110.36	16.81 152.58	18.49 232.32	21.06 413.59	24.32 782.36	27.68 1391.3	31.49 2473.0	35.97 4484.8	41.25 8294.2	46.87 14724
0.03400 1/ 29	14.12 63.866	15.14 86.658	16.11 113.84	17.34 157.40	19.07 239.64	21.73 426.59	25.08 806.89	28.54 1434.8	32.47 2550.3	37.09 4624.8	42.54 8552.6	48.33 15182
0.03600 1/ 28	14.54 65.769	15.59 89.237	16.58 117.23	17.85 162.07	19.64 246.74	22.37 439.21	25.82 830.71	29.39 1477.1	33.43 2625.3	38.18 4760.6	43.78 8803.5	49.74 15627
0.03800 1/ 26	14.95 67.621	16.02 91.746	17.05 120.52	18.35 166.62	20.18 253.65	22.99 451.48	26.54 853.88	30.20 1518.2	34.36 2698.3	39.24 4892.8	45.00 9047.5	51.12 16059
0.04000 1/ 25	15.35 69.425	16.45 94.191	17.50 123.73	18.84 171.04	20.72 260.38	23.60 463.44	27.24 876.45	31.00 1558.3	35.26 2769.4	40.27 5021.5	46.18 9285.2	52.46 16481
0.04200 1/ 24	15.74 71.185	16.87 96.576	17.95 126.86	19.31 175.36	21.24 266.95	24.20 475.11	27.93 898.47	31.78 1597.4	36.14 2838.7	41.28 5147.1	47.33 9517.0	53.77 16892
0.04400 1/ 23	16.12 72.904	17.27 98.905	18.38 129.91	19.78 179.58	21.75 273.36	24.78 486.50	28.60 919.97	32.54 1635.5	37.01 2906.4	42.26 5269.7	48.46 9743.4	55.05 17293
0.04600 1/ 22	16.49 74.584	17.67 101.18	18.80 132.90	20.23 183.71	22.25 279.63	25.34 497.63	29.25 940.98	33.28 1672.9	37.85 2972.6	43.22 5389.5	49.56 9964.7	56.30 17686
0.04800 1/ 21	16.85 76.229	18.06 103.41	19.22 135.82	20.68 187.75	22.74 285.77	25.90 508.53	29.89 961.55	34.01 1709.4	38.67 3037.4	44.16 5506.8	50.64 10181	57.52 18070
0.05000 1/ 20	17.21 77.840	18.44 105.59	19.62 138.69	21.11 191.70	23.22 291.78	26.44 519.21	30.52 981.70	34.72 1745.1	39.48 3100.8	45.09 5621.7	51.69 10393	58.71 18446
0.05500 1/ 18	18.07 81.733	19.36 110.87	20.60 145.61	22.17 201.25	24.37 306.30	27.76 545.00	32.03 1030.4	36.44 1831.5	41.43 3254.1	47.31 5899.2	54.24 10906	61.61 19354
0.06000 1/ 17	18.89 85.453	20.24 115.91	21.53 152.22	23.17 210.38	25.48 320.18	29.01 569.64	33.47 1076.9	38.08 1914.0	43.30 3400.6	49.44 6164.3	56.68 11395	64.37 20222
0.06500 1/ 15	19.68 89.023	21.09 120.74	22.43 158.57	24.14 219.14	26.54 333.49	30.22 593.29	34.86 1121.5	39.65 1993.2	45.09 3541.1	51.48 6418.7	59.01 11865	67.02 21055
0.07000 1/ 14	20.44 92.459	21.90 125.40	23.30 164.67	25.07 227.57	27.56 346.30	31.37 616.04	36.20 1164.4	41.17 2069.4	46.81 3676.3	53.44 6663.5	61.26 12317	69.57 21856
0.07500 1/ 13	21.17 95.774	22.69 129.89	24.13 170.57	25.96 235.71	28.54 358.66	32.49 637.99	37.48 1205.9	42.63 2143.0	48.47 3806.7	55.33 6899.6	63.43 12753	72.03 22629
	2800	3100	3400	3900	4600	5500	7500	9000	11000	15000	18000	23000

θ_{medial} for part-full circular pipes.

k_s = 0.030 mm
S = 0.00010 to 0.00048

Water (or sewage) at 15°C;
full bore conditions.

ie hydraulic gradient =
1 in 10000 to 1 in 2083

velocities in ms^{-1}
discharges in m^3s^{-1}

Gradient	(Equivalent) Pipe diameters in m											
	2.400	2.700	3.000	3.400	4.000	5.000	6.400	8.000	10.00	12.60	16.00	20.00
0.00010 1/10000	0.643 2.9067	0.692 3.9624	0.739 5.2265	0.800 7.2603	0.885 11.120	1.016 19.951	1.183 38.053	1.356 68.154	1.553 121.97	1.786 222.67	2.062 414.52	2.356 740.13
0.00011 1/ 9091	0.676 3.0598	0.728 4.1707	0.778 5.5009	0.842 7.6407	0.931 11.701	1.069 20.991	1.244 40.029	1.426 71.681	1.633 128.26	1.878 234.13	2.167 435.77	2.476 777.97
0.00012 1/ 8333	0.709 3.2065	0.763 4.3703	0.815 5.7637	0.882 8.0050	0.975 12.258	1.120 21.986	1.303 41.920	1.493 75.057	1.710 134.28	1.966 245.09	2.268 456.10	2.592 814.17
0.00013 1/ 7692	0.740 3.3476	0.797 4.5622	0.851 6.0163	0.920 8.3552	1.018 12.793	1.168 22.943	1.360 43.737	1.558 78.300	1.783 140.07	2.050 255.61	2.366 475.63	2.702 848.93
0.00014 1/ 7143	0.770 3.4836	0.829 4.7472	0.886 6.2599	0.957 8.6929	1.059 13.308	1.215 23.864	1.414 45.488	1.620 81.426	1.854 145.64	2.131 265.75	2.459 494.45	2.809 882.41
0.00015 1/ 6667	0.799 3.6151	0.860 4.9261	0.919 6.4954	0.993 9.0193	1.099 13.807	1.261 24.755	1.467 47.181	1.680 84.446	1.923 151.03	2.210 275.55	2.550 512.61	2.912 914.75
0.00016 1/ 6250	0.827 3.7425	0.891 5.0994	0.951 6.7236	1.028 9.3355	1.137 14.290	1.305 25.618	1.518 48.820	1.738 87.370	1.989 156.24	2.286 285.03	2.637 530.20	3.011 946.05
0.00017 1/ 5882	0.855 3.8662	0.920 5.2676	0.983 6.9451	1.062 9.6424	1.174 14.758	1.347 26.456	1.567 50.410	1.795 90.207	2.054 161.30	2.360 294.23	2.722 547.26	3.108 976.40
0.00018 1/ 5556	0.881 3.9865	0.949 5.4312	1.013 7.1604	1.095 9.9409	1.211 15.214	1.389 27.270	1.615 51.956	1.849 92.965	2.116 166.22	2.431 303.18	2.804 563.84	3.202 1005.9
0.00019 1/ 5263	0.907 4.1037	0.976 5.5906	1.043 7.3702	1.127 10.232	1.246 15.658	1.429 28.063	1.662 53.461	1.903 95.650	2.177 171.00	2.501 311.88	2.885 579.98	3.293 1034.6
0.00020 1/ 5000	0.932 4.2180	1.004 5.7460	1.072 7.5748	1.158 10.515	1.280 16.091	1.469 28.836	1.707 54.929	1.955 98.268	2.237 175.67	2.569 320.36	2.963 595.71	3.382 1062.6
0.00022 1/ 4545	0.981 4.4388	1.056 6.0462	1.127 7.9698	1.218 11.062	1.347 16.926	1.545 30.328	1.796 57.761	2.055 103.32	2.351 184.67	2.701 336.73	3.114 626.05	3.554 1116.6
0.00024 1/ 4167	1.028 4.6502	1.106 6.3336	1.181 8.3481	1.276 11.586	1.411 17.726	1.617 31.757	1.880 60.472	2.152 108.15	2.461 193.28	2.826 352.39	3.258 655.07	3.718 1168.2
0.00026 1/ 3846	1.073 4.8534	1.154 6.6099	1.232 8.7116	1.332 12.090	1.472 18.495	1.687 33.129	1.961 63.076	2.244 112.80	2.566 201.55	2.947 367.42	3.397 682.93	3.876 1217.7
0.00028 1/ 3571	1.116 5.0492	1.201 6.8761	1.282 9.0620	1.385 12.575	1.531 19.235	1.755 34.451	2.039 65.585	2.333 117.27	2.668 209.52	3.063 381.90	3.530 709.77	4.028 1265.4
0.00030 1/ 3333	1.158 5.2386	1.246 7.1335	1.330 9.4006	1.437 13.044	1.588 19.951	1.820 35.729	2.114 68.008	2.419 121.59	2.766 217.22	3.175 395.89	3.659 735.67	4.175 1311.5
0.00032 1/ 3125	1.199 5.4220	1.289 7.3828	1.376 9.7285	1.487 13.498	1.643 20.644	1.883 36.966	2.187 70.355	2.502 125.77	2.861 224.67	3.284 409.42	3.784 760.75	4.317 1356.1
0.00034 1/ 2941	1.238 5.6000	1.332 7.6247	1.421 10.047	1.535 13.939	1.696 21.316	1.944 38.166	2.258 72.631	2.583 129.83	2.953 231.89	3.389 422.55	3.905 785.07	4.454 1399.3
0.00036 1/ 2778	1.276 5.7731	1.373 7.8599	1.465 10.356	1.582 14.368	1.748 21.970	2.003 39.333	2.326 74.843	2.661 133.77	3.042 238.91	3.491 435.30	4.022 808.69	4.588 1441.3
0.00038 1/ 2632	1.313 5.9416	1.413 8.0890	1.508 10.658	1.628 14.785	1.799 22.606	2.061 40.468	2.393 76.996	2.738 137.61	3.129 245.74	3.591 447.71	4.136 831.68	4.718 1482.2
0.00040 1/ 2500	1.350 6.1059	1.452 8.3123	1.549 10.951	1.673 15.192	1.848 23.227	2.117 41.575	2.459 79.095	2.812 141.35	3.214 252.40	3.688 459.81	4.248 854.09	4.845 1522.0
0.00042 1/ 2381	1.385 6.2664	1.490 8.5303	1.590 11.238	1.717 15.589	1.897 23.832	2.172 42.656	2.522 81.144	2.885 144.99	3.296 258.90	3.782 471.61	4.357 875.95	4.968 1560.9
0.00044 1/ 2273	1.420 6.4232	1.527 8.7435	1.630 11.518	1.760 15.977	1.944 24.424	2.226 43.712	2.585 83.146	2.956 148.56	3.377 265.25	3.875 483.14	4.463 897.30	5.089 1598.8
0.00046 1/ 2174	1.454 6.5767	1.564 8.9520	1.668 11.793	1.802 16.356	1.990 25.003	2.279 44.745	2.645 85.104	3.025 152.05	3.456 271.46	3.965 494.42	4.567 918.18	5.207 1635.9
0.00048 1/ 2083	1.487 6.7270	1.599 9.1562	1.706 12.061	1.842 16.728	2.035 25.570	2.330 45.756	2.705 87.020	3.093 155.46	3.534 277.54	4.054 505.45	4.668 938.62	5.323 1672.2
	550	600	650	750	900	1100	1400	1800	2200	2800	3600	4400

θ_{medial} for part-full circular pipes.

k_s = 0.030 mm S = 0.00010 to 0.00048

k_s = 0.030 mm
S = 0.00050 to 0.00280

ie hydraulic gradient =
1 in 2000 to 1 in 357

Water (or sewage) at 15°C;
full bore conditions.

velocities in ms^{-1}
discharges in m^{3}s^{-1}

(Equivalent) Pipe diameters in m

Gradient	2.400	2.700	3.000	3.400	4.000	5.000	6.400	8.000	10.00	12.60	16.00	20.00
0.00050	1.520	1.634	1.744	1.883	2.079	2.381	2.763	3.159	3.610	4.140	4.768	5.436
1/ 2000	6.8743	9.3563	12.324	17.093	26.126	46.747	88.899	158.81	283.49	516.27	958.64	1707.8
0.00055	1.598	1.719	1.834	1.980	2.186	2.503	2.905	3.320	3.793	4.350	5.009	5.710
1/ 1818	7.2307	9.8405	12.961	17.974	27.470	49.144	93.441	166.90	297.89	542.41	1007.0	1793.8
0.00060	1.674	1.800	1.920	2.073	2.288	2.620	3.040	3.474	3.968	4.551	5.239	5.972
1/ 1667	7.5718	10.304	13.570	18.817	28.755	51.437	97.785	174.63	311.65	567.41	1053.3	1876.0
0.00065	1.746	1.877	2.003	2.162	2.387	2.732	3.169	3.622	4.136	4.743	5.460	6.223
1/ 1538	7.8995	10.749	14.156	19.627	29.990	53.638	101.96	182.06	324.87	591.40	1097.7	1954.9
0.00070	1.816	1.952	2.082	2.248	2.481	2.840	3.294	3.764	4.298	4.928	5.672	6.465
1/ 1429	8.2153	11.178	14.720	20.408	31.180	55.759	105.97	189.21	337.60	614.50	1140.5	2030.9
0.00075	1.883	2.025	2.159	2.331	2.573	2.944	3.415	3.902	4.455	5.107	5.878	6.698
1/ 1333	8.5204	11.592	15.264	21.162	32.329	57.807	109.85	196.12	349.88	636.80	1181.8	2104.2
0.00080	1.949	2.095	2.234	2.411	2.661	3.045	3.531	4.035	4.606	5.280	6.076	6.924
1/ 1250	8.8158	11.993	15.792	21.891	33.441	59.789	113.61	202.80	361.77	658.37	1221.7	2175.1
0.00085	2.012	2.163	2.306	2.489	2.747	3.143	3.645	4.164	4.753	5.448	6.269	7.143
1/ 1176	9.1024	12.383	16.303	22.599	34.520	61.712	117.25	209.28	373.30	679.30	1260.4	2243.9
0.00090	2.074	2.229	2.377	2.565	2.830	3.238	3.755	4.289	4.896	5.611	6.456	7.355
1/ 1111	9.3809	12.761	16.801	23.287	35.568	63.580	120.79	215.58	384.50	699.62	1298.0	2310.7
0.00095	2.134	2.293	2.445	2.639	2.912	3.331	3.862	4.411	5.034	5.769	6.638	7.562
1/ 1053	9.6521	13.129	17.285	23.957	36.588	65.399	124.23	221.70	395.40	719.40	1334.6	2375.7
0.00100	2.192	2.356	2.512	2.711	2.991	3.421	3.966	4.529	5.170	5.924	6.815	7.764
1/ 1000	9.9165	13.488	17.756	24.610	37.583	67.171	127.58	227.67	406.01	738.66	1370.3	2439.0
0.00110	2.305	2.477	2.641	2.849	3.143	3.595	4.167	4.759	5.430	6.222	7.157	8.152
1/ 909	10.427	14.181	18.667	25.869	39.502	70.589	134.05	239.19	426.49	775.82	1439.0	2561.1
0.00120	2.413	2.592	2.764	2.982	3.289	3.762	4.359	4.978	5.680	6.507	7.484	8.524
1/ 833	10.915	14.843	19.538	27.074	41.337	73.858	140.24	250.20	446.07	811.34	1504.7	2677.8
0.00130	2.516	2.704	2.882	3.109	3.430	3.921	4.544	5.188	5.919	6.780	7.797	8.880
1/ 769	11.384	15.480	20.374	28.230	43.099	76.997	146.18	260.76	464.86	845.42	1567.8	2789.7
0.00140	2.616	2.811	2.996	3.232	3.565	4.075	4.722	5.390	6.149	7.043	8.099	9.223
1/ 714	11.835	16.092	21.179	29.344	44.795	80.019	151.90	270.94	482.95	878.23	1628.5	2897.5
0.00150	2.713	2.914	3.106	3.351	3.695	4.224	4.893	5.586	6.371	7.297	8.391	9.554
1/ 667	12.271	16.684	21.957	30.420	46.434	82.936	157.42	280.76	500.40	909.90	1687.0	3001.5
0.00160	2.806	3.014	3.213	3.465	3.821	4.368	5.060	5.775	6.586	7.543	8.673	9.874
1/ 625	12.693	17.257	22.710	31.461	48.019	85.760	162.76	290.26	517.30	940.54	1743.7	3102.1
0.00170	2.896	3.111	3.316	3.576	3.944	4.507	5.221	5.958	6.795	7.781	8.946	10.18
1/ 588	13.103	17.813	23.440	32.471	49.557	88.498	167.94	299.47	533.68	970.24	1798.7	3199.6
0.00180	2.984	3.205	3.416	3.684	4.063	4.643	5.377	6.136	6.998	8.013	9.211	10.49
1/ 556	13.501	18.353	24.150	33.452	51.051	91.158	172.98	308.42	549.58	999.09	1852.0	3294.4
0.00190	3.070	3.297	3.514	3.790	4.178	4.774	5.529	6.309	7.195	8.238	9.469	10.78
1/ 526	13.888	18.879	24.840	34.407	52.505	93.746	177.87	317.13	565.06	1027.2	1903.9	3386.5
0.00200	3.153	3.387	3.609	3.892	4.291	4.903	5.677	6.478	7.387	8.457	9.721	11.07
1/ 500	14.265	19.391	25.513	35.337	53.921	96.268	182.64	325.61	580.14	1054.5	1954.5	3476.3
0.00220	3.314	3.559	3.793	4.090	4.508	5.151	5.963	6.803	7.757	8.880	10.21	11.62
1/ 455	14.994	20.379	26.811	37.132	56.654	101.13	191.84	341.97	609.21	1107.2	2052.0	3649.3
0.00240	3.468	3.724	3.969	4.279	4.716	5.387	6.237	7.114	8.111	9.284	10.67	12.14
1/ 417	15.690	21.324	28.053	38.848	59.267	105.78	200.64	357.61	637.00	1157.6	2145.1	3814.7
0.00260	3.616	3.883	4.137	4.460	4.916	5.615	6.499	7.413	8.450	9.672	11.11	12.65
1/ 385	16.359	22.231	29.244	40.495	61.774	110.25	209.08	372.61	663.67	1206.0	2234.5	3973.4
0.00280	3.758	4.035	4.300	4.635	5.108	5.834	6.752	7.700	8.777	10.04	11.54	13.13
1/ 357	17.002	23.104	30.391	42.081	64.188	114.54	217.20	387.06	689.33	1252.5	2320.6	4126.1
	900	1000	1100	1300	1500	1900	2400	3000	3800	4800	6000	7500

θ_{medial} for part-full circular pipes.

k_s = 0.030 mm   S = 0.00050 to 0.00280

$k_s = 0.030$ mm
$S = 0.0030$ to 0.014

Water (or sewage) at 15°C;
full bore conditions.

ie hydraulic gradient =
1 in 333 to 1 in 71

velocities in ms^{-1}
discharges in m^3s^{-1}

Gradient	(Equivalent) Pipe diameters in m											
	2.400	2.700	3.000	3.400	4.000	5.000	6.400	8.000	10.00	12.60	16.00	20.00
0.00300 1/ 333	3.896 17.624	4.183 23.948	4.456 31.499	4.804 43.612	5.293 66.519	6.045 118.69	6.995 225.04	7.978 401.00	9.092 714.10	10.41 1297.4	11.95 2403.6	13.60 4273.4
0.00320 1/ 313	4.029 18.226	4.325 24.764	4.608 32.571	4.967 45.094	5.473 68.774	6.249 122.70	7.231 232.63	8.246 414.48	9.397 738.07	10.75 1340.8	12.35 2483.9	14.06 4416.0
0.00340 1/ 294	4.158 18.809	4.463 25.555	4.755 33.611	5.125 46.531	5.647 70.961	6.447 126.59	7.460 239.99	8.506 427.56	9.693 761.30	11.09 1383.0	12.74 2561.8	14.50 4554.2
0.00360 1/ 278	4.283 19.376	4.598 26.324	4.898 34.620	5.279 47.926	5.816 73.085	6.640 130.37	7.682 247.13	8.759 440.26	9.980 783.86	11.42 1423.9	13.12 2637.4	14.92 4688.3
0.00380 1/ 263	4.405 19.928	4.728 27.072	5.037 35.603	5.428 49.284	5.980 75.152	6.827 134.05	7.898 254.08	9.004 452.61	10.26 805.80	11.74 1463.6	13.48 2710.9	15.34 4818.8
0.00400 1/ 250	4.524 20.465	4.856 27.801	5.172 36.560	5.574 50.607	6.141 77.165	7.009 137.63	8.109 260.85	9.244 464.64	10.53 827.18	12.05 1502.4	13.84 2782.6	15.74 4946.0
0.00420 1/ 238	4.640 20.989	4.980 28.513	5.304 37.494	5.716 51.898	6.297 79.129	7.187 141.12	8.314 267.46	9.477 476.38	10.80 848.04	12.35 1540.2	14.19 2852.4	16.14 5070.0
0.00440 1/ 227	4.753 21.502	5.101 29.207	5.433 38.406	5.855 53.159	6.450 81.048	7.361 144.54	8.514 273.91	9.705 487.85	11.06 868.40	12.65 1577.1	14.53 2920.7	16.52 5191.1
0.00460 1/ 217	4.864 22.003	5.220 29.887	5.560 39.299	5.991 54.392	6.599 82.924	7.531 147.87	8.710 280.21	9.928 499.06	11.31 888.31	12.94 1613.2	14.86 2987.4	16.90 5309.5
0.00480 1/ 208	4.972 22.493	5.336 30.551	5.683 40.172	6.124 55.598	6.745 84.760	7.697 151.14	8.902 286.39	10.15 510.02	11.56 907.80	13.22 1648.5	15.18 3052.7	17.27 5425.3
0.00500 1/ 200	5.078 22.973	5.450 31.203	5.804 41.027	6.254 56.780	6.888 86.558	7.860 154.34	9.090 292.43	10.36 520.77	11.80 926.88	13.50 1683.1	15.50 3116.6	17.63 5538.7
0.00550 1/ 182	5.335 24.134	5.725 32.777	6.097 43.094	6.569 59.637	7.234 90.904	8.254 162.07	9.544 307.04	10.88 546.73	12.39 972.99	14.17 1766.7	16.27 3271.1	18.50 5812.9
0.00600 1/ 167	5.580 25.244	5.988 34.282	6.376 45.070	6.869 62.367	7.565 95.059	8.630 169.46	9.978 321.00	11.37 571.54	12.95 1017.1	14.81 1846.5	17.00 3418.7	19.34 6074.8
0.00650 1/ 154	5.816 26.309	6.240 35.727	6.644 46.967	7.158 64.988	7.882 99.045	8.991 176.55	10.39 334.40	11.84 595.34	13.49 1059.3	15.42 1923.2	17.71 3560.3	20.14 6326.1
0.00700 1/ 143	6.042 27.334	6.483 37.117	6.903 48.792	7.436 67.510	8.187 102.88	9.339 183.37	10.80 347.29	12.30 618.25	14.01 1100.0	16.01 1996.9	18.39 3696.6	20.91 6567.9
0.00750 1/ 133	6.261 28.324	6.717 38.459	7.152 50.554	7.704 69.944	8.482 106.58	9.674 189.95	11.18 359.73	12.74 640.35	14.51 1139.3	16.59 2068.0	19.04 3828.1	21.65 6801.2
0.00800 1/ 125	6.473 29.282	6.944 39.757	7.393 52.259	7.963 72.299	8.767 110.17	9.999 196.32	11.56 371.77	13.16 661.74	14.99 1177.3	17.14 2136.9	19.67 3955.3	22.37 7026.9
0.00850 1/ 118	6.678 30.210	7.164 41.016	7.627 53.911	8.215 74.582	9.043 113.64	10.31 202.50	11.92 383.44	13.58 682.46	15.46 1214.1	17.67 2203.6	20.29 4078.6	23.06 7245.6
0.00900 1/ 111	6.877 31.112	7.377 42.239	7.854 55.516	8.459 76.800	9.312 117.01	10.62 208.50	12.27 394.77	13.98 702.59	15.91 1249.8	18.19 2268.4	20.88 4198.3	23.74 7458.0
0.00950 1/ 105	7.071 31.989	7.585 43.428	8.075 57.078	8.696 78.957	9.573 120.29	10.92 214.33	12.61 405.79	14.37 722.17	16.36 1284.6	18.70 2331.4	21.46 4314.7	24.40 7664.6
0.01000 1/ 100	7.260 32.844	7.787 44.587	8.290 58.599	8.928 81.058	9.827 123.49	11.21 220.01	12.95 416.53	14.75 741.24	16.79 1318.5	19.19 2392.7	22.02 4428.2	25.04 7865.8
0.01100 1/ 91	7.624 34.492	8.178 46.821	8.705 61.532	9.374 85.110	10.32 129.65	11.76 230.97	13.59 437.22	15.48 778.01	17.62 1383.8	20.14 2511.1	23.11 4646.8	26.27 8253.8
0.01200 1/ 83	7.973 36.068	8.551 48.957	9.102 64.336	9.801 88.983	10.79 135.54	12.30 241.44	14.21 457.00	16.18 813.15	18.41 1446.2	21.05 2624.1	24.15 4855.7	27.45 8624.5
0.01300 1/ 77	8.307 37.579	8.909 51.006	9.482 67.025	10.21 92.699	11.24 141.19	12.81 251.49	14.80 475.98	16.85 846.86	19.18 1506.0	21.91 2732.6	25.15 5056.2	28.58 8980.0
0.01400 1/ 71	8.628 39.034	9.253 52.978	9.848 69.614	10.60 96.274	11.67 146.63	13.30 261.15	15.36 494.24	17.49 879.30	19.91 1563.6	22.75 2836.9	26.11 5249.0	29.67 9322.1
	1500	1700	1900	2200	2600	3200	4100	5000	6500	8000	10000	13000

θ_{medial} **for part-full circular pipes.**

$k_s = 0.030$ mm $S = 0.0030$ to 0.014

$k_s = 0.030$ mm
$S = 0.015$ to 0.075

ie hydraulic gradient =
1 in 66 to 1 in 13

Water (or sewage) at 15°C;
full bore conditions.

velocities in ms^{-1}
discharges in m^{3}s^{-1}

| Gradient | (Equivalent) Pipe diameters in m | | | | | | | | | | | |
	2.400	2.700	3.000	3.400	4.000	5.000	6.400	8.000	10.00	12.60	16.00	20.00
0.01500	8.939	9.585	10.20	10.98	12.09	13.78	15.91	18.12	20.62	23.56	27.03	30.72
1/ 67	40.438	54.882	72.113	99.725	151.88	270.48	511.86	910.60	1619.2	2937.6	5435.1	9652.3
0.01600	9.239	9.907	10.54	11.35	12.49	14.24	16.44	18.72	21.30	24.34	27.93	31.74
1/ 62	41.796	56.723	74.529	103.06	156.96	279.51	528.91	940.87	1672.9	3035.0	5615.1	9971.6
0.01700	9.530	10.22	10.88	11.71	12.88	14.68	16.95	19.30	21.96	25.10	28.79	32.73
1/ 59	43.113	58.508	76.872	106.30	161.88	288.26	545.43	970.22	1725.1	3129.4	5789.5	10281
0.01800	9.813	10.52	11.20	12.05	13.26	15.11	17.45	19.87	22.61	25.83	29.64	33.68
1/ 56	44.392	60.241	79.148	109.44	166.66	296.75	561.47	998.71	1775.7	3221.1	5958.9	10582
0.01900	10.09	10.82	11.51	12.39	13.63	15.53	17.94	20.42	23.23	26.55	30.46	34.61
1/ 53	45.636	61.928	81.361	112.50	171.31	305.01	577.08	1026.4	1824.9	3310.3	6123.7	10874
0.02000	10.36	11.10	11.82	12.72	13.99	15.94	18.41	20.96	23.85	27.24	31.26	35.52
1/ 50	46.848	63.570	83.517	115.48	175.83	313.06	592.28	1053.4	1872.8	3397.1	6284.2	11159
0.02200	10.87	11.66	12.40	13.35	14.69	16.73	19.32	21.99	25.02	28.59	32.79	37.27
1/ 45	49.185	66.737	87.674	121.22	184.56	328.58	621.58	1105.5	1965.2	3564.6	6593.6	11707
0.02400	11.37	12.18	12.97	13.96	15.35	17.49	20.19	22.98	26.15	29.87	34.26	38.94
1/ 42	51.418	69.764	91.646	126.70	192.91	343.40	649.58	1155.2	2053.5	3724.6	6889.2	12232
0.02600	11.84	12.69	13.50	14.54	15.99	18.21	21.03	23.93	27.22	31.10	35.67	40.54
1/ 38	53.560	72.668	95.457	131.97	200.91	357.63	676.45	1202.9	2138.2	3878.0	7172.7	12735
0.02800	12.30	13.18	14.02	15.09	16.60	18.91	21.83	24.84	28.26	32.29	37.03	42.08
1/ 36	55.622	75.462	99.124	137.03	208.61	371.31	702.29	1248.8	2219.7	4025.7	7445.6	13219
0.03000	12.73	13.65	14.52	15.63	17.19	19.58	22.61	25.73	29.26	33.43	38.34	43.56
1/ 33	57.611	78.158	102.66	141.92	216.04	384.52	727.23	1293.1	2298.4	4168.1	7708.8	13686
0.03200	13.16	14.11	15.01	16.15	17.76	20.23	23.36	26.58	30.23	34.53	39.61	45.00
1/ 31	59.536	80.767	106.09	146.65	223.23	397.30	751.36	1336.0	2374.4	4306.0	7963.5	14138
0.03400	13.57	14.55	15.48	16.66	18.32	20.86	24.08	27.40	31.17	35.60	40.83	46.40
1/ 29	61.402	83.295	109.40	151.23	230.20	409.68	774.74	1377.5	2448.2	4439.5	8210.3	14575
0.03600	13.97	14.98	15.93	17.15	18.86	21.48	24.79	28.21	32.08	36.64	42.03	47.75
1/ 28	63.213	85.751	112.63	155.68	236.96	421.70	797.45	1417.8	2519.8	4569.2	8449.9	15001
0.03800	14.36	15.39	16.38	17.62	19.38	22.07	25.48	28.99	32.97	37.66	43.19	49.06
1/ 26	64.976	88.139	115.76	160.01	243.54	433.40	819.53	1457.0	2589.4	4695.4	8683.0	15414
0.04000	14.74	15.80	16.81	18.09	19.89	22.65	26.14	29.75	33.83	38.64	44.31	50.35
1/ 25	66.692	90.465	118.81	164.22	249.95	444.79	841.04	1495.2	2657.2	4818.2	8910.0	15817
0.04200	15.11	16.20	17.23	18.54	20.39	23.22	26.80	30.49	34.67	39.60	45.42	51.60
1/ 24	68.366	92.734	121.79	168.34	256.21	455.90	862.02	1532.5	2723.4	4938.1	9131.4	16210
0.04400	15.47	16.58	17.64	18.98	20.87	23.77	27.43	31.21	35.50	40.54	46.49	52.82
1/ 23	70.001	94.950	124.70	172.35	262.31	466.75	882.51	1568.9	2788.0	5055.1	9347.6	16593
0.04600	15.83	16.96	18.04	19.42	21.35	24.31	28.06	31.92	36.30	41.46	47.54	54.01
1/ 22	71.600	97.116	127.54	176.28	268.28	477.35	902.54	1604.4	2851.1	5169.5	9559.0	16968
0.04800	16.17	17.33	18.44	19.84	21.81	24.84	28.66	32.61	37.09	42.36	48.57	55.18
1/ 21	73.164	99.236	130.32	180.12	274.12	487.73	922.14	1639.2	2912.9	5281.4	9765.8	17335
0.05000	16.51	17.69	18.82	20.25	22.27	25.36	29.26	33.29	37.86	43.24	49.58	56.32
1/ 20	74.696	101.31	133.05	183.88	279.84	497.90	941.33	1673.3	2973.4	5391.0	9968.3	17694
0.05500	17.33	18.57	19.75	21.25	23.37	26.61	30.70	34.93	39.72	45.36	52.01	59.09
1/ 18	78.397	106.33	139.63	192.97	293.66	522.45	987.70	1755.7	3119.6	5655.9	10458	18562
0.06000	18.11	19.41	20.64	22.21	24.42	27.80	32.08	36.49	41.50	47.39	54.34	61.73
1/ 17	81.934	111.12	145.92	201.66	306.86	545.92	1032.0	1834.4	3259.3	5908.9	10925	19391
0.06500	18.86	20.21	21.50	23.13	25.43	28.95	33.40	37.99	43.20	49.34	56.57	64.26
1/ 15	85.326	115.72	151.95	209.99	319.53	568.42	1074.5	1909.8	3393.2	6151.6	11374	20187
0.07000	19.58	20.98	22.32	24.01	26.40	30.05	34.67	39.44	44.85	51.21	58.71	66.69
1/ 14	88.591	120.14	157.76	218.00	331.72	590.08	1115.4	1982.4	3522.2	6385.2	11805	20952
0.07500	20.28	21.73	23.11	24.86	27.33	31.12	35.90	40.83	46.43	53.02	60.78	69.04
1/ 13	91.741	124.41	163.36	225.74	343.48	610.98	1154.9	2052.5	3646.6	6610.5	12222	21691
	2700	3000	3400	3800	4500	5500	7000	9000	11000	14000	18000	23000

θ_{medial} for part-full circular pipes.

$k_s = 0.030$ mm $S = 0.0015$ to 0.075

$k_s = 0.060$ mm
$S = 0.00010$ to 0.00048
ie hydraulic gradient =
1 in 10000 to 1 in 2083

Water (or sewage) at 15°C;
full bore conditions.

velocities in ms⁻¹
discharges in m³s⁻¹

Gradient	\(Equivalent) Pipe diameters in m 2.400	2.700	3.000	3.400	4.000	5.000	6.400	8.000	10.00	12.60	16.00	20.00
0.00010 1/10000	0.634 2.8661	0.682 3.9053	0.728 5.1492	0.787 7.1494	0.871 10.943	0.999 19.617	1.162 37.378	1.331 66.882	1.523 119.58	1.749 218.10	2.017 405.60	2.303 723.54
0.00011 1/ 9091	0.667 3.0156	0.718 4.1086	0.766 5.4168	0.828 7.5202	0.916 11.509	1.051 20.628	1.222 39.297	1.399 70.305	1.600 125.68	1.838 229.19	2.120 426.16	2.420 760.11
0.00012 1/ 8333	0.698 3.1588	0.752 4.3033	0.803 5.6730	0.867 7.8752	0.959 12.051	1.100 21.596	1.279 41.133	1.464 73.579	1.674 131.51	1.923 239.80	2.217 445.82	2.531 795.08
0.00013 1/ 7692	0.729 3.2964	0.784 4.4903	0.837 5.9191	0.905 8.2161	1.000 12.571	1.147 22.525	1.333 42.896	1.526 76.722	1.746 137.11	2.005 249.98	2.311 464.69	2.638 828.64
0.00014 1/ 7143	0.758 3.4289	0.816 4.6706	0.871 6.1563	0.941 8.5446	1.040 13.073	1.193 23.420	1.386 44.595	1.587 79.750	1.814 142.51	2.083 259.78	2.402 482.86	2.740 860.95
0.00015 1/ 6667	0.786 3.5570	0.846 4.8447	0.903 6.3854	0.976 8.8620	1.079 13.557	1.237 24.285	1.437 46.235	1.645 82.674	1.881 147.72	2.159 269.25	2.489 500.40	2.840 892.14
0.00016 1/ 6250	0.814 3.6811	0.876 5.0133	0.935 6.6073	1.010 9.1693	1.116 14.026	1.279 25.122	1.487 47.823	1.701 85.504	1.945 152.76	2.233 278.41	2.573 517.37	2.936 922.32
0.00017 1/ 5882	0.840 3.8015	0.904 5.1770	0.965 6.8226	1.043 9.4675	1.152 14.481	1.321 25.934	1.534 49.363	1.756 88.248	2.007 157.65	2.304 287.29	2.655 533.83	3.029 951.58
0.00018 1/ 5556	0.866 3.9186	0.932 5.3361	0.995 7.0319	1.075 9.7573	1.188 14.923	1.361 26.723	1.581 50.859	1.809 90.915	2.068 162.40	2.373 295.92	2.735 549.81	3.119 980.00
0.00019 1/ 5263	0.891 4.0325	0.959 5.4909	1.024 7.2356	1.106 10.039	1.222 15.353	1.400 27.491	1.626 52.315	1.860 93.510	2.127 167.02	2.441 304.32	2.812 565.37	3.207 1007.7
0.00020 1/ 5000	0.916 4.1436	0.985 5.6419	1.052 7.4342	1.136 10.314	1.255 15.773	1.438 28.239	1.670 53.734	1.911 96.039	2.184 171.52	2.506 312.50	2.887 580.52	3.293 1034.6
0.00022 1/ 4545	0.963 4.3581	1.036 5.9333	1.106 7.8175	1.194 10.845	1.320 16.582	1.512 29.683	1.755 56.472	2.008 100.92	2.294 180.21	2.633 328.28	3.033 609.75	3.459 1086.5
0.00024 1/ 4167	1.009 4.5633	1.085 6.2121	1.158 8.1843	1.250 11.353	1.381 17.357	1.582 31.065	1.837 59.091	2.100 105.58	2.400 188.51	2.754 343.36	3.172 637.69	3.617 1136.2
0.00026 1/ 3846	1.052 4.7604	1.132 6.4799	1.208 8.5365	1.304 11.840	1.440 18.100	1.650 32.391	1.915 61.605	2.190 110.06	2.502 196.48	2.870 357.84	3.305 664.50	3.768 1183.9
0.00028 1/ 3571	1.094 4.9503	1.177 6.7379	1.256 8.8758	1.356 12.310	1.497 18.816	1.715 33.668	1.990 64.025	2.275 114.37	2.599 204.16	2.982 371.78	3.433 690.31	3.914 1229.7
0.00030 1/ 3333	1.135 5.1337	1.220 6.9871	1.302 9.2035	1.406 12.764	1.552 19.508	1.778 34.902	2.063 66.363	2.358 118.53	2.694 211.56	3.089 385.23	3.557 715.22	4.055 1274.0
0.00032 1/ 3125	1.174 5.3113	1.262 7.2284	1.347 9.5207	1.454 13.203	1.606 20.177	1.838 36.095	2.133 68.624	2.438 122.56	2.785 218.73	3.194 398.24	3.677 739.32	4.192 1316.8
0.00034 1/ 2941	1.212 5.4836	1.303 7.4624	1.390 9.8284	1.501 13.629	1.657 20.826	1.897 37.253	2.201 70.818	2.516 126.47	2.873 225.68	3.295 410.87	3.793 762.69	4.324 1358.4
0.00036 1/ 2778	1.249 5.6511	1.343 7.6899	1.433 10.127	1.547 14.042	1.708 21.457	1.955 38.378	2.268 72.948	2.591 130.26	2.959 232.44	3.393 423.12	3.906 785.38	4.452 1398.7
0.00038 1/ 2632	1.285 5.8141	1.382 7.9112	1.474 10.418	1.591 14.445	1.756 22.071	2.010 39.472	2.332 75.022	2.665 133.95	3.043 239.00	3.489 435.05	4.016 807.46	4.577 1437.9
0.00040 1/ 2500	1.320 5.9729	1.419 8.1270	1.514 10.702	1.634 14.838	1.804 22.669	2.065 40.539	2.395 77.042	2.736 137.55	3.125 245.40	3.582 446.67	4.123 828.97	4.699 1476.1
0.00042 1/ 2381	1.355 6.1280	1.456 8.3375	1.553 10.979	1.676 15.221	1.850 23.253	2.118 41.579	2.456 79.013	2.806 141.06	3.204 251.65	3.673 458.00	4.227 849.95	4.817 1513.4
0.00044 1/ 2273	1.388 6.2795	1.492 8.5433	1.591 11.249	1.718 15.595	1.896 23.824	2.169 42.596	2.516 80.938	2.874 144.49	3.282 257.75	3.762 469.07	4.329 870.44	4.933 1549.8
0.00046 1/ 2174	1.421 6.4276	1.527 8.7445	1.629 11.514	1.758 15.961	1.940 24.381	2.220 43.590	2.574 82.821	2.941 147.84	3.358 263.71	3.849 479.89	4.429 890.48	5.047 1585.4
0.00048 1/ 2083	1.453 6.5727	1.562 8.9415	1.666 11.773	1.797 16.319	1.984 24.927	2.270 44.563	2.632 84.664	3.006 151.12	3.432 269.54	3.934 490.49	4.526 910.08	5.157 1620.2
	550	600	650	750	900	1100	1400	1800	2200	2800	3500	4400

θ_{medial} for part-full circular pipes.

$k_s = 0.060$ mm $S = 0.00010$ to 0.00048

k_s = 0.060 mm
S = 0.00050 to 0.00280

ie hydraulic gradient =
1 in 2000 to 1 in 357

Water (or sewage) at 15°C;
full bore conditions.

velocities in ms^{-1}
discharges in m^{3}s^{-1}

| Gradient | (Equivalent) Pipe diameters in m | | | | | | | | | | | |
	2.400	2.700	3.000	3.400	4.000	5.000	6.400	8.000	10.00	12.60	16.00	20.00
0.00050	1.484	1.595	1.701	1.836	2.026	2.318	2.688	3.070	3.505	4.017	4.622	5.266
1/ 2000	6.7149	9.1345	12.027	16.670	25.462	45.516	86.469	154.33	275.26	500.86	929.29	1654.4
0.00055	1.560	1.677	1.788	1.930	2.129	2.435	2.824	3.225	3.681	4.218	4.853	5.528
1/ 1818	7.0585	9.6011	12.640	17.519	26.755	47.821	90.832	162.10	289.08	525.94	975.69	1736.8
0.00060	1.633	1.755	1.871	2.019	2.228	2.548	2.953	3.372	3.849	4.410	5.073	5.779
1/ 1667	7.3873	10.047	13.227	18.331	27.992	50.024	95.003	169.52	302.28	549.90	1020.0	1815.6
0.00065	1.703	1.830	1.951	2.105	2.322	2.655	3.078	3.514	4.010	4.595	5.285	6.020
1/ 1538	7.7029	10.476	13.790	19.110	29.179	52.138	99.006	176.64	314.95	572.89	1062.6	1891.2
0.00070	1.770	1.902	2.028	2.187	2.413	2.759	3.197	3.651	4.165	4.772	5.489	6.251
1/ 1429	8.0068	10.889	14.332	19.860	30.321	54.174	102.86	183.50	327.15	595.02	1103.6	1963.9
0.00075	1.835	1.971	2.102	2.267	2.501	2.859	3.313	3.782	4.315	4.943	5.685	6.475
1/ 1333	8.3003	11.287	14.856	20.584	31.424	56.139	106.58	190.12	338.92	616.38	1143.1	2034.1
0.00080	1.898	2.039	2.173	2.344	2.586	2.956	3.425	3.910	4.460	5.109	5.875	6.691
1/ 1250	8.5843	11.672	15.362	21.284	32.492	58.040	110.18	196.52	350.30	637.04	1181.3	2102.0
0.00085	1.958	2.104	2.243	2.419	2.668	3.050	3.533	4.033	4.601	5.270	6.060	6.901
1/ 1176	8.8598	12.046	15.854	21.964	33.527	59.883	113.67	202.73	361.34	657.07	1218.4	2167.9
0.00090	2.018	2.167	2.310	2.492	2.748	3.141	3.639	4.153	4.737	5.426	6.239	7.104
1/ 1111	9.1274	12.410	16.331	22.624	34.532	61.674	117.06	208.76	372.06	676.52	1254.4	2231.8
0.00095	2.075	2.229	2.376	2.563	2.826	3.230	3.741	4.270	4.870	5.577	6.413	7.302
1/ 1053	9.3878	12.763	16.795	23.266	35.510	63.416	120.35	214.62	382.49	695.44	1289.4	2294.0
0.00100	2.131	2.289	2.440	2.632	2.902	3.316	3.841	4.383	4.999	5.725	6.582	7.495
1/ 1000	9.6416	13.107	17.248	23.892	36.463	65.113	123.56	220.33	392.65	713.88	1323.5	2354.5
0.00110	2.239	2.405	2.564	2.764	3.048	3.483	4.033	4.603	5.249	6.010	6.909	7.866
1/ 909	10.131	13.772	18.121	25.099	38.301	68.386	129.76	231.35	412.24	749.41	1389.2	2471.3
0.00120	2.343	2.516	2.682	2.892	3.188	3.642	4.217	4.812	5.487	6.283	7.222	8.222
1/ 833	10.599	14.407	18.955	26.253	40.058	71.515	135.68	241.88	430.96	783.38	1452.1	2582.9
0.00130	2.442	2.623	2.795	3.013	3.322	3.795	4.394	5.013	5.716	6.544	7.522	8.562
1/ 769	11.048	15.016	19.756	27.360	41.744	74.517	141.36	251.98	448.92	815.96	1512.3	2689.9
0.00140	2.538	2.725	2.904	3.131	3.451	3.942	4.564	5.206	5.936	6.795	7.810	8.890
1/ 714	11.481	15.603	20.527	28.426	43.367	77.406	146.82	261.70	466.20	847.31	1570.3	2792.9
0.00150	2.630	2.824	3.009	3.244	3.576	4.084	4.728	5.393	6.148	7.038	8.089	9.207
1/ 667	11.898	16.170	21.271	29.455	44.934	80.195	152.10	271.09	482.88	877.57	1626.3	2892.3
0.00160	2.719	2.920	3.111	3.354	3.696	4.222	4.887	5.574	6.354	7.273	8.358	9.513
1/ 625	12.302	16.718	21.991	30.450	46.450	82.893	157.20	280.16	499.02	906.84	1680.4	2988.5
0.00170	2.806	3.013	3.210	3.460	3.813	4.355	5.040	5.749	6.553	7.500	8.619	9.809
1/ 588	12.694	17.249	22.689	31.416	47.919	85.508	162.15	288.96	514.66	935.21	1732.9	3081.7
0.00180	2.890	3.103	3.306	3.563	3.927	4.484	5.190	5.919	6.746	7.721	8.872	10.10
1/ 556	13.074	17.765	23.367	32.353	49.346	88.049	166.95	297.51	529.85	962.76	1783.9	3172.1
0.00190	2.972	3.191	3.399	3.664	4.037	4.610	5.335	6.084	6.934	7.936	9.119	10.38
1/ 526	13.444	18.268	24.027	33.265	50.734	90.520	171.63	305.82	544.62	989.55	1833.5	3260.2
0.00200	3.052	3.276	3.490	3.762	4.145	4.733	5.477	6.245	7.118	8.145	9.359	10.65
1/ 500	13.805	18.757	24.669	34.153	52.087	92.927	176.18	313.91	559.02	1015.7	1881.7	3345.9
0.00220	3.205	3.441	3.665	3.950	4.352	4.969	5.749	6.556	7.471	8.549	9.822	11.18
1/ 455	14.501	19.700	25.909	35.866	54.695	97.569	184.96	329.52	586.76	1066.0	1974.8	3511.2
0.00240	3.352	3.598	3.833	4.131	4.551	5.195	6.010	6.852	7.808	8.935	10.26	11.68
1/ 417	15.165	20.602	27.093	37.504	57.187	102.00	193.35	344.44	613.28	1114.1	2063.8	3669.1
0.00260	3.493	3.749	3.994	4.304	4.741	5.412	6.260	7.137	8.132	9.305	10.69	12.16
1/ 385	15.803	21.468	28.230	39.075	59.578	106.26	201.40	358.75	638.71	1160.2	2149.1	3820.6
0.00280	3.629	3.895	4.148	4.470	4.924	5.620	6.501	7.411	8.444	9.661	11.10	12.63
1/ 357	16.418	22.301	29.324	40.587	61.880	110.36	209.14	372.52	663.19	1204.6	2231.2	3966.4
	900	1000	1100	1300	1500	1900	2400	3000	3700	4700	6000	7500

θ_{medial} for part-full circular pipes.

k_s = 0.060 mm   S = 0.00050 to 0.00280

A36
continued

k_s = 0.060 mm
S = 0.0030 to 0.014

Water (or sewage) at 15°C;
full bore conditions.

ie hydraulic gradient =
1 in 333 to 1 in 71

velocities in ms⁻¹
discharges in m³s⁻¹

Gradient	(Equivalent) Pipe diameters in m											
	2.400	2.700	3.000	3.400	4.000	5.000	6.400	8.000	10.00	12.60	16.00	20.00
0.00300 1/ 333	3.760 17.010	4.035 23.105	4.298 30.380	4.631 42.047	5.101 64.101	5.822 114.31	6.734 216.62	7.676 385.81	8.745 686.81	10.00 1247.4	11.49 2310.4	13.07 4107.0
0.00320 1/ 313	3.887 17.584	4.171 23.882	4.442 31.401	4.787 43.458	5.272 66.250	6.016 118.13	6.958 223.85	7.931 398.67	9.036 709.65	10.34 1288.9	11.87 2387.0	13.51 4243.1
0.00340 1/ 294	4.010 18.140	4.303 24.636	4.582 32.392	4.937 44.827	5.438 68.333	6.205 121.84	7.176 230.86	8.179 411.13	9.318 731.80	10.66 1329.0	12.24 2461.3	13.93 4375.0
0.00360 1/ 278	4.129 18.679	4.431 25.369	4.719 33.353	5.084 46.156	5.599 70.356	6.389 125.44	7.388 237.66	8.420 423.23	9.591 753.30	10.97 1368.0	12.60 2533.4	14.33 4503.0
0.00380 1/ 263	4.245 19.205	4.555 26.081	4.851 34.289	5.226 47.449	5.755 72.324	6.567 128.94	7.593 244.28	8.654 435.00	9.858 774.22	11.28 1405.9	12.95 2603.6	14.73 4627.6
0.00400 1/ 250	4.358 19.716	4.676 26.775	4.980 35.200	5.365 48.709	5.908 74.241	6.741 132.35	7.794 250.73	8.882 446.46	10.12 794.59	11.57 1442.9	13.29 2671.9	15.12 4748.9
0.00420 1/ 238	4.469 20.215	4.795 27.452	5.106 36.089	5.500 49.937	6.057 76.110	6.910 135.68	7.989 257.02	9.104 457.64	10.37 814.46	11.86 1478.9	13.62 2738.6	15.49 4867.2
0.00440 1/ 227	4.576 20.703	4.910 28.113	5.228 36.957	5.632 51.137	6.202 77.936	7.075 138.93	8.180 263.16	9.322 468.56	10.62 833.87	12.14 1514.1	13.94 2803.6	15.86 4982.7
0.00460 1/ 217	4.682 21.180	5.023 28.759	5.348 37.806	5.762 52.310	6.344 79.721	7.237 142.10	8.367 269.17	9.534 479.23	10.86 852.84	12.42 1548.5	14.26 2867.3	16.22 5095.7
0.00480 1/ 208	4.785 21.646	5.133 29.392	5.466 38.637	5.888 53.458	6.483 81.468	7.395 145.21	8.550 275.04	9.742 489.68	11.09 871.40	12.69 1582.2	14.57 2929.5	16.57 5206.2
0.00500 1/ 200	4.886 22.103	5.242 30.011	5.581 39.450	6.012 54.582	6.619 83.179	7.550 148.25	8.729 280.80	9.945 499.91	11.33 889.58	12.95 1615.1	14.87 2990.5	16.92 5314.4
0.00550 1/ 182	5.130 23.206	5.503 31.508	5.859 41.416	6.311 57.299	6.948 87.312	7.925 155.61	9.161 294.70	10.44 524.62	11.89 933.50	13.59 1694.8	15.61 3137.8	17.75 5575.9
0.00600 1/ 167	5.363 24.261	5.753 32.939	6.125 43.294	6.597 59.895	7.262 91.263	8.283 162.63	9.574 307.98	10.91 548.24	12.42 975.47	14.20 1770.9	16.31 3278.5	18.54 5825.8
0.00650 1/ 154	5.587 25.273	5.993 34.311	6.380 45.096	6.871 62.385	7.564 95.052	8.626 169.38	9.970 320.73	11.36 570.90	12.93 1015.7	14.79 1843.9	16.98 3413.5	19.31 6065.4
0.00700 1/ 143	5.802 26.247	6.223 35.632	6.625 46.831	7.135 64.782	7.854 98.699	8.957 175.86	10.35 332.99	11.79 592.69	13.43 1054.5	15.35 1914.1	17.62 3543.4	20.04 6296.0
0.00750 1/ 133	6.010 27.187	6.446 36.907	6.862 48.504	7.390 67.094	8.134 102.22	9.275 182.12	10.72 344.82	12.21 613.73	13.90 1091.8	15.89 1981.9	18.25 3668.7	20.75 6518.5
0.00800 1/ 125	6.211 28.097	6.661 38.140	7.091 50.123	7.636 69.331	8.405 105.62	9.584 188.18	11.07 356.27	12.61 634.07	14.36 1128.0	16.42 2047.4	18.85 3789.9	21.43 6733.7
0.00850 1/ 118	6.406 28.978	6.870 39.335	7.313 51.693	7.875 71.500	8.668 108.92	9.883 194.05	11.42 367.36	13.01 653.79	14.81 1163.0	16.93 2111.0	19.43 3907.4	22.10 6942.3
0.00900 1/ 111	6.595 29.834	7.073 40.495	7.529 53.216	8.107 73.605	8.923 112.12	10.17 199.75	11.75 378.14	13.39 672.94	15.24 1197.1	17.42 2172.7	20.00 4021.5	22.74 7144.9
0.00950 1/ 105	6.779 30.666	7.270 41.624	7.738 54.699	8.333 75.654	9.171 115.24	10.46 205.29	12.08 388.62	13.76 691.57	15.66 1230.1	17.91 2232.7	20.55 4132.5	23.37 7341.9
0.01000 1/ 100	6.958 31.477	7.462 42.724	7.943 56.143	8.552 77.649	9.412 118.28	10.73 210.69	12.40 398.82	14.12 709.71	16.07 1262.4	18.37 2291.1	21.09 4240.6	23.98 7533.8
0.01100 1/ 91	7.304 33.041	7.832 44.844	8.336 58.926	8.976 81.495	9.878 124.13	11.26 221.10	13.01 418.50	14.81 744.68	16.86 1324.5	19.28 2403.8	22.13 4448.9	25.16 7903.6
0.01200 1/ 83	7.634 34.535	8.186 46.870	8.713 61.586	9.381 85.170	10.32 129.72	11.77 231.04	13.59 437.30	15.48 778.10	17.62 1383.9	20.14 2511.4	23.12 4648.0	26.28 8257.1
0.01300 1/ 77	7.951 35.968	8.525 48.813	9.074 64.137	9.769 88.695	10.75 135.08	12.25 240.58	14.15 455.33	16.12 810.15	18.35 1440.8	20.97 2614.7	24.07 4838.9	27.36 8596.0
0.01400 1/ 71	8.256 37.347	8.852 50.683	9.421 66.593	10.14 92.088	11.16 140.24	12.72 249.76	14.69 472.68	16.73 840.99	19.04 1495.6	21.77 2714.0	24.98 5022.7	28.40 8922.2
	1500	1700	1900	2100	2500	3100	4000	5000	6500	8000	10000	13000

θ_{medial} for part-full circular pipes.

k_s = 0.060 mm   S = 0.0030 to 0.014

$k_s = 0.060$ mm
$S = 0.015$ to 0.075

ie hydraulic gradient = 1 in 66 to 1 in 13

Water (or sewage) at 15°C; full bore conditions.

velocities in ms⁻¹
discharges in m³s⁻¹

Gradient	2.400	2.700	3.000	3.400	4.000	5.000	6.400	8.000	10.00	12.60	16.00	20.00
0.01500 1/67	8.550 / 38.678	9.167 / 52.488	9.756 / 68.962	10.50 / 95.361	11.56 / 145.22	13.17 / 258.62	15.21 / 489.43	17.32 / 870.75	19.72 / 1548.5	22.54 / 2809.9	25.86 / 5200.0	29.40 / 9236.9
0.01600 1/62	8.834 / 39.966	9.472 / 54.233	10.08 / 71.253	10.85 / 98.528	11.94 / 150.04	13.61 / 267.19	15.72 / 505.62	17.90 / 899.53	20.37 / 1599.7	23.28 / 2902.7	26.72 / 5371.4	30.37 / 9541.4
0.01700 1/59	9.110 / 41.213	9.768 / 55.925	10.39 / 73.475	11.19 / 101.60	12.31 / 154.71	14.03 / 275.49	16.21 / 521.32	18.45 / 927.43	21.00 / 1649.2	24.00 / 2992.5	27.54 / 5537.6	31.31 / 9836.4
0.01800 1/56	9.378 / 42.425	10.05 / 57.568	10.70 / 75.632	11.52 / 104.58	12.67 / 159.25	14.44 / 283.56	16.68 / 536.56	18.99 / 954.52	21.61 / 1697.4	24.70 / 3079.8	28.34 / 5699.0	32.22 / 10123
0.01900 1/53	9.638 / 43.603	10.33 / 59.166	11.00 / 77.730	11.84 / 107.48	13.02 / 163.65	14.84 / 291.40	17.14 / 551.38	19.51 / 980.87	22.21 / 1744.2	25.38 / 3164.7	29.13 / 5856.0	33.11 / 10401
0.02000 1/50	9.892 / 44.751	10.61 / 60.722	11.29 / 79.773	12.15 / 110.30	13.37 / 167.95	15.23 / 299.04	17.59 / 565.82	20.02 / 1006.5	22.79 / 1789.8	26.04 / 3247.3	29.89 / 6008.8	33.97 / 10673
0.02200 1/45	10.38 / 46.964	11.13 / 63.723	11.84 / 83.712	12.75 / 115.74	14.02 / 176.23	15.98 / 313.77	18.45 / 593.66	21.01 / 1056.0	23.91 / 1877.7	27.32 / 3406.7	31.35 / 6303.5	35.64 / 11196
0.02400 1/42	10.85 / 49.079	11.63 / 66.590	12.38 / 87.476	13.32 / 120.94	14.65 / 184.14	16.70 / 327.84	19.28 / 620.25	21.95 / 1103.3	24.98 / 1961.7	28.54 / 3559.0	32.75 / 6585.1	37.23 / 11696
0.02600 1/38	11.30 / 51.108	12.11 / 69.340	12.89 / 91.087	13.87 / 125.93	15.26 / 191.73	17.38 / 341.34	20.07 / 645.76	22.85 / 1148.6	26.00 / 2042.2	29.71 / 3705.0	34.09 / 6855.2	38.76 / 12175
0.02800 1/36	11.73 / 53.060	12.57 / 71.987	13.38 / 94.561	14.40 / 130.73	15.84 / 199.03	18.05 / 354.32	20.84 / 670.31	23.72 / 1192.2	26.99 / 2119.7	30.84 / 3845.6	35.39 / 7115.1	40.22 / 12637
0.03000 1/33	12.15 / 54.943	13.02 / 74.540	13.85 / 97.914	14.91 / 135.36	16.40 / 206.08	18.68 / 366.86	21.57 / 693.99	24.56 / 1234.3	27.94 / 2194.5	31.93 / 3981.2	36.63 / 7365.8	41.64 / 13082
0.03200 1/31	12.55 / 56.765	13.45 / 77.010	14.31 / 101.16	15.40 / 139.84	16.94 / 212.90	19.30 / 378.98	22.28 / 716.90	25.37 / 1275.0	28.86 / 2266.9	32.98 / 4112.3	37.84 / 7608.3	43.01 / 13512
0.03400 1/29	12.94 / 58.530	13.87 / 79.404	14.76 / 104.30	15.88 / 144.18	17.47 / 219.50	19.90 / 390.72	22.98 / 739.11	26.15 / 1314.5	29.76 / 2337.0	34.00 / 4239.4	39.01 / 7843.4	44.34 / 13930
0.03600 1/28	13.32 / 60.245	14.27 / 81.729	15.19 / 107.35	16.35 / 148.40	17.98 / 225.92	20.48 / 402.13	23.65 / 760.67	26.91 / 1352.8	30.62 / 2405.1	34.99 / 4362.9	40.14 / 8071.6	45.63 / 14335
0.03800 1/26	13.69 / 61.913	14.67 / 83.990	15.61 / 110.32	16.80 / 152.50	18.47 / 232.15	21.05 / 413.23	24.30 / 781.63	27.66 / 1390.1	31.47 / 2471.3	35.95 / 4482.9	41.25 / 8293.6	46.88 / 14729
0.04000 1/25	14.04 / 63.537	15.05 / 86.192	16.02 / 113.21	17.24 / 156.50	18.96 / 238.23	21.60 / 424.03	24.93 / 802.06	28.38 / 1426.4	32.29 / 2535.8	36.89 / 4599.9	42.32 / 8509.8	48.11 / 15113
0.04200 1/24	14.39 / 65.121	15.43 / 88.340	16.41 / 116.03	17.67 / 160.39	19.43 / 244.16	22.13 / 434.57	25.55 / 821.98	29.08 / 1461.8	33.09 / 2598.7	37.81 / 4713.9	43.37 / 8720.7	49.30 / 15487
0.04400 1/23	14.74 / 66.668	15.80 / 90.437	16.80 / 118.78	18.08 / 164.20	19.89 / 249.94	22.66 / 444.87	26.16 / 841.43	29.77 / 1496.4	33.87 / 2660.1	38.70 / 4825.3	44.40 / 8926.6	50.46 / 15853
0.04600 1/22	15.07 / 68.180	16.15 / 92.487	17.19 / 121.48	18.49 / 167.91	20.34 / 255.60	23.17 / 454.93	26.75 / 860.44	30.44 / 1530.2	34.63 / 2720.1	39.57 / 4934.1	45.40 / 9127.9	51.60 / 16210
0.04800 1/21	15.40 / 69.660	16.50 / 94.493	17.56 / 124.11	18.90 / 171.55	20.78 / 261.14	23.67 / 464.77	27.33 / 879.05	31.10 / 1563.2	35.38 / 2778.9	40.43 / 5040.6	46.38 / 9324.9	52.71 / 16560
0.05000 1/20	15.72 / 71.110	16.85 / 96.458	17.92 / 126.69	19.29 / 175.12	21.21 / 266.56	24.16 / 474.41	27.89 / 897.27	31.74 / 1595.6	36.11 / 2836.4	41.26 / 5144.9	47.34 / 9517.8	53.80 / 16902
0.05500 1/18	16.49 / 74.611	17.68 / 101.21	18.80 / 132.92	20.24 / 183.73	22.25 / 279.65	25.35 / 497.70	29.26 / 941.29	33.30 / 1673.8	37.88 / 2975.4	43.28 / 5396.9	49.66 / 9983.7	56.43 / 17729
0.06000 1/17	17.23 / 77.956	18.47 / 105.74	19.65 / 138.87	21.14 / 191.95	23.25 / 292.17	26.48 / 519.96	30.57 / 983.35	34.79 / 1748.6	39.58 / 3108.2	45.21 / 5637.7	51.87 / 10429	58.95 / 18520
0.06500 1/15	17.94 / 81.165	19.23 / 110.09	20.45 / 144.59	22.01 / 199.84	24.21 / 304.17	27.57 / 541.31	31.82 / 1023.7	36.21 / 1820.3	41.20 / 3235.6	47.07 / 5868.7	53.99 / 10856	61.36 / 19278
0.07000 1/14	18.62 / 84.253	19.96 / 114.28	21.23 / 150.08	22.85 / 207.44	25.12 / 315.72	28.61 / 561.85	33.03 / 1062.5	37.59 / 1889.3	42.76 / 3358.2	48.85 / 6090.9	56.04 / 11267	63.68 / 20007
0.07500 1/13	19.28 / 87.232	20.66 / 118.32	21.98 / 155.38	23.65 / 214.76	26.01 / 326.87	29.62 / 581.66	34.19 / 1100.0	38.91 / 1955.9	44.26 / 3476.5	50.57 / 6305.4	58.01 / 11663	65.93 / 20711
	2600	2900	3300	3700	4400	5500	7000	8500	11000	14000	17000	22000

θ_{medial} for part-full circular pipes.

$k_s = 0.060$ mm $S = 0.0015$ to 0.075

k$_s$ = 0.150 mm
S = 0.00010 to 0.00048

Water (or sewage) at 15°C;
full bore conditions.

ie hydraulic gradient =
1 in 10000 to 1 in 2083

velocities in ms^{-1}
discharges in m^{3}s^{-1}

Gradient	(Equivalent) Pipe diameters in m											
	2.400	2.700	3.000	3.400	4.000	5.000	6.400	8.000	10.00	12.60	16.00	20.00
0.00010 1/10000	0.613 2.7716	0.659 3.7739	0.704 4.9729	0.760 6.8995	0.840 10.551	0.962 18.889	1.117 35.941	1.278 64.235	1.461 114.72	1.676 209.00	1.931 388.27	2.203 691.99
0.00011 1/ 9091	0.644 2.9137	0.693 3.9669	0.739 5.2267	0.799 7.2510	0.882 11.087	1.011 19.846	1.174 37.755	1.342 67.468	1.534 120.47	1.760 219.46	2.027 407.65	2.312 726.46
0.00012 1/ 8333	0.674 3.0495	0.725 4.1515	0.774 5.4695	0.836 7.5871	0.923 11.599	1.057 20.760	1.228 39.489	1.404 70.557	1.604 125.98	1.840 229.46	2.120 426.18	2.417 759.40
0.00013 1/ 7692	0.703 3.1799	0.756 4.3287	0.807 5.7025	0.871 7.9097	0.962 12.091	1.102 21.638	1.279 41.153	1.463 73.522	1.671 131.26	1.917 239.06	2.208 443.95	2.518 791.01
0.00014 1/ 7143	0.731 3.3055	0.786 4.4993	0.838 5.9269	0.905 8.2204	1.000 12.565	1.145 22.483	1.329 42.755	1.519 76.376	1.736 136.34	1.991 248.29	2.293 461.06	2.615 821.43
0.00015 1/ 6667	0.757 3.4268	0.815 4.6641	0.869 6.1436	0.938 8.5202	1.036 13.022	1.187 23.299	1.377 44.301	1.574 79.130	1.798 141.24	2.063 257.20	2.375 477.56	2.708 850.78
0.00016 1/ 6250	0.783 3.5442	0.842 4.8235	0.899 6.3532	0.970 8.8105	1.071 13.465	1.227 24.088	1.424 45.797	1.627 81.795	1.859 145.99	2.132 265.82	2.455 493.53	2.798 879.17
0.00017 1/ 5882	0.809 3.6580	0.869 4.9781	0.928 6.5566	1.001 9.0919	1.106 13.894	1.266 24.853	1.469 47.247	1.679 84.378	1.917 150.59	2.199 274.17	2.532 509.01	2.886 906.69
0.00018 1/ 5556	0.833 3.7686	0.896 5.1283	0.956 6.7541	1.032 9.3652	1.139 14.311	1.304 25.596	1.512 48.656	1.729 86.887	1.974 155.05	2.264 282.29	2.606 524.04	2.971 933.42
0.00019 1/ 5263	0.857 3.8762	0.921 5.2745	0.983 6.9462	1.061 9.6312	1.171 14.716	1.340 26.319	1.555 50.026	1.777 89.328	2.030 159.40	2.327 290.18	2.679 538.66	3.054 959.42
0.00020 1/ 5000	0.880 3.9810	0.946 5.4169	1.009 7.1335	1.089 9.8903	1.203 15.111	1.376 27.024	1.597 51.361	1.824 91.706	2.083 163.63	2.389 297.87	2.750 552.91	3.135 984.74
0.00022 1/ 4545	0.925 4.1832	0.994 5.6915	1.060 7.4946	1.144 10.390	1.263 15.873	1.445 28.382	1.677 53.935	1.916 96.290	2.187 171.79	2.508 312.69	2.887 580.37	3.290 1033.6
0.00024 1/ 4167	0.967 4.3766	1.040 5.9541	1.109 7.8398	1.197 10.868	1.321 16.601	1.512 29.681	1.753 56.395	2.003 100.67	2.287 179.59	2.621 326.86	3.017 606.61	3.438 1080.2
0.00026 1/ 3846	1.008 4.5622	1.084 6.2061	1.156 8.1711	1.247 11.326	1.377 17.300	1.575 30.926	1.826 58.756	2.086 104.88	2.382 187.07	2.730 340.45	3.142 631.78	3.581 1125.0
0.00028 1/ 3571	1.048 4.7408	1.126 6.4487	1.201 8.4901	1.296 11.768	1.430 17.973	1.636 32.126	1.897 61.027	2.167 108.92	2.474 194.28	2.835 353.53	3.263 656.00	3.718 1168.0
0.00030 1/ 3333	1.086 4.9133	1.167 6.6829	1.245 8.7979	1.343 12.194	1.482 18.622	1.695 33.283	1.965 63.220	2.245 112.83	2.562 201.22	2.936 366.15	3.379 679.38	3.850 1209.6
0.00032 1/ 3125	1.123 5.0802	1.207 6.9095	1.287 9.0958	1.388 12.606	1.532 19.250	1.752 34.403	2.031 65.341	2.320 116.60	2.648 207.95	3.034 378.36	3.491 701.99	3.978 1249.8
0.00034 1/ 2941	1.159 5.2420	1.245 7.1292	1.328 9.3846	1.432 13.005	1.580 19.859	1.807 35.488	2.095 67.397	2.393 120.26	2.731 214.46	3.129 390.19	3.600 723.90	4.102 1288.7
0.00036 1/ 2778	1.193 5.3991	1.282 7.3426	1.367 9.6651	1.475 13.393	1.627 20.450	1.861 36.542	2.157 69.394	2.463 123.82	2.811 220.79	3.221 401.68	3.706 745.19	4.223 1326.6
0.00038 1/ 2632	1.227 5.5520	1.319 7.5502	1.406 9.9380	1.517 13.771	1.673 21.026	1.913 37.568	2.217 71.336	2.532 127.28	2.890 226.94	3.311 412.86	3.809 765.89	4.340 1363.4
0.00040 1/ 2500	1.260 5.7010	1.354 7.7525	1.444 10.204	1.557 14.139	1.718 21.586	1.964 38.566	2.276 73.228	2.599 130.65	2.966 232.94	3.398 423.74	3.909 786.05	4.454 1399.2
0.00042 1/ 2381	1.292 5.8463	1.388 7.9498	1.480 10.463	1.597 14.498	1.761 22.133	2.014 39.541	2.334 75.074	2.664 133.93	3.040 238.79	3.484 434.36	4.007 805.71	4.565 1434.2
0.00044 1/ 2273	1.324 5.9883	1.422 8.1425	1.516 10.716	1.635 14.848	1.804 22.667	2.062 40.492	2.390 76.876	2.728 137.14	3.113 244.50	3.567 444.73	4.103 824.92	4.674 1468.3
0.00046 1/ 2174	1.354 6.1271	1.455 8.3309	1.551 10.964	1.673 15.191	1.845 23.189	2.110 41.423	2.444 78.638	2.791 140.28	3.184 250.08	3.648 454.87	4.196 843.69	4.780 1501.7
0.00048 1/ 2083	1.384 6.2629	1.487 8.5153	1.585 11.206	1.710 15.526	1.886 23.700	2.156 42.333	2.498 80.362	2.852 143.35	3.254 255.54	3.728 464.79	4.288 862.06	4.884 1534.3
	500	600	650	750	850	1100	1400	1700	2200	2700	3500	4300

θ_{medial} for part-full circular pipes.

k$_s$ = 0.150 mm   S = 0.00010 to 0.00048

k_s = 0.150 mm
S = 0.00050 to 0.00280

Water (or sewage) at 15°C;
full bore conditions.

ie hydraulic gradient =
1 in 2000 to 1 in 357

velocities in ms⁻¹
discharges in m³s⁻¹

| Gradient | (Equivalent) Pipe diameters in m ||||||||||||
	2.400	2.700	3.000	3.400	4.000	5.000	6.400	8.000	10.00	12.60	16.00	20.00
0.00050	1.414	1.519	1.619	1.746	1.926	2.201	2.551	2.912	3.322	3.805	4.377	4.986
1/ 2000	6.3959	8.6959	11.444	15.854	24.200	43.225	82.051	146.35	260.89	474.50	880.05	1566.3
0.00055	1.485	1.595	1.700	1.834	2.022	2.311	2.677	3.056	3.486	3.994	4.593	5.232
1/ 1818	6.7175	9.1324	12.018	16.648	25.409	45.379	86.131	153.62	273.82	497.98	923.52	1643.6
0.00060	1.553	1.668	1.778	1.917	2.114	2.416	2.799	3.194	3.644	4.174	4.800	5.467
1/ 1667	7.0248	9.5496	12.566	17.406	26.565	47.439	90.031	160.56	286.17	520.41	965.06	1717.4
0.00065	1.618	1.738	1.852	1.997	2.202	2.517	2.915	3.327	3.795	4.346	4.998	5.692
1/ 1538	7.3197	9.9499	13.092	18.134	27.673	49.414	93.771	167.22	298.02	541.92	1004.9	1788.2
0.00070	1.681	1.805	1.924	2.074	2.287	2.613	3.027	3.454	3.940	4.512	5.189	5.909
1/ 1429	7.6035	10.335	13.598	18.834	28.740	51.315	97.371	173.63	309.43	562.62	1043.2	1856.4
0.00075	1.741	1.870	1.993	2.149	2.369	2.707	3.135	3.577	4.080	4.672	5.373	6.118
1/ 1333	7.8774	10.707	14.087	19.510	29.770	53.149	100.84	179.81	320.43	582.60	1080.2	1922.1
0.00080	1.800	1.933	2.060	2.221	2.448	2.797	3.239	3.696	4.215	4.827	5.551	6.321
1/ 1250	8.1424	11.067	14.559	20.164	30.766	54.924	104.20	185.79	331.07	601.92	1116.0	1985.7
0.00085	1.857	1.994	2.125	2.291	2.525	2.885	3.340	3.812	4.347	4.978	5.723	6.517
1/ 1176	8.3993	11.415	15.018	20.798	31.731	56.644	107.46	191.59	341.39	620.65	1150.7	2047.3
0.00090	1.912	2.053	2.188	2.358	2.600	2.970	3.439	3.924	4.474	5.123	5.891	6.707
1/ 1111	8.6487	11.754	15.463	21.413	32.669	58.314	110.62	197.22	351.40	638.84	1184.4	2107.2
0.00095	1.965	2.110	2.249	2.424	2.672	3.053	3.534	4.032	4.598	5.265	6.053	6.893
1/ 1053	8.8914	12.083	15.895	22.012	33.581	59.939	113.70	202.69	361.14	656.53	1217.1	2165.4
0.00100	2.018	2.166	2.308	2.489	2.743	3.133	3.628	4.139	4.719	5.403	6.212	7.073
1/ 1000	9.1278	12.404	16.317	22.595	34.469	61.522	116.70	208.03	370.63	673.76	1249.0	2222.1
0.00110	2.118	2.275	2.423	2.612	2.879	3.289	3.807	4.343	4.952	5.670	6.518	7.421
1/ 909	9.5836	13.023	17.130	23.719	36.182	64.573	122.47	218.31	388.93	706.97	1310.5	2331.4
0.00120	2.215	2.378	2.533	2.731	3.009	3.437	3.979	4.539	5.175	5.924	6.810	7.754
1/ 833	10.019	13.614	17.907	24.794	37.818	67.488	127.99	228.13	406.41	738.71	1369.3	2435.9
0.00130	2.307	2.477	2.639	2.844	3.134	3.580	4.143	4.726	5.388	6.168	7.091	8.072
1/ 769	10.437	14.181	18.652	25.824	39.388	70.285	133.29	237.56	423.18	769.15	1425.6	2536.0
0.00140	2.396	2.572	2.740	2.954	3.255	3.717	4.302	4.906	5.593	6.403	7.360	8.379
1/ 714	10.839	14.727	19.369	26.816	40.899	72.976	138.38	246.62	439.31	798.44	1479.9	2632.4
0.00150	2.482	2.664	2.838	3.059	3.371	3.849	4.454	5.081	5.792	6.630	7.621	8.675
1/ 667	11.227	15.254	20.061	27.773	42.357	75.573	143.30	255.38	454.88	826.70	1532.2	2725.5
0.00160	2.565	2.753	2.933	3.161	3.483	3.977	4.602	5.249	5.983	6.849	7.872	8.962
1/ 625	11.603	15.763	20.731	28.699	43.767	78.085	148.05	263.84	469.94	854.04	1582.8	2815.4
0.00170	2.645	2.839	3.025	3.260	3.592	4.101	4.746	5.412	6.169	7.062	8.116	9.239
1/ 588	11.967	16.257	21.380	29.597	45.134	80.520	152.66	272.04	484.53	880.54	1631.9	2902.6
0.00180	2.723	2.923	3.114	3.356	3.697	4.221	4.885	5.571	6.350	7.268	8.353	9.509
1/ 556	12.320	16.737	22.010	30.468	46.461	82.884	157.14	280.01	498.70	906.27	1679.5	2987.3
0.00190	2.799	3.005	3.200	3.449	3.800	4.338	5.020	5.725	6.525	7.469	8.584	9.771
1/ 526	12.664	17.203	22.623	31.316	47.752	85.183	161.49	287.76	512.49	931.29	1725.9	3069.6
0.00200	2.873	3.084	3.285	3.540	3.900	4.452	5.152	5.875	6.696	7.664	8.808	10.03
1/ 500	12.999	17.658	23.220	32.141	49.010	87.423	165.73	295.30	525.91	955.66	1771.0	3149.8
0.00220	3.016	3.237	3.448	3.715	4.093	4.672	5.406	6.164	7.026	8.041	9.241	10.52
1/ 455	13.644	18.533	24.371	33.733	51.434	91.741	173.91	309.85	551.79	1002.6	1858.0	3304.4
0.00240	3.152	3.383	3.603	3.883	4.277	4.882	5.649	6.441	7.341	8.401	9.654	10.99
1/ 417	14.261	19.371	25.471	35.254	53.750	95.867	181.72	323.75	576.52	1047.5	1941.1	3452.2
0.00260	3.283	3.523	3.753	4.044	4.454	5.084	5.882	6.706	7.643	8.747	10.05	11.44
1/ 385	14.853	20.174	26.526	36.713	55.972	99.825	189.21	337.08	600.24	1090.6	2020.8	3593.9
0.00280	3.409	3.658	3.896	4.198	4.624	5.278	6.106	6.961	7.933	9.079	10.43	11.87
1/ 357	15.422	20.946	27.541	38.117	58.110	103.63	196.42	349.91	623.07	1132.0	2097.6	3730.2
	850	1000	1100	1200	1400	1800	2300	2900	3600	4600	6000	7000

θ_{medial} for part-full circular pipes.

k_s = 0.150 mm   S = 0.00050 to 0.00280

k_s = 0.150 mm
S = 0.0030 to 0.014

Water (or sewage) at 15°C;
full bore conditions.

ie hydraulic gradient =
1 in 333 to 1 in 71

velocities in ms^{-1}
discharges in m^3s^{-1}

Gradient	(Equivalent) Pipe diameters in m											
	2.400	2.700	3.000	3.400	4.000	5.000	6.400	8.000	10.00	12.60	16.00	20.00
0.00300 1/ 333	3.531 15.972	3.789 21.692	4.035 28.521	4.347 39.471	4.788 60.173	5.465 107.31	6.322 203.37	7.208 362.29	8.214 645.09	9.399 1172.0	10.80 2171.6	12.29 3861.8
0.00320 1/ 313	3.648 16.503	3.915 22.413	4.169 29.468	4.492 40.782	4.947 62.169	5.646 110.86	6.531 210.10	7.446 374.27	8.485 666.39	9.710 1210.7	11.16 2243.2	12.70 3989.0
0.00340 1/ 294	3.762 17.018	4.037 23.112	4.299 30.386	4.632 42.052	5.101 64.103	5.822 114.31	6.734 216.62	7.677 385.87	8.748 687.04	10.01 1248.2	11.50 2312.6	13.09 4112.3
0.00360 1/ 278	3.872 17.519	4.155 23.791	4.425 31.278	4.767 43.285	5.251 65.981	5.992 117.65	6.931 222.95	7.901 397.14	9.003 707.09	10.30 1284.6	11.84 2379.9	13.47 4232.1
0.00380 1/ 263	3.980 18.005	4.271 24.451	4.548 32.146	4.900 44.484	5.396 67.808	6.158 120.91	7.122 229.11	8.119 408.10	9.251 726.58	10.59 1320.0	12.16 2445.5	13.84 4348.6
0.00400 1/ 250	4.085 18.479	4.383 25.094	4.667 32.991	5.028 45.653	5.538 69.587	6.319 124.07	7.308 235.11	8.331 418.77	9.493 745.58	10.86 1354.4	12.48 2509.3	14.20 4462.0
0.00420 1/ 238	4.187 18.941	4.492 25.722	4.784 33.815	5.154 46.792	5.676 71.323	6.476 127.17	7.490 240.96	8.538 429.19	9.729 764.10	11.13 1388.1	12.79 2571.6	14.56 4572.7
0.00440 1/ 227	4.287 19.393	4.599 26.334	4.898 34.619	5.276 47.905	5.811 73.017	6.630 130.18	7.668 246.67	8.741 439.35	9.959 782.18	11.40 1420.9	13.09 2632.4	14.90 4680.7
0.00460 1/ 217	4.384 19.834	4.704 26.933	5.009 35.406	5.396 48.993	5.942 74.674	6.780 133.13	7.841 252.26	8.938 449.29	10.18 799.86	11.65 1453.0	13.39 2691.8	15.24 4786.3
0.00480 1/ 208	4.480 20.266	4.806 27.519	5.118 36.176	5.513 50.057	6.071 76.295	6.927 136.02	8.011 257.72	9.132 459.01	10.40 817.16	11.90 1484.4	13.68 2749.9	15.56 4889.6
0.00500 1/ 200	4.573 20.689	4.907 28.092	5.224 36.930	5.628 51.099	6.198 77.882	7.071 138.85	8.178 263.08	9.321 468.54	10.62 834.11	12.15 1515.1	13.96 2806.9	15.89 4990.8
0.00550 1/ 182	4.799 21.710	5.149 29.479	5.482 38.751	5.906 53.618	6.503 81.717	7.419 145.68	8.580 276.00	9.779 491.55	11.14 875.04	12.75 1589.4	14.64 2944.4	16.66 5235.3
0.00600 1/ 167	5.015 22.687	5.380 30.803	5.728 40.491	6.171 56.024	6.794 85.382	7.752 152.21	8.964 288.36	10.22 513.53	11.64 914.15	13.32 1660.4	15.30 3075.9	17.41 5468.9
0.00650 1/ 154	5.222 23.623	5.602 32.074	5.964 42.160	6.425 58.332	7.074 88.897	8.071 158.47	9.332 300.21	10.64 534.62	12.12 951.66	13.86 1728.5	15.93 3202.0	18.12 5693.0
0.00700 1/ 143	5.421 24.524	5.815 33.297	6.192 43.767	6.669 60.553	7.343 92.279	8.377 164.49	9.686 311.61	11.04 554.91	12.58 987.76	14.39 1794.1	16.53 3323.3	18.81 5908.6
0.00750 1/ 133	5.613 25.393	6.021 34.476	6.411 45.317	6.905 62.696	7.603 95.543	8.673 170.30	10.03 322.61	11.43 574.49	13.02 1022.6	14.90 1857.3	17.11 3440.3	19.47 6116.6
0.00800 1/ 125	5.799 26.234	6.221 35.617	6.623 46.816	7.134 64.769	7.854 98.700	8.960 175.92	10.36 333.25	11.81 593.42	13.45 1056.3	15.39 1918.4	17.67 3553.6	20.11 6317.9
0.00850 1/ 118	5.979 27.049	6.414 36.723	6.829 48.269	7.355 66.778	8.098 101.76	9.237 181.37	10.68 343.56	12.17 611.78	13.86 1088.9	15.86 1977.7	18.22 3663.3	20.73 6512.9
0.00900 1/ 111	6.154 27.841	6.601 37.797	7.028 49.679	7.570 68.729	8.334 104.73	9.507 186.66	10.99 353.58	12.53 629.60	14.27 1120.6	16.32 2035.3	18.75 3769.8	21.33 6702.2
0.00950 1/ 105	6.324 28.611	6.784 38.842	7.222 51.052	7.779 70.626	8.564 107.62	9.769 191.81	11.29 363.32	12.87 646.93	14.66 1151.5	16.77 2091.2	19.26 3873.5	21.92 6886.4
0.01000 1/ 100	6.490 29.360	6.962 39.859	7.411 52.388	7.982 72.474	8.788 110.43	10.02 196.82	11.59 372.80	13.21 663.81	15.04 1181.5	17.21 2145.7	19.77 3974.4	22.49 7065.8
0.01100 1/ 91	6.810 30.806	7.304 41.820	7.776 54.965	8.375 76.037	9.220 115.86	10.52 206.48	12.16 391.09	13.85 696.35	15.78 1239.4	18.05 2250.8	20.73 4169.0	23.59 7411.5
0.01200 1/ 83	7.115 32.187	7.631 43.694	8.124 57.426	8.750 79.441	9.632 121.04	10.99 215.71	12.70 408.56	14.47 727.45	16.48 1294.7	18.86 2351.2	21.66 4354.9	24.64 7741.9
0.01300 1/ 77	7.408 33.512	7.945 45.491	8.458 59.788	9.109 82.706	10.03 126.01	11.44 224.57	13.22 425.32	15.07 757.27	17.16 1347.7	19.63 2447.6	22.55 4533.2	25.65 8058.8
0.01400 1/ 71	7.689 34.786	8.247 47.221	8.780 62.060	9.455 85.847	10.41 130.80	11.87 233.09	13.72 441.45	15.64 785.97	17.81 1398.8	20.37 2540.2	23.40 4704.7	26.62 8363.7
	1400	1600	1800	2000	2400	3000	3800	4800	6000	7500	9500	12000

θ_{medial} **for part-full circular pipes.**

k_s = 0.150 mm   S = 0.0030 to 0.014

k_s = 0.150 mm S = 0.015 to 0.075	Water (or sewage) at 15°C; full bore conditions.

A37
continued

ie hydraulic gradient =
1 in 66 to 1 in 13

velocities in ms^{-1}
discharges in m^{3}s^{-1}

Gradient	(Equivalent) Pipe diameters in m											
	2.400	2.700	3.000	3.400	4.000	5.000	6.400	8.000	10.00	12.60	16.00	20.00
0.01500 1/ 67	7.961 36.016	8.539 48.890	9.090 64.252	9.789 88.879	10.78 135.41	12.29 241.31	14.21 457.01	16.19 813.66	18.44 1448.1	21.09 2629.6	24.22 4870.3	27.56 8658.0
0.01600 1/ 62	8.224 37.206	8.821 50.504	9.390 66.372	10.11 91.811	11.13 139.88	12.69 249.26	14.67 472.06	16.72 840.44	19.04 1495.7	21.78 2716.1	25.02 5030.4	28.46 8942.5
0.01700 1/ 59	8.479 38.359	9.094 52.068	9.681 68.428	10.43 94.652	11.48 144.21	13.09 256.97	15.13 486.65	17.24 866.40	19.63 1541.9	22.46 2799.9	25.79 5185.6	29.34 9218.3
0.01800 1/ 56	8.727 39.478	9.359 53.587	9.963 70.423	10.73 97.412	11.81 148.41	13.47 264.45	15.57 500.81	17.74 891.60	20.20 1586.7	23.11 2881.3	26.54 5336.3	30.20 9486.1
0.01900 1/ 53	8.967 40.567	9.617 55.065	10.24 72.364	11.02 100.10	12.14 152.50	13.84 271.73	16.00 514.58	18.23 916.11	20.76 1630.3	23.74 2960.5	27.27 5482.8	31.02 9746.6
0.02000 1/ 50	9.202 41.628	9.869 56.504	10.50 74.255	11.31 102.71	12.45 156.48	14.20 278.82	16.41 528.00	18.70 939.99	21.30 1672.8	24.36 3037.6	27.98 5625.6	31.83 10000
0.02200 1/ 45	9.654 43.672	10.35 59.277	11.02 77.899	11.87 107.75	13.06 164.15	14.90 292.48	17.22 553.86	19.62 986.01	22.34 1754.7	25.55 3186.2	29.35 5900.7	33.39 10489
0.02400 1/ 42	10.09 45.626	10.82 61.928	11.51 81.381	12.40 112.56	13.65 171.48	15.56 305.54	17.99 578.58	20.49 1030.0	23.34 1832.9	26.69 3328.2	30.66 6163.7	34.88 10957
0.02600 1/ 38	10.50 47.499	11.26 64.470	11.99 84.721	12.91 117.18	14.21 178.51	16.20 318.06	18.72 602.28	21.33 1072.2	24.29 1907.9	27.78 3464.4	31.91 6415.8	36.30 11405
0.02800 1/ 36	10.90 49.302	11.69 66.916	12.44 87.934	13.40 121.62	14.74 185.28	16.81 330.11	19.43 625.09	22.14 1112.8	25.21 1980.1	28.84 3595.5	33.12 6658.5	37.67 11836
0.03000 1/ 33	11.28 51.042	12.10 69.277	12.88 91.035	13.87 125.91	15.26 191.81	17.40 341.74	20.11 647.09	22.92 1151.9	26.10 2049.8	29.85 3721.9	34.28 6892.6	39.00 12252
0.03200 1/ 31	11.65 52.724	12.50 71.560	13.30 94.034	14.32 130.06	15.77 198.12	17.98 352.98	20.78 668.38	23.67 1189.8	26.96 2117.2	30.83 3844.3	35.41 7119.0	40.28 12654
0.03400 1/ 29	12.02 54.355	12.88 73.772	13.71 96.941	14.77 134.08	16.25 204.24	18.53 363.88	21.42 689.01	24.40 1226.5	27.79 2182.5	31.78 3962.8	36.50 7338.5	41.52 13044
0.03600 1/ 28	12.37 55.938	13.26 75.921	14.11 99.763	15.20 137.98	16.73 210.19	19.07 374.47	22.04 709.04	25.11 1262.1	28.60 2245.9	32.70 4077.9	37.56 7551.6	42.73 13423
0.03800 1/ 26	12.71 57.479	13.62 78.010	14.50 102.51	15.62 141.77	17.19 215.96	19.60 384.76	22.65 728.52	25.80 1296.8	29.38 2307.6	33.60 4189.9	38.59 7758.9	43.90 13792
0.04000 1/ 25	13.04 58.979	13.98 80.046	14.88 105.18	16.02 145.47	17.63 221.59	20.11 394.78	23.24 747.49	26.47 1330.6	30.15 2367.6	34.48 4298.9	39.59 7960.8	45.04 14150
0.04200 1/ 24	13.36 60.442	14.33 82.031	15.25 107.79	16.42 149.08	18.07 227.09	20.60 404.56	23.81 766.00	27.13 1363.5	30.89 2426.2	35.33 4405.2	40.57 8157.7	46.16 14500
0.04400 1/ 23	13.68 61.870	14.67 83.969	15.61 110.34	16.81 152.60	18.50 232.45	21.09 414.11	24.37 784.07	27.77 1395.7	31.62 2483.4	36.16 4509.1	41.53 8349.9	47.24 14842
0.04600 1/ 22	13.98 63.267	15.00 85.864	15.96 112.83	17.19 156.04	18.91 237.69	21.57 423.44	24.92 801.74	28.39 1427.1	32.33 2539.3	36.98 4610.6	42.46 8537.8	48.31 15176
0.04800 1/ 21	14.29 64.633	15.32 87.718	16.31 115.26	17.56 159.41	19.32 242.82	22.03 432.58	25.46 819.02	29.00 1457.9	33.03 2594.0	37.77 4709.9	43.38 8721.7	49.35 15503
0.05000 1/ 20	14.58 65.971	15.64 89.533	16.64 117.65	17.92 162.71	19.72 247.84	22.49 441.52	25.99 835.95	29.60 1488.0	33.71 2647.6	38.55 4807.2	44.27 8901.8	50.37 15823
0.05500 1/ 18	15.30 69.204	16.40 93.920	17.46 123.41	18.80 170.67	20.69 259.97	23.59 463.13	27.26 876.85	31.05 1560.8	35.36 2777.1	40.44 5042.2	46.44 9336.9	52.83 16596
0.06000 1/ 17	15.98 72.293	17.14 98.111	18.24 128.91	19.64 178.29	21.61 271.56	24.64 483.77	28.47 915.92	32.43 1630.3	36.93 2900.8	42.24 5266.7	48.51 9752.6	55.18 17335
0.06500 1/ 15	16.64 75.256	17.84 102.13	18.98 134.20	20.44 185.59	22.50 282.68	25.65 503.57	29.64 953.40	33.76 1697.0	38.44 3019.4	43.97 5482.1	50.49 10151	57.43 18043
0.07000 1/ 14	17.27 78.107	18.51 106.00	19.70 139.28	21.21 192.61	23.35 293.38	26.62 522.63	30.76 989.46	35.04 1761.2	39.90 3133.6	45.63 5689.3	52.40 10535	59.60 18725
0.07500 1/ 13	17.87 80.858	19.17 109.73	20.40 144.18	21.96 199.39	24.17 303.71	27.55 541.01	31.84 1024.3	36.27 1823.1	41.30 3243.7	47.23 5889.3	54.24 10905	61.70 19383
	2400	2700	3000	3400	4000	5000	6500	8000	10000	13000	16000	20000

θ_{medial} for part-full circular pipes.

k_s = 0.150 mm   S = 0.0015 to 0.075

$k_s = 0.30$ mm
$S = 0.00010$ to 0.00048

Water (or sewage) at 15°C;
full bore conditions.

ie hydraulic gradient =
1 in 10000 to 1 in 2083

velocities in ms⁻¹
discharges in m³s⁻¹

velocities in ms^{-1}
discharges in m^3s^{-1}

| Gradient | (Equivalent) Pipe diameters in m | | | | | | | | | | | |
	2.400	2.700	3.000	3.400	4.000	5.000	6.400	8.000	10.00	12.60	16.00	20.00
0.00010	0.589	0.633	0.676	0.729	0.806	0.922	1.071	1.224	1.399	1.605	1.848	2.108
1/10000	2.6636	3.6254	4.7755	6.6230	10.123	18.113	34.444	61.532	109.85	200.07	371.58	662.14
0.00011	0.619	0.665	0.710	0.766	0.846	0.969	1.124	1.285	1.468	1.684	1.940	2.212
1/ 9091	2.7982	3.8082	5.0159	6.9558	10.631	19.018	36.162	64.594	115.30	209.99	389.97	694.86
0.00012	0.647	0.696	0.742	0.801	0.885	1.013	1.175	1.343	1.534	1.760	2.027	2.311
1/ 8333	2.9267	3.9829	5.2456	7.2739	11.116	19.884	37.804	67.521	120.52	219.47	407.54	726.12
0.00013	0.674	0.725	0.773	0.835	0.922	1.055	1.224	1.399	1.598	1.833	2.111	2.407
1/ 7692	3.0501	4.1505	5.4661	7.5792	11.581	20.715	39.379	70.328	125.52	228.56	424.40	756.11
0.00014	0.700	0.753	0.803	0.867	0.957	1.096	1.271	1.453	1.659	1.903	2.191	2.499
1/ 7143	3.1689	4.3119	5.6783	7.8729	12.029	21.514	40.895	73.030	130.33	237.31	440.61	784.97
0.00015	0.726	0.780	0.832	0.898	0.992	1.135	1.317	1.505	1.719	1.971	2.269	2.587
1/ 6667	3.2835	4.4676	5.8831	8.1565	12.462	22.286	42.358	75.637	134.98	245.75	456.26	812.82
0.00016	0.750	0.807	0.860	0.929	1.025	1.173	1.361	1.555	1.776	2.036	2.345	2.673
1/ 6250	3.3944	4.6182	6.0812	8.4307	12.880	23.032	43.773	78.158	139.47	253.92	471.40	839.75
0.00017	0.774	0.832	0.887	0.958	1.057	1.210	1.403	1.604	1.831	2.100	2.418	2.756
1/ 5882	3.5020	4.7643	6.2733	8.6966	13.285	23.755	45.144	80.603	143.82	261.83	486.08	865.86
0.00018	0.797	0.857	0.914	0.986	1.089	1.246	1.445	1.651	1.885	2.161	2.488	2.837
1/ 5556	3.6064	4.9061	6.4599	8.9548	13.679	24.458	46.476	82.976	148.05	269.52	500.32	891.20
0.00019	0.820	0.881	0.940	1.014	1.119	1.280	1.485	1.697	1.937	2.221	2.557	2.915
1/ 5263	3.7080	5.0441	6.6413	9.2060	14.062	25.141	47.772	85.285	152.16	276.99	514.18	915.86
0.00020	0.842	0.904	0.965	1.041	1.149	1.314	1.524	1.741	1.988	2.280	2.624	2.992
1/ 5000	3.8069	5.1786	6.8181	9.4507	14.435	25.806	49.033	87.534	156.17	284.27	527.68	939.87
0.00022	0.884	0.950	1.013	1.093	1.206	1.380	1.600	1.828	2.087	2.392	2.754	3.139
1/ 4545	3.9977	5.4377	7.1589	9.9224	15.155	27.090	51.466	91.870	163.89	298.31	553.70	986.17
0.00024	0.924	0.993	1.059	1.143	1.261	1.442	1.672	1.910	2.181	2.500	2.878	3.280
1/ 4167	4.1801	5.6855	7.4846	10.373	15.842	28.316	53.791	96.012	171.27	311.72	578.56	1030.4
0.00026	0.963	1.035	1.103	1.190	1.313	1.502	1.741	1.989	2.271	2.603	2.996	3.415
1/ 3846	4.3551	5.9231	7.7971	10.806	16.501	29.492	56.021	99.986	178.35	324.59	602.41	1072.8
0.00028	1.000	1.074	1.146	1.236	1.364	1.560	1.808	2.065	2.358	2.702	3.110	3.545
1/ 3571	4.5235	6.1518	8.0978	11.222	17.136	30.624	58.167	103.81	185.16	336.97	625.36	1113.7
0.00030	1.036	1.113	1.187	1.280	1.412	1.615	1.872	2.139	2.441	2.798	3.220	3.670
1/ 3333	4.6860	6.3725	8.3880	11.624	17.748	31.716	60.238	107.50	191.73	348.91	647.50	1153.0
0.00032	1.071	1.150	1.226	1.323	1.460	1.669	1.935	2.210	2.522	2.891	3.327	3.792
1/ 3125	4.8432	6.5860	8.6688	12.012	18.341	32.773	62.241	111.07	198.09	360.47	668.92	1191.2
0.00034	1.104	1.186	1.265	1.365	1.505	1.721	1.995	2.278	2.601	2.981	3.430	3.909
1/ 2941	4.9955	6.7930	8.9409	12.389	18.915	33.797	64.182	114.53	204.25	371.67	689.67	1228.1
0.00036	1.137	1.222	1.302	1.405	1.550	1.772	2.054	2.345	2.677	3.068	3.530	4.023
1/ 2778	5.1435	6.9940	9.2051	12.755	19.473	34.792	66.068	117.89	210.24	382.54	709.83	1263.9
0.00038	1.169	1.256	1.339	1.444	1.593	1.821	2.111	2.410	2.751	3.153	3.628	4.134
1/ 2632	5.2875	7.1895	9.4622	13.110	20.015	35.759	67.901	121.16	216.06	393.12	729.44	1298.8
0.00040	1.200	1.289	1.374	1.482	1.635	1.869	2.166	2.474	2.823	3.235	3.723	4.242
1/ 2500	5.4277	7.3799	9.7125	13.457	20.543	36.701	69.687	124.34	221.72	403.42	748.53	1332.8
0.00042	1.230	1.321	1.409	1.519	1.676	1.916	2.220	2.535	2.893	3.316	3.816	4.348
1/ 2381	5.5645	7.5656	9.9568	13.795	21.059	37.620	71.429	127.44	227.25	413.47	767.15	1365.9
0.00044	1.260	1.353	1.442	1.556	1.716	1.962	2.273	2.596	2.962	3.395	3.906	4.451
1/ 2273	5.6980	7.7470	10.195	14.125	21.562	38.518	73.130	130.47	232.65	423.28	785.34	1398.3
0.00046	1.288	1.384	1.475	1.591	1.755	2.006	2.325	2.655	3.029	3.472	3.994	4.552
1/ 2174	5.8286	7.9244	10.428	14.447	22.054	39.395	74.793	133.44	237.93	432.87	803.12	1429.9
0.00048	1.317	1.414	1.508	1.626	1.793	2.050	2.376	2.712	3.095	3.547	4.081	4.650
1/ 2083	5.9564	8.0979	10.657	14.763	22.535	40.253	76.421	136.34	243.09	442.25	820.51	1460.8
	500	550	650	700	850	1000	1300	1700	2100	2600	3400	4200

θ_{medial} for part-full circular pipes.

$k_s = 0.30$ mm $S = 0.00010$ to 0.00048

$k_s = 0.30$ mm
$S = 0.00050$ to 0.00280

Water (or sewage) at 15°C; full bore conditions.

ie hydraulic gradient =
1 in 2000 to 1 in 357

velocities in ms^{-1}
discharges in m^{3}s^{-1}

Gradient	\multicolumn{12}{c}{(Equivalent) Pipe diameters in m}
	2.400	2.700	3.000	3.400	4.000	5.000	6.400	8.000	10.00	12.60	16.00	20.00
0.00050 1/ 2000	1.344 6.0815	1.444 8.2678	1.539 10.880	1.660 15.072	1.831 23.006	2.093 41.094	2.425 78.014	2.769 139.17	3.160 248.15	3.621 451.44	4.166 837.55	4.746 1491.2
0.00055 1/ 1818	1.411 6.3839	1.516 8.6784	1.616 11.420	1.742 15.820	1.921 24.145	2.196 43.125	2.545 81.864	2.905 146.03	3.315 260.37	3.799 473.65	4.370 878.70	4.980 1564.4
0.00060 1/ 1667	1.475 6.6728	1.584 9.0708	1.689 11.936	1.821 16.533	2.008 25.234	2.295 45.066	2.659 85.543	3.036 152.59	3.464 272.04	3.969 494.86	4.566 918.03	5.202 1634.3
0.00065 1/ 1538	1.536 6.9500	1.650 9.4472	1.759 12.431	1.896 17.218	2.091 26.278	2.390 46.928	2.769 89.072	3.161 158.88	3.606 283.24	4.132 515.21	4.754 955.75	5.416 1701.4
0.00070 1/ 1429	1.595 7.2167	1.713 9.8094	1.826 12.907	1.969 17.877	2.171 27.282	2.481 48.719	2.874 92.468	3.281 164.93	3.743 294.01	4.289 534.80	4.934 992.04	5.621 1766.0
0.00075 1/ 1333	1.652 7.4741	1.774 10.159	1.891 13.366	2.039 18.513	2.248 28.251	2.569 50.448	2.976 95.744	3.397 170.76	3.876 304.41	4.441 553.69	5.108 1027.1	5.820 1828.3
0.00080 1/ 1250	1.707 7.7231	1.833 10.497	1.954 13.811	2.107 19.128	2.323 29.189	2.654 52.120	3.075 98.913	3.510 176.41	4.004 314.47	4.587 571.96	5.277 1060.9	6.011 1888.6
0.00085 1/ 1176	1.761 7.9644	1.891 10.825	2.015 14.242	2.172 19.724	2.395 30.098	2.737 53.740	3.170 101.98	3.618 181.88	4.128 324.21	4.729 589.68	5.440 1093.8	6.197 1947.0
0.00090 1/ 1111	1.812 8.1987	1.946 11.143	2.074 14.660	2.236 20.303	2.465 30.980	2.817 55.314	3.263 104.97	3.724 187.20	4.248 333.68	4.867 606.88	5.598 1125.6	6.378 2003.7
0.00095 1/ 1053	1.863 8.4266	2.000 11.452	2.132 15.067	2.298 20.866	2.534 31.839	2.895 56.845	3.353 107.87	3.827 192.37	4.366 342.88	5.001 623.60	5.753 1156.6	6.553 2058.8
0.00100 1/ 1000	1.912 8.6486	2.053 11.754	2.188 15.463	2.359 21.415	2.600 32.675	2.971 58.335	3.441 110.69	3.927 197.40	4.480 351.85	5.132 639.90	5.903 1186.8	6.724 2112.5
0.00110 1/ 909	2.006 9.0766	2.154 12.335	2.296 16.227	2.475 22.472	2.728 34.286	3.117 61.209	3.610 116.14	4.120 207.10	4.700 369.13	5.384 671.30	6.192 1245.1	7.054 2216.1
0.00120 1/ 833	2.097 9.4855	2.251 12.890	2.399 16.957	2.586 23.482	2.851 35.826	3.257 63.955	3.772 121.34	4.305 216.38	4.910 385.64	5.624 701.31	6.469 1300.7	7.369 2315.0
0.00130 1/ 769	2.183 9.8778	2.344 13.423	2.498 17.657	2.693 24.451	2.968 37.303	3.391 66.589	3.927 126.34	4.482 225.27	5.112 401.48	5.855 730.10	6.734 1354.0	7.671 2410.0
0.00140 1/ 714	2.267 10.255	2.434 13.936	2.593 18.331	2.796 25.384	3.082 38.724	3.520 69.123	4.076 131.14	4.652 233.83	5.306 416.72	6.077 757.79	6.990 1405.4	7.962 2501.3
0.00150 1/ 667	2.347 10.620	2.520 14.430	2.685 18.981	2.895 26.283	3.191 40.096	3.645 71.569	4.221 135.77	4.816 242.08	5.493 431.43	6.292 784.52	7.236 1454.9	8.242 2589.4
0.00160 1/ 625	2.425 10.972	2.604 14.909	2.774 19.610	2.991 27.153	3.296 41.422	3.765 73.934	4.360 140.26	4.975 250.07	5.674 445.65	6.499 810.36	7.474 1502.8	8.514 2674.6
0.00170 1/ 588	2.501 11.313	2.685 15.372	2.861 20.220	3.084 27.997	3.399 42.708	3.882 76.226	4.495 144.60	5.129 257.81	5.850 459.44	6.700 835.42	7.705 1549.2	8.776 2757.2
0.00180 1/ 556	2.574 11.645	2.763 15.822	2.944 20.812	3.174 28.816	3.498 43.956	3.996 78.452	4.626 148.82	5.279 265.33	6.020 472.82	6.895 859.74	7.929 1594.3	9.032 2837.4
0.00190 1/ 526	2.645 11.967	2.840 16.260	3.026 21.387	3.262 29.612	3.595 45.170	4.106 80.617	4.754 152.92	5.424 272.64	6.186 485.84	7.085 883.40	8.148 1638.2	9.280 2915.4
0.00200 1/ 500	2.715 12.281	2.914 16.687	3.105 21.948	3.347 30.388	3.689 46.353	4.213 82.726	4.878 156.92	5.566 279.76	6.347 498.52	7.270 906.44	8.360 1680.9	9.522 2991.4
0.00220 1/ 455	2.849 12.887	3.058 17.509	3.258 23.029	3.512 31.884	3.870 48.633	4.420 86.791	5.117 164.63	5.839 293.48	6.659 522.96	7.626 950.86	8.769 1763.2	9.988 3137.9
0.00240 1/ 417	2.977 13.466	3.195 18.295	3.404 24.062	3.669 33.313	4.043 50.811	4.618 90.675	5.346 171.99	6.100 306.60	6.956 546.32	7.966 993.31	9.161 1841.9	10.43 3277.8
0.00260 1/ 385	3.099 14.021	3.327 19.048	3.544 25.052	3.820 34.684	4.210 52.900	4.808 94.401	5.566 179.05	6.350 319.18	7.241 568.72	8.293 1034.0	9.536 1917.3	10.86 3412.0
0.00280 1/ 357	3.217 14.555	3.454 19.773	3.679 26.005	3.965 36.003	4.370 54.911	4.990 97.986	5.777 185.84	6.591 331.28	7.516 590.28	8.607 1073.2	9.897 1989.9	11.27 3541.2
	800	950	1000	1200	1400	1700	2200	2700	3400	4300	5500	7000

θ_{medial} for part-full circular pipes.

$k_s = 0.30$ mm $S = 0.00050$ to 0.00280

$k_s = 0.30$ mm
$S = 0.0030$ to 0.014

Water (or sewage) at 15°C;
full bore conditions.

ie hydraulic gradient =
1 in 333 to 1 in 71

velocities in ms^{-1}
discharges in m^{3}s^{-1}

Gradient	(Equivalent) Pipe diameters in m											
	2.400	2.700	3.000	3.400	4.000	5.000	6.400	8.000	10.00	12.60	16.00	20.00
0.00300 1/ 333	3.331 15.070	3.576 20.473	3.809 26.925	4.106 37.275	4.524 56.851	5.167 101.44	5.981 192.40	6.823 342.96	7.780 611.08	8.910 1111.0	10.25 2060.0	11.67 3665.8
0.00320 1/ 313	3.441 15.568	3.694 21.150	3.935 27.815	4.241 38.506	4.673 58.727	5.337 104.79	6.178 198.74	7.048 354.26	8.037 631.19	9.203 1147.5	10.58 2127.7	12.05 3786.3
0.00340 1/ 294	3.548 16.051	3.808 21.806	4.057 28.677	4.373 39.699	4.818 60.545	5.502 108.03	6.369 204.88	7.266 365.20	8.285 650.69	9.487 1183.0	10.91 2193.4	12.42 3903.1
0.00360 1/ 278	3.652 16.520	3.920 22.443	4.175 29.514	4.500 40.858	4.959 62.311	5.662 111.18	6.554 210.85	7.477 375.83	8.526 669.62	9.763 1217.4	11.23 2257.2	12.79 4016.6
0.00380 1/ 263	3.753 16.977	4.028 23.062	4.291 30.328	4.624 41.985	5.095 64.029	5.818 114.24	6.735 216.65	7.683 386.17	8.760 688.03	10.03 1250.8	11.53 2319.2	13.14 4126.9
0.00400 1/ 250	3.851 17.421	4.133 23.665	4.403 31.122	4.745 43.082	5.228 65.701	5.970 117.23	6.910 222.31	7.883 396.24	8.989 705.97	10.29 1283.4	11.84 2379.6	13.48 4234.3
0.00420 1/ 238	3.947 17.854	4.236 24.254	4.512 31.895	4.863 44.153	5.358 67.333	6.118 120.14	7.082 227.82	8.078 406.07	9.211 723.46	10.55 1315.2	12.13 2438.5	13.81 4339.1
0.00440 1/ 227	4.040 18.278	4.336 24.828	4.619 32.651	4.978 45.198	5.485 68.926	6.263 122.98	7.249 233.20	8.269 415.66	9.429 740.54	10.80 1346.2	12.41 2496.0	14.14 4441.5
0.00460 1/ 217	4.132 18.691	4.435 25.390	4.724 33.389	5.091 46.219	5.609 70.483	6.405 125.75	7.413 238.46	8.456 425.03	9.641 757.23	11.04 1376.6	12.69 2552.2	14.46 4541.5
0.00480 1/ 208	4.221 19.096	4.531 25.940	4.826 34.111	5.201 47.219	5.730 72.007	6.543 128.47	7.573 243.61	8.638 434.21	9.849 773.57	11.28 1406.3	12.97 2607.3	14.77 4639.4
0.00500 1/ 200	4.309 19.492	4.625 26.478	4.926 34.819	5.309 48.198	5.849 73.499	6.678 131.13	7.729 248.66	8.817 443.19	10.05 789.57	11.51 1435.3	13.24 2661.2	15.07 4735.2
0.00550 1/ 182	4.520 20.450	4.852 27.778	5.168 36.528	5.569 50.564	6.136 77.105	7.006 137.56	8.108 260.84	9.249 464.89	10.55 828.22	12.07 1505.6	13.88 2791.3	15.81 4966.8
0.00600 1/ 167	4.723 21.365	5.069 29.021	5.399 38.162	5.818 52.824	6.410 80.549	7.319 143.70	8.470 272.48	9.661 485.63	11.02 865.15	12.61 1572.7	14.50 2915.7	16.51 5188.1
0.00650 1/ 154	4.917 22.243	5.277 30.213	5.620 39.728	6.057 54.991	6.673 83.853	7.619 149.59	8.817 283.65	10.06 505.52	11.47 900.57	13.13 1637.1	15.10 3035.0	17.19 5400.3
0.00700 1/ 143	5.103 23.088	5.477 31.359	5.834 41.236	6.287 57.077	6.926 87.033	7.907 155.26	9.151 294.39	10.44 524.66	11.90 934.66	13.63 1699.0	15.67 3149.8	17.84 5604.5
0.00750 1/ 133	5.284 23.903	5.670 32.466	6.039 42.690	6.508 59.090	7.170 90.100	8.186 160.73	9.473 304.75	10.81 543.13	12.32 967.54	14.11 1758.8	16.22 3260.6	18.47 5801.6
0.00800 1/ 125	5.458 24.691	5.857 33.536	6.238 44.097	6.723 61.036	7.406 93.067	8.455 166.02	9.785 314.78	11.16 560.99	12.72 999.35	14.57 1816.6	16.75 3367.7	19.07 5992.2
0.00850 1/ 118	5.627 25.455	6.038 34.573	6.431 45.461	6.930 62.923	7.635 95.943	8.717 171.15	10.09 324.50	11.50 578.30	13.12 1030.2	15.02 1872.6	17.27 3471.6	19.66 6176.9
0.00900 1/ 111	5.791 26.196	6.214 35.581	6.619 46.785	7.132 64.755	7.857 98.735	8.970 176.13	10.38 333.93	11.84 595.11	13.50 1060.1	15.45 1927.0	17.77 3572.4	20.23 6356.2
0.00950 1/ 105	5.950 26.918	6.385 36.560	6.801 48.072	7.328 66.537	8.073 101.45	9.217 180.97	10.67 343.11	12.16 611.46	13.87 1089.2	15.88 1979.9	18.26 3670.4	20.79 6530.6
0.01000 1/ 100	6.105 27.621	6.552 37.514	6.978 49.327	7.520 68.272	8.284 104.10	9.457 185.69	10.94 352.05	12.48 627.38	14.23 1117.6	16.29 2031.4	18.73 3765.9	21.33 6700.5
0.01100 1/ 91	6.405 28.975	6.873 39.353	7.320 51.744	7.888 71.618	8.689 109.19	9.920 194.78	11.48 369.28	13.09 658.08	14.93 1172.2	17.09 2130.8	19.65 3950.0	22.37 7028.1
0.01200 1/ 83	6.691 30.270	7.180 41.111	7.647 54.054	8.240 74.814	9.077 114.07	10.36 203.46	11.99 385.74	13.68 687.41	15.59 1224.5	17.85 2225.7	20.52 4126.0	23.37 7341.0
0.01300 1/ 77	6.965 31.511	7.475 42.796	7.961 56.270	8.578 77.880	9.449 118.74	10.79 211.80	12.48 401.53	14.24 715.54	16.23 1274.6	18.58 2316.7	21.36 4294.7	24.32 7641.1
0.01400 1/ 71	7.229 32.705	7.758 44.418	8.262 58.402	8.903 80.830	9.807 123.24	11.19 219.81	12.95 416.72	14.77 742.60	16.84 1322.8	19.28 2404.3	22.17 4457.0	25.24 7929.9
	1300	1500	1600	1900	2200	2700	3500	4400	5500	7000	8500	11000

θ_{medial} for part-full circular pipes.

$k_s = 0.30$ mm $S = 0.0030$ to 0.014

k_s = 0.30 mm
S = 0.015 to 0.075

Water (or sewage) at 15°C;
full bore conditions.

A38
continued

ie hydraulic gradient =
1 in 66 to 1 in 13

velocities in ms^{-1}
discharges in m^{3}s^{-1}

Gradient	(Equivalent) Pipe diameters in m											
	2.400	2.700	3.000	3.400	4.000	5.000	6.400	8.000	10.00	12.60	16.00	20.00
0.01500 1/ 67	7.484 33.858	8.031 45.983	8.553 60.459	9.216 83.676	10.15 127.57	11.59 227.55	13.41 431.38	15.29 768.72	17.43 1369.3	19.96 2488.8	22.95 4613.7	26.13 8208.6
0.01600 1/ 62	7.731 34.973	8.296 47.497	8.835 62.449	9.519 86.429	10.49 131.77	11.97 235.03	13.85 445.56	15.80 793.98	18.01 1414.3	20.62 2570.6	23.70 4765.2	26.99 8478.1
0.01700 1/ 59	7.970 36.053	8.552 48.964	9.108 64.377	9.813 89.098	10.81 135.84	12.34 242.28	14.28 459.31	16.28 818.46	18.56 1457.9	21.25 2649.8	24.43 4912.0	27.82 8739.3
0.01800 1/ 56	8.201 37.102	8.801 50.388	9.372 66.250	10.10 91.688	11.12 139.79	12.70 249.32	14.69 472.65	16.76 842.24	19.10 1500.2	21.87 2726.7	25.14 5054.6	28.63 8993.0
0.01900 1/ 53	8.427 38.123	9.043 51.774	9.630 68.071	10.38 94.208	11.43 143.63	13.05 256.17	15.10 485.63	17.22 865.36	19.63 1541.4	22.47 2801.6	25.83 5193.3	29.41 9239.7
0.02000 1/ 50	8.647 39.117	9.278 53.123	9.881 69.845	10.65 96.663	11.73 147.37	13.39 262.84	15.49 498.27	17.66 887.88	20.14 1581.5	23.05 2874.4	26.50 5328.4	30.18 9479.9
0.02200 1/ 45	9.070 41.032	9.733 55.724	10.36 73.264	11.17 101.39	12.30 154.58	14.04 275.70	16.25 522.64	18.53 931.29	21.12 1658.8	24.18 3014.9	27.80 5588.7	31.65 9943.1
0.02400 1/ 42	9.475 42.863	10.17 58.210	10.83 76.531	11.67 105.91	12.85 161.47	14.67 287.98	16.97 545.92	19.35 972.76	22.06 1732.7	25.26 3149.1	29.03 5837.5	33.06 10386
0.02600 1/ 38	9.863 44.619	10.58 60.593	11.27 79.665	12.14 110.25	13.38 168.08	15.27 299.77	17.66 568.25	20.14 1012.5	22.96 1803.5	26.29 3277.9	30.22 6076.1	34.41 10810
0.02800 1/ 36	10.24 46.308	10.98 62.887	11.70 82.680	12.60 114.42	13.88 174.44	15.84 311.11	18.33 589.74	20.91 1050.8	23.83 1871.7	27.28 3401.8	31.36 6305.7	35.71 11219
0.03000 1/ 33	10.60 47.938	11.37 65.100	12.11 85.589	13.05 118.45	14.37 180.57	16.40 322.05	18.98 610.47	21.64 1087.8	24.67 1937.4	28.24 3521.3	32.46 6527.3	36.96 11613
0.03200 1/ 31	10.95 49.515	11.74 67.241	12.51 88.403	13.47 122.34	14.84 186.51	16.94 332.63	19.60 630.52	22.35 1123.5	25.48 2001.1	29.17 3636.9	33.53 6741.5	38.18 11994
0.03400 1/ 29	11.28 51.043	12.11 69.316	12.89 91.130	13.89 126.12	15.30 192.26	17.46 342.88	20.20 649.96	23.04 1158.1	26.26 2062.7	30.07 3748.9	34.56 6949.2	39.35 12363
0.03600 1/ 28	11.61 52.526	12.46 71.330	13.27 93.779	14.29 129.78	15.74 197.84	17.97 352.84	20.79 668.83	23.71 1191.7	27.03 2122.6	30.94 3857.7	35.57 7150.8	40.50 12722
0.03800 1/ 26	11.93 53.969	12.80 73.290	13.63 96.354	14.69 133.34	16.18 203.28	18.46 362.52	21.36 687.18	24.36 1224.4	27.77 2180.8	31.79 3963.6	36.54 7346.9	41.61 13071
0.04000 1/ 25	12.24 55.375	13.13 75.198	13.99 98.863	15.07 136.81	16.60 208.57	18.94 371.96	21.92 705.06	24.99 1256.3	28.49 2237.5	32.61 4066.6	37.49 7538.0	42.69 13411
0.04200 1/ 24	12.54 56.746	13.46 77.059	14.33 101.31	15.44 140.20	17.01 213.73	19.41 381.16	22.46 722.49	25.61 1287.3	29.19 2292.9	33.42 4167.1	38.42 7724.3	43.74 13742
0.04400 1/ 23	12.84 58.084	13.78 78.877	14.67 103.70	15.81 143.51	17.41 218.76	19.87 390.14	22.99 739.52	26.21 1317.7	29.88 2346.9	34.21 4265.3	39.32 7906.2	44.77 14066
0.04600 1/ 22	13.13 59.393	14.09 80.653	15.00 106.03	16.16 146.74	17.80 223.69	20.32 398.92	23.51 756.16	26.80 1347.3	30.55 2399.7	34.98 4361.2	40.21 8084.0	45.78 14382
0.04800 1/ 21	13.41 60.673	14.39 82.392	15.32 108.32	16.51 149.90	18.18 228.51	20.75 407.52	24.01 772.45	27.38 1376.3	31.21 2451.3	35.73 4455.1	41.07 8258.0	46.76 14691
0.05000 1/ 20	13.69 61.927	14.69 84.094	15.64 110.56	16.85 153.00	18.56 233.23	21.18 415.93	24.51 788.39	27.95 1404.7	31.86 2501.9	36.47 4547.1	41.92 8428.4	47.73 14995
0.05500 1/ 18	14.36 64.956	15.41 88.207	16.41 115.96	17.68 160.48	19.47 244.63	22.22 436.26	25.70 826.92	29.31 1473.4	33.41 2624.2	38.25 4769.2	43.97 8840.1	50.06 15727
0.06000 1/ 17	15.00 67.851	16.09 92.137	17.14 121.13	18.46 167.62	20.33 255.53	23.21 455.69	26.85 863.74	30.62 1539.0	34.90 2740.9	39.95 4981.4	45.92 9233.5	52.29 16427
0.06500 1/ 15	15.61 70.627	16.75 95.906	17.84 126.08	19.22 174.48	21.17 265.98	24.16 474.32	27.95 899.04	31.87 1601.9	36.33 2853.0	41.58 5185.0	47.80 9610.7	54.42 17098
0.07000 1/ 14	16.20 73.298	17.38 99.533	18.51 130.85	19.94 181.08	21.97 276.03	25.07 492.25	29.00 933.02	33.07 1662.4	37.70 2960.8	43.15 5380.9	49.61 9973.8	56.48 17744
0.07500 1/ 13	16.77 75.875	18.00 103.03	19.16 135.45	20.65 187.44	22.74 285.73	25.95 509.54	30.02 965.80	34.23 1720.8	39.02 3064.8	44.67 5569.9	51.35 10324	58.46 18367
	2100	2300	2600	2900	3500	4300	5500	7000	8500	11000	14000	17000

θ_{medial} for part-full circular pipes.

k_s = 0.60 mm
S = 0.00010 to 0.00048

Water (or sewage) at 15°C;
full bore conditions

ie hydraulic gradient =
1 in 10000 to 1 in 2083

velocities in ms^{-1}
discharges in m^{3}s^{-1}

Gradient	(Equivalent) Pipe diameters in m											
	2.400	2.700	3.000	3.400	4.000	5.000	6.400	8.000	10.00	12.60	16.00	20.00
0.00010 1/10000	0.559 2.5270	0.601 3.4391	0.641 4.5298	0.692 6.2820	0.764 9.6015	0.875 17.180	1.016 32.674	1.161 58.378	1.327 104.24	1.523 189.91	1.755 352.82	2.002 628.93
0.00011 1/ 9091	0.586 2.6530	0.631 3.6104	0.673 4.7552	0.726 6.5943	0.802 10.078	0.918 18.031	1.066 34.289	1.219 61.261	1.393 109.38	1.598 199.26	1.841 370.18	2.100 659.85
0.00012 1/ 8333	0.613 2.7735	0.659 3.7741	0.703 4.9707	0.759 6.8926	0.838 10.533	0.960 18.844	1.114 35.833	1.274 64.015	1.455 114.29	1.670 208.20	1.924 386.76	2.194 689.38
0.00013 1/ 7692	0.639 2.8890	0.687 3.9312	0.732 5.1773	0.791 7.1788	0.873 10.970	0.999 19.624	1.160 37.314	1.326 66.657	1.515 119.00	1.738 216.77	2.003 402.67	2.285 717.72
0.00014 1/ 7143	0.663 3.0002	0.713 4.0823	0.761 5.3762	0.821 7.4543	0.906 11.390	1.038 20.375	1.204 38.739	1.377 69.199	1.573 123.54	1.805 225.02	2.079 417.98	2.371 744.98
0.00015 1/ 6667	0.687 3.1075	0.738 4.2282	0.788 5.5680	0.850 7.7200	0.939 11.796	1.075 21.099	1.247 40.114	1.425 71.652	1.629 127.91	1.868 232.98	2.152 432.75	2.455 771.28
0.00016 1/ 6250	0.710 3.2113	0.763 4.3693	0.814 5.7536	0.879 7.9771	0.970 12.188	1.110 21.800	1.288 41.444	1.473 74.025	1.682 132.14	1.930 240.68	2.223 447.04	2.536 796.73
0.00017 1/ 5882	0.732 3.3120	0.787 4.5060	0.839 5.9336	0.906 8.2263	1.000 12.569	1.145 22.479	1.328 42.733	1.518 76.324	1.735 136.24	1.990 248.14	2.292 460.88	2.615 821.39
0.00018 1/ 5556	0.754 3.4097	0.810 4.6388	0.864 6.1083	0.933 8.4683	1.030 12.938	1.178 23.138	1.367 43.985	1.563 78.557	1.785 140.22	2.048 255.38	2.359 474.33	2.691 845.33
0.00019 1/ 5263	0.775 3.5047	0.833 4.7680	0.888 6.2782	0.959 8.7036	1.058 13.297	1.211 23.780	1.405 45.202	1.606 80.729	1.835 144.10	2.105 262.43	2.424 487.41	2.765 868.62
0.00020 1/ 5000	0.795 3.5973	0.855 4.8938	0.912 6.4437	0.984 8.9328	1.086 13.647	1.243 24.404	1.442 46.388	1.648 82.844	1.883 147.87	2.160 269.29	2.488 500.14	2.837 891.30
0.00022 1/ 4545	0.835 3.7757	0.897 5.1363	0.957 6.7628	1.033 9.3748	1.140 14.321	1.304 25.609	1.513 48.674	1.729 86.922	1.975 155.14	2.266 282.52	2.610 524.69	2.976 935.02
0.00024 1/ 4167	0.872 3.9462	0.938 5.3681	1.000 7.0677	1.079 9.7970	1.191 14.966	1.363 26.759	1.581 50.858	1.807 90.819	2.064 162.09	2.367 295.17	2.726 548.15	3.109 976.80
0.00026 1/ 3846	0.908 4.1098	0.976 5.5904	1.041 7.3601	1.124 10.202	1.240 15.584	1.419 27.863	1.646 52.953	1.881 94.556	2.149 168.75	2.464 307.29	2.838 570.65	3.237 1016.9
0.00028 1/ 3571	0.943 4.2672	1.014 5.8043	1.081 7.6416	1.167 10.592	1.287 16.178	1.473 28.925	1.709 54.969	1.953 98.152	2.230 175.16	2.558 318.96	2.946 592.31	3.360 1055.4
0.00030 1/ 3333	0.977 4.4191	1.050 6.0107	1.119 7.9131	1.208 10.968	1.333 16.752	1.525 29.950	1.769 56.914	2.022 101.62	2.309 181.35	2.648 330.22	3.050 613.20	3.478 1092.6
0.00032 1/ 3125	1.009 4.5660	1.085 6.2103	1.157 8.1758	1.248 11.332	1.377 17.307	1.576 30.941	1.828 58.795	2.088 104.98	2.385 187.34	2.736 341.10	3.150 633.40	3.593 1128.6
0.00034 1/ 2941	1.041 4.7084	1.118 6.4039	1.193 8.4304	1.287 11.684	1.420 17.845	1.625 31.902	1.884 60.619	2.153 108.23	2.459 193.14	2.820 351.66	3.248 652.99	3.704 1163.5
0.00036 1/ 2778	1.071 4.8467	1.151 6.5918	1.228 8.6776	1.325 12.027	1.462 18.368	1.672 32.835	1.939 62.389	2.216 111.39	2.531 198.77	2.902 361.91	3.342 672.00	3.811 1197.4
0.00038 1/ 2632	1.101 4.9812	1.183 6.7745	1.262 8.9180	1.361 12.360	1.502 18.876	1.718 33.742	1.993 64.112	2.277 114.46	2.601 204.25	2.982 371.87	3.434 690.50	3.916 1230.3
0.00040 1/ 2500	1.130 5.1122	1.214 6.9526	1.295 9.1522	1.397 12.684	1.541 19.371	1.763 34.626	2.045 65.789	2.337 117.45	2.668 209.58	3.060 381.58	3.524 708.51	4.018 1262.4
0.00042 1/ 2381	1.158 5.2399	1.245 7.1262	1.327 9.3806	1.432 13.000	1.580 19.854	1.807 35.488	2.096 67.425	2.395 120.37	2.735 214.78	3.136 391.05	3.611 726.08	4.118 1293.7
0.00044 1/ 2273	1.186 5.3647	1.274 7.2957	1.359 9.6037	1.466 13.309	1.617 20.325	1.850 36.329	2.146 69.022	2.451 123.22	2.799 219.87	3.210 400.29	3.697 743.24	4.215 1324.2
0.00046 1/ 2174	1.213 5.4867	1.303 7.4615	1.389 9.8217	1.499 13.611	1.654 20.786	1.892 37.152	2.194 70.584	2.507 126.01	2.863 224.83	3.283 409.33	3.780 760.01	4.310 1354.1
0.00048 1/ 2083	1.239 5.6060	1.332 7.6237	1.420 10.035	1.532 13.907	1.690 21.236	1.933 37.957	2.242 72.112	2.561 128.73	2.925 229.69	3.354 418.17	3.862 776.42	4.403 1383.3
	470	550	600	650	800	1000	1300	1600	2000	2500	3200	3900

θ_{medial} for part-full circular pipes.

k_s = 0.60 mm   S = 0.00010 to 0.00048

$k_s = 0.60$ mm
$S = 0.00050$ to 0.00280

Water (or sewage) at 15°C;
full bore conditions.

ie hydraulic gradient =
1 in 2000 to 1 in 357

velocities in ms⁻¹
discharges in m³s⁻¹

Gradient	(Equivalent) Pipe diameters in m											
	2.400	2.700	3.000	3.400	4.000	5.000	6.400	8.000	10.00	12.60	16.00	20.00
0.00050 1/ 2000	1.265 5.7229	1.359 7.7825	1.449 10.244	1.564 14.196	1.725 21.678	1.973 38.745	2.288 73.608	2.614 131.40	2.985 234.45	3.423 426.83	3.942 792.49	4.494 1411.9
0.00055 1/ 1818	1.327 6.0053	1.426 8.1662	1.521 10.749	1.641 14.895	1.810 22.745	2.070 40.650	2.400 77.224	2.742 137.85	3.132 245.95	3.591 447.76	4.135 831.31	4.714 1481.1
0.00060 1/ 1667	1.387 6.2751	1.490 8.5329	1.589 11.231	1.714 15.563	1.891 23.764	2.163 42.470	2.508 80.678	2.865 144.01	3.271 256.94	3.751 467.75	4.319 868.41	4.925 1547.1
0.00065 1/ 1538	1.444 6.5339	1.552 8.8846	1.654 11.694	1.785 16.204	1.969 24.742	2.252 44.216	2.611 83.991	2.983 149.92	3.406 267.48	3.905 486.92	4.496 903.99	5.126 1610.5
0.00070 1/ 1429	1.499 6.7829	1.611 9.2230	1.717 12.139	1.853 16.820	2.044 25.682	2.337 45.896	2.710 87.180	3.096 155.61	3.535 277.62	4.053 505.37	4.666 938.23	5.320 1671.5
0.00075 1/ 1333	1.552 7.0232	1.668 9.5495	1.778 12.569	1.918 17.415	2.116 26.590	2.420 47.516	2.806 90.256	3.205 161.10	3.659 287.40	4.196 523.17	4.831 971.26	5.508 1730.3
0.00080 1/ 1250	1.604 7.2556	1.723 9.8654	1.837 12.984	1.982 17.991	2.186 27.468	2.500 49.084	2.898 93.231	3.311 166.41	3.780 296.87	4.334 540.39	4.990 1003.2	5.689 1787.2
0.00085 1/ 1176	1.654 7.4809	1.777 10.171	1.894 13.387	2.043 18.548	2.254 28.319	2.577 50.603	2.988 96.115	3.413 171.55	3.897 306.04	4.468 557.08	5.144 1034.2	5.864 1842.3
0.00090 1/ 1111	1.702 7.6996	1.828 10.469	1.949 13.778	2.103 19.090	2.319 29.145	2.652 52.079	3.075 98.915	3.512 176.54	4.010 314.94	4.598 573.28	5.293 1064.2	6.035 1895.9
0.00095 1/ 1053	1.749 7.9123	1.879 10.758	2.003 14.158	2.161 19.617	2.383 29.949	2.725 53.514	3.159 101.64	3.609 181.40	4.120 323.61	4.724 589.04	5.439 1093.5	6.201 1948.0
0.00100 1/ 1000	1.795 8.1196	1.928 11.039	2.055 14.529	2.217 20.130	2.446 30.732	2.797 54.911	3.242 104.29	3.703 186.13	4.228 332.04	4.847 604.39	5.580 1122.0	6.362 1998.7
0.00110 1/ 909	1.883 8.5190	2.023 11.582	2.156 15.243	2.326 21.118	2.566 32.240	2.934 57.605	3.401 109.40	3.884 195.26	4.435 348.31	5.084 633.98	5.853 1176.9	6.673 2096.5
0.00120 1/ 833	1.967 8.9007	2.113 12.101	2.253 15.925	2.430 22.063	2.680 33.682	3.065 60.179	3.553 114.29	4.058 203.97	4.633 363.84	5.311 662.26	6.114 1229.4	6.971 2189.9
0.00130 1/ 769	2.048 9.2668	2.200 12.598	2.345 16.579	2.530 22.970	2.790 35.065	3.191 62.648	3.698 118.98	4.224 212.33	4.822 378.75	5.529 689.37	6.365 1279.7	7.256 2279.5
0.00140 1/ 714	2.126 9.6190	2.284 13.077	2.435 17.209	2.626 23.842	2.896 36.395	3.312 65.024	3.839 123.49	4.384 220.37	5.005 393.09	5.738 715.47	6.605 1328.1	7.530 2365.8
0.00150 1/ 667	2.201 9.9589	2.365 13.539	2.521 17.817	2.719 24.683	2.998 37.679	3.428 67.317	3.974 127.84	4.539 228.13	5.181 406.93	5.940 740.64	6.838 1374.8	7.795 2449.0
0.00160 1/ 625	2.274 10.288	2.443 13.986	2.604 18.404	2.808 25.497	3.097 38.921	3.541 69.534	4.105 132.04	4.688 235.64	5.352 420.31	6.135 764.99	7.062 1420.0	8.051 2529.4
0.00170 1/ 588	2.345 10.606	2.518 14.419	2.684 18.974	2.895 26.286	3.193 40.125	3.651 71.683	4.231 136.12	4.833 242.92	5.517 433.28	6.324 788.60	7.280 1463.8	8.300 2607.4
0.00180 1/ 556	2.413 10.916	2.592 14.839	2.762 19.527	2.979 27.051	3.286 41.293	3.757 73.769	4.354 140.08	4.973 249.98	5.677 445.88	6.508 811.51	7.492 1506.3	8.541 2683.2
0.00190 1/ 526	2.479 11.217	2.663 15.248	2.839 20.065	3.062 27.796	3.376 42.430	3.860 75.798	4.474 143.93	5.110 256.85	5.833 458.13	6.687 833.80	7.698 1547.7	8.775 2756.8
0.00200 1/ 500	2.544 11.510	2.733 15.646	2.913 20.589	3.141 28.522	3.465 43.537	3.961 77.775	4.591 147.69	5.243 263.54	5.985 470.06	6.861 855.51	7.898 1588.0	9.004 2828.5
0.00220 1/ 455	2.669 12.075	2.867 16.414	3.056 21.599	3.296 29.921	3.634 45.671	4.155 81.585	4.816 154.92	5.500 276.44	6.278 493.06	7.197 897.36	8.284 1665.6	9.444 2966.8
0.00240 1/ 417	2.788 12.614	2.995 17.148	3.192 22.564	3.443 31.257	3.797 47.710	4.341 85.226	5.030 161.83	5.745 288.77	6.558 515.04	7.517 937.34	8.653 1739.8	9.864 3099.0
0.00260 1/ 385	2.903 13.132	3.118 17.851	3.323 23.489	3.584 32.539	3.952 49.665	4.518 88.718	5.236 168.45	5.980 300.59	6.826 536.12	7.825 975.69	9.007 1811.0	10.27 3225.7
0.00280 1/ 357	3.013 13.631	3.236 18.528	3.449 24.380	3.720 33.772	4.102 51.547	4.689 92.078	5.435 174.83	6.206 311.96	7.084 556.40	8.121 1012.6	9.348 1879.5	10.66 3347.6
	750	850	950	1100	1200	1600	2000	2500	3100	3900	5000	6000

θ_{medial} for part-full circular pipes.

$k_s = 0.60$ mm $S = 0.00050$ to 0.00280

$k_s = 0.60$ mm
$S = 0.0030$ to 0.014

Water (or sewage) at 15°C;
full bore conditions.

ie hydraulic gradient =
1 in 333 to 1 in 71

velocities in ms⁻¹
discharges in m³s⁻¹

velocities in ms^{-1}
discharges in m^3s^{-1}

Gradient	(Equivalent) Pipe diameters in m											
	2.400	2.700	3.000	3.400	4.000	5.000	6.400	8.000	10.00	12.60	16.00	20.00
0.00300 1/ 333	3.119 14.111	3.350 19.181	3.571 25.239	3.851 34.962	4.246 53.363	4.855 95.320	5.626 180.98	6.425 322.94	7.333 575.97	8.406 1048.2	9.676 1945.5	11.03 3465.3
0.00320 1/ 313	3.222 14.576	3.461 19.813	3.688 26.070	3.978 36.113	4.386 55.119	5.014 98.456	5.811 186.94	6.636 333.56	7.574 594.90	8.683 1082.6	9.994 2009.4	11.39 3579.1
0.00340 1/ 294	3.322 15.027	3.567 20.426	3.802 26.876	4.100 37.229	4.522 56.821	5.169 101.49	5.990 192.70	6.841 343.84	7.808 613.24	8.950 1116.0	10.30 2071.4	11.74 3689.4
0.00360 1/ 278	3.418 15.464	3.671 21.020	3.913 27.658	4.220 38.312	4.653 58.474	5.319 104.45	6.164 198.30	7.039 353.83	8.035 631.05	9.210 1148.4	10.60 2131.5	12.08 3796.5
0.00380 1/ 263	3.512 15.890	3.772 21.599	4.020 28.419	4.336 39.365	4.781 60.082	5.466 107.32	6.334 203.75	7.233 363.55	8.255 648.38	9.463 1179.9	10.89 2190.0	12.42 3900.7
0.00400 1/ 250	3.604 16.305	3.871 22.162	4.125 29.160	4.449 40.392	4.906 61.647	5.608 110.11	6.499 209.06	7.421 373.02	8.470 665.25	9.709 1210.6	11.18 2247.0	12.74 4002.3
0.00420 1/ 238	3.693 16.709	3.967 22.711	4.228 29.883	4.559 41.392	5.027 63.174	5.747 112.84	6.659 214.23	7.605 382.25	8.680 681.71	9.949 1240.6	11.45 2302.5	13.05 4101.1
0.00440 1/ 227	3.781 17.104	4.060 23.248	4.327 30.588	4.667 42.370	5.146 64.665	5.882 115.50	6.816 219.28	7.784 391.26	8.884 697.78	10.18 1269.8	11.72 2356.8	13.36 4197.7
0.00460 1/ 217	3.866 17.490	4.152 23.772	4.425 31.278	4.772 43.325	5.262 66.123	6.015 118.10	6.970 224.22	7.959 400.07	9.084 713.49	10.41 1298.4	11.99 2409.8	13.66 4292.1
0.00480 1/ 208	3.949 17.867	4.242 24.285	4.520 31.953	4.875 44.260	5.375 67.549	6.145 120.65	7.120 229.06	8.131 408.69	9.280 728.86	10.64 1326.4	12.24 2461.7	13.96 4384.6
0.00500 1/ 200	4.031 18.237	4.329 24.788	4.614 32.614	4.976 45.175	5.487 68.946	6.272 123.14	7.267 233.79	8.299 417.13	9.472 743.92	10.86 1353.8	12.50 2512.5	14.24 4475.1
0.00550 1/ 182	4.229 19.130	4.541 26.002	4.840 34.211	5.219 47.387	5.755 72.320	6.578 129.17	7.623 245.22	8.704 437.53	9.935 780.28	11.39 1419.9	13.11 2635.3	14.94 4693.7
0.00600 1/ 167	4.417 19.984	4.744 27.162	5.056 35.737	5.452 49.500	6.012 75.544	6.872 134.92	7.962 256.15	9.092 457.02	10.38 815.03	11.89 1483.2	13.69 2752.6	15.61 4902.7
0.00650 1/ 154	4.598 20.803	4.938 28.274	5.263 37.200	5.675 51.526	6.258 78.637	7.153 140.45	8.288 266.63	9.464 475.71	10.80 848.36	12.38 1543.8	14.25 2865.2	16.24 5103.0
0.00700 1/ 143	4.773 21.591	5.125 29.345	5.462 38.609	5.890 53.477	6.494 81.612	7.423 145.76	8.602 276.71	9.822 493.70	11.21 880.43	12.85 1602.1	14.79 2973.4	16.86 5295.9
0.00750 1/ 133	4.941 22.351	5.306 30.378	5.654 39.968	6.097 55.358	6.723 84.483	7.684 150.88	8.904 286.44	10.17 511.05	11.60 911.37	13.30 1658.4	15.31 3077.9	17.45 5481.9
0.00800 1/ 125	5.103 23.086	5.480 31.377	5.840 41.282	6.298 57.178	6.944 87.260	7.937 155.84	9.197 295.85	10.50 527.84	11.99 941.30	13.74 1712.9	15.81 3179.0	18.02 5661.8
0.00850 1/ 118	5.261 23.798	5.649 32.345	6.020 42.556	6.492 58.942	7.158 89.951	8.182 160.65	9.480 304.97	10.82 544.11	12.35 970.31	14.16 1765.7	16.30 3276.9	18.58 5836.2
0.00900 1/ 111	5.414 24.490	5.813 33.285	6.195 43.792	6.681 60.655	7.366 92.565	8.419 165.31	9.755 313.83	11.14 559.90	12.71 998.47	14.57 1816.9	16.77 3372.0	19.12 6005.6
0.00950 1/ 105	5.562 25.163	5.973 34.200	6.366 44.995	6.864 62.321	7.568 95.106	8.650 169.85	10.02 322.44	11.44 575.26	13.06 1025.9	14.97 1866.8	17.23 3464.5	19.64 6170.3
0.01000 1/ 100	5.707 25.819	6.129 35.091	6.531 46.167	7.043 63.943	7.765 97.582	8.876 174.27	10.28 330.83	11.74 590.23	13.40 1052.5	15.36 1915.3	17.68 3554.5	20.15 6330.7
0.01100 1/ 91	5.986 27.082	6.429 36.807	6.851 48.425	7.387 67.071	8.145 102.35	9.310 182.79	10.79 347.00	12.32 619.07	14.06 1104.0	16.11 2008.9	18.54 3728.2	21.14 6639.9
0.01200 1/ 83	6.253 28.290	6.715 38.448	7.156 50.583	7.716 70.059	8.508 106.91	9.724 190.93	11.27 362.45	12.86 646.64	14.68 1153.1	16.83 2098.3	19.37 3894.1	22.08 6935.4
0.01300 1/ 77	6.509 29.447	6.990 40.021	7.449 52.653	8.032 72.926	8.856 111.29	10.12 198.74	11.73 377.27	13.39 673.07	15.28 1200.3	17.52 2184.0	20.16 4053.2	22.98 7218.8
0.01400 1/ 71	6.756 30.562	7.254 41.535	7.731 54.645	8.336 75.684	9.191 115.50	10.50 206.26	12.17 391.53	13.90 698.51	15.86 1245.6	18.18 2266.6	20.92 4206.4	23.85 7491.5
	1100	1300	1400	1600	1900	2300	3000	3800	4700	6000	7500	9500

θ_{medial} for part-full circular pipes.

$k_s = 0.60$ mm $\quad S = 0.0030$ to 0.014

k_s = 0.60 mm
S = 0.015 to 0.075

Water (or sewage) at 15°C; full bore conditions.

ie hydraulic gradient = 1 in 66 to 1 in 13

velocities in ms^{-1}
discharges in m^{3}s^{-1}

Gradient	(Equivalent) Pipe diameters in m											
	2.400	2.700	3.000	3.400	4.000	5.000	6.400	8.000	10.00	12.60	16.00	20.00
0.01500 1/ 67	6.993 31.637	7.510 42.996	8.003 56.567	8.629 78.345	9.514 119.56	10.87 213.51	12.60 405.29	14.38 723.05	16.42 1289.4	18.82 2346.2	21.66 4354.1	24.68 7754.6
0.01600 1/ 62	7.223 32.676	7.756 44.409	8.265 58.425	8.913 80.919	9.826 123.48	11.23 220.52	13.01 418.60	14.86 746.79	16.96 1331.7	19.43 2423.2	22.37 4497.0	25.49 8009.0
0.01700 1/ 59	7.446 33.684	7.995 45.779	8.520 60.227	9.187 83.413	10.13 127.29	11.58 227.31	13.41 431.50	15.31 769.80	17.48 1372.7	20.03 2497.8	23.06 4635.5	26.28 8255.7
0.01800 1/ 56	7.662 34.663	8.228 47.108	8.768 61.976	9.454 85.836	10.42 130.98	11.91 233.91	13.80 444.02	15.76 792.14	17.99 1412.5	20.61 2570.3	23.72 4770.0	27.04 8495.1
0.01900 1/ 53	7.873 35.614	8.454 48.402	9.008 63.677	9.714 88.192	10.71 134.58	12.24 240.33	14.18 456.20	16.19 813.86	18.48 1451.3	21.18 2640.8	24.37 4900.8	27.78 8728.1
0.02000 1/ 50	8.077 36.541	8.674 49.661	9.243 65.334	9.966 90.486	10.99 138.08	12.56 246.58	14.55 468.07	16.61 835.03	18.96 1489.0	21.73 2709.4	25.01 5028.1	28.50 8954.9
0.02200 1/ 45	8.472 38.328	9.098 52.089	9.695 68.528	10.45 94.909	11.53 144.83	13.17 258.63	15.26 490.94	17.42 875.82	19.88 1561.8	22.79 2841.8	26.23 5273.7	29.90 9392.2
0.02400 1/ 42	8.850 40.036	9.503 54.409	10.13 71.580	10.92 99.136	12.04 151.28	13.76 270.14	15.94 512.79	18.20 914.80	20.77 1631.3	23.80 2968.2	27.40 5508.4	31.23 9810.1
0.02600 1/ 38	9.212 41.673	9.892 56.635	10.54 74.507	11.37 103.19	12.53 157.46	14.32 281.19	16.59 533.75	18.94 952.18	21.62 1697.9	24.78 3089.5	28.52 5733.4	32.50 10211
0.02800 1/ 36	9.560 43.249	10.27 58.776	10.94 77.324	11.80 107.09	13.00 163.41	14.86 291.81	17.22 553.91	19.66 988.16	22.44 1762.1	25.71 3206.2	29.59 5949.9	33.73 10596
0.03000 1/ 33	9.896 44.769	10.63 60.842	11.32 80.041	12.21 110.85	13.46 169.16	15.38 302.07	17.82 573.37	20.35 1022.9	23.22 1823.9	26.62 3318.8	30.63 6158.9	34.91 10969
0.03200 1/ 31	10.22 46.240	10.98 62.840	11.70 82.670	12.61 114.49	13.90 174.71	15.89 311.98	18.41 592.19	21.02 1056.4	23.99 1883.8	27.49 3427.7	31.64 6361.0	36.06 11328
0.03400 1/ 29	10.54 47.665	11.31 64.776	12.06 85.218	13.00 118.02	14.33 180.09	16.38 321.59	18.98 610.43	21.66 1089.0	24.72 1941.8	28.34 3533.2	32.61 6556.8	37.17 11677
0.03600 1/ 28	10.84 49.049	11.64 66.657	12.41 87.691	13.38 121.45	14.75 185.32	16.85 330.93	19.53 628.14	22.29 1120.6	25.44 1998.1	29.16 3635.7	33.56 6747.0	38.25 12016
0.03800 1/ 26	11.14 50.395	11.96 68.486	12.75 90.097	13.74 124.78	15.15 190.40	17.32 340.00	20.06 645.37	22.90 1151.3	26.14 2052.9	29.96 3735.4	34.48 6932.0	39.30 12345
0.04000 1/ 25	11.43 51.706	12.27 70.267	13.08 92.440	14.10 128.02	15.55 195.35	17.77 348.84	20.58 662.15	23.50 1181.2	26.82 2106.3	30.74 3832.5	35.37 7112.1	40.32 12666
0.04200 1/ 24	11.71 52.984	12.58 72.005	13.40 94.726	14.45 131.19	15.93 200.18	18.21 357.46	21.09 678.51	24.08 1210.4	27.48 2158.3	31.50 3927.2	36.25 7287.9	41.31 12979
0.04400 1/ 23	11.99 54.233	12.87 73.701	13.72 96.958	14.79 134.28	16.31 204.90	18.63 365.88	21.59 694.49	24.65 1238.9	28.13 2209.2	32.24 4019.6	37.10 7459.4	42.29 13285
0.04600 1/ 22	12.26 55.453	13.16 75.360	14.03 99.139	15.12 137.30	16.67 209.51	19.05 374.11	22.07 710.11	25.20 1266.8	28.76 2258.8	32.96 4110.0	37.93 7627.1	43.24 13583
0.04800 1/ 21	12.52 56.647	13.45 76.982	14.33 101.27	15.45 140.26	17.03 214.02	19.46 382.17	22.55 725.39	25.74 1294.0	29.38 2307.4	33.67 4198.5	38.75 7791.3	44.17 13876
0.05000 1/ 20	12.78 57.817	13.72 78.571	14.62 103.36	15.77 143.15	17.38 218.43	19.87 390.05	23.01 740.36	26.28 1320.7	29.99 2355.0	34.37 4285.1	39.55 7952.0	45.08 14162
0.05500 1/ 18	13.40 60.642	14.39 82.411	15.34 108.41	16.54 150.14	18.23 229.11	20.84 409.10	24.14 776.52	27.56 1385.2	31.45 2470.0	36.04 4494.3	41.48 8340.3	47.28 14853
0.06000 1/ 17	14.00 63.342	15.03 86.079	16.02 113.24	17.27 156.83	19.04 239.30	21.76 427.31	25.21 811.07	28.78 1446.9	32.85 2579.9	37.65 4694.3	43.33 8711.3	49.38 15514
0.06500 1/ 15	14.57 65.931	15.65 89.597	16.67 117.87	17.98 163.24	19.82 249.08	22.65 444.77	26.24 844.21	29.96 1506.0	34.19 2685.3	39.19 4886.0	45.10 9067.1	51.40 16148
0.07000 1/ 14	15.12 68.422	16.24 92.983	17.31 122.32	18.66 169.40	20.57 258.49	23.51 461.57	27.23 876.10	31.09 1562.9	35.48 2786.7	40.67 5070.5	46.80 9409.5	53.34 16757
0.07500 1/ 13	15.66 70.826	16.81 96.250	17.91 126.62	19.31 175.35	21.29 267.57	24.33 477.78	28.19 906.86	32.18 1617.7	36.73 2884.6	42.09 5248.6	48.44 9739.8	55.21 17346
	1600	1900	2100	2300	2700	3400	4400	5500	7000	8500	11000	14000

θ_{medial} for part-full circular pipes.

k_s = 0.60 mm   S = 0.0015 to 0.075

A40

$k_s = 1.50$ mm
S — 0.00010 to 0.00048

Water (or sewage) at 15°C;
full bore conditions.

ie hydraulic gradient =
1 in 10000 to 1 in 2083

velocities in ms^{-1}
discharges in m^{3}s^{-1}

Gradient	(Equivalent) Pipe diameters in m											
	2.400	2.700	3.000	3.400	4.000	5.000	6.400	8.000	10.00	12.60	16.00	20.00
0.00010	0.512	0.551	0.588	0.635	0.702	0.804	0.935	1.070	1.224	1.406	1.622	1.853
1/10000	2.3153	3.1526	4.1544	5.7646	8.8172	15.793	30.072	53.788	96.150	175.37	326.20	582.13
0.00011	0.537	0.578	0.617	0.666	0.736	0.844	0.981	1.123	1.284	1.475	1.702	1.944
1/ 9091	2.4295	3.3080	4.3591	6.0485	9.2511	16.570	31.549	56.428	100.87	183.97	342.18	610.64
0.00012	0.561	0.604	0.644	0.696	0.769	0.882	1.025	1.173	1.342	1.541	1.778	2.030
1/ 8333	2.5387	3.4566	4.5547	6.3197	9.6656	17.311	32.960	58.950	105.37	192.18	357.45	637.87
0.00013	0.584	0.629	0.671	0.725	0.801	0.918	1.067	1.221	1.397	1.604	1.851	2.114
1/ 7692	2.6434	3.5990	4.7423	6.5799	10.063	18.023	34.313	61.369	109.69	200.06	372.10	664.00
0.00014	0.607	0.653	0.696	0.752	0.831	0.953	1.107	1.267	1.450	1.665	1.921	2.194
1/ 7143	2.7441	3.7361	4.9229	6.8302	10.446	18.707	35.616	63.697	113.85	207.64	386.19	689.13
0.00015	0.628	0.676	0.721	0.779	0.861	0.986	1.146	1.312	1.501	1.724	1.988	2.271
1/ 6667	2.8413	3.8684	5.0971	7.0718	10.815	19.368	36.872	65.943	117.86	214.95	399.78	713.39
0.00016	0.649	0.698	0.745	0.805	0.889	1.019	1.184	1.355	1.550	1.781	2.054	2.345
1/ 6250	2.9354	3.9963	5.2655	7.3054	11.172	20.007	38.088	68.115	121.74	222.02	412.93	736.84
0.00017	0.669	0.720	0.768	0.830	0.917	1.050	1.221	1.397	1.598	1.836	2.117	2.418
1/ 5882	3.0265	4.1203	5.4288	7.5318	11.518	20.626	39.266	70.221	125.50	228.88	425.68	759.58
0.00018	0.689	0.741	0.790	0.854	0.943	1.081	1.256	1.438	1.644	1.889	2.179	2.488
1/ 5556	3.1150	4.2407	5.5874	7.7517	11.854	21.228	40.410	72.265	129.16	235.54	438.06	781.65
0.00019	0.708	0.761	0.812	0.877	0.969	1.111	1.291	1.477	1.690	1.941	2.239	2.556
1/ 5263	3.2010	4.3578	5.7416	7.9656	12.181	21.812	41.522	74.253	132.71	242.01	450.09	803.12
0.00020	0.726	0.781	0.834	0.900	0.995	1.140	1.324	1.516	1.734	1.991	2.297	2.623
1/ 5000	3.2849	4.4719	5.8919	8.1739	12.499	22.382	42.606	76.190	136.17	248.32	461.82	824.04
0.00022	0.762	0.819	0.874	0.945	1.044	1.196	1.389	1.590	1.819	2.089	2.409	2.751
1/ 4545	3.4465	4.6918	6.1814	8.5754	13.113	23.480	44.695	79.923	142.84	260.47	484.42	864.35
0.00024	0.796	0.856	0.914	0.987	1.090	1.249	1.451	1.661	1.900	2.182	2.517	2.874
1/ 4167	3.6009	4.9018	6.4581	8.9590	13.699	24.529	46.690	83.490	149.21	272.09	506.01	902.87
0.00026	0.829	0.891	0.951	1.027	1.135	1.301	1.511	1.729	1.978	2.271	2.620	2.992
1/ 3846	3.7490	5.1034	6.7235	9.3270	14.262	25.536	48.605	86.912	155.32	283.23	526.72	939.81
0.00028	0.860	0.925	0.987	1.066	1.178	1.350	1.568	1.795	2.052	2.357	2.719	3.105
1/ 3571	3.8915	5.2973	6.9788	9.6811	14.803	26.504	50.447	90.204	161.20	293.95	546.65	975.36
0.00030	0.891	0.958	1.022	1.104	1.220	1.397	1.623	1.858	2.125	2.440	2.814	3.214
1/ 3333	4.0290	5.4843	7.2252	10.023	15.325	27.438	52.224	93.380	166.88	304.29	565.88	1009.7
0.00032	0.920	0.989	1.056	1.140	1.260	1.443	1.677	1.919	2.195	2.521	2.907	3.319
1/ 3125	4.1620	5.6653	7.4635	10.353	15.830	28.342	53.943	96.452	172.36	314.30	584.48	1042.8
0.00034	0.948	1.020	1.089	1.176	1.299	1.488	1.729	1.978	2.262	2.598	2.997	3.422
1/ 2941	4.2909	5.8407	7.6945	10.673	16.320	29.218	55.609	99.430	177.68	323.99	602.50	1075.0
0.00036	0.976	1.050	1.120	1.210	1.336	1.531	1.779	2.036	2.328	2.674	3.084	3.521
1/ 2778	4.4160	6.0110	7.9188	10.984	16.795	30.068	57.227	102.32	182.85	333.41	620.00	1106.2
0.00038	1.003	1.079	1.151	1.243	1.373	1.573	1.828	2.092	2.392	2.747	3.168	3.618
1/ 2632	4.5378	6.1766	8.1369	11.287	17.257	30.895	58.800	105.13	187.87	342.56	637.03	1136.6
0.00040	1.029	1.107	1.181	1.276	1.409	1.615	1.875	2.146	2.454	2.819	3.251	3.712
1/ 2500	4.6563	6.3379	8.3494	11.582	17.707	31.701	60.332	107.87	192.76	351.48	653.61	1166.1
0.00042	1.055	1.134	1.211	1.307	1.444	1.655	1.922	2.199	2.515	2.889	3.331	3.804
1/ 2381	4.7720	6.4953	8.5566	11.869	18.146	32.486	61.827	110.54	197.54	360.18	669.78	1195.0
0.00044	1.080	1.161	1.239	1.338	1.478	1.694	1.967	2.251	2.574	2.957	3.410	3.893
1/ 2273	4.8849	6.6489	8.7589	12.149	18.575	33.254	63.287	113.15	202.20	368.67	685.57	1223.2
0.00046	1.104	1.188	1.267	1.368	1.512	1.732	2.012	2.302	2.632	3.023	3.486	3.981
1/ 2174	4.9953	6.7991	8.9567	12.424	18.994	34.004	64.713	115.70	206.75	376.98	701.00	1250.7
0.00048	1.128	1.213	1.294	1.398	1.544	1.769	2.055	2.351	2.689	3.088	3.562	4.067
1/ 2083	5.1033	6.9461	9.1502	12.692	19.404	34.738	66.109	118.20	211.21	385.10	716.10	1277.6
	400	450	500	550	650	850	1100	1300	1700	2100	2700	3400

θ_{medial} for part-full circular pipes.

$k_s = 1.50$ mm   S = 0.00010 to 0.00048

k_s = 1.50 mm
S = 0.00050 to 0.00280

Water (or sewage) at 15°C; full bore conditions.

ie hydraulic gradient = 1 in 2000 to 1 in 357

velocities in ms⁻¹
discharges in m³s⁻¹

velocities in ms^{-1}
discharges in m^3s^{-1}

Gradient	(Equivalent) Pipe diameters in m											
	2.400	2.700	3.000	3.400	4.000	5.000	6.400	8.000	10.00	12.60	16.00	20.00
0.00050 1/ 2000	1.151 5.2091	1.238 7.0900	1.321 9.3398	1.427 12.955	1.576 19.806	1.806 35.456	2.097 67.476	2.400 120.64	2.745 215.57	3.152 393.06	3.635 730.90	4.151 1304.0
0.00055 1/ 1818	1.208 5.4646	1.299 7.4377	1.386 9.7977	1.497 13.590	1.653 20.776	1.894 37.193	2.200 70.779	2.518 126.54	2.879 226.12	3.306 412.28	3.813 766.63	4.354 1367.8
0.00060 1/ 1667	1.262 5.7088	1.357 7.7699	1.448 10.235	1.564 14.196	1.727 21.704	1.979 38.852	2.298 73.935	2.630 132.18	3.007 236.19	3.454 430.65	3.983 800.77	4.548 1428.7
0.00065 1/ 1538	1.314 5.9430	1.413 8.0886	1.507 10.655	1.628 14.778	1.798 22.593	2.060 40.443	2.392 76.962	2.737 137.59	3.130 245.86	3.595 448.26	4.146 833.52	4.733 1487.1
0.00070 1/ 1429	1.364 6.1683	1.466 8.3952	1.564 11.059	1.689 15.338	1.866 23.449	2.138 41.974	2.483 79.874	2.841 142.80	3.249 255.16	3.731 465.21	4.302 865.03	4.912 1543.3
0.00075 1/ 1333	1.412 6.3858	1.518 8.6910	1.620 11.448	1.749 15.878	1.932 24.274	2.213 43.451	2.570 82.685	2.941 147.82	3.363 264.13	3.862 481.56	4.454 895.43	5.085 1597.5
0.00080 1/ 1250	1.458 6.5961	1.568 8.9772	1.673 11.825	1.806 16.401	1.995 25.073	2.286 44.880	2.655 85.403	3.037 152.68	3.473 272.81	3.989 497.38	4.600 924.84	5.252 1650.0
0.00085 1/ 1176	1.503 6.7999	1.616 9.2545	1.725 12.190	1.862 16.907	2.057 25.847	2.356 46.265	2.737 88.037	3.131 157.39	3.581 281.22	4.112 512.71	4.742 953.34	5.414 1700.8
0.00090 1/ 1111	1.547 6.9978	1.663 9.5238	1.775 12.545	1.916 17.399	2.117 26.598	2.425 47.610	2.816 90.595	3.222 161.96	3.685 289.38	4.231 527.60	4.879 981.02	5.571 1750.2
0.00095 1/ 1053	1.589 7.1903	1.709 9.7857	1.824 12.890	1.969 17.877	2.175 27.329	2.491 48.917	2.893 93.083	3.311 166.41	3.786 297.33	4.347 542.08	5.013 1007.9	5.724 1798.2
0.00100 1/ 1000	1.631 7.3778	1.754 10.041	1.871 13.226	2.020 18.343	2.231 28.041	2.556 50.191	2.969 95.506	3.397 170.74	3.884 305.06	4.461 556.18	5.143 1034.1	5.873 1844.9
0.00110 1/ 909	1.711 7.7392	1.840 10.533	1.963 13.873	2.119 19.241	2.341 29.413	2.681 52.647	3.114 100.18	3.563 179.08	4.074 319.98	4.679 583.36	5.395 1084.7	6.160 1935.1
0.00120 1/ 833	1.787 8.0845	1.922 11.002	2.050 14.492	2.214 20.099	2.445 30.725	2.801 54.993	3.253 104.64	3.721 187.06	4.255 334.22	4.887 609.34	5.635 1133.0	6.434 2021.2
0.00130 1/ 769	1.860 8.4158	2.000 11.453	2.134 15.086	2.304 20.922	2.545 31.982	2.915 57.243	3.386 108.92	3.874 194.71	4.429 347.89	5.087 634.25	5.865 1179.3	6.697 2103.8
0.00140 1/ 714	1.931 8.7345	2.076 11.887	2.215 15.657	2.392 21.714	2.641 33.192	3.026 59.409	3.514 113.04	4.020 202.07	4.597 361.04	5.279 658.22	6.087 1223.8	6.950 2183.3
0.00150 1/ 667	1.999 9.0420	2.149 12.305	2.293 16.208	2.476 22.478	2.734 34.360	3.132 61.498	3.637 117.01	4.161 209.18	4.758 373.73	5.464 681.35	6.301 1266.8	7.194 2260.0
0.00160 1/ 625	2.064 9.3394	2.220 12.710	2.368 16.741	2.557 23.217	2.824 35.489	3.235 63.519	3.757 120.86	4.298 216.05	4.915 386.00	5.644 703.72	6.508 1308.4	7.430 2334.2
0.00170 1/ 588	2.128 9.6277	2.288 13.102	2.441 17.257	2.636 23.933	2.911 36.584	3.335 65.477	3.873 124.58	4.431 222.71	5.066 397.90	5.818 725.40	6.708 1348.7	7.659 2406.1
0.00180 1/ 556	2.190 9.9076	2.355 13.483	2.512 17.759	2.713 24.629	2.996 37.647	3.432 67.379	3.985 128.20	4.559 229.17	5.213 409.45	5.986 746.45	6.903 1387.9	7.881 2475.9
0.00190 1/ 526	2.250 10.180	2.420 13.853	2.581 18.246	2.787 25.305	3.078 38.680	3.526 69.229	4.094 131.72	4.684 235.46	5.356 420.68	6.151 766.93	7.092 1425.9	8.097 2543.8
0.00200 1/ 500	2.309 10.445	2.483 14.214	2.649 18.722	2.860 25.964	3.158 39.687	3.618 71.030	4.201 135.15	4.806 241.58	5.496 431.62	6.311 786.87	7.276 1463.0	8.308 2609.9
0.00220 1/ 455	2.422 10.956	2.604 14.909	2.778 19.637	3.000 27.234	3.313 41.628	3.794 74.503	4.406 141.75	5.041 253.39	5.764 452.71	6.619 825.31	7.632 1534.5	8.713 2737.4
0.00240 1/ 417	2.530 11.444	2.720 15.574	2.902 20.513	3.133 28.447	3.460 43.482	3.963 77.821	4.603 148.06	5.265 264.67	6.021 472.86	6.914 862.04	7.971 1602.8	9.101 2859.2
0.00260 1/ 385	2.633 11.913	2.831 16.211	3.021 21.352	3.261 29.611	3.602 45.261	4.125 81.004	4.791 154.12	5.481 275.49	6.267 492.19	7.196 897.28	8.297 1668.3	9.473 2976.0
0.00280 1/ 357	2.733 12.364	2.939 16.825	3.135 22.160	3.385 30.731	3.738 46.973	4.281 84.066	4.972 159.94	5.688 285.90	6.504 510.79	7.468 931.18	8.611 1731.3	9.831 3088.5
	600	650	750	850	950	1200	1600	2000	2400	3100	3900	4900

θ_{medial} for part-full circular pipes.

k_s = 1.50 mm   S = 0.00050 to 0.00280

$k_s = 1.50$ mm
$S = 0.0030$ to 0.014

Water (or sewage) at 15°C;
full bore conditions.

ie hydraulic gradient =
1 in 333 to 1 in 71

velocities in ms⁻¹
discharges in m³s⁻¹

Gradient	(Equivalent) Pipe diameters in m											
	2.400	2.700	3.000	3.400	4.000	5.000	6.400	8.000	10.00	12.60	16.00	20.00
0.00300 1/ 333	2.829 12.799	3.042 17.416	3.245 22.939	3.504 31.811	3.869 48.624	4.432 87.021	5.147 165.56	5.888 295.95	6.732 528.73	7.730 963.89	8.913 1792.1	10.18 3196.9
0.00320 1/ 313	2.922 13.219	3.142 17.989	3.352 23.692	3.619 32.856	3.996 50.221	4.577 89.878	5.316 171.00	6.081 305.66	6.953 546.09	7.984 995.52	9.206 1850.9	10.51 3301.8
0.00340 1/ 294	3.012 13.627	3.239 18.543	3.455 24.423	3.730 33.869	4.120 51.769	4.719 92.648	5.479 176.27	6.268 315.08	7.167 562.91	8.230 1026.2	9.489 1907.9	10.83 3403.5
0.00360 1/ 278	3.100 14.023	3.333 19.082	3.555 25.132	3.839 34.853	4.239 53.272	4.856 95.338	5.638 181.38	6.450 324.22	7.375 579.24	8.469 1056.0	9.764 1963.3	11.15 3502.2
0.00380 1/ 263	3.185 14.408	3.424 19.606	3.653 25.822	3.944 35.809	4.356 54.734	4.989 97.953	5.793 186.36	6.627 333.12	7.577 595.13	8.701 1084.9	10.03 2017.1	11.45 3598.3
0.00400 1/ 250	3.268 14.783	3.513 20.116	3.748 26.494	4.047 36.741	4.469 56.158	5.118 100.50	5.944 191.21	6.799 341.78	7.774 610.60	8.927 1113.1	10.29 2069.5	11.75 3691.8
0.00420 1/ 238	3.349 15.148	3.600 20.614	3.841 27.149	4.147 37.650	4.579 57.546	5.245 102.99	6.091 195.93	6.968 350.23	7.967 625.69	9.148 1140.6	10.55 2120.7	12.04 3783.0
0.00440 1/ 227	3.427 15.506	3.685 21.099	3.931 27.789	4.245 38.537	4.687 58.902	5.369 105.41	6.234 200.55	7.132 358.47	8.154 640.43	9.363 1167.5	10.80 2170.6	12.33 3872.1
0.00460 1/ 217	3.505 15.855	3.768 21.574	4.020 28.415	4.340 39.404	4.793 60.228	5.489 107.78	6.374 205.06	7.292 366.54	8.338 654.83	9.574 1193.7	11.04 2219.4	12.60 3959.1
0.00480 1/ 208	3.580 16.196	3.849 22.039	4.106 29.027	4.434 40.253	4.896 61.525	5.608 110.10	6.512 209.47	7.449 374.43	8.517 668.92	9.780 1219.4	11.28 2267.2	12.87 4044.3
0.00500 1/ 200	3.654 16.531	3.929 22.494	4.191 29.626	4.525 41.084	4.997 62.795	5.723 112.38	6.646 213.80	7.603 382.16	8.693 682.73	9.981 1244.6	11.51 2313.9	13.14 4127.8
0.00550 1/ 182	3.833 17.339	4.121 23.594	4.396 31.074	4.746 43.092	5.241 65.864	6.003 117.87	6.971 224.24	7.974 400.82	9.117 716.08	10.47 1305.4	12.07 2426.9	13.78 4329.3
0.00600 1/ 167	4.003 18.111	4.304 24.645	4.592 32.458	4.958 45.011	5.475 68.796	6.270 123.12	7.281 234.22	8.329 418.66	9.523 747.94	10.93 1363.5	12.61 2534.9	14.39 4521.9
0.00650 1/ 154	4.167 18.852	4.480 25.652	4.780 33.785	5.160 46.851	5.698 71.608	6.526 128.15	7.578 243.79	8.669 435.77	9.912 778.50	11.38 1419.2	13.12 2638.5	14.98 4706.7
0.00700 1/ 143	4.325 19.565	4.650 26.622	4.960 35.062	5.355 48.622	5.914 74.314	6.773 132.99	7.865 253.01	8.997 452.23	10.29 807.90	11.81 1472.8	13.62 2738.1	15.55 4884.4
0.00750 1/ 133	4.477 20.252	4.813 27.558	5.135 36.294	5.543 50.330	6.122 76.925	7.011 137.66	8.141 261.89	9.313 468.11	10.65 836.28	12.23 1524.5	14.10 2834.3	16.09 5055.9
0.00800 1/ 125	4.624 20.917	4.971 28.463	5.303 37.486	5.725 51.983	6.322 79.451	7.241 142.18	8.408 270.49	9.618 483.47	11.00 863.72	12.63 1574.5	14.56 2927.2	16.62 5221.8
0.00850 1/ 118	4.766 21.562	5.124 29.340	5.467 38.641	5.902 53.584	6.517 81.898	7.464 146.56	8.667 278.82	9.915 498.36	11.34 890.31	13.02 1623.0	15.01 3017.4	17.13 5382.5
0.00900 1/ 111	4.905 22.188	5.273 30.191	5.625 39.762	6.073 55.139	6.706 84.275	7.681 150.81	8.918 286.91	10.20 512.82	11.66 916.14	13.39 1670.1	15.44 3104.9	17.63 5538.7
0.00950 1/ 105	5.039 22.796	5.418 31.020	5.780 40.853	6.240 56.652	6.890 86.586	7.891 154.95	9.163 294.77	10.48 526.88	11.98 941.26	13.76 1715.8	15.87 3190.0	18.11 5690.5
0.01000 1/ 100	5.170 23.389	5.559 31.826	5.930 41.916	6.402 58.125	7.069 88.838	8.097 158.98	9.401 302.44	10.75 540.58	12.30 965.72	14.12 1760.4	16.28 3272.9	18.58 5838.4
0.01100 1/ 91	5.423 24.532	5.830 33.381	6.220 43.964	6.715 60.965	7.415 93.177	8.492 166.74	9.860 317.21	11.28 566.98	12.90 1012.9	14.81 1846.4	17.07 3432.7	19.49 6123.4
0.01200 1/ 83	5.664 25.624	6.090 34.867	6.496 45.920	7.014 63.678	7.745 97.324	8.870 174.16	10.30 331.32	11.78 592.20	13.47 1057.9	15.47 1928.5	17.83 3585.4	20.36 6395.8
0.01300 1/ 77	5.896 26.672	6.339 36.292	6.762 47.797	7.300 66.280	8.061 101.30	9.232 181.28	10.72 344.86	12.26 616.39	14.02 1101.2	16.10 2007.3	18.56 3731.9	21.19 6657.0
0.01400 1/ 71	6.119 27.680	6.578 37.664	7.017 49.603	7.576 68.784	8.366 105.13	9.581 188.12	11.12 357.88	12.73 639.67	14.55 1142.7	16.71 2083.1	19.26 3872.8	21.99 6908.4
	800	900	1000	1100	1300	1700	2100	2600	3300	4200	5500	6500

θ_{medial} for part-full circular pipes.

$k_s = 1.50$ mm $S = 0.0030$ to 0.014

k_s = 1.50 mm
S = 0.015 to 0.075

ie hydraulic gradient =
1 in 66 to 1 in 13

Water (or sewage) at 15°C;
full bore conditions.

velocities in ms^{-1}
discharges in m^{3}s^{-1}

Gradient	(Equivalent) Pipe diameters in m											
	2.400	2.700	3.000	3.400	4.000	5.000	6.400	8.000	10.00	12.60	16.00	20.00
0.01500 1/ 67	6.334 28.652	6.809 38.987	7.264 51.346	7.842 71.201	8.660 108.82	9.918 194.73	11.52 370.45	13.17 662.14	15.06 1182.9	17.29 2156.3	19.94 4008.8	22.76 7150.9
0.01600 1/ 62	6.541 29.593	7.033 40.267	7.502 53.031	8.100 73.537	8.944 112.39	10.24 201.12	11.89 382.61	13.60 683.86	15.55 1221.7	17.86 2227.0	20.59 4140.3	23.51 7385.5
0.01700 1/ 59	6.743 30.504	7.249 41.507	7.733 54.664	8.349 75.802	9.219 115.85	10.56 207.32	12.26 394.39	14.02 704.92	16.03 1259.3	18.41 2295.6	21.23 4267.7	24.23 7612.9
0.01800 1/ 56	6.939 31.390	7.460 42.711	7.958 56.250	8.591 78.002	9.487 119.21	10.86 213.33	12.62 405.83	14.43 725.36	16.50 1295.8	18.94 2362.1	21.84 4391.5	24.94 7833.6
0.01900 1/ 53	7.129 32.251	7.664 43.883	8.176 57.793	8.827 80.140	9.747 122.48	11.16 219.18	12.96 416.96	14.83 745.25	16.95 1331.3	19.46 2426.9	22.44 4511.9	25.62 8048.3
0.02000 1/ 50	7.314 33.089	7.864 45.024	8.389 59.295	9.056 82.224	10.00 125.67	11.45 224.87	13.30 427.79	15.21 764.62	17.39 1365.9	19.97 2489.9	23.02 4629.1	26.28 8257.5
0.02200 1/ 45	7.672 34.705	8.248 47.223	8.798 62.192	9.499 86.240	10.49 131.80	12.01 235.86	13.95 448.68	15.95 801.95	18.24 1432.6	20.94 2611.5	24.15 4855.1	27.57 8660.6
0.02400 1/ 42	8.013 36.250	8.615 49.324	9.190 64.959	9.921 90.077	10.96 137.67	12.55 246.35	14.57 468.64	16.66 837.62	19.05 1496.4	21.88 2727.7	25.22 5071.0	28.79 9045.8
0.02600 1/ 38	8.340 37.731	8.967 51.340	9.565 67.613	10.33 93.757	11.40 143.29	13.06 256.41	15.16 487.79	17.34 871.84	19.83 1557.5	22.77 2839.1	26.25 5278.2	29.97 9415.2
0.02800 1/ 36	8.655 39.157	9.305 53.279	9.927 70.167	10.72 97.299	11.83 148.71	13.55 266.10	15.74 506.21	18.00 904.76	20.58 1616.3	23.63 2946.3	27.24 5477.4	31.10 9770.7
0.03000 1/ 33	8.960 40.532	9.632 55.150	10.28 72.632	11.09 100.72	12.25 153.93	14.03 275.44	16.29 523.98	18.63 936.53	21.30 1673.0	24.46 3049.7	28.20 5669.7	32.19 10114
0.03200 1/ 31	9.254 41.862	9.948 56.960	10.61 75.015	11.46 104.02	12.65 158.98	14.49 284.48	16.82 541.17	19.24 967.25	22.00 1727.9	25.26 3149.8	29.12 5855.7	33.25 10445
0.03400 1/ 29	9.539 43.151	10.25 58.714	10.94 77.325	11.81 107.22	13.04 163.87	14.93 293.24	17.34 557.83	19.84 997.03	22.68 1781.1	26.04 3246.7	30.02 6036.0	34.27 10767
0.03600 1/ 28	9.815 44.403	10.55 60.417	11.26 79.568	12.15 110.33	13.42 168.63	15.37 301.74	17.84 574.01	20.41 1025.9	23.34 1832.8	26.79 3340.9	30.89 6211.0	35.27 11079
0.03800 1/ 26	10.08 45.621	10.84 62.074	11.57 81.749	12.49 113.36	13.79 173.25	15.79 310.01	18.33 589.75	20.97 1054.1	23.98 1883.0	27.53 3432.4	31.74 6381.2	36.23 11383
0.04000 1/ 25	10.35 46.806	11.12 63.687	11.87 83.874	12.81 116.31	14.15 177.75	16.20 318.07	18.81 605.07	21.51 1081.5	24.60 1931.9	28.24 3521.6	32.56 6547.0	37.17 11679
0.04200 1/ 24	10.60 47.963	11.40 65.261	12.16 85.947	13.13 119.18	14.49 182.14	16.60 325.93	19.27 620.02	22.05 1108.2	25.21 1979.6	28.94 3608.6	33.37 6708.8	38.09 11967
0.04400 1/ 23	10.85 49.092	11.67 66.798	12.45 87.970	13.44 121.98	14.84 186.43	16.99 333.60	19.73 634.61	22.57 1134.3	25.80 2026.2	29.62 3693.6	34.15 6866.7	38.99 12249
0.04600 1/ 22	11.10 50.196	11.93 68.300	12.73 89.948	13.74 124.73	15.17 190.62	17.37 341.10	20.17 648.88	23.07 1159.8	26.38 2071.8	30.29 3776.6	34.92 7021.0	39.87 12524
0.04800 1/ 21	11.33 51.276	12.19 69.769	13.00 91.884	14.03 127.41	15.50 194.72	17.75 348.44	20.60 662.84	23.57 1184.7	26.95 2116.4	30.94 3857.8	35.67 7172.0	40.72 12793
0.05000 1/ 20	11.57 52.334	12.44 71.209	13.27 93.779	14.32 130.04	15.82 198.74	18.11 355.63	21.03 676.51	24.06 1209.1	27.50 2160.0	31.58 3937.4	36.41 7320.0	41.56 13057
0.05500 1/ 18	12.13 54.890	13.04 74.686	13.91 98.359	15.02 136.39	16.59 208.45	19.00 372.99	22.06 709.54	25.23 1268.2	28.84 2265.5	33.12 4129.6	38.18 7677.3	43.59 13695
0.06000 1/ 17	12.67 57.332	13.62 78.009	14.53 102.73	15.69 142.46	17.33 217.72	19.84 389.58	23.04 741.10	26.35 1324.6	30.13 2366.2	34.59 4313.3	39.88 8018.7	45.53 14304
0.06500 1/ 15	13.19 59.674	14.18 81.195	15.13 106.93	16.33 148.28	18.03 226.61	20.65 405.50	23.98 771.37	27.43 1378.7	31.36 2462.9	36.00 4489.4	41.51 8346.2	47.39 14888
0.07000 1/ 14	13.69 61.928	14.72 84.262	15.70 110.97	16.95 153.88	18.71 235.17	21.43 420.81	24.88 800.50	28.46 1430.7	32.54 2555.9	37.36 4658.9	43.08 8661.3	49.18 15450
0.07500 1/ 13	14.17 64.103	15.23 87.221	16.25 114.87	17.54 159.28	19.37 243.42	22.18 435.58	25.76 828.60	29.46 1481.0	33.68 2645.6	38.68 4822.5	44.59 8965.4	50.91 15992
	1000	1100	1300	1400	1700	2100	2700	3400	4200	5500	6500	8500

θ_{medial} for part-full circular pipes.

k_s = 1.50 mm   S = 0.0015 to 0.075

k_s = 3.00 mm
S = 0.00010 to 0.00048

Water (or sewage) at 15°C;
full bore conditions.

ie hydraulic gradient =
1 in 10000 to 1 in 2083

velocities in ms^{-1}
discharges in m^{3}s^{-1}

Gradient	(Equivalent) Pipe diameters in m											
	2.400	2.700	3.000	3.400	4.000	5.000	6.400	8.000	10.00	12.60	16.00	20.00
0.00010 1/10000	0.473 2.1413	0.510 2.9181	0.544 3.8482	0.589 5.3444	0.651 8.1835	0.748 14.680	0.870 27.996	0.998 50.145	1.143 89.760	1.315 163.94	1.519 305.36	1.737 545.61
0.00011 1/ 9091	0.497 2.2465	0.535 3.0614	0.571 4.0371	0.618 5.6065	0.683 8.5847	0.784 15.399	0.913 29.367	1.046 52.600	1.199 94.153	1.379 171.96	1.593 320.29	1.822 572.29
0.00012 1/ 8333	0.519 2.3469	0.559 3.1982	0.597 4.2175	0.645 5.8570	0.714 8.9681	0.819 16.086	0.954 30.677	1.093 54.946	1.252 98.350	1.441 179.62	1.664 334.56	1.903 597.78
0.00013 1/ 7692	0.540 2.4433	0.582 3.3295	0.621 4.3906	0.672 6.0973	0.743 9.3358	0.853 16.746	0.993 31.934	1.138 57.196	1.303 102.38	1.500 186.97	1.732 348.25	1.981 622.23
0.00014 1/ 7143	0.561 2.5361	0.604 3.4558	0.645 4.5571	0.697 6.3285	0.771 9.6897	0.885 17.380	1.030 33.143	1.181 59.360	1.353 106.25	1.556 194.05	1.798 361.42	2.055 645.75
0.00015 1/ 6667	0.580 2.6255	0.625 3.5777	0.667 4.7178	0.722 6.5515	0.798 10.031	0.916 17.992	1.067 34.310	1.222 61.449	1.400 109.99	1.611 200.87	1.861 374.12	2.128 668.45
0.00016 1/ 6250	0.599 2.7121	0.645 3.6956	0.689 4.8732	0.745 6.7673	0.825 10.361	0.946 18.584	1.102 35.438	1.263 63.469	1.446 113.60	1.664 207.47	1.922 386.41	2.198 690.40
0.00017 1/ 5882	0.618 2.7960	0.665 3.8099	0.711 5.0238	0.768 6.9764	0.850 10.681	0.976 19.158	1.136 36.532	1.302 65.427	1.491 117.10	1.715 213.87	1.981 398.32	2.265 711.68
0.00018 1/ 5556	0.636 2.8774	0.685 3.9208	0.731 5.1701	0.791 7.1794	0.875 10.992	1.004 19.715	1.169 37.594	1.339 67.329	1.534 120.51	1.765 220.08	2.039 409.89	2.331 732.34
0.00019 1/ 5263	0.654 2.9566	0.704 4.0287	0.752 5.3123	0.813 7.3769	0.899 11.294	1.032 20.257	1.201 38.627	1.376 69.178	1.576 123.81	1.813 226.12	2.095 421.14	2.395 752.43
0.00020 1/ 5000	0.671 3.0338	0.722 4.1338	0.771 5.4509	0.834 7.5692	0.922 11.589	1.059 20.784	1.232 39.632	1.412 70.979	1.617 127.04	1.861 232.00	2.149 432.09	2.457 772.00
0.00022 1/ 4545	0.703 3.1825	0.757 4.3364	0.809 5.7179	0.875 7.9400	0.967 12.156	1.110 21.802	1.292 41.572	1.481 74.451	1.697 133.25	1.952 243.34	2.254 453.21	2.577 809.73
0.00024 1/ 4167	0.735 3.3246	0.791 4.5300	0.845 5.9731	0.914 8.2943	1.011 12.698	1.160 22.774	1.350 43.424	1.547 77.768	1.772 139.19	2.039 254.18	2.354 473.39	2.692 845.78
0.00026 1/ 3846	0.765 3.4609	0.824 4.7157	0.880 6.2179	0.951 8.6340	1.052 13.218	1.207 23.706	1.405 45.202	1.610 80.949	1.845 144.88	2.122 264.58	2.451 492.75	2.802 880.35
0.00028 1/ 3571	0.794 3.5921	0.855 4.8943	0.913 6.4534	0.987 8.9610	1.092 13.719	1.253 24.603	1.458 46.911	1.671 84.011	1.914 150.36	2.202 274.58	2.543 511.37	2.908 913.62
0.00030 1/ 3333	0.822 3.7186	0.885 5.0667	0.945 6.6807	1.022 9.2765	1.130 14.202	1.297 25.469	1.510 48.561	1.730 86.965	1.982 155.64	2.279 284.23	2.633 529.34	3.010 945.72
0.00032 1/ 3125	0.849 3.8410	0.914 5.2334	0.976 6.9004	1.055 9.5816	1.167 14.669	1.340 26.306	1.559 50.157	1.787 89.822	2.047 160.75	2.354 293.56	2.719 546.72	3.109 976.77
0.00034 1/ 2941	0.875 3.9596	0.942 5.3950	1.006 7.1135	1.088 9.8773	1.203 15.121	1.381 27.118	1.607 51.704	1.842 92.591	2.110 165.71	2.427 302.61	2.803 563.57	3.205 1006.9
0.00036 1/ 2778	0.901 4.0748	0.970 5.5519	1.036 7.3203	1.120 10.164	1.238 15.561	1.421 27.905	1.654 53.206	1.896 95.280	2.171 170.52	2.497 311.39	2.884 579.92	3.298 1036.1
0.00038 1/ 2632	0.925 4.1869	0.996 5.7046	1.064 7.5215	1.150 10.444	1.272 15.988	1.460 28.672	1.699 54.666	1.948 97.895	2.231 175.20	2.566 319.94	2.963 595.83	3.388 1064.5
0.00040 1/ 2500	0.950 4.2960	1.022 5.8532	1.092 7.7174	1.180 10.716	1.305 16.404	1.498 29.418	1.744 56.089	1.998 100.44	2.289 179.76	2.633 328.26	3.040 611.33	3.476 1092.2
0.00042 1/ 2381	0.973 4.4024	1.048 5.9982	1.119 7.9085	1.209 10.981	1.338 16.810	1.535 30.146	1.787 57.476	2.048 102.93	2.345 184.20	2.698 336.37	3.116 626.44	3.562 1119.2
0.00044 1/ 2273	0.996 4.5063	1.072 6.1397	1.145 8.0951	1.238 11.240	1.369 17.207	1.572 30.857	1.829 58.831	2.096 105.35	2.401 188.54	2.761 344.30	3.189 641.19	3.646 1145.5
0.00046 1/ 2174	1.019 4.6079	1.097 6.2781	1.171 8.2776	1.266 11.493	1.400 17.595	1.607 31.552	1.870 60.156	2.143 107.72	2.455 192.78	2.823 352.04	3.261 655.62	3.728 1171.3
0.00048 1/ 2083	1.041 4.7073	1.120 6.4135	1.196 8.4560	1.293 11.741	1.430 17.974	1.642 32.232	1.910 61.452	2.189 110.04	2.507 196.94	2.884 359.62	3.331 669.73	3.809 1196.5
	320	360	400	460	550	650	850	1100	1300	1700	2100	2700

θ_{medial} for part-full circular pipes.

k_s = 3.00 mm   S = 0.00010 to 0.00048

$k_s = 3.00$ mm
$S = 0.00050$ to 0.00280

ie hydraulic gradient =
1 in 2000 to 1 in 357

Water (or sewage) at 15°C;
full bore conditions.

velocities in ms^{-1}
discharges in m^{3}s^{-1}

Gradient	(Equivalent) Pipe diameters in m											
	2.400	2.700	3.000	3.400	4.000	5.000	6.400	8.000	10.00	12.60	16.00	20.00
0.00050 1/ 2000	1.062 4.8046	1.143 6.5461	1.221 8.6309	1.320 11.984	1.460 18.345	1.675 32.897	1.950 62.721	2.234 112.32	2.559 201.00	2.944 367.05	3.400 683.56	3.887 1221.2
0.00055 1/ 1818	1.114 5.0398	1.199 6.8664	1.281 9.0532	1.384 12.570	1.531 19.242	1.757 34.506	2.045 65.787	2.344 117.80	2.684 210.82	3.088 384.98	3.566 716.95	4.077 1280.9
0.00060 1/ 1667	1.164 5.2645	1.253 7.1725	1.338 9.4567	1.446 13.130	1.599 20.100	1.836 36.043	2.136 68.716	2.448 123.05	2.804 220.21	3.225 402.12	3.725 748.86	4.259 1337.8
0.00065 1/ 1538	1.211 5.4800	1.304 7.4661	1.393 9.8437	1.505 13.667	1.665 20.922	1.911 37.517	2.223 71.526	2.548 128.08	2.918 229.21	3.357 418.55	3.877 779.46	4.433 1392.5
0.00070 1/ 1429	1.257 5.6874	1.353 7.7486	1.445 10.216	1.562 14.184	1.728 21.713	1.983 38.936	2.307 74.230	2.644 132.92	3.029 237.87	3.484 434.36	4.023 808.91	4.600 1445.1
0.00075 1/ 1333	1.301 5.8875	1.401 8.0212	1.496 10.575	1.617 14.683	1.789 22.477	2.053 40.304	2.389 76.839	2.737 137.59	3.135 246.23	3.606 449.62	4.164 837.32	4.762 1495.9
0.00080 1/ 1250	1.344 6.0810	1.447 8.2848	1.545 10.923	1.670 15.166	1.847 23.215	2.120 41.628	2.467 79.362	2.827 142.11	3.238 254.31	3.724 464.38	4.301 864.80	4.918 1545.0
0.00085 1/ 1176	1.386 6.2686	1.492 8.5403	1.593 11.260	1.722 15.633	1.904 23.931	2.185 42.911	2.543 81.808	2.914 146.49	3.338 262.15	3.839 478.69	4.434 891.43	5.069 1592.5
0.00090 1/ 1111	1.426 6.4507	1.535 8.7884	1.639 11.587	1.772 16.087	1.960 24.626	2.249 44.157	2.617 84.182	2.999 150.74	3.435 269.75	3.950 492.57	4.562 917.29	5.216 1638.7
0.00095 1/ 1053	1.465 6.6279	1.577 9.0297	1.684 11.905	1.821 16.529	2.013 25.301	2.311 45.368	2.689 86.492	3.081 154.87	3.529 277.15	4.059 506.08	4.687 942.45	5.359 1683.7
0.00100 1/ 1000	1.503 6.8004	1.618 9.2647	1.728 12.215	1.868 16.959	2.066 25.960	2.371 46.549	2.759 88.741	3.161 158.90	3.621 284.36	4.164 519.24	4.809 966.95	5.499 1727.4
0.00110 1/ 909	1.577 7.1330	1.697 9.7178	1.813 12.812	1.959 17.788	2.167 27.229	2.487 48.823	2.893 93.077	3.316 166.66	3.797 298.25	4.368 544.60	5.044 1014.2	5.767 1811.8
0.00120 1/ 833	1.647 7.4508	1.773 10.151	1.893 13.383	2.046 18.580	2.263 28.441	2.597 50.997	3.022 97.220	3.463 174.08	3.966 311.52	4.562 568.83	5.269 1059.3	6.024 1892.4
0.00130 1/ 769	1.714 7.7556	1.845 10.566	1.971 13.930	2.130 19.340	2.356 29.604	2.703 53.082	3.146 101.19	3.605 181.20	4.129 324.25	4.748 592.08	5.484 1102.6	6.270 1969.7
0.00140 1/ 714	1.779 8.0489	1.915 10.965	2.045 14.457	2.211 20.071	2.445 30.723	2.806 55.088	3.264 105.02	3.741 188.04	4.284 336.50	4.928 614.44	5.691 1144.2	6.507 2044.1
0.00150 1/ 667	1.842 8.3319	1.982 11.351	2.117 14.965	2.288 20.777	2.531 31.803	2.904 57.024	3.379 108.71	3.872 194.65	4.435 348.32	5.101 636.02	5.891 1184.4	6.735 2115.9
0.00160 1/ 625	1.902 8.6056	2.048 11.724	2.187 15.456	2.364 21.459	2.614 32.847	3.000 58.896	3.490 112.28	3.999 201.04	4.581 359.75	5.268 656.89	6.084 1223.3	6.956 2185.3
0.00170 1/ 588	1.961 8.8708	2.111 12.085	2.254 15.933	2.436 22.120	2.694 33.859	3.092 60.710	3.598 115.73	4.123 207.23	4.722 370.83	5.430 677.12	6.271 1260.9	7.170 2252.6
0.00180 1/ 556	2.018 9.1284	2.172 12.436	2.319 16.395	2.507 22.762	2.773 34.842	3.182 62.472	3.702 119.09	4.242 213.24	4.859 381.59	5.588 696.76	6.453 1297.5	7.378 2317.9
0.00190 1/ 526	2.073 9.3789	2.232 12.777	2.383 16.845	2.576 23.387	2.849 35.797	3.269 64.185	3.803 122.36	4.359 219.09	4.992 392.05	5.741 715.87	6.630 1333.1	7.580 2381.5
0.00200 1/ 500	2.127 9.6229	2.290 13.109	2.445 17.283	2.643 23.995	2.923 36.728	3.354 65.854	3.902 125.54	4.472 224.78	5.122 402.24	5.890 734.47	6.802 1367.7	7.777 2443.4
0.00220 1/ 455	2.231 10.093	2.402 13.750	2.565 18.128	2.772 25.168	3.066 38.523	3.518 69.072	4.093 131.67	4.690 235.76	5.372 421.89	6.178 770.34	7.135 1434.5	8.157 2562.7
0.00240 1/ 417	2.330 10.543	2.508 14.362	2.679 18.935	2.895 26.288	3.202 40.238	3.674 72.146	4.275 137.53	4.899 246.25	5.611 440.66	6.453 804.61	7.452 1498.3	8.520 2676.7
0.00260 1/ 385	2.426 10.974	2.611 14.950	2.788 19.709	3.014 27.362	3.333 41.882	3.825 75.094	4.450 143.15	5.099 256.31	5.840 458.66	6.717 837.48	7.756 1559.5	8.868 2786.0
0.00280 1/ 357	2.517 11.389	2.710 15.515	2.894 20.454	3.128 28.396	3.459 43.465	3.969 77.931	4.618 148.56	5.292 266.00	6.060 475.98	6.970 869.11	8.049 1618.4	9.203 2891.2
	430	480	550	600	700	900	1100	1400	1800	2200	2900	3600

θ_{medial} for part-full circular pipes.

$k_s = 3.00$ mm $S = 0.00050$ to 0.00280

$k_s = 3.00$ mm
$S = 0.0030$ to 0.014

Water (or sewage) at 15°C;
full bore conditions

ie hydraulic gradient =
1 in 333 to 1 in 71

velocities in ms^{-1}
discharges in m^{3}s^{-1}

Gradient	(Equivalent) Pipe diameters in m											
	2.400	2.700	3.000	3.400	4.000	5.000	6.400	8.000	10.00	12.60	16.00	20.00
0.00300 1/ 333	2.606 11.789	2.805 16.060	2.995 21.172	3.238 29.394	3.580 44.992	4.108 80.668	4.780 153.78	5.478 275.34	6.273 492.70	7.215 899.63	8.332 1675.2	9.526 2992.7
0.00320 1/ 313	2.691 12.176	2.897 16.587	3.094 21.867	3.344 30.359	3.698 46.468	4.243 83.316	4.937 158.82	5.657 284.37	6.479 508.87	7.452 929.14	8.605 1730.2	9.839 3090.9
0.00340 1/ 294	2.774 12.551	2.986 17.098	3.189 22.541	3.447 31.294	3.812 47.900	4.374 85.882	5.089 163.71	5.832 293.13	6.679 524.53	7.681 957.75	8.870 1783.5	10.14 3186.0
0.00360 1/ 278	2.855 12.915	3.073 17.594	3.281 23.195	3.547 32.202	3.922 49.289	4.501 88.373	5.237 168.46	6.001 301.63	6.872 539.75	7.904 985.53	9.128 1835.2	10.44 3278.4
0.00380 1/ 263	2.933 13.269	3.157 18.077	3.371 23.831	3.644 33.085	4.030 50.641	4.624 90.797	5.380 173.08	6.165 309.90	7.061 554.54	8.121 1012.5	9.378 1885.5	10.72 3368.3
0.00400 1/ 250	3.009 13.615	3.239 18.547	3.459 24.451	3.739 33.946	4.135 51.958	4.744 93.157	5.520 177.58	6.326 317.96	7.244 568.96	8.332 1038.9	9.621 1934.5	11.00 3455.8
0.00420 1/ 238	3.084 13.951	3.319 19.005	3.545 25.055	3.831 34.785	4.237 53.242	4.862 95.459	5.656 181.97	6.482 325.81	7.423 583.01	8.537 1064.5	9.859 1982.3	11.27 3541.2
0.00440 1/ 227	3.157 14.280	3.398 19.453	3.628 25.645	3.921 35.604	4.337 54.496	4.976 97.707	5.790 186.25	6.634 333.48	7.598 596.74	8.738 1089.6	10.09 2029.0	11.54 3624.6
0.00460 1/ 217	3.228 14.601	3.474 19.890	3.710 26.222	4.010 36.405	4.434 55.721	5.088 99.904	5.920 190.44	6.784 340.98	7.769 610.16	8.935 1114.1	10.32 2074.6	11.80 3706.0
0.00480 1/ 208	3.297 14.915	3.549 20.319	3.790 26.787	4.096 37.188	4.530 56.920	5.198 102.05	6.047 194.54	6.930 348.32	7.936 623.28	9.127 1138.0	10.54 2119.2	12.05 3785.8
0.00500 1/ 200	3.365 15.223	3.622 20.738	3.868 27.339	4.180 37.956	4.623 58.095	5.305 104.16	6.172 198.55	7.073 355.50	8.100 636.14	9.315 1161.5	10.76 2162.9	12.30 3863.8
0.00550 1/ 182	3.529 15.967	3.799 21.751	4.057 28.675	4.385 39.809	4.849 60.932	5.564 109.25	6.473 208.25	7.418 372.86	8.495 667.20	9.770 1218.2	11.28 2268.5	12.90 4052.5
0.00600 1/ 167	3.687 16.677	3.968 22.719	4.237 29.951	4.580 41.581	5.065 63.644	5.811 114.11	6.761 217.51	7.748 389.45	8.873 696.88	10.20 1272.4	11.78 2369.4	13.47 4232.7
0.00650 1/ 154	3.837 17.359	4.130 23.647	4.410 31.175	4.767 43.280	5.272 66.244	6.049 118.77	7.038 226.40	8.064 405.36	9.235 725.35	10.62 1324.4	12.27 2466.2	14.02 4405.6
0.00700 1/ 143	3.982 18.015	4.286 24.541	4.577 32.352	4.947 44.915	5.471 68.746	6.277 123.25	7.303 234.95	8.369 420.67	9.584 752.74	11.02 1374.4	12.73 2559.3	14.55 4571.9
0.00750 1/ 133	4.122 18.648	4.437 25.403	4.738 33.489	5.121 46.492	5.663 71.160	6.498 127.58	7.560 243.20	8.663 435.44	9.921 779.17	11.41 1422.7	13.18 2649.2	15.06 4732.4
0.00800 1/ 125	4.257 19.260	4.582 26.236	4.893 34.588	5.289 48.018	5.849 73.495	6.711 131.77	7.808 251.18	8.947 449.72	10.25 804.73	11.78 1469.3	13.61 2736.1	15.56 4887.7
0.00850 1/ 118	4.388 19.853	4.723 27.044	5.044 35.653	5.452 49.496	6.029 75.758	6.918 135.83	8.048 258.91	9.222 463.57	10.56 829.50	12.15 1514.6	14.03 2820.3	16.04 5038.1
0.00900 1/ 111	4.516 20.429	4.860 27.829	5.190 36.687	5.610 50.932	6.204 77.956	7.118 139.77	8.282 266.42	9.490 477.01	10.87 853.56	12.50 1558.5	14.43 2902.1	16.50 5184.2
0.00950 1/ 105	4.640 20.989	4.994 28.592	5.332 37.693	5.764 52.329	6.374 80.093	7.313 143.60	8.509 273.72	9.750 490.09	11.17 876.95	12.84 1601.2	14.83 2981.6	16.95 5326.3
0.01000 1/ 100	4.760 21.535	5.124 29.335	5.471 38.673	5.913 53.689	6.539 82.175	7.503 147.33	8.730 280.84	10.00 502.82	11.46 899.74	13.18 1642.8	15.21 3059.1	17.39 5464.7
0.01100 1/ 91	4.993 22.586	5.374 30.768	5.738 40.561	6.202 56.311	6.859 86.187	7.870 154.52	9.156 294.55	10.49 527.37	12.02 943.67	13.82 1723.0	15.96 3208.4	18.24 5731.5
0.01200 1/ 83	5.215 23.591	5.613 32.137	5.994 42.366	6.478 58.816	7.164 90.021	8.220 161.40	9.563 307.65	10.96 550.83	12.55 985.64	14.43 1799.6	16.67 3351.1	19.06 5986.4
0.01300 1/ 77	5.428 24.555	5.842 33.450	6.238 44.097	6.743 61.218	7.456 93.699	8.556 167.99	9.954 320.22	11.41 573.33	13.06 1025.9	15.02 1873.1	17.35 3488.0	19.83 6230.9
0.01400 1/ 71	5.633 25.483	6.063 34.713	6.474 45.762	6.997 63.530	7.738 97.237	8.879 174.33	10.33 332.31	11.84 594.97	13.56 1064.6	15.59 1943.9	18.00 3619.7	20.58 6466.1
	550	600	650	750	900	1100	1400	1800	2200	2800	3500	4400

θ_{medial} for part-full circular pipes.

$k_s = 3.00$ mm $S = 0.0030$ to 0.014

k_s = 3.00 mm
S = 0.015 to 0.075

ie hydraulic gradient =
1 in 66 to 1 in 13

Water (or sewage) at 15°C;
full bore conditions.

velocities in ms⁻¹
discharges in m³s⁻¹

| Gradient | (Equivalent) Pipe diameters in m | | | | | | | | | | | |
	2.400	2.700	3.000	3.400	4.000	5.000	6.400	8.000	10.00	12.60	16.00	20.00
0.01500 1/ 67	5.831 26.377	6.276 35.932	6.701 47.369	7.243 65.761	8.010 100.65	9.190 180.45	10.69 343.97	12.25 615.86	14.03 1102.0	16.14 2012.1	18.63 3746.7	21.30 6693.1
0.01600 1/ 62	6.022 27.243	6.482 37.111	6.921 48.923	7.481 67.919	8.272 103.95	9.492 186.37	11.04 355.26	12.65 636.07	14.49 1138.2	16.67 2078.1	19.25 3869.6	22.00 6912.6
0.01700 1/ 59	6.207 28.082	6.681 38.254	7.134 50.430	7.711 70.010	8.527 107.15	9.784 192.11	11.38 366.19	13.04 655.65	14.94 1173.2	17.18 2142.1	19.84 3988.7	22.68 7125.4
0.01800 1/ 56	6.387 28.896	6.875 39.363	7.341 51.892	7.935 72.040	8.774 110.26	10.07 197.68	11.71 376.81	13.42 674.66	15.37 1207.2	17.68 2204.2	20.41 4104.4	23.34 7332.0
0.01900 1/ 53	6.563 29.689	7.063 40.442	7.543 53.315	8.152 74.015	9.015 113.28	10.34 203.10	12.03 387.14	13.79 693.15	15.79 1240.3	18.16 2264.6	20.97 4216.9	23.98 7532.9
0.02000 1/ 50	6.733 30.460	7.247 41.494	7.739 54.700	8.364 75.939	9.249 116.23	10.61 208.38	12.35 397.20	14.15 711.16	16.20 1272.5	18.63 2323.4	21.52 4326.4	24.60 7728.7
0.02200 1/ 45	7.062 31.948	7.601 43.520	8.116 57.371	8.772 79.647	9.701 121.90	11.13 218.55	12.95 416.59	14.84 745.88	16.99 1334.6	19.54 2436.9	22.57 4537.6	25.80 8105.9
0.02400 1/ 42	7.376 33.369	7.939 45.456	8.477 59.923	9.163 83.189	10.13 127.33	11.63 228.27	13.53 435.12	15.50 779.05	17.75 1394.0	20.41 2545.2	23.57 4739.4	26.95 8466.4
0.02600 1/ 38	7.677 34.732	8.263 47.312	8.824 62.371	9.537 86.587	10.55 132.53	12.10 237.60	14.08 452.89	16.13 810.87	18.47 1450.9	21.25 2649.2	24.53 4933.0	28.05 8812.2
0.02800 1/ 36	7.967 36.043	8.575 49.099	9.157 64.726	9.897 89.857	10.94 137.53	12.56 246.57	14.61 469.99	16.74 841.48	19.17 1505.7	22.05 2749.2	25.46 5119.2	29.11 9144.9
0.03000 1/ 33	8.247 37.309	8.877 50.823	9.478 66.999	10.24 93.012	11.33 142.36	13.00 255.23	15.12 486.49	17.33 871.02	19.84 1558.6	22.82 2845.7	26.35 5298.9	30.13 9465.9
0.03200 1/ 31	8.518 38.533	9.168 52.490	9.789 69.197	10.58 96.063	11.70 147.03	13.42 263.60	15.62 502.45	17.90 899.60	20.50 1609.7	23.57 2939.0	27.22 5472.7	31.12 9776.3
0.03400 1/ 29	8.780 39.719	9.450 54.106	10.09 71.327	10.91 99.021	12.06 151.55	13.84 271.71	16.10 517.92	18.45 927.29	21.13 1659.2	24.30 3029.5	28.06 5641.2	32.08 10077
0.03600 1/ 28	9.035 40.871	9.724 55.675	10.38 73.396	11.22 101.89	12.41 155.95	14.24 279.59	16.57 532.94	18.98 954.17	21.74 1707.4	25.00 3117.3	28.87 5804.8	33.01 10369
0.03800 1/ 26	9.282 41.992	9.991 57.201	10.67 75.408	11.53 104.68	12.75 160.22	14.63 287.25	17.02 547.54	19.50 980.33	22.33 1754.1	25.69 3202.8	29.66 5963.8	33.91 10654
0.04000 1/ 25	9.523 43.083	10.25 58.688	10.95 77.367	11.83 107.41	13.08 164.39	15.01 294.72	17.46 561.77	20.01 1005.8	22.91 1799.7	26.35 3286.0	30.43 6118.8	34.79 10930
0.04200 1/ 24	9.759 44.147	10.50 60.138	11.22 79.278	12.12 110.06	13.40 168.45	15.38 302.00	17.89 575.65	20.50 1030.6	23.48 1844.2	27.00 3367.1	31.18 6269.9	35.65 11200
0.04400 1/ 23	9.988 45.186	10.75 61.553	11.48 81.144	12.41 112.65	13.72 172.41	15.74 309.11	18.32 589.19	20.99 1054.9	24.03 1887.6	27.64 3446.4	31.92 6417.5	36.49 11464
0.04600 1/ 22	10.21 46.202	10.99 62.937	11.74 82.969	12.69 115.18	14.03 176.29	16.10 316.05	18.73 602.44	21.46 1078.6	24.57 1930.0	28.26 3523.9	32.64 6561.7	37.31 11722
0.04800 1/ 21	10.43 47.196	11.23 64.291	11.99 84.753	12.96 117.66	14.33 180.08	16.44 322.85	19.13 615.40	21.92 1101.8	25.10 1971.5	28.87 3599.7	33.34 6702.8	38.11 11974
0.05000 1/ 20	10.65 48.170	11.46 65.617	12.24 86.502	13.23 120.09	14.63 183.80	16.78 329.51	19.52 628.09	22.37 1124.5	25.62 2012.2	29.46 3673.9	34.02 6841.1	38.90 12221
0.05500 1/ 18	11.17 50.522	12.02 68.821	12.83 90.725	13.87 125.95	15.34 192.77	17.60 345.60	20.48 658.75	23.46 1179.4	26.87 2110.4	30.90 3853.2	35.69 7175.0	40.80 12817
0.06000 1/ 17	11.66 52.769	12.55 71.882	13.41 94.760	14.49 131.55	16.02 201.34	18.38 360.97	21.39 688.05	24.51 1231.9	28.07 2204.3	32.28 4024.6	37.27 7494.1	42.61 13387
0.06500 1/ 15	12.14 54.924	13.07 74.818	13.95 98.630	15.08 136.92	16.68 209.56	19.13 375.71	22.26 716.15	25.51 1282.2	29.21 2294.3	33.59 4188.9	38.79 7800.1	44.35 13934
0.07000 1/ 14	12.60 56.998	13.56 77.643	14.48 102.35	15.65 142.09	17.31 217.48	19.86 389.89	23.10 743.18	26.47 1330.6	30.31 2380.9	34.86 4347.1	40.26 8094.6	46.03 14460
0.07500 1/ 13	13.04 58.999	14.04 80.368	14.99 105.95	16.20 147.08	17.91 225.11	20.55 403.58	23.91 769.27	27.40 1377.3	31.38 2464.5	36.09 4499.7	41.67 8378.7	47.64 14967
	600	700	800	900	1000	1300	1700	2100	2600	3300	4100	5000

θ_{medial} for part-full circular pipes.

k_s = 3.00 mm   S = 0.0015 to 0.075

$k_s = 6.00$ mm
S = 0.00010 to 0.00048

Water (or sewage) at 15°C;
full bore conditions.

ie hydraulic gradient =
1 in 10000 to 1 in 2083

velocities in ms⁻¹
discharges in m³s⁻¹

Gradient	(Equivalent) Pipe diameters in m											
	2.400	2.700	3.000	3.400	4.000	5.000	6.400	8.000	10.00	12.60	16.00	20.00
0.00010 1/10000	0.434 1.9611	0.467 2.6757	0.500 3.5323	0.541 4.9115	0.599 7.5321	0.690 13.538	0.804 25.874	0.924 46.430	1.060 83.256	1.222 152.33	1.414 284.22	1.619 508.63
0.00011 1/ 9091	0.455 2.0572	0.490 2.8068	0.524 3.7052	0.567 5.1519	0.629 7.9007	0.723 14.201	0.844 27.140	0.969 48.700	1.112 87.326	1.281 159.77	1.483 298.11	1.698 533.47
0.00012 1/ 8333	0.475 2.1489	0.512 2.9320	0.548 3.8704	0.593 5.3816	0.657 8.2528	0.755 14.833	0.881 28.349	1.012 50.869	1.161 91.214	1.338 166.88	1.549 311.38	1.774 557.22
0.00013 1/ 7692	0.494 2.2370	0.533 3.0520	0.570 4.0289	0.617 5.6019	0.684 8.5906	0.786 15.440	0.917 29.508	1.053 52.949	1.209 94.943	1.393 173.71	1.612 324.10	1.846 579.99
0.00014 1/ 7143	0.513 2.3217	0.553 3.1676	0.592 4.1814	0.640 5.8139	0.709 8.9156	0.816 16.024	0.952 30.624	1.093 54.951	1.255 98.532	1.446 180.27	1.673 336.35	1.916 601.90
0.00015 1/ 6667	0.531 2.4034	0.573 3.2790	0.612 4.3285	0.663 6.0184	0.734 9.2292	0.845 16.588	0.985 31.701	1.132 56.882	1.299 101.99	1.497 186.60	1.732 348.16	1.983 623.04
0.00016 1/ 6250	0.549 2.4824	0.592 3.3869	0.632 4.4708	0.685 6.2162	0.759 9.5325	0.873 17.133	1.018 32.742	1.169 58.750	1.341 105.34	1.546 192.73	1.788 359.59	2.048 643.49
0.00017 1/ 5882	0.566 2.5590	0.610 3.4914	0.652 4.6087	0.706 6.4080	0.782 9.8264	0.899 17.661	1.049 33.751	1.205 60.561	1.383 108.59	1.593 198.67	1.844 370.67	2.111 663.31
0.00018 1/ 5556	0.582 2.6334	0.628 3.5928	0.671 4.7427	0.726 6.5941	0.805 10.112	0.926 18.174	1.080 34.731	1.240 62.319	1.423 111.74	1.640 204.43	1.897 381.42	2.173 682.56
0.00019 1/ 5263	0.598 2.7058	0.645 3.6915	0.689 4.8729	0.746 6.7752	0.827 10.389	0.951 18.673	1.109 35.684	1.274 64.028	1.462 114.81	1.684 210.04	1.949 391.88	2.232 701.27
0.00020 1/ 5000	0.614 2.7763	0.662 3.7876	0.707 4.9998	0.766 6.9516	0.848 10.660	0.976 19.158	1.138 36.612	1.307 65.694	1.500 117.79	1.728 215.50	2.000 402.07	2.290 719.50
0.00022 1/ 4545	0.644 2.9121	0.694 3.9729	0.742 5.2443	0.803 7.2915	0.890 11.181	1.023 20.095	1.194 38.402	1.371 68.904	1.573 123.55	1.813 226.03	2.097 421.71	2.402 754.65
0.00024 1/ 4167	0.672 3.0419	0.725 4.1500	0.775 5.4780	0.839 7.6164	0.929 11.679	1.069 20.990	1.247 40.111	1.432 71.971	1.643 129.05	1.893 236.09	2.191 440.48	2.509 788.22
0.00026 1/ 3846	0.700 3.1664	0.754 4.3198	0.807 5.7021	0.873 7.9279	0.967 12.157	1.113 21.848	1.298 41.751	1.490 74.913	1.710 134.32	1.971 245.74	2.280 458.48	2.612 820.43
0.00028 1/ 3571	0.726 3.2862	0.783 4.4832	0.837 5.9178	0.906 8.2277	1.004 12.616	1.155 22.674	1.347 43.329	1.547 77.744	1.775 139.39	2.045 255.02	2.366 475.80	2.710 851.42
0.00030 1/ 3333	0.752 3.4017	0.811 4.6408	0.867 6.1259	0.938 8.5170	1.039 13.060	1.195 23.471	1.394 44.852	1.601 80.475	1.837 144.29	2.117 263.98	2.450 492.51	2.805 881.32
0.00032 1/ 3125	0.777 3.5135	0.837 4.7933	0.895 6.3271	0.969 8.7968	1.073 13.489	1.235 24.242	1.440 46.324	1.654 83.117	1.897 149.03	2.187 272.64	2.530 508.67	2.897 910.24
0.00034 1/ 2941	0.801 3.6219	0.863 4.9411	0.923 6.5222	0.999 9.0679	1.106 13.905	1.273 24.989	1.484 47.751	1.704 85.677	1.956 153.62	2.254 281.04	2.608 524.33	2.987 938.26
0.00036 1/ 2778	0.824 3.7271	0.888 5.0846	0.949 6.7116	1.028 9.3312	1.139 14.308	1.310 25.714	1.527 49.137	1.754 88.164	2.013 158.07	2.319 289.19	2.683 539.54	3.073 965.48
0.00038 1/ 2632	0.846 3.8294	0.912 5.2242	0.976 6.8958	1.056 9.5873	1.170 14.701	1.346 26.419	1.569 50.485	1.802 90.582	2.068 162.41	2.383 297.12	2.757 554.34	3.157 991.95
0.00040 1/ 2500	0.869 3.9290	0.936 5.3601	1.001 7.0752	1.083 9.8367	1.200 15.083	1.381 27.107	1.610 51.798	1.849 92.937	2.122 166.63	2.445 304.84	2.829 568.74	3.240 1017.7
0.00042 1/ 2381	0.890 4.0262	0.959 5.4927	1.026 7.2502	1.110 10.080	1.230 15.456	1.415 27.777	1.650 53.078	1.895 95.234	2.174 170.75	2.505 312.38	2.899 582.80	3.320 1042.9
0.00044 1/ 2273	0.911 4.1212	0.982 5.6222	1.050 7.4211	1.136 10.317	1.259 15.820	1.448 28.431	1.689 54.328	1.939 97.477	2.225 174.77	2.564 319.73	2.967 596.52	3.398 1067.4
0.00046 1/ 2174	0.931 4.2139	1.004 5.7487	1.073 7.5881	1.162 10.550	1.287 16.176	1.481 29.071	1.727 55.550	1.983 99.669	2.275 178.70	2.622 326.92	3.034 609.93	3.474 1091.4
0.00048 1/ 2083	0.952 4.3047	1.026 5.8725	1.097 7.7515	1.187 10.777	1.315 16.525	1.512 29.697	1.764 56.746	2.026 101.81	2.324 182.55	2.678 333.96	3.099 623.06	3.549 1114.9
	230	260	290	320	380	480	600	750	950	1200	1500	1900

θ_{medial} **for part-full circular pipes.**

$k_s = 6.00$ mm   S = 0.00010 to 0.00048

$k_s = 6.00$ mm
$S = 0.00050$ to 0.00280

Water (or sewage) at 15°C; full bore conditions.

ie hydraulic gradient = 1 in 2000 to 1 in 357

velocities in ms^{-1}
discharges in m^{3}s^{-1}

Gradient	(Equivalent) Pipe diameters in m											
	2.400	2.700	3.000	3.400	4.000	5.000	6.400	8.000	10.00	12.60	16.00	20.00
0.00050 1/ 2000	0.971 4.3936	1.047 5.9938	1.119 7.9116	1.211 10.999	1.342 16.866	1.544 30.310	1.800 57.917	2.067 103.92	2.372 186.31	2.734 340.85	3.163 635.91	3.622 1137.9
0.00055 1/ 1818	1.019 4.6084	1.098 6.2868	1.174 8.2983	1.271 11.537	1.408 17.690	1.619 31.790	1.888 60.747	2.168 108.99	2.488 195.41	2.867 357.49	3.317 666.96	3.799 1193.5
0.00060 1/ 1667	1.064 4.8136	1.147 6.5667	1.226 8.6677	1.327 12.051	1.470 18.477	1.691 33.205	1.972 63.450	2.265 113.84	2.599 204.11	2.995 373.40	3.465 696.64	3.968 1246.6
0.00065 1/ 1538	1.108 5.0105	1.194 6.8352	1.276 9.0221	1.382 12.543	1.530 19.233	1.760 34.562	2.053 66.043	2.357 118.49	2.705 212.45	3.117 388.65	3.606 725.09	4.130 1297.5
0.00070 1/ 1429	1.149 5.1999	1.239 7.0936	1.325 9.3631	1.434 13.017	1.588 19.959	1.827 35.868	2.130 68.538	2.446 122.97	2.807 220.47	3.235 403.33	3.743 752.48	4.286 1346.5
0.00075 1/ 1333	1.190 5.3826	1.282 7.3429	1.371 9.6921	1.484 13.475	1.644 20.661	1.891 37.128	2.205 70.945	2.532 127.29	2.906 228.21	3.348 417.49	3.874 778.90	4.436 1393.8
0.00080 1/ 1250	1.229 5.5594	1.325 7.5840	1.416 10.010	1.533 13.917	1.698 21.339	1.953 38.347	2.278 73.273	2.615 131.46	3.001 235.70	3.458 431.19	4.001 804.45	4.582 1439.5
0.00085 1/ 1176	1.267 5.7307	1.365 7.8176	1.460 10.319	1.580 14.346	1.750 21.996	2.013 39.528	2.348 75.530	2.696 135.51	3.093 242.96	3.565 444.47	4.124 829.22	4.723 1483.8
0.00090 1/ 1111	1.304 5.8970	1.405 8.0445	1.502 10.618	1.626 14.762	1.801 22.634	2.072 40.675	2.416 77.721	2.774 139.44	3.183 250.01	3.668 457.36	4.244 853.27	4.860 1526.8
0.00095 1/ 1053	1.339 6.0588	1.444 8.2652	1.543 10.909	1.671 15.167	1.851 23.255	2.128 41.790	2.482 79.852	2.850 143.27	3.270 256.86	3.769 469.90	4.360 876.66	4.993 1568.7
0.00100 1/ 1000	1.374 6.2164	1.481 8.4801	1.584 11.193	1.714 15.561	1.899 23.860	2.184 42.877	2.547 81.928	2.924 146.99	3.355 263.54	3.866 482.11	4.473 899.44	5.123 1609.5
0.00110 1/ 909	1.441 6.5201	1.553 8.8945	1.661 11.740	1.798 16.321	1.991 25.025	2.290 44.971	2.671 85.929	3.067 154.17	3.519 276.41	4.055 505.65	4.692 943.36	5.373 1688.0
0.00120 1/ 833	1.505 6.8103	1.623 9.2904	1.735 12.262	1.878 17.048	2.080 26.139	2.392 46.972	2.790 89.752	3.204 161.03	3.676 288.70	4.236 528.14	4.901 985.32	5.612 1763.1
0.00130 1/ 769	1.567 7.0887	1.689 9.6701	1.806 12.764	1.954 17.744	2.165 27.207	2.490 48.891	2.904 93.419	3.334 167.61	3.826 300.50	4.409 549.72	5.101 1025.6	5.841 1835.1
0.00140 1/ 714	1.626 7.3566	1.753 10.035	1.874 13.246	2.028 18.415	2.247 28.235	2.584 50.738	3.014 96.947	3.460 173.94	3.971 311.84	4.575 570.48	5.293 1064.3	6.062 1904.4
0.00150 1/ 667	1.683 7.6150	1.814 10.388	1.940 13.711	2.099 19.062	2.326 29.226	2.675 52.520	3.119 100.35	3.582 180.04	4.110 322.79	4.736 590.50	5.479 1101.7	6.275 1971.3
0.00160 1/ 625	1.739 7.8650	1.874 10.729	2.003 14.161	2.168 19.687	2.402 30.186	2.763 54.243	3.222 103.64	3.699 185.95	4.245 333.38	4.891 609.88	5.659 1137.8	6.481 2035.9
0.00170 1/ 588	1.792 8.1072	1.932 11.059	2.065 14.597	2.235 20.294	2.476 31.115	2.848 55.913	3.321 106.84	3.813 191.68	4.375 343.65	5.042 628.65	5.833 1172.8	6.680 2098.6
0.00180 1/ 556	1.844 8.3425	1.988 11.380	2.125 15.021	2.300 20.882	2.548 32.018	2.930 57.535	3.417 109.93	3.924 197.23	4.502 353.61	5.188 646.88	6.002 1206.8	6.874 2159.5
0.00190 1/ 526	1.895 8.5713	2.042 11.692	2.183 15.433	2.363 21.455	2.618 32.896	3.011 59.113	3.511 112.95	4.031 202.64	4.626 363.31	5.330 664.62	6.167 1239.9	7.062 2218.7
0.00200 1/ 500	1.944 8.7941	2.095 11.996	2.240 15.834	2.425 22.013	2.686 33.751	3.089 60.649	3.602 115.88	4.136 207.91	4.746 372.75	5.469 681.89	6.327 1272.1	7.246 2276.3
0.00220 1/ 455	2.039 9.2237	2.198 12.582	2.349 16.607	2.543 23.088	2.817 35.399	3.240 63.611	3.778 121.54	4.338 218.06	4.978 390.95	5.736 715.18	6.636 1334.2	7.599 2387.4
0.00240 1/ 417	2.130 9.6341	2.295 13.142	2.454 17.346	2.656 24.115	2.942 36.974	3.384 66.440	3.946 126.95	4.531 227.76	5.199 408.34	5.991 746.99	6.931 1393.6	7.937 2493.6
0.00260 1/ 385	2.217 10.028	2.389 13.679	2.554 18.055	2.765 25.100	3.062 38.484	3.522 69.155	4.107 132.13	4.716 237.06	5.411 425.01	6.235 777.50	7.214 1450.5	8.262 2595.5
0.00280 1/ 357	2.300 10.407	2.479 14.196	2.651 18.737	2.869 26.048	3.178 39.938	3.655 71.766	4.263 137.12	4.894 246.01	5.616 441.06	6.471 806.85	7.487 1505.3	8.574 2693.5
	280	310	350	390	460	600	750	950	1200	1500	1900	2300

θ_{medial} **for part-full circular pipes.**

$k_s = 6.00$ mm $S = 0.00050$ to 0.00280

A42
continued

k_s = 6.00 mm
S = 0.0030 to 0.014

Water (or sewage) at 15°C;
full bore conditions.

ie hydraulic gradient =
1 in 333 to 1 in 71

velocities in ms⁻¹
discharges in m³s⁻¹

Gradient	(Equivalent) Pipe diameters in m											
	2.400	2.700	3.000	3.400	4.000	5.000	6.400	8.000	10.00	12.60	16.00	20.00
0.00300 1/ 333	2.381 10.772	2.566 14.694	2.744 19.395	2.970 26.963	3.290 41.340	3.783 74.286	4.412 141.94	5.066 254.65	5.813 456.55	6.698 835.18	7.749 1558.1	8.874 2788.0
0.00320 1/ 313	2.459 11.126	2.651 15.177	2.834 20.031	3.067 27.848	3.398 42.697	3.907 76.724	4.557 146.60	5.232 263.00	6.004 471.52	6.918 862.58	8.004 1609.2	9.166 2879.5
0.00340 1/ 294	2.535 11.468	2.732 15.644	2.921 20.648	3.162 28.705	3.502 44.011	4.028 79.086	4.697 151.11	5.393 271.10	6.188 486.04	7.131 889.13	8.250 1658.7	9.448 2968.1
0.00360 1/ 278	2.609 11.801	2.812 16.098	3.006 21.247	3.253 29.538	3.604 45.288	4.145 81.379	4.833 155.49	5.550 278.96	6.368 500.13	7.337 914.91	8.489 1706.8	9.722 3054.2
0.00380 1/ 263	2.680 12.124	2.889 16.539	3.088 21.830	3.343 30.348	3.703 46.529	4.258 83.610	4.966 159.75	5.702 286.61	6.542 513.84	7.539 939.99	8.722 1753.6	9.988 3137.9
0.00400 1/ 250	2.750 12.440	2.964 16.969	3.169 22.397	3.429 31.136	3.799 47.739	4.369 85.783	5.095 163.90	5.850 294.06	6.712 527.19	7.734 964.41	8.948 1799.2	10.25 3219.4
0.00420 1/ 238	2.818 12.747	3.037 17.388	3.247 22.950	3.514 31.906	3.893 48.918	4.477 87.902	5.221 167.95	5.995 301.32	6.878 540.22	7.926 988.23	9.169 1843.6	10.50 3298.9
0.00440 1/ 227	2.884 13.047	3.108 17.798	3.323 23.491	3.597 32.657	3.984 50.070	4.582 89.971	5.344 171.91	6.136 308.41	7.040 552.93	8.112 1011.5	9.385 1887.0	10.75 3376.5
0.00460 1/ 217	2.949 13.340	3.178 18.198	3.398 24.019	3.678 33.391	4.074 51.195	4.685 91.994	5.464 175.77	6.274 315.35	7.198 565.36	8.294 1034.2	9.596 1929.4	10.99 3452.4
0.00480 1/ 208	3.012 13.628	3.247 18.589	3.471 24.536	3.757 34.109	4.162 52.297	4.786 93.973	5.581 179.55	6.409 322.13	7.353 577.52	8.473 1056.5	9.803 1970.9	11.23 3526.7
0.00500 1/ 200	3.075 13.909	3.314 18.973	3.543 25.042	3.834 34.813	4.248 53.376	4.885 95.912	5.697 183.26	6.541 328.78	7.505 589.44	8.648 1078.3	10.00 2011.6	11.46 3599.4
0.00550 1/ 182	3.225 14.588	3.476 19.899	3.716 26.265	4.022 36.513	4.455 55.982	5.123 100.59	5.975 192.20	6.860 344.83	7.871 618.21	9.070 1130.9	10.49 2109.8	12.02 3775.1
0.00600 1/ 167	3.368 15.237	3.630 20.785	3.881 27.433	4.200 38.137	4.653 58.472	5.351 105.07	6.240 200.75	7.165 360.16	8.221 645.71	9.473 1181.2	10.96 2203.6	12.55 3943.0
0.00650 1/ 154	3.506 15.859	3.778 21.634	4.040 28.554	4.372 39.695	4.843 60.860	5.570 109.36	6.495 208.95	7.458 374.87	8.557 672.08	9.860 1229.4	11.41 2293.6	13.06 4104.1
0.00700 1/ 143	3.638 16.458	3.921 22.451	4.192 29.632	4.537 41.194	5.026 63.158	5.780 113.49	6.741 216.84	7.739 389.03	8.880 697.45	10.23 1275.9	11.84 2380.2	13.56 4259.0
0.00750 1/ 133	3.766 17.036	4.059 23.239	4.339 30.672	4.696 42.640	5.202 65.376	5.983 117.47	6.977 224.45	8.011 402.68	9.192 721.94	10.59 1320.6	12.25 2463.8	14.03 4408.5
0.00800 1/ 125	3.889 17.595	4.192 24.001	4.482 31.679	4.851 44.039	5.373 67.521	6.179 121.33	7.206 231.82	8.274 415.89	9.493 745.62	10.94 1364.0	12.66 2544.6	14.49 4553.1
0.00850 1/ 118	4.009 18.137	4.321 24.740	4.620 32.654	5.000 45.395	5.539 69.599	6.369 125.06	7.428 238.95	8.529 428.69	9.786 768.57	11.28 1405.9	13.05 2622.9	14.94 4693.2
0.00900 1/ 111	4.125 18.663	4.446 25.458	4.754 33.601	5.145 46.711	5.699 71.618	6.554 128.69	7.643 245.88	8.776 441.12	10.07 790.85	11.60 1446.7	13.42 2698.9	15.37 4829.3
0.00950 1/ 105	4.238 19.174	4.568 26.156	4.884 34.522	5.286 47.992	5.855 73.581	6.734 132.22	7.853 252.62	9.016 453.21	10.35 812.53	11.92 1486.4	13.79 2772.9	15.79 4961.7
0.01000 1/ 100	4.349 19.673	4.687 26.835	5.011 35.419	5.423 49.239	6.008 75.493	6.909 135.65	8.057 259.19	9.251 464.99	10.61 833.64	12.23 1525.0	14.15 2845.0	16.20 5090.6
0.01100 1/ 91	4.561 20.633	4.916 28.146	5.255 37.148	5.688 51.643	6.301 79.178	7.246 142.27	8.450 271.84	9.702 487.69	11.13 874.33	12.83 1599.4	14.84 2983.8	16.99 5339.1
0.01200 1/ 83	4.764 21.551	5.134 29.398	5.489 38.801	5.941 53.940	6.581 82.700	7.568 148.60	8.826 283.93	10.13 509.38	11.63 913.22	13.40 1670.5	15.50 3116.5	17.75 5576.5
0.01300 1/ 77	4.958 22.431	5.344 30.598	5.713 40.386	6.184 56.143	6.850 86.078	7.877 154.67	9.186 295.52	10.55 530.18	12.10 950.51	13.94 1738.8	16.13 3243.8	18.48 5804.2
0.01400 1/ 71	5.146 23.278	5.546 31.754	5.929 41.910	6.417 58.263	7.108 89.328	8.175 160.51	9.533 306.68	10.95 550.20	12.56 986.40	14.47 1804.4	16.74 3366.2	19.17 6023.3
	320	360	400	450	550	650	850	1100	1300	1700	2100	2700

θ_{medial} **for part-full circular pipes.**

k_s = 6.00 mm   S = 0.0030 to 0.014

$k_s = 6.00$ mm
S = 0.015 to 0.075

ie hydraulic gradient =
1 in 66 to 1 in 13

Water (or sewage) at 15°C;
full bore conditions.

velocities in ms^{-1}
discharges in m^{3}s^{-1}

Gradient	(Equivalent) Pipe diameters in m											
	2.400	2.700	3.000	3.400	4.000	5.000	6.400	8.000	10.00	12.60	16.00	20.00
0.01500 1/ 67	5.326 24.096	5.741 32.869	6.137 43.382	6.642 60.308	7.358 92.463	8.462 166.15	9.868 317.45	11.33 569.51	13.00 1021.0	14.98 1867.8	17.33 3484.4	19.85 6234.8
0.01600 1/ 62	5.501 24.886	5.929 33.947	6.339 44.805	6.860 62.286	7.599 95.497	8.739 171.60	10.19 327.86	11.70 588.19	13.43 1054.5	15.47 1929.0	17.90 3598.7	20.50 6439.3
0.01700 1/ 59	5.670 25.652	6.112 34.992	6.534 46.184	7.072 64.204	7.833 98.436	9.008 176.88	10.51 337.95	12.06 606.30	13.84 1087.0	15.95 1988.4	18.45 3709.5	21.13 6637.4
0.01800 1/ 56	5.835 26.396	6.289 36.007	6.723 47.523	7.277 66.066	8.060 101.29	9.270 182.01	10.81 347.75	12.41 623.88	14.24 1118.5	16.41 2046.0	18.98 3817.0	21.74 6829.9
0.01900 1/ 53	5.995 27.120	6.461 36.993	6.907 48.826	7.476 67.876	8.281 104.07	9.524 187.00	11.11 357.28	12.75 640.97	14.63 1149.1	16.86 2102.1	19.50 3921.6	22.34 7017.1
0.02000 1/ 50	6.151 27.824	6.629 37.955	7.087 50.095	7.670 69.640	8.497 106.77	9.771 191.85	11.39 366.56	13.08 657.63	15.01 1179.0	17.30 2156.7	20.01 4023.5	22.92 7199.4
0.02200 1/ 45	6.451 29.183	6.953 39.808	7.433 52.540	8.045 73.040	8.911 111.98	10.25 201.22	11.95 384.46	13.72 689.73	15.74 1236.5	18.14 2262.0	20.99 4219.9	24.03 7550.8
0.02400 1/ 42	6.738 30.481	7.262 41.578	7.764 54.877	8.403 76.288	9.308 116.96	10.70 210.17	12.48 401.55	14.33 720.40	16.44 1291.5	18.95 2362.6	21.92 4407.5	25.10 7886.6
0.02600 1/ 38	7.013 31.726	7.558 43.276	8.081 57.118	8.746 79.404	9.688 121.74	11.14 218.75	12.99 417.95	14.92 749.82	17.12 1344.3	19.72 2459.1	22.82 4587.5	26.13 8208.6
0.02800 1/ 36	7.278 32.923	7.844 44.910	8.386 59.275	9.076 82.402	10.05 126.34	11.56 227.01	13.48 433.73	15.48 778.13	17.76 1395.0	20.47 2551.9	23.68 4760.7	27.12 8518.5
0.03000 1/ 33	7.533 34.079	8.119 46.487	8.680 61.356	9.394 85.294	10.41 130.77	11.97 234.98	13.96 448.96	16.02 805.44	18.39 1444.0	21.18 2641.5	24.51 4927.8	28.07 8817.5
0.03200 1/ 31	7.780 35.197	8.386 48.012	8.965 63.368	9.703 88.092	10.75 135.06	12.36 242.69	14.41 463.68	16.55 831.86	18.99 1491.4	21.88 2728.1	25.31 5089.4	28.99 9106.7
0.03400 1/ 29	8.020 36.281	8.644 49.490	9.241 65.319	10.00 90.804	11.08 139.22	12.74 250.16	14.86 477.96	17.06 857.46	19.57 1537.3	22.55 2812.1	26.09 5246.1	29.88 9387.0
0.03600 1/ 28	8.252 37.333	8.894 50.925	9.509 67.213	10.29 93.437	11.40 143.25	13.11 257.41	15.29 491.81	17.55 882.33	20.14 1581.8	23.21 2893.6	26.85 5398.2	30.75 9659.1
0.03800 1/ 26	8.479 38.356	9.138 52.320	9.769 69.055	10.57 95.997	11.71 147.18	13.47 264.46	15.71 505.29	18.03 906.51	20.69 1625.2	23.84 2972.9	27.58 5546.1	31.59 9923.8
0.04000 1/ 25	8.699 39.352	9.375 53.680	10.02 70.849	10.85 98.492	12.02 151.00	13.82 271.33	16.12 518.42	18.50 930.06	21.23 1667.4	24.46 3050.1	28.30 5690.2	32.41 10182
0.04200 1/ 24	8.914 40.324	9.607 55.006	10.27 72.599	11.12 100.92	12.31 154.73	14.16 278.04	16.51 531.22	18.96 953.03	21.75 1708.6	25.07 3125.5	29.00 5830.7	33.21 10433
0.04400 1/ 23	9.123 41.274	9.833 56.300	10.51 74.308	11.38 103.30	12.60 158.38	14.49 284.58	16.90 543.73	19.41 975.46	22.27 1748.8	25.66 3199.0	29.68 5968.0	33.99 10679
0.04600 1/ 22	9.329 42.201	10.05 57.566	10.75 75.978	11.63 105.62	12.89 161.94	14.82 290.98	17.28 555.95	19.84 997.38	22.77 1788.1	26.23 3270.9	30.35 6102.1	34.76 10919
0.04800 1/ 21	9.529 43.109	10.27 58.804	10.98 77.612	11.88 107.89	13.16 165.42	15.14 297.24	17.65 567.90	20.27 1018.8	23.26 1826.6	26.80 3341.3	31.00 6233.3	35.50 11153
0.05000 1/ 20	9.726 43.998	10.48 60.017	11.21 79.213	12.13 110.12	13.44 168.83	15.45 303.37	18.02 579.62	20.69 1039.8	23.74 1864.2	27.35 3410.2	31.64 6361.9	36.23 11383
0.05500 1/ 18	10.20 46.146	10.99 62.947	11.75 83.080	12.72 115.49	14.09 177.07	16.20 318.17	18.90 607.91	21.70 1090.6	24.89 1955.2	28.68 3576.6	33.19 6672.4	38.00 11939
0.06000 1/ 17	10.65 48.198	11.48 65.746	12.28 86.774	13.29 120.63	14.72 184.95	16.93 332.32	19.74 634.94	22.66 1139.1	26.00 2042.2	29.96 3735.7	34.66 6969.1	39.69 12470
0.06500 1/ 15	11.09 50.167	11.95 68.431	12.78 90.318	13.83 125.56	15.32 192.50	17.62 345.89	20.54 660.87	23.59 1185.6	27.06 2125.6	31.18 3888.2	36.08 7253.7	41.31 12979
0.07000 1/ 14	11.51 52.061	12.40 71.014	13.26 93.728	14.35 130.30	15.90 199.77	18.28 358.95	21.32 685.82	24.48 1230.4	28.09 2205.8	32.36 4035.0	37.44 7527.5	42.87 13469
0.07500 1/ 13	11.91 53.888	12.84 73.507	13.73 97.018	14.85 134.87	16.45 206.78	18.92 371.55	22.07 709.89	25.34 1273.6	29.07 2283.2	33.50 4176.7	38.75 7791.8	44.38 13942
	350	390	440	500	600	750	950	1200	1500	1800	2300	2900

θ_{medial} for part-full circular pipes.

$k_s = 6.00$ mm   S = 0.0015 to 0.075

73

A43

k_s = 15.0 mm
S = 0.00010 to 0.00048

Water (or sewage) at 15°C;
full bore conditions.

ie hydraulic gradient =
1 in 10000 to 1 in 2083

velocities in ms^{-1}
discharges in m^{3}s^{-1}

Gradient	(Equivalent) Pipe diameters in m											
	2.400	2.700	3.000	3.400	4.000	5.000	6.400	8.000	10.00	12.60	16.00	20.00
0.00010	0.380	0.410	0.440	0.477	0.530	0.612	0.716	0.825	0.950	1.097	1.274	1.462
1/10000	1.7182	2.3493	3.1071	4.3297	6.6577	12.009	23.035	41.465	74.574	136.84	256.06	459.39
0.00011	0.398	0.430	0.461	0.500	0.556	0.641	0.751	0.865	0.996	1.151	1.336	1.534
1/ 9091	1.8022	2.4642	3.2590	4.5412	6.9830	12.595	24.160	43.490	78.216	143.52	268.56	481.82
0.00012	0.416	0.450	0.482	0.522	0.580	0.670	0.784	0.904	1.040	1.202	1.395	1.602
1/ 8333	1.8824	2.5739	3.4041	4.7434	7.2938	13.156	25.236	45.425	81.696	149.91	280.51	503.26
0.00013	0.433	0.468	0.501	0.544	0.604	0.697	0.817	0.941	1.083	1.251	1.452	1.667
1/ 7692	1.9594	2.6791	3.5433	4.9373	7.5920	13.694	26.267	47.281	85.034	156.03	291.97	523.82
0.00014	0.449	0.486	0.520	0.564	0.627	0.724	0.847	0.976	1.124	1.299	1.507	1.730
1/ 7143	2.0335	2.7804	3.6772	5.1239	7.8788	14.211	27.259	49.067	88.246	161.93	302.99	543.60
0.00015	0.465	0.503	0.538	0.584	0.649	0.749	0.877	1.010	1.163	1.344	1.560	1.791
1/ 6667	2.1050	2.8781	3.8064	5.3040	8.1556	14.710	28.217	50.791	91.344	167.61	313.63	562.68
0.00016	0.481	0.519	0.556	0.603	0.670	0.774	0.906	1.044	1.201	1.388	1.611	1.850
1/ 6250	2.1741	2.9726	3.9314	5.4781	8.4234	15.193	29.143	52.457	94.342	173.11	323.92	581.14
0.00017	0.495	0.535	0.573	0.622	0.691	0.798	0.934	1.076	1.238	1.431	1.661	1.907
1/ 5882	2.2411	3.0642	4.0525	5.6469	8.6828	15.661	30.040	54.073	97.247	178.44	333.89	599.03
0.00018	0.510	0.551	0.590	0.640	0.711	0.821	0.961	1.107	1.274	1.473	1.709	1.962
1/ 5556	2.3061	3.1531	4.1701	5.8107	8.9348	16.115	30.912	55.641	100.07	183.62	343.58	616.41
0.00019	0.524	0.566	0.606	0.658	0.731	0.843	0.987	1.137	1.309	1.513	1.756	2.016
1/ 5263	2.3694	3.2396	4.2845	5.9701	9.1798	16.557	31.759	57.167	102.81	188.65	352.99	633.30
0.00020	0.537	0.581	0.622	0.675	0.750	0.865	1.013	1.167	1.343	1.552	1.801	2.068
1/ 5000	2.4310	3.3239	4.3959	6.1253	9.4185	16.988	32.585	58.653	105.48	193.55	362.17	649.76
0.00022	0.564	0.609	0.652	0.708	0.786	0.907	1.062	1.224	1.409	1.628	1.889	2.169
1/ 4545	2.5498	3.4863	4.6107	6.4246	9.8786	17.818	34.176	61.517	110.63	203.00	379.85	681.48
0.00024	0.589	0.636	0.681	0.739	0.821	0.948	1.110	1.278	1.471	1.700	1.973	2.266
1/ 4167	2.6633	3.6415	4.8159	6.7105	10.318	18.610	35.697	64.254	115.56	212.03	396.75	711.79
0.00026	0.613	0.662	0.709	0.769	0.855	0.987	1.155	1.331	1.531	1.770	2.054	2.358
1/ 3846	2.7722	3.7903	5.0128	6.9847	10.740	19.371	37.155	66.879	120.28	220.69	412.95	740.87
0.00028	0.636	0.687	0.736	0.798	0.887	1.024	1.199	1.381	1.589	1.837	2.131	2.447
1/ 3571	2.8769	3.9335	5.2021	7.2486	11.146	20.102	38.559	69.405	124.82	229.03	428.55	768.84
0.00030	0.658	0.711	0.762	0.826	0.918	1.060	1.241	1.429	1.645	1.901	2.206	2.533
1/ 3333	2.9780	4.0717	5.3849	7.5032	11.537	20.808	39.913	71.842	129.20	237.07	443.59	795.83
0.00032	0.680	0.734	0.787	0.854	0.948	1.095	1.281	1.476	1.699	1.964	2.279	2.616
1/ 3125	3.0758	4.2053	5.5616	7.7495	11.916	21.491	41.222	74.199	133.44	244.85	458.14	821.94
0.00034	0.701	0.757	0.811	0.880	0.977	1.128	1.321	1.522	1.751	2.024	2.349	2.697
1/ 2941	3.1705	4.3349	5.7329	7.9882	12.283	22.153	42.491	76.483	137.55	252.39	472.25	847.24
0.00036	0.721	0.779	0.835	0.905	1.006	1.161	1.359	1.566	1.802	2.083	2.417	2.775
1/ 2778	3.2625	4.4607	5.8992	8.2199	12.639	22.796	43.724	78.702	141.54	259.70	485.94	871.81
0.00038	0.741	0.800	0.857	0.930	1.033	1.193	1.396	1.609	1.851	2.140	2.483	2.851
1/ 2632	3.3520	4.5830	6.0610	8.4453	12.985	23.421	44.923	80.859	145.42	266.82	499.26	895.70
0.00040	0.760	0.821	0.880	0.954	1.060	1.224	1.433	1.650	1.900	2.195	2.548	2.925
1/ 2500	3.4391	4.7021	6.2186	8.6648	13.323	24.029	46.090	82.960	149.20	273.76	512.23	918.98
0.00042	0.779	0.842	0.901	0.978	1.086	1.254	1.468	1.691	1.947	2.250	2.611	2.997
1/ 2381	3.5241	4.8183	6.3723	8.8790	13.652	24.623	47.229	85.010	152.88	280.52	524.89	941.68
0.00044	0.797	0.861	0.923	1.001	1.112	1.284	1.503	1.731	1.992	2.303	2.672	3.068
1/ 2273	3.6071	4.9318	6.5223	9.0880	13.974	25.203	48.341	87.011	156.48	287.12	537.24	963.84
0.00046	0.815	0.881	0.943	1.023	1.137	1.312	1.536	1.770	2.037	2.354	2.732	3.137
1/ 2174	3.6882	5.0427	6.6690	9.2924	14.288	25.770	49.428	88.967	160.00	293.58	549.32	985.51
0.00048	0.833	0.900	0.964	1.046	1.161	1.341	1.570	1.808	2.081	2.405	2.791	3.204
1/ 2083	3.7676	5.1513	6.8125	9.4924	14.595	26.324	50.491	90.881	163.44	299.89	561.14	1006.7
	120	140	150	170	210	260	330	410	500	650	800	1000

θ_{medial} for part-full circular pipes.

k_s = 15.0 mm   S = 0.00010 to 0.00048

k_s = 15.0 mm
S = 0.00050 to 0.00280

ie hydraulic gradient =
1 in 2000 to 1 in 357

Water (or sewage) at 15°C;
full bore conditions.

velocities in ms^{-1}
discharges in m^{3}s^{-1}

A43

continued

Gradient	(Equivalent) Pipe diameters in m											
	2.400	2.700	3.000	3.400	4.000	5.000	6.400	8.000	10.00	12.60	16.00	20.00
0.00050 1/ 2000	0.850 3.8454	0.918 5.2576	0.984 6.9531	1.067 9.6882	1.185 14.896	1.368 26.867	1.602 51.533	1.845 92.756	2.124 166.81	2.455 306.08	2.848 572.71	3.271 1027.5
0.00055 1/ 1818	0.892 4.0332	0.963 5.5143	1.032 7.2927	1.119 10.161	1.243 15.624	1.435 28.179	1.680 54.049	1.935 97.285	2.228 174.96	2.575 321.02	2.987 600.67	3.430 1077.6
0.00060 1/ 1667	0.931 4.2127	1.006 5.7597	1.078 7.6172	1.169 10.613	1.299 16.319	1.499 29.433	1.755 56.453	2.022 101.61	2.327 182.74	2.689 335.30	3.120 627.38	3.583 1125.6
0.00065 1/ 1538	0.969 4.3848	1.047 5.9950	1.122 7.9284	1.217 11.047	1.352 16.986	1.560 30.635	1.827 58.759	2.104 105.76	2.422 190.20	2.799 348.99	3.248 653.01	3.729 1171.5
0.00070 1/ 1429	1.006 4.5504	1.087 6.2215	1.164 8.2278	1.263 11.464	1.403 17.627	1.619 31.792	1.895 60.978	2.184 109.76	2.513 197.38	2.905 362.17	3.370 677.66	3.870 1215.8
0.00075 1/ 1333	1.041 4.7102	1.125 6.4400	1.205 8.5167	1.307 11.867	1.452 18.246	1.676 32.908	1.962 63.119	2.260 113.61	2.601 204.31	3.007 374.88	3.489 701.45	4.006 1258.4
0.00080 1/ 1250	1.075 4.8648	1.162 6.6513	1.244 8.7962	1.350 12.256	1.500 18.845	1.731 33.988	2.026 65.189	2.334 117.34	2.687 211.01	3.105 387.18	3.603 724.46	4.137 1299.7
0.00085 1/ 1176	1.108 5.0146	1.197 6.8561	1.283 9.0671	1.391 12.634	1.546 19.425	1.784 35.034	2.089 67.196	2.406 120.95	2.769 217.51	3.201 399.10	3.714 746.76	4.264 1339.7
0.00090 1/ 1111	1.141 5.1601	1.232 7.0549	1.320 9.3300	1.432 13.000	1.591 19.988	1.836 36.050	2.149 69.145	2.476 124.46	2.850 223.82	3.294 410.67	3.822 768.41	4.388 1378.6
0.00095 1/ 1053	1.172 5.3015	1.266 7.2484	1.356 9.5858	1.471 13.356	1.634 20.536	1.886 37.038	2.208 71.040	2.544 127.87	2.928 229.95	3.384 421.93	3.927 789.47	4.508 1416.3
0.00100 1/ 1000	1.202 5.4393	1.299 7.4368	1.391 9.8350	1.509 13.704	1.677 21.070	1.935 38.001	2.266 72.886	2.610 131.19	3.004 235.93	3.472 432.89	4.029 809.98	4.625 1453.1
0.00110 1/ 909	1.261 5.7050	1.362 7.7999	1.459 10.315	1.583 14.373	1.759 22.099	2.030 39.856	2.376 76.445	2.737 137.59	3.151 247.44	3.641 454.02	4.225 849.52	4.851 1524.1
0.00120 1/ 833	1.317 5.9588	1.423 8.1469	1.524 10.774	1.653 15.012	1.837 23.082	2.120 41.629	2.482 79.845	2.859 143.71	3.291 258.45	3.803 474.22	4.413 887.30	5.067 1591.9
0.00130 1/ 769	1.371 6.2022	1.481 8.4797	1.586 11.214	1.721 15.625	1.912 24.024	2.207 43.329	2.583 83.106	2.976 149.58	3.425 269.00	3.958 493.58	4.593 923.54	5.274 1656.9
0.00140 1/ 714	1.423 6.4364	1.537 8.7999	1.646 11.638	1.786 16.215	1.984 24.932	2.290 44.965	2.681 86.244	3.088 155.23	3.554 279.16	4.108 512.22	4.767 958.41	5.473 1719.4
0.00150 1/ 667	1.473 6.6624	1.591 9.1089	1.704 12.046	1.849 16.785	2.054 25.807	2.370 46.544	2.775 89.271	3.197 160.68	3.679 288.96	4.252 530.20	4.934 992.05	5.665 1779.8
0.00160 1/ 625	1.521 6.8810	1.643 9.4078	1.760 12.441	1.909 17.335	2.121 26.654	2.448 48.071	2.866 92.200	3.302 165.95	3.800 298.44	4.392 547.59	5.096 1024.6	5.851 1838.1
0.00170 1/ 588	1.568 7.0929	1.694 9.6974	1.814 12.825	1.968 17.869	2.186 27.474	2.524 49.551	2.954 95.038	3.403 171.06	3.917 307.63	4.527 564.44	5.253 1056.1	6.031 1894.7
0.00180 1/ 556	1.613 7.2986	1.743 9.9787	1.867 13.196	2.025 18.387	2.250 28.271	2.597 50.988	3.040 97.794	3.502 176.02	4.030 316.54	4.658 580.81	5.405 1086.7	6.206 1949.7
0.00190 1/ 526	1.658 7.4987	1.791 10.252	1.918 13.558	2.081 18.891	2.311 29.046	2.668 52.385	3.123 100.47	3.598 180.84	4.141 325.22	4.786 596.73	5.553 1116.5	6.376 2003.1
0.00200 1/ 500	1.701 7.6935	1.837 10.519	1.968 13.911	2.135 19.382	2.371 29.801	2.737 53.746	3.204 103.08	3.691 185.54	4.248 333.67	4.910 612.23	5.697 1145.5	6.542 2055.1
0.00220 1/ 455	1.784 8.0692	1.927 11.032	2.064 14.590	2.239 20.328	2.487 31.255	2.871 56.370	3.361 108.12	3.871 194.60	4.456 349.96	5.150 642.12	5.976 1201.5	6.861 2155.5
0.00240 1/ 417	1.863 8.4281	2.013 11.523	2.156 15.239	2.339 21.232	2.598 32.646	2.999 58.877	3.510 112.93	4.044 203.26	4.654 365.52	5.379 670.67	6.241 1254.9	7.166 2251.3
0.00260 1/ 385	1.939 8.7724	2.095 11.994	2.244 15.861	2.434 22.100	2.704 33.979	3.121 61.282	3.654 117.54	4.209 211.56	4.844 380.45	5.598 698.06	6.496 1306.1	7.459 2343.2
0.00280 1/ 357	2.012 9.1036	2.174 12.446	2.329 16.460	2.526 22.934	2.806 35.262	3.239 63.596	3.792 121.98	4.368 219.54	5.027 394.81	5.810 724.42	6.741 1355.4	7.740 2431.7
	140	150	170	190	230	280	360	450	550	700	900	1100

θ_{medial} **for part-full circular pipes.**

k_s = 15.0 mm   S = 0.00050 to 0.00280

$k_s = 15.0$ mm
$S = 0.0030$ to 0.014

Water (or sewage) at 15°C;
full bore conditions.

ie hydraulic gradient =
1 in 333 to 1 in 71

velocities in ms^{-1}
discharges in m^{3}s^{-1}

Gradient	(Equivalent) Pipe diameters in m											
	2.400	2.700	3.000	3.400	4.000	5.000	6.400	8.000	10.00	12.60	16.00	20.00
0.00300 1/ 333	2.083 9.4232	2.250 12.883	2.410 17.038	2.615 23.739	2.905 36.500	3.353 65.828	3.925 126.26	4.521 227.25	5.203 408.67	6.014 749.85	6.978 1403.0	8.012 2517.1
0.00320 1/ 313	2.151 9.7324	2.324 13.306	2.489 17.597	2.700 24.518	3.000 37.697	3.463 67.987	4.053 130.40	4.669 234.70	5.374 422.08	6.211 774.44	7.207 1449.0	8.275 2599.6
0.00340 1/ 294	2.218 10.032	2.396 13.716	2.566 18.138	2.784 25.273	3.092 38.858	3.569 70.080	4.178 134.41	4.813 241.93	5.539 435.07	6.402 798.28	7.429 1493.6	8.530 2679.6
0.00360 1/ 278	2.282 10.323	2.465 14.113	2.640 18.664	2.864 26.006	3.182 39.984	3.673 72.112	4.299 138.31	4.953 248.94	5.700 447.68	6.588 821.42	7.644 1536.9	8.777 2757.3
0.00380 1/ 263	2.344 10.606	2.533 14.500	2.713 19.176	2.943 26.718	3.269 41.080	3.773 74.089	4.417 142.10	5.088 255.77	5.856 459.95	6.768 843.93	7.854 1579.1	9.017 2832.9
0.00400 1/ 250	2.405 10.881	2.598 14.877	2.783 19.674	3.019 27.413	3.354 42.148	3.871 76.014	4.532 145.79	5.220 262.41	6.008 471.90	6.944 865.86	8.058 1620.1	9.252 2906.5
0.00420 1/ 238	2.465 11.150	2.663 15.244	2.852 20.160	3.094 28.090	3.437 43.189	3.967 77.891	4.644 149.39	5.349 268.89	6.157 483.56	7.116 887.24	8.257 1660.1	9.480 2978.3
0.00440 1/ 227	2.523 11.413	2.725 15.603	2.919 20.635	3.167 28.751	3.518 44.205	4.060 79.724	4.753 152.91	5.475 275.22	6.302 494.94	7.283 908.12	8.451 1699.2	9.703 3048.4
0.00460 1/ 217	2.579 11.669	2.786 15.954	2.985 21.099	3.238 29.397	3.597 45.199	4.152 81.516	4.860 156.35	5.598 281.41	6.443 506.06	7.447 928.54	8.641 1737.4	9.921 3116.9
0.00480 1/ 208	2.635 11.920	2.846 16.297	3.049 21.552	3.308 30.030	3.674 46.171	4.241 83.270	4.965 159.71	5.719 287.46	6.582 516.95	7.607 948.51	8.827 1774.7	10.13 3183.9
0.00500 1/ 200	2.689 12.166	2.905 16.633	3.112 21.997	3.376 30.649	3.750 47.123	4.328 84.987	5.067 163.00	5.837 293.39	6.718 527.61	7.764 968.07	9.009 1811.3	10.34 3249.6
0.00550 1/ 182	2.821 12.760	3.047 17.445	3.264 23.071	3.541 32.145	3.933 49.424	4.540 89.136	5.314 170.96	6.122 307.71	7.046 553.36	8.143 1015.3	9.449 1899.7	10.85 3408.2
0.00600 1/ 167	2.946 13.328	3.182 18.221	3.409 24.097	3.698 33.575	4.108 51.622	4.742 93.100	5.551 178.56	6.394 321.39	7.359 577.97	8.505 1060.5	9.869 1984.2	11.33 3559.7
0.00650 1/ 154	3.066 13.872	3.312 18.966	3.548 25.081	3.849 34.946	4.276 53.730	4.935 96.902	5.777 185.85	6.655 334.52	7.659 601.57	8.852 1103.8	10.27 2065.3	11.79 3705.1
0.00700 1/ 143	3.182 14.396	3.437 19.682	3.682 26.028	3.994 36.265	4.437 55.758	5.122 100.56	5.995 192.87	6.906 347.15	7.949 624.28	9.186 1145.4	10.66 2143.2	12.24 3845.0
0.00750 1/ 133	3.294 14.901	3.558 20.372	3.811 26.942	4.135 37.538	4.593 57.716	5.301 104.09	6.206 199.64	7.149 359.33	8.228 646.19	9.509 1185.7	11.03 2218.4	12.67 3979.9
0.00800 1/ 125	3.402 15.390	3.675 21.041	3.936 27.825	4.270 38.770	4.744 59.609	5.475 107.50	6.409 206.19	7.383 371.12	8.497 667.39	9.821 1224.5	11.40 2291.2	13.08 4110.5
0.00850 1/ 118	3.507 15.864	3.788 21.688	4.058 28.682	4.402 39.963	4.890 61.444	5.644 110.81	6.607 212.54	7.610 382.54	8.759 687.93	10.12 1262.2	11.75 2361.7	13.49 4237.0
0.00900 1/ 111	3.608 16.324	3.898 22.317	4.175 29.514	4.529 41.122	5.031 63.225	5.807 114.03	6.798 218.70	7.831 393.63	9.013 707.88	10.42 1298.8	12.09 2430.2	13.88 4359.8
0.00950 1/ 105	3.707 16.771	4.005 22.929	4.290 30.322	4.653 42.249	5.169 64.958	5.966 117.15	6.985 224.69	8.046 404.42	9.260 727.27	10.70 1334.4	12.42 2496.8	14.26 4479.3
0.01000 1/ 100	3.804 17.207	4.109 23.525	4.401 31.110	4.774 43.346	5.303 66.646	6.121 120.20	7.166 230.53	8.255 414.93	9.501 746.17	10.98 1369.1	12.74 2561.7	14.63 4595.7
0.01100 1/ 91	3.989 18.047	4.309 24.673	4.616 32.629	5.007 45.462	5.562 69.899	6.420 126.06	7.516 241.78	8.658 435.18	9.964 782.59	11.52 1435.9	13.36 2686.7	15.34 4820.0
0.01200 1/ 83	4.167 18.849	4.501 25.770	4.821 34.080	5.230 47.484	5.810 73.007	6.706 131.67	7.850 252.53	9.043 454.53	10.41 817.39	12.03 1499.8	13.96 2806.2	16.02 5034.3
0.01300 1/ 77	4.337 19.619	4.685 26.823	5.018 35.472	5.444 49.423	6.047 75.989	6.980 137.05	8.171 262.85	9.412 473.09	10.83 850.77	12.52 1561.0	14.53 2920.8	16.68 5239.9
0.01400 1/ 71	4.500 20.360	4.862 27.835	5.208 36.811	5.649 51.289	6.275 78.858	7.243 142.22	8.479 272.77	9.767 490.95	11.24 882.89	12.99 1619.9	15.08 3031.0	17.31 5437.7
	150	160	180	210	240	300	390	480	600	750	950	1200

θ_{medial} **for part-full circular pipes.**

$k_s = 15.0$ mm $S = 0.0030$ to 0.014

k_s = 15.0 mm
S = 0.015 to 0.075

ie hydraulic gradient =
1 in 66 to 1 in 13

Water (or sewage) at 15°C;
full bore conditions.

velocities in ms^{-1}
discharges in m^{3}s^{-1}

Gradient	(Equivalent) Pipe diameters in m											
	2.400	2.700	3.000	3.400	4.000	5.000	6.400	8.000	10.00	12.60	16.00	20.00
0.01500	4.658	5.032	5.390	5.847	6.496	7.497	8.777	10.11	11.64	13.45	15.60	17.92
1/ 67	21.074	28.813	38.103	53.090	81.626	147.21	282.34	508.18	913.88	1676.8	3137.4	5628.6
0.01600	4.811	5.197	5.567	6.039	6.709	7.743	9.064	10.44	12.02	13.89	16.12	18.50
1/ 62	21.766	29.758	39.353	54.831	84.303	152.04	291.60	524.85	943.85	1731.8	3240.3	5813.2
0.01700	4.959	5.357	5.739	6.225	6.915	7.982	9.343	10.76	12.39	14.32	16.61	19.07
1/ 59	22.436	30.673	40.564	56.518	86.898	156.72	300.58	541.01	972.90	1785.1	3340.0	5992.1
0.01800	5.103	5.513	5.905	6.406	7.116	8.213	9.614	11.08	12.75	14.73	17.09	19.63
1/ 56	23.086	31.563	41.740	58.157	89.417	161.26	309.29	556.69	1001.1	1836.9	3436.9	6165.8
0.01900	5.243	5.664	6.067	6.581	7.311	8.438	9.878	11.38	13.10	15.14	17.56	20.16
1/ 53	23.719	32.428	42.884	59.751	91.868	165.68	317.77	571.95	1028.5	1887.2	3531.1	6334.8
0.02000	5.379	5.811	6.224	6.752	7.501	8.657	10.13	11.67	13.44	15.53	18.02	20.69
1/ 50	24.335	33.270	43.998	61.303	94.254	169.99	326.03	586.81	1055.3	1936.2	3622.8	6499.3
0.02200	5.642	6.095	6.528	7.082	7.867	9.080	10.63	12.24	14.09	16.29	18.90	21.70
1/ 45	25.523	34.894	46.146	64.296	98.855	178.28	341.94	615.45	1106.8	2030.7	3799.6	6816.6
0.02400	5.893	6.366	6.819	7.397	8.216	9.484	11.10	12.79	14.72	17.01	19.74	22.66
1/ 42	26.658	36.446	48.198	67.155	103.25	186.21	357.15	642.82	1156.0	2121.0	3968.6	7119.7
0.02600	6.133	6.625	7.097	7.699	8.552	9.871	11.56	13.31	15.32	17.71	20.54	23.59
1/ 38	27.747	37.935	50.166	69.897	107.47	193.82	371.73	669.07	1203.2	2207.6	4130.6	7410.4
0.02800	6.365	6.876	7.365	7.989	8.875	10.24	11.99	13.81	15.90	18.37	21.32	24.48
1/ 36	28.794	39.367	52.060	72.536	111.52	201.13	385.76	694.32	1248.6	2291.0	4286.6	7690.1
0.03000	6.588	7.117	7.624	8.270	9.186	10.60	12.41	14.30	16.46	19.02	22.07	25.34
1/ 33	29.805	40.748	53.888	75.082	115.44	208.19	399.30	718.69	1292.4	2371.4	4437.0	7960.1
0.03200	6.804	7.350	7.874	8.541	9.488	10.95	12.82	14.77	17.00	19.64	22.79	26.17
1/ 31	30.782	42.085	55.655	77.545	119.23	215.02	412.40	742.26	1334.8	2449.2	4582.5	8221.1
0.03400	7.014	7.577	8.116	8.804	9.780	11.29	13.21	15.22	17.52	20.25	23.49	26.97
1/ 29	31.730	43.380	57.368	79.932	122.89	221.64	425.09	765.11	1375.9	2524.5	4723.6	8474.1
0.03600	7.217	7.796	8.351	9.059	10.06	11.62	13.60	15.66	18.03	20.83	24.17	27.76
1/ 28	32.650	44.638	59.031	82.249	126.46	228.07	437.42	787.29	1415.8	2597.7	4860.5	8719.8
0.03800	7.415	8.010	8.580	9.307	10.34	11.93	13.97	16.09	18.52	21.40	24.84	28.52
1/ 26	33.545	45.861	60.649	84.503	129.92	234.31	449.40	808.87	1454.6	2668.9	4993.7	8958.8
0.04000	7.608	8.218	8.803	9.549	10.61	12.24	14.33	16.51	19.00	21.96	25.48	29.26
1/ 25	34.416	47.053	62.225	86.698	133.30	240.40	461.08	829.88	1492.4	2738.3	5123.4	9191.5
0.04200	7.796	8.421	9.020	9.785	10.87	12.55	14.69	16.92	19.47	22.50	26.11	29.98
1/ 24	35.266	48.215	63.762	88.840	136.59	246.34	472.46	850.38	1529.2	2805.9	5250.0	9418.5
0.04400	7.979	8.619	9.233	10.02	11.13	12.84	15.03	17.32	19.93	23.03	26.73	30.69
1/ 23	36.096	49.350	65.262	90.930	139.81	252.14	483.58	870.39	1565.2	2871.9	5373.5	9640.1
0.04600	8.158	8.813	9.440	10.24	11.38	13.13	15.37	17.71	20.38	23.55	27.33	31.38
1/ 22	36.907	50.459	66.729	92.974	142.95	257.80	494.45	889.95	1600.4	2936.5	5494.3	9856.8
0.04800	8.334	9.002	9.643	10.46	11.62	13.41	15.70	18.09	20.82	24.06	27.91	32.05
1/ 21	37.701	51.544	68.164	94.974	146.02	263.35	505.09	909.09	1634.8	2999.6	5612.5	10069
0.05000	8.506	9.188	9.842	10.68	11.86	13.69	16.02	18.46	21.24	24.55	28.49	32.71
1/ 20	38.479	52.607	69.570	96.932	149.03	268.78	515.50	927.84	1668.5	3061.5	5728.2	10276
0.05500	8.921	9.637	10.32	11.20	12.44	14.36	16.81	19.36	22.28	25.75	29.88	34.31
1/ 18	40.357	55.175	72.966	101.66	156.31	281.90	540.67	973.13	1750.0	3210.9	6007.8	10778
0.06000	9.318	10.07	10.78	11.70	12.99	15.00	17.55	20.22	23.27	26.90	31.21	35.83
1/ 17	42.152	57.629	76.211	106.18	163.26	294.43	564.71	1016.4	1827.8	3353.7	6274.9	11257
0.06500	9.698	10.48	11.22	12.17	13.52	15.61	18.27	21.05	24.22	27.99	32.48	37.30
1/ 15	43.873	59.982	79.323	110.52	169.93	306.46	587.77	1057.9	1902.4	3490.6	6531.2	11717
0.07000	10.06	10.87	11.65	12.63	14.03	16.20	18.96	21.84	25.14	29.05	33.71	38.70
1/ 14	45.529	62.246	82.317	114.69	176.34	318.03	609.96	1097.8	1974.3	3622.4	6777.7	12159
0.07500	10.42	11.25	12.05	13.08	14.53	16.77	19.63	22.61	26.02	30.07	34.89	40.06
1/ 13	47.127	64.431	85.206	118.72	182.53	329.19	631.36	1136.4	2043.6	3749.5	7015.6	12586
	150	170	190	210	250	320	400	500	650	800	1000	1300

θ_{medial} for part-full circular pipes.

k_s = 15.0 mm   S = 0.0015 to 0.075

$k_s = 30.0$ mm
$S = 0.00010$ to 0.00048

Water (or sewage) at 15°C;
full bore conditions

ie hydraulic gradient =
1 in 10000 to 1 in 2083

velocities in ms^{-1}
discharges in m^{3}s^{-1}

| Gradient | (Equivalent) Pipe diameters in m | | | | | | | | | | | |
	2.400	2.700	3.000	3.400	4.000	5.000	6.400	8.000	10.00	12.60	16.00	20.00
0.00010	0.339	0.367	0.394	0.428	0.477	0.552	0.649	0.750	0.865	1.003	1.167	1.343
1/10000	1.5327	2.1002	2.7829	3.8860	5.9914	10.844	20.875	37.689	67.975	125.08	234.67	422.03
0.00011	0.355	0.385	0.413	0.449	0.500	0.579	0.681	0.786	0.908	1.052	1.224	1.409
1/ 9091	1.6076	2.2028	2.9188	4.0758	6.2840	11.374	21.894	39.530	71.294	131.18	246.13	442.63
0.00012	0.371	0.402	0.431	0.469	0.522	0.605	0.711	0.821	0.948	1.099	1.279	1.472
1/ 8333	1.6791	2.3008	3.0487	4.2572	6.5636	11.880	22.868	41.288	74.466	137.02	257.08	462.32
0.00013	0.386	0.418	0.449	0.488	0.544	0.630	0.740	0.855	0.987	1.144	1.331	1.532
1/ 7692	1.7477	2.3949	3.1732	4.4312	6.8317	12.365	23.803	42.975	77.507	142.61	267.58	481.20
0.00014	0.401	0.434	0.466	0.506	0.564	0.654	0.768	0.887	1.024	1.187	1.381	1.590
1/ 7143	1.8138	2.4853	3.2931	4.5985	7.0898	12.832	24.702	44.598	80.434	148.00	277.68	499.37
0.00015	0.415	0.449	0.482	0.524	0.584	0.676	0.795	0.918	1.060	1.229	1.430	1.645
1/ 6667	1.8775	2.5726	3.4088	4.7600	7.3387	13.283	25.569	46.163	83.258	153.20	287.43	516.90
0.00016	0.429	0.464	0.498	0.541	0.603	0.699	0.821	0.949	1.095	1.269	1.476	1.699
1/ 6250	1.9391	2.6570	3.5206	4.9162	7.5796	13.718	26.408	47.678	85.989	158.22	296.86	533.85
0.00017	0.442	0.478	0.513	0.558	0.622	0.720	0.846	0.978	1.129	1.308	1.522	1.752
1/ 5882	1.9988	2.7389	3.6290	5.0676	7.8129	14.141	27.221	49.146	88.636	163.09	305.99	550.29
0.00018	0.455	0.492	0.528	0.574	0.640	0.741	0.871	1.006	1.161	1.346	1.566	1.802
1/ 5556	2.0568	2.8183	3.7343	5.2146	8.0396	14.551	28.010	50.571	91.207	167.82	314.87	566.24
0.00019	0.467	0.506	0.543	0.590	0.657	0.761	0.895	1.034	1.193	1.383	1.609	1.852
1/ 5263	2.1132	2.8956	3.8367	5.3576	8.2600	14.950	28.778	51.957	93.707	172.42	323.50	581.76
0.00020	0.479	0.519	0.557	0.605	0.674	0.781	0.918	1.061	1.224	1.419	1.651	1.900
1/ 5000	2.1681	2.9709	3.9364	5.4968	8.4747	15.338	29.526	53.307	96.142	176.90	331.90	596.88
0.00022	0.503	0.544	0.584	0.635	0.707	0.819	0.963	1.112	1.284	1.488	1.731	1.993
1/ 4545	2.2740	3.1160	4.1287	5.7653	8.8885	16.087	30.967	55.910	100.84	185.54	348.11	626.01
0.00024	0.525	0.568	0.610	0.663	0.739	0.856	1.005	1.162	1.341	1.554	1.808	2.081
1/ 4167	2.3752	3.2546	4.3124	6.0218	9.2839	16.803	32.345	58.397	105.32	193.79	363.59	653.86
0.00026	0.546	0.592	0.635	0.690	0.769	0.891	1.047	1.209	1.396	1.618	1.882	2.166
1/ 3846	2.4722	3.3876	4.4885	6.2678	9.6631	17.489	33.666	60.782	109.62	201.70	378.44	680.56
0.00028	0.567	0.614	0.659	0.716	0.798	0.924	1.086	1.255	1.448	1.679	1.953	2.248
1/ 3571	2.5656	3.5155	4.6581	6.5045	10.028	18.150	34.937	63.077	113.76	209.32	392.72	706.25
0.00030	0.587	0.636	0.682	0.742	0.826	0.957	1.124	1.299	1.499	1.738	2.022	2.327
1/ 3333	2.6557	3.6390	4.8216	6.7329	10.380	18.787	36.164	65.291	117.75	216.67	406.51	731.04
0.00032	0.606	0.656	0.705	0.766	0.853	0.988	1.161	1.342	1.548	1.795	2.088	2.403
1/ 3125	2.7429	3.7584	4.9798	6.9538	10.721	19.403	37.350	67.433	121.62	223.77	419.84	755.02
0.00034	0.625	0.677	0.726	0.789	0.879	1.019	1.197	1.383	1.596	1.850	2.152	2.477
1/ 2941	2.8273	3.8741	5.1331	7.1678	11.051	20.001	38.500	69.509	125.36	230.66	432.77	778.26
0.00036	0.643	0.696	0.747	0.812	0.905	1.048	1.231	1.423	1.642	1.904	2.215	2.549
1/ 2778	2.9093	3.9865	5.2820	7.3757	11.371	20.581	39.616	71.524	129.00	237.35	445.32	800.83
0.00038	0.661	0.715	0.768	0.835	0.930	1.077	1.265	1.462	1.687	1.956	2.276	2.619
1/ 2632	2.9891	4.0957	5.4268	7.5779	11.683	21.145	40.702	73.485	132.53	243.86	457.52	822.78
0.00040	0.678	0.734	0.788	0.856	0.954	1.105	1.298	1.500	1.731	2.007	2.335	2.687
1/ 2500	3.0668	4.2022	5.5679	7.7749	11.987	21.694	41.760	75.394	135.97	250.19	469.41	844.15
0.00042	0.695	0.752	0.807	0.877	0.977	1.132	1.330	1.537	1.774	2.056	2.392	2.753
1/ 2381	3.1425	4.3060	5.7054	7.9669	12.283	22.230	42.791	77.256	139.33	256.37	481.00	865.00
0.00044	0.711	0.770	0.826	0.898	1.000	1.159	1.361	1.573	1.816	2.104	2.449	2.818
1/ 2273	3.2165	4.4074	5.8397	8.1545	12.572	22.753	43.799	79.075	142.61	262.41	492.32	885.36
0.00046	0.727	0.787	0.845	0.918	1.023	1.185	1.392	1.609	1.857	2.152	2.504	2.882
1/ 2174	3.2888	4.5065	5.9710	8.3378	12.854	23.265	44.783	80.852	145.82	268.30	503.39	905.26
0.00048	0.743	0.804	0.863	0.938	1.045	1.210	1.422	1.643	1.897	2.198	2.558	2.944
1/ 2083	3.3596	4.6034	6.0995	8.5172	13.131	23.765	45.747	82.592	148.96	274.08	514.22	924.73
	70	80	85	100	120	150	190	230	290	370	460	600

θ_{medial} for part-full circular pipes.

$k_s = 30.0$ mm $S = 0.00010$ to 0.00048

$k_s = 30.0$ mm
$S = 0.00050$ to 0.00280

Water (or sewage) at 15°C; full bore conditions.

ie hydraulic gradient = 1 in 2000 to 1 in 357

velocities in ms⁻¹
discharges in m³s⁻¹

Gradient	(Equivalent) Pipe diameters in m											
	2.400	2.700	3.000	3.400	4.000	5.000	6.400	8.000	10.00	12.60	16.00	20.00
0.00050 1/ 2000	0.758 3.4289	0.821 4.6984	0.881 6.2253	0.957 8.6929	1.066 13.402	1.235 24.255	1.451 46.690	1.677 84.295	1.936 152.03	2.243 279.73	2.610 524.82	3.004 943.80
0.00055 1/ 1818	0.795 3.5963	0.861 4.9278	0.924 6.5293	1.004 9.1173	1.119 14.056	1.296 25.440	1.522 48.970	1.759 88.410	2.030 159.45	2.353 293.38	2.738 550.44	3.151 989.87
0.00060 1/ 1667	0.830 3.7563	0.899 5.1470	0.965 6.8197	1.049 9.5228	1.168 14.681	1.353 26.571	1.590 51.148	1.837 92.342	2.120 166.54	2.458 306.43	2.859 574.92	3.291 1033.9
0.00065 1/ 1538	0.864 3.9098	0.936 5.3572	1.004 7.0983	1.092 9.9118	1.216 15.281	1.409 27.656	1.655 53.236	1.912 96.113	2.207 173.34	2.558 318.94	2.976 598.40	3.425 1076.1
0.00070 1/ 1429	0.897 4.0574	0.971 5.5595	1.042 7.3663	1.133 10.286	1.262 15.858	1.462 28.701	1.717 55.246	1.984 99.742	2.290 179.89	2.654 330.99	3.089 620.99	3.555 1116.7
0.00075 1/ 1333	0.928 4.1999	1.005 5.7547	1.079 7.6249	1.173 10.647	1.306 16.415	1.513 29.708	1.778 57.186	2.054 103.24	2.371 186.20	2.748 342.60	3.197 642.79	3.679 1155.9
0.00080 1/ 1250	0.959 4.3376	1.038 5.9435	1.114 7.8750	1.211 10.996	1.349 16.953	1.563 30.683	1.836 59.062	2.121 106.63	2.449 192.31	2.838 353.84	3.302 663.87	3.800 1193.9
0.00085 1/ 1176	0.988 4.4712	1.070 6.1265	1.148 8.1175	1.248 11.335	1.391 17.475	1.611 31.627	1.892 60.880	2.187 109.91	2.524 198.23	2.925 364.73	3.403 684.30	3.917 1230.6
0.00090 1/ 1111	1.017 4.6008	1.101 6.3041	1.182 8.3529	1.285 11.664	1.431 17.982	1.657 32.544	1.947 62.645	2.250 113.10	2.597 203.97	3.010 375.31	3.502 704.14	4.031 1266.3
0.00095 1/ 1053	1.045 4.7270	1.131 6.4769	1.214 8.5818	1.320 11.983	1.470 18.475	1.703 33.436	2.001 64.362	2.312 116.20	2.668 209.56	3.092 385.59	3.598 723.44	4.141 1301.0
0.00100 1/ 1000	1.072 4.8498	1.161 6.6453	1.246 8.8048	1.354 12.295	1.508 18.955	1.747 34.305	2.053 66.034	2.372 119.22	2.738 215.01	3.173 395.61	3.692 742.23	4.249 1334.8
0.00110 1/ 909	1.124 5.0866	1.217 6.9697	1.306 9.2347	1.420 12.895	1.582 19.880	1.832 35.980	2.153 69.258	2.488 125.04	2.871 225.50	3.328 414.92	3.872 778.46	4.456 1399.9
0.00120 1/ 833	1.174 5.3128	1.271 7.2797	1.365 9.6454	1.483 13.468	1.652 20.764	1.914 37.580	2.249 72.338	2.598 130.60	2.999 235.53	3.476 433.37	4.044 813.08	4.654 1462.2
0.00130 1/ 769	1.222 5.5298	1.323 7.5770	1.420 10.039	1.544 14.019	1.720 21.612	1.992 39.114	2.340 75.292	2.704 135.93	3.121 245.15	3.618 451.07	4.209 846.29	4.844 1521.9
0.00140 1/ 714	1.269 5.7386	1.373 7.8631	1.474 10.418	1.602 14.548	1.785 22.428	2.067 40.591	2.429 78.134	2.806 141.06	3.239 254.41	3.754 468.10	4.368 878.23	5.027 1579.3
0.00150 1/ 667	1.313 5.9401	1.422 8.1391	1.526 10.784	1.659 15.059	1.847 23.215	2.140 42.016	2.514 80.877	2.905 146.01	3.353 263.34	3.886 484.53	4.521 909.06	5.204 1634.8
0.00160 1/ 625	1.356 6.1349	1.468 8.4061	1.576 11.138	1.713 15.552	1.908 23.977	2.210 43.394	2.597 83.530	3.000 150.80	3.463 271.97	4.013 500.42	4.670 938.88	5.374 1688.4
0.00170 1/ 588	1.398 6.3238	1.513 8.6649	1.624 11.481	1.766 16.031	1.967 24.715	2.278 44.730	2.676 86.101	3.092 155.45	3.569 280.34	4.137 515.82	4.813 967.77	5.540 1740.4
0.00180 1/ 556	1.438 6.5072	1.557 8.9161	1.671 11.814	1.817 16.496	2.024 25.432	2.344 46.027	2.754 88.597	3.182 159.95	3.673 288.47	4.257 530.78	4.953 995.83	5.700 1790.8
0.00190 1/ 526	1.478 6.6855	1.600 9.1605	1.717 12.137	1.867 16.948	2.079 26.129	2.408 47.288	2.830 91.025	3.269 164.34	3.774 296.38	4.373 545.33	5.089 1023.1	5.857 1839.9
0.00200 1/ 500	1.516 6.8592	1.642 9.3985	1.762 12.453	1.915 17.389	2.133 26.808	2.471 48.517	2.903 93.390	3.354 168.61	3.872 304.08	4.487 559.49	5.221 1049.7	6.009 1887.7
0.00220 1/ 455	1.590 7.1941	1.722 9.8573	1.848 13.061	2.009 18.237	2.237 28.116	2.592 50.885	3.045 97.949	3.518 176.84	4.061 318.92	4.706 586.80	5.476 1100.9	6.302 1979.8
0.00240 1/ 417	1.661 7.5140	1.798 10.296	1.930 13.642	2.098 19.048	2.337 29.367	2.707 53.148	3.180 102.30	3.674 184.70	4.241 333.10	4.915 612.90	5.719 1149.9	6.582 2067.9
0.00260 1/ 385	1.729 7.8209	1.872 10.716	2.009 14.199	2.184 19.826	2.432 30.566	2.817 55.319	3.310 106.48	3.825 192.24	4.414 346.71	5.116 637.93	5.953 1196.9	6.851 2152.3
0.00280 1/ 357	1.794 8.1162	1.942 11.121	2.085 14.735	2.266 20.575	2.524 31.720	2.924 57.407	3.435 110.50	3.969 199.50	4.581 359.80	5.309 662.01	6.177 1242.0	7.110 2233.6
	75	85	90	100	120	150	200	250	310	390	490	600

θ_{medial} for part-full circular pipes.

$k_s = 30.0$ mm $S = 0.00050$ to 0.00280

k_s = 30.0 mm
S = 0.0030 to 0.014

Water (or sewage) at 15°C;
full bore conditions.

ie hydraulic gradient =
1 in 333 to 1 in 71

velocities in ms⁻¹
discharges in m³s⁻¹

velocities in ms^{-1}
discharges in m^3s^{-1}

Gradient	(Equivalent) Pipe diameters in m											
	2.400	2.700	3.000	3.400	4.000	5.000	6.400	8.000	10.00	12.60	16.00	20.00
0.00300 1/ 333	1.857 8.4011	2.010 11.511	2.158 15.252	2.346 21.297	2.613 32.833	3.026 59.422	3.556 114.38	4.108 206.50	4.742 372.42	5.496 685.25	6.394 1285.6	7.359 2312.0
0.00320 1/ 313	1.918 8.6767	2.076 11.889	2.228 15.752	2.423 21.996	2.698 33.910	3.126 61.371	3.672 118.13	4.243 213.28	4.897 384.64	5.676 707.72	6.604 1327.8	7.601 2387.8
0.00340 1/ 294	1.977 8.9437	2.140 12.255	2.297 16.237	2.497 22.673	2.782 34.954	3.222 63.260	3.785 121.77	4.374 219.84	5.048 396.48	5.851 729.50	6.807 1368.7	7.835 2461.3
0.00360 1/ 278	2.034 9.2031	2.202 12.610	2.364 16.708	2.570 23.330	2.862 35.967	3.315 65.094	3.895 125.30	4.500 226.21	5.194 407.97	6.020 750.65	7.005 1408.3	8.062 2532.6
0.00380 1/ 263	2.090 9.4553	2.263 12.956	2.428 17.166	2.640 23.970	2.941 36.953	3.406 66.878	4.002 128.73	4.624 232.41	5.337 419.15	6.185 771.22	7.196 1446.9	8.283 2602.1
0.00400 1/ 250	2.144 9.7010	2.322 13.292	2.492 17.612	2.709 24.592	3.017 37.913	3.495 68.616	4.106 132.08	4.744 238.45	5.475 430.04	6.346 791.26	7.383 1484.5	8.498 2669.7
0.00420 1/ 238	2.197 9.9406	2.379 13.621	2.553 18.047	2.776 25.200	3.092 38.850	3.581 70.310	4.207 135.34	4.861 244.34	5.611 440.66	6.503 810.80	7.566 1521.2	8.708 2735.6
0.00440 1/ 227	2.249 10.175	2.435 13.941	2.613 18.472	2.841 25.793	3.164 39.764	3.665 71.965	4.306 138.52	4.975 250.09	5.743 451.03	6.656 829.88	7.744 1557.0	8.913 2800.0
0.00460 1/ 217	2.300 10.403	2.490 14.254	2.672 18.887	2.905 26.372	3.235 40.658	3.748 73.583	4.403 141.64	5.087 255.71	5.872 461.17	6.805 848.53	7.918 1592.0	9.113 2862.9
0.00480 1/ 208	2.349 10.627	2.543 14.561	2.729 19.293	2.967 26.940	3.305 41.532	3.828 75.165	4.498 144.69	5.197 261.21	5.998 471.09	6.952 866.78	8.088 1626.2	9.309 2924.5
0.00500 1/ 200	2.398 10.846	2.596 14.861	2.786 19.691	3.028 27.495	3.373 42.389	3.907 76.716	4.590 147.67	5.304 266.60	6.122 480.80	7.095 884.66	8.255 1659.8	9.501 2984.8
0.00550 1/ 182	2.515 11.376	2.722 15.587	2.922 20.652	3.176 28.837	3.538 44.458	4.098 80.460	4.814 154.88	5.563 279.61	6.421 504.27	7.441 927.84	8.658 1740.8	9.965 3130.5
0.00600 1/ 167	2.626 11.882	2.843 16.280	3.052 21.570	3.317 30.120	3.695 46.435	4.280 84.038	5.028 161.76	5.810 292.04	6.706 526.70	7.772 969.10	9.043 1818.2	10.41 3269.7
0.00650 1/ 154	2.734 12.367	2.960 16.945	3.176 22.451	3.453 31.350	3.846 48.331	4.455 87.470	5.234 168.37	6.047 303.97	6.980 548.21	8.089 1008.7	9.412 1892.4	10.83 3403.2
0.00700 1/ 143	2.837 12.834	3.071 17.585	3.296 23.299	3.583 32.533	3.991 50.156	4.623 90.772	5.431 174.73	6.276 315.45	7.243 568.90	8.395 1046.7	9.767 1963.9	11.24 3531.6
0.00750 1/ 133	2.936 13.284	3.179 18.202	3.412 24.117	3.709 33.675	4.131 51.916	4.785 93.958	5.622 180.86	6.496 326.52	7.498 588.87	8.689 1083.5	10.11 2032.8	11.64 3655.6
0.00800 1/ 125	3.033 13.720	3.283 18.799	3.524 24.908	3.831 34.780	4.267 53.619	4.942 97.040	5.806 186.79	6.709 337.23	7.744 608.18	8.974 1119.0	10.44 2099.5	12.02 3775.5
0.00850 1/ 118	3.126 14.142	3.384 19.377	3.632 25.674	3.949 35.850	4.398 55.269	5.094 100.03	5.985 192.54	6.915 347.61	7.982 626.90	9.251 1153.5	10.76 2164.1	12.39 3891.7
0.00900 1/ 111	3.217 14.552	3.483 19.939	3.737 26.419	4.063 36.890	4.526 56.872	5.242 102.93	6.159 198.12	7.116 357.68	8.213 645.08	9.519 1186.9	11.08 2226.8	12.75 4004.5
0.00950 1/ 105	3.305 14.951	3.578 20.486	3.840 27.143	4.174 37.901	4.650 58.430	5.386 105.75	6.327 203.55	7.311 367.49	8.438 662.75	9.780 1219.4	11.38 2287.8	13.10 4114.3
0.01000 1/ 100	3.391 15.339	3.671 21.018	3.940 27.848	4.283 38.885	4.771 59.948	5.526 108.49	6.492 208.84	7.501 377.03	8.658 679.97	10.03 1251.1	11.67 2347.3	13.44 4221.1
0.01100 1/ 91	3.556 16.088	3.850 22.044	4.132 29.207	4.492 40.784	5.003 62.874	5.795 113.79	6.809 219.03	7.867 395.44	9.080 713.16	10.52 1312.2	12.24 2461.9	14.09 4427.2
0.01200 1/ 83	3.714 16.804	4.021 23.024	4.316 30.506	4.692 42.597	5.226 65.670	6.053 118.85	7.111 228.77	8.217 413.02	9.484 744.87	10.99 1370.5	12.79 2571.3	14.72 4624.0
0.01300 1/ 77	3.866 17.490	4.186 23.964	4.492 31.752	4.883 44.337	5.439 68.352	6.300 123.70	7.402 238.11	8.552 429.89	9.871 775.29	11.44 1426.5	13.31 2676.3	15.32 4812.9
0.01400 1/ 71	4.012 18.150	4.344 24.869	4.662 32.951	5.068 46.010	5.645 70.932	6.538 128.37	7.681 247.10	8.875 446.11	10.24 804.56	11.87 1480.3	13.81 2777.4	15.90 4994.5
	75	85	95	110	130	160	200	250	320	400	500	630

θ_{medial} for part-full circular pipes.

k_s = 30.0 mm   S = 0.0030 to 0.014

$k_s = 30.0$ mm
$S = 0.015$ to 0.075

ie hydraulic gradient =
1 in 66 to 1 in 13

Water (or sewage) at 15°C;
full bore conditions.

velocities in ms^{-1}
discharges in m^{3}s^{-1}

| Gradient | (Equivalent) Pipe diameters in m | | | | | | | | | | | |
	2.400	2.700	3.000	3.400	4.000	5.000	6.400	8.000	10.00	12.60	16.00	20.00
0.01500 1/ 67	4.153 18.787	4.496 25.742	4.825 34.107	5.246 47.625	5.843 73.422	6.768 132.88	7.951 255.78	9.187 461.77	10.60 832.80	12.29 1532.3	14.30 2874.8	16.46 5169.8
0.01600 1/ 62	4.289 19.403	4.643 26.586	4.983 35.226	5.418 49.187	6.034 75.830	6.989 137.24	8.212 264.17	9.488 476.92	10.95 860.11	12.69 1582.6	14.77 2969.1	17.00 5339.4
0.01700 1/ 59	4.421 20.001	4.786 27.405	5.137 36.310	5.584 50.701	6.220 78.164	7.205 141.46	8.464 272.30	9.780 491.60	11.29 886.58	13.08 1631.3	15.22 3060.5	17.52 5503.7
0.01800 1/ 56	4.549 20.580	4.925 28.199	5.286 37.363	5.746 52.171	6.400 80.430	7.413 145.56	8.710 280.19	10.06 505.85	11.62 912.28	13.46 1678.6	15.66 3149.2	18.03 5663.3
0.01900 1/ 53	4.674 21.144	5.060 28.972	5.431 38.387	5.904 53.601	6.576 82.634	7.617 149.55	8.948 287.87	10.34 519.71	11.93 937.28	13.83 1724.6	16.09 3235.5	18.52 5818.5
0.02000 1/ 50	4.795 21.694	5.192 29.725	5.572 39.384	6.057 54.993	6.747 84.781	7.814 153.44	9.181 295.35	10.61 533.21	12.24 961.63	14.19 1769.4	16.51 3319.6	19.00 5969.6
0.02200 1/ 45	5.029 22.753	5.445 31.175	5.844 41.306	6.353 57.678	7.076 88.919	8.196 160.93	9.629 309.76	11.13 559.24	12.84 1008.6	14.88 1855.7	17.32 3481.6	19.93 6261.0
0.02400 1/ 42	5.253 23.764	5.687 32.562	6.103 43.143	6.635 60.243	7.391 92.873	8.560 168.08	10.06 323.54	11.62 584.10	13.41 1053.4	15.54 1938.2	18.09 3636.4	20.82 6539.4
0.02600 1/ 38	5.468 24.735	5.919 33.891	6.353 44.905	6.906 62.703	7.692 96.666	8.910 174.95	10.47 336.75	12.09 607.96	13.96 1096.4	16.18 2017.4	18.82 3784.9	21.67 6806.5
0.02800 1/ 36	5.674 25.669	6.143 35.171	6.593 46.600	7.167 65.070	7.983 100.32	9.246 181.55	10.86 349.46	12.55 630.91	14.49 1137.8	16.79 2093.5	19.54 3927.8	22.48 7063.4
0.03000 1/ 33	5.873 26.570	6.358 36.405	6.824 48.236	7.418 67.354	8.263 103.84	9.571 187.92	11.24 361.73	12.99 653.05	15.00 1177.8	17.38 2167.0	20.22 4065.7	23.27 7311.3
0.03200 1/ 31	6.066 27.441	6.567 37.599	7.048 49.818	7.662 69.563	8.534 107.24	9.885 194.09	11.61 373.59	13.42 674.47	15.49 1216.4	17.95 2238.1	20.88 4199.0	24.04 7551.1
0.03400 1/ 29	6.252 28.286	6.769 38.757	7.265 51.351	7.898 71.704	8.797 110.54	10.19 200.06	11.97 385.09	13.83 695.23	15.96 1253.8	18.50 2307.0	21.53 4328.2	24.78 7783.5
0.03600 1/ 28	6.434 29.106	6.965 39.880	7.475 52.840	8.127 73.782	9.052 113.75	10.48 205.86	12.32 396.25	14.23 715.38	16.43 1290.2	19.04 2373.9	22.15 4453.7	25.49 8009.1
0.03800 1/ 26	6.610 29.903	7.156 40.973	7.680 54.288	8.349 75.804	9.300 116.86	10.77 211.50	12.66 407.11	14.62 734.99	16.88 1325.5	19.56 2438.9	22.76 4575.8	26.19 8228.6
0.04000 1/ 25	6.782 30.680	7.342 42.038	7.880 55.698	8.566 77.774	9.541 119.90	11.05 217.00	12.98 417.69	15.00 754.08	17.32 1360.0	20.07 2502.3	23.35 4694.6	26.87 8442.4
0.04200 1/ 24	6.949 31.438	7.523 43.076	8.074 57.074	8.778 79.694	9.777 122.86	11.32 222.35	13.30 428.00	15.37 772.70	17.74 1393.6	20.56 2564.1	23.93 4810.6	27.54 8650.9
0.04400 1/ 23	7.113 32.178	7.700 44.090	8.264 58.417	8.984 81.570	10.01 125.75	11.59 227.59	13.62 438.08	15.73 790.89	18.16 1426.3	21.05 2624.4	24.49 4923.8	28.18 8854.5
0.04600 1/ 22	7.273 32.901	7.874 45.080	8.450 59.730	9.186 83.403	10.23 128.58	11.85 232.70	13.92 447.92	16.09 808.66	18.57 1458.4	21.52 2683.4	25.04 5034.4	28.82 9053.5
0.04800 1/ 21	7.429 33.609	8.043 46.050	8.632 61.014	9.384 85.197	10.45 131.34	12.11 237.71	14.22 457.56	16.43 826.06	18.97 1489.8	21.98 2741.1	25.58 5142.7	29.44 9248.2
0.05000 1/ 20	7.582 34.302	8.209 47.000	8.810 62.273	9.577 86.954	10.67 134.05	12.36 242.61	14.52 466.99	16.77 843.09	19.36 1520.5	22.44 2797.6	26.11 5248.8	30.04 9438.9
0.05500 1/ 18	7.952 35.976	8.609 49.294	9.240 65.312	10.04 91.198	11.19 140.60	12.96 254.45	15.22 489.78	17.59 884.24	20.30 1594.7	23.53 2934.2	27.38 5505.0	31.51 9899.6
0.06000 1/ 17	8.306 37.576	8.992 51.486	9.651 68.216	10.49 95.253	11.69 146.85	13.54 265.77	15.90 511.56	18.37 923.56	21.21 1665.6	24.58 3064.6	28.60 5749.7	32.91 10340
0.06500 1/ 15	8.645 39.110	9.359 53.588	10.04 71.002	10.92 99.143	12.16 152.84	14.09 276.62	16.55 532.45	19.12 961.27	22.07 1733.6	25.58 3189.8	29.76 5984.5	34.26 10762
0.07000 1/ 14	8.972 40.586	9.713 55.611	10.42 73.682	11.33 102.89	12.62 158.61	14.62 287.06	17.18 552.55	19.85 997.56	22.91 1799.1	26.55 3310.2	30.89 6210.4	35.55 11168
0.07500 1/ 13	9.286 42.011	10.05 57.563	10.79 76.269	11.73 106.50	13.07 164.18	15.13 297.14	17.78 571.95	20.54 1032.6	23.71 1862.2	27.48 3426.4	31.97 6428.4	36.80 11560
	80	90	95	110	130	160	210	260	320	410	500	650

θ_{medial} for part-full circular pipes.

$k_s = 30.0$ mm $S = 0.0015$ to 0.075

A45

k_s = 60.0 mm
S = 0.00010 to 0.00048

ie hydraulic gradient =
1 in 10000 to 1 in 2083

Water (or sewage) at 15°C;
full bore conditions.

velocities in ms^{-1}
discharges in m^{3}s^{-1}

Gradient	\(Equivalent\) Pipe diameters in m											
	2.400	2.700	3.000	3.400	4.000	5.000	6.400	8.000	10.00	12.60	16.00	20.00
0.00010 1/10000	0.298 1.3465	0.323 1.8502	0.348 2.4575	0.379 3.4411	0.424 5.3232	0.493 9.6764	0.582 18.710	0.675 33.906	0.781 61.365	0.909 113.29	1.061 213.26	1.224 384.61
0.00011 1/ 9091	0.312 1.4123	0.339 1.9406	0.365 2.5775	0.398 3.6091	0.444 5.5831	0.517 10.149	0.610 19.623	0.707 35.562	0.819 64.361	0.953 118.83	1.112 223.67	1.284 403.39
0.00012 1/ 8333	0.326 1.4751	0.354 2.0269	0.381 2.6921	0.415 3.7696	0.464 5.8314	0.540 10.600	0.637 20.496	0.739 37.143	0.856 67.223	0.995 124.11	1.162 233.62	1.341 421.33
0.00013 1/ 7692	0.339 1.5354	0.368 2.1097	0.396 2.8021	0.432 3.9236	0.483 6.0696	0.562 11.033	0.663 21.333	0.769 38.660	0.891 69.969	1.036 129.18	1.209 243.16	1.396 438.54
0.00014 1/ 7143	0.352 1.5933	0.382 2.1894	0.411 2.9079	0.448 4.0718	0.501 6.2988	0.583 11.450	0.688 22.139	0.798 40.120	0.925 72.610	1.075 134.06	1.255 252.34	1.449 455.09
0.00015 1/ 6667	0.365 1.6493	0.396 2.2663	0.426 3.0100	0.464 4.2147	0.519 6.5199	0.604 11.852	0.712 22.916	0.826 41.528	0.957 75.159	1.113 138.76	1.299 261.20	1.499 471.07
0.00016 1/ 6250	0.377 1.7034	0.409 2.3406	0.440 3.1088	0.479 4.3530	0.536 6.7338	0.623 12.241	0.736 23.668	0.853 42.890	0.988 77.625	1.149 143.31	1.342 269.76	1.549 486.52
0.00017 1/ 5882	0.388 1.7559	0.421 2.4127	0.453 3.2045	0.494 4.4870	0.552 6.9411	0.643 12.617	0.758 24.396	0.880 44.211	1.019 80.014	1.185 147.72	1.383 278.07	1.596 501.49
0.00018 1/ 5556	0.399 1.8068	0.434 2.4827	0.466 3.2974	0.509 4.6171	0.568 7.1424	0.661 12.983	0.780 25.104	0.905 45.493	1.048 82.334	1.219 152.01	1.423 286.13	1.643 516.03
0.00019 1/ 5263	0.410 1.8563	0.445 2.5507	0.479 3.3878	0.522 4.7437	0.584 7.3382	0.679 13.339	0.802 25.792	0.930 46.740	1.077 84.591	1.252 156.17	1.462 293.97	1.688 530.17
0.00020 1/ 5000	0.421 1.9045	0.457 2.6170	0.492 3.4759	0.536 4.8670	0.599 7.5289	0.697 13.686	0.823 26.462	0.954 47.954	1.105 86.788	1.285 160.23	1.500 301.61	1.731 543.95
0.00022 1/ 4545	0.442 1.9975	0.479 2.7448	0.516 3.6456	0.562 5.1046	0.628 7.8964	0.731 14.354	0.863 27.754	1.001 50.295	1.159 91.025	1.348 168.05	1.573 316.33	1.816 570.50
0.00024 1/ 4167	0.461 2.0864	0.501 2.8669	0.539 3.8077	0.587 5.3316	0.656 8.2477	0.764 14.992	0.901 28.988	1.045 52.532	1.211 95.073	1.408 175.53	1.643 330.40	1.897 595.87
0.00026 1/ 3846	0.480 2.1716	0.521 2.9840	0.561 3.9632	0.611 5.5494	0.683 8.5845	0.795 15.605	0.938 30.172	1.088 54.677	1.260 98.956	1.465 182.69	1.710 343.89	1.974 620.20
0.00028 1/ 3571	0.498 2.2536	0.541 3.0966	0.582 4.1129	0.634 5.7589	0.709 8.9086	0.825 16.194	0.973 31.311	1.129 56.741	1.308 102.69	1.521 189.59	1.775 356.87	2.049 643.62
0.00030 1/ 3333	0.516 2.3327	0.560 3.2054	0.602 4.2573	0.657 5.9611	0.734 9.2214	0.854 16.762	1.007 32.410	1.168 58.733	1.353 106.30	1.574 196.25	1.837 369.40	2.121 666.21
0.00032 1/ 3125	0.533 2.4093	0.578 3.3105	0.622 4.3969	0.678 6.1566	0.758 9.5239	0.882 17.312	1.041 33.473	1.207 60.660	1.398 109.78	1.625 202.68	1.897 381.51	2.190 688.06
0.00034 1/ 2941	0.549 2.4834	0.596 3.4124	0.641 4.5323	0.699 6.3462	0.781 9.8170	0.909 17.845	1.073 34.504	1.244 62.527	1.441 113.16	1.676 208.92	1.956 393.26	2.258 709.24
0.00036 1/ 2778	0.565 2.5555	0.613 3.5114	0.660 4.6637	0.719 6.5302	0.804 10.102	0.935 18.362	1.104 35.504	1.280 64.340	1.483 116.44	1.724 214.98	2.013 404.66	2.323 729.80
0.00038 1/ 2632	0.580 2.6255	0.630 3.6076	0.678 4.7916	0.739 6.7092	0.826 10.379	0.961 18.866	1.134 36.477	1.315 66.103	1.523 119.63	1.771 220.87	2.068 415.75	2.387 749.80
0.00040 1/ 2500	0.595 2.6937	0.646 3.7014	0.695 4.9161	0.758 6.8835	0.847 10.648	0.986 19.356	1.163 37.425	1.349 67.820	1.563 122.74	1.817 226.61	2.121 426.55	2.449 769.28
0.00042 1/ 2381	0.610 2.7603	0.662 3.7928	0.713 5.0375	0.777 7.0535	0.868 10.911	1.010 19.834	1.192 38.349	1.383 69.495	1.601 125.77	1.862 232.20	2.174 437.08	2.509 788.28
0.00044 1/ 2273	0.625 2.8252	0.678 3.8821	0.729 5.1561	0.795 7.2195	0.889 11.168	1.034 20.301	1.220 39.252	1.415 71.131	1.639 128.73	1.906 237.67	2.225 447.37	2.568 806.83
0.00046 1/ 2174	0.639 2.8888	0.693 3.9693	0.746 5.2720	0.813 7.3818	0.909 11.419	1.057 20.757	1.248 40.134	1.447 72.730	1.676 131.63	1.949 243.01	2.275 457.43	2.626 824.96
0.00048 1/ 2083	0.652 2.9509	0.708 4.0547	0.762 5.3854	0.831 7.5406	0.928 11.665	1.080 21.204	1.274 40.997	1.478 74.294	1.712 134.46	1.991 248.24	2.324 467.27	2.682 842.71
	38	42	46	55	60	80	100	120	160	200	250	310

θ_{medial} for part-full circular pipes.

k_s = 60.0 mm   S = 0.00010 to 0.00048

$k_s = 60.0$ mm
$S = 0.00050$ to 0.00280

Water (or sewage) at 15°C; full bore conditions.

A45 continued

ie hydraulic gradient = 1 in 2000 to 1 in 357

velocities in ms^{-1}
discharges in m^{3}s^{-1}

Gradient	(Equivalent) Pipe diameters in m											
	2.400	2.700	3.000	3.400	4.000	5.000	6.400	8.000	10.00	12.60	16.00	20.00
0.00050 1/ 2000	0.666 3.0118	0.723 4.1383	0.778 5.4964	0.848 7.6961	0.947 11.905	1.102 21.641	1.301 41.843	1.509 75.826	1.747 137.23	2.032 253.36	2.372 476.90	2.738 860.08
0.00055 1/ 1818	0.698 3.1588	0.758 4.3404	0.816 5.7648	0.889 8.0719	0.994 12.486	1.156 22.697	1.364 43.885	1.582 79.528	1.833 143.93	2.131 265.73	2.488 500.18	2.871 902.07
0.00060 1/ 1667	0.729 3.2993	0.792 4.5334	0.852 6.0212	0.929 8.4308	1.038 13.042	1.207 23.707	1.425 45.837	1.653 83.064	1.914 150.33	2.226 277.54	2.598 522.42	2.999 942.18
0.00065 1/ 1538	0.759 3.4340	0.824 4.7186	0.887 6.2671	0.967 8.7751	1.080 13.574	1.257 24.675	1.483 47.709	1.720 86.456	1.992 156.47	2.317 288.88	2.704 543.76	3.122 980.65
0.00070 1/ 1429	0.788 3.5637	0.855 4.8967	0.920 6.5037	1.003 9.1065	1.121 14.087	1.304 25.606	1.539 49.510	1.785 89.720	2.067 162.38	2.404 299.78	2.807 564.28	3.239 1017.7
0.00075 1/ 1333	0.815 3.6888	0.885 5.0686	0.952 6.7320	1.038 9.4261	1.160 14.581	1.350 26.505	1.593 51.248	1.848 92.870	2.140 168.08	2.489 310.30	2.905 584.09	3.353 1053.4
0.00080 1/ 1250	0.842 3.8098	0.914 5.2349	0.984 6.9528	1.072 9.7353	1.198 15.060	1.394 27.374	1.645 52.929	1.908 95.916	2.210 173.59	2.570 320.48	3.000 603.25	3.463 1087.9
0.00085 1/ 1176	0.868 3.9271	0.942 5.3960	1.014 7.1668	1.105 10.035	1.235 15.523	1.437 28.217	1.696 54.558	1.967 98.868	2.278 178.93	2.649 330.34	3.093 621.81	3.570 1121.4
0.00090 1/ 1111	0.893 4.0409	0.970 5.5525	1.043 7.3746	1.137 10.326	1.271 15.973	1.479 29.035	1.745 56.140	2.024 101.73	2.344 184.12	2.726 339.92	3.182 639.84	3.673 1153.9
0.00095 1/ 1053	0.918 4.1517	0.996 5.7047	1.072 7.5767	1.168 10.609	1.306 16.411	1.519 29.831	1.793 57.678	2.079 104.52	2.409 189.17	2.801 349.24	3.270 657.37	3.774 1185.6
0.00100 1/ 1000	0.942 4.2596	1.022 5.8529	1.100 7.7736	1.199 10.885	1.340 16.837	1.559 30.606	1.840 59.177	2.133 107.24	2.471 194.08	2.874 358.31	3.354 674.45	3.872 1216.4
0.00110 1/ 909	0.988 4.4675	1.072 6.1386	1.153 8.1531	1.257 11.416	1.405 17.659	1.635 32.100	1.929 62.065	2.238 112.47	2.592 203.55	3.014 375.80	3.518 707.37	4.061 1275.7
0.00120 1/ 833	1.031 4.6662	1.120 6.4116	1.205 8.5157	1.313 11.924	1.468 18.445	1.708 33.527	2.015 64.826	2.337 117.47	2.707 212.60	3.148 392.51	3.675 738.83	4.241 1332.5
0.00130 1/ 769	1.074 4.8567	1.166 6.6734	1.254 8.8634	1.367 12.411	1.528 19.198	1.777 34.897	2.097 67.473	2.433 122.27	2.818 221.29	3.276 408.54	3.825 769.00	4.415 1386.9
0.00140 1/ 714	1.114 5.0401	1.210 6.9254	1.301 9.1980	1.419 12.879	1.585 19.923	1.844 36.214	2.177 70.020	2.524 126.89	2.924 229.64	3.400 423.96	3.969 798.03	4.581 1439.2
0.00150 1/ 667	1.153 5.2170	1.252 7.1685	1.347 9.5209	1.468 13.331	1.641 20.622	1.909 37.485	2.253 72.478	2.613 131.34	3.027 237.70	3.519 438.84	4.108 826.04	4.742 1489.7
0.00160 1/ 625	1.191 5.3881	1.293 7.4036	1.391 9.8332	1.516 13.768	1.695 21.298	1.972 38.715	2.327 74.855	2.699 135.65	3.126 245.50	3.635 453.24	4.243 853.13	4.898 1538.6
0.00170 1/ 588	1.228 5.5540	1.333 7.6315	1.434 10.136	1.563 14.192	1.747 21.954	2.032 39.906	2.398 77.159	2.782 139.82	3.222 253.05	3.747 467.19	4.374 879.39	5.048 1586.0
0.00180 1/ 556	1.263 5.7150	1.372 7.8528	1.476 10.430	1.608 14.604	1.798 22.590	2.091 41.063	2.468 79.396	2.862 143.88	3.315 260.39	3.855 480.73	4.501 904.88	5.195 1631.9
0.00190 1/ 526	1.298 5.8717	1.409 8.0680	1.516 10.716	1.653 15.004	1.847 23.210	2.149 42.189	2.536 81.572	2.941 147.82	3.406 267.53	3.961 493.90	4.624 929.68	5.337 1676.7
0.00200 1/ 500	1.332 6.0242	1.446 8.2776	1.555 10.994	1.695 15.394	1.895 23.813	2.204 43.285	2.602 83.691	3.017 151.66	3.495 274.48	4.064 506.74	4.744 953.83	5.476 1720.2
0.00220 1/ 455	1.397 6.3183	1.516 8.6816	1.631 11.531	1.778 16.145	1.987 24.975	2.312 45.397	2.729 87.776	3.164 159.06	3.665 287.87	4.262 531.47	4.976 1000.4	5.743 1804.2
0.00240 1/ 417	1.459 6.5993	1.584 9.0677	1.704 12.043	1.857 16.863	2.076 26.086	2.415 47.416	2.850 91.679	3.305 166.14	3.828 300.67	4.452 555.10	5.197 1044.9	5.998 1884.4
0.00260 1/ 385	1.518 6.8688	1.648 9.4380	1.773 12.535	1.933 17.552	2.161 27.151	2.514 49.352	2.966 95.423	3.440 172.92	3.985 312.95	4.634 577.77	5.409 1087.5	6.243 1961.4
0.00280 1/ 357	1.576 7.1281	1.711 9.7943	1.840 13.008	2.006 18.214	2.242 28.176	2.608 51.216	3.078 99.025	3.570 179.45	4.135 324.77	4.809 599.58	5.613 1128.6	6.479 2035.4
	38	44	48	55	65	80	100	130	160	200	260	320

θ_{medial} for part-full circular pipes.

$k_s = 60.0$ mm $S = 0.00050$ to 0.00280

$k_s = 60.0$ mm
$S = 0.0030$ to 0.014

Water (or sewage) at 15°C;
full bore conditions.

ie hydraulic gradient =
1 in 333 to 1 in 71

velocities in ms^{-1}
discharges in m^{3}s^{-1}

Gradient	(Equivalent) Pipe diameters in m											
	2.400	2.700	3.000	3.400	4.000	5.000	6.400	8.000	10.00	12.60	16.00	20.00
0.00300 1/ 333	1.631 7.3783	1.771 10.138	1.905 13.465	2.077 18.854	2.321 29.165	2.700 53.013	3.186 102.50	3.695 185.75	4.280 336.16	4.977 620.63	5.810 1168.2	6.706 2106.8
0.00320 1/ 313	1.684 7.6203	1.829 10.471	1.967 13.907	2.145 19.472	2.397 30.121	2.788 54.752	3.291 105.86	3.817 191.84	4.421 347.19	5.141 640.98	6.001 1206.5	6.926 2175.9
0.00340 1/ 294	1.736 7.8548	1.885 10.793	2.028 14.335	2.211 20.071	2.471 31.048	2.874 56.437	3.392 109.12	3.934 197.74	4.557 357.88	5.299 660.71	6.185 1243.7	7.139 2242.9
0.00360 1/ 278	1.787 8.0826	1.940 11.106	2.087 14.750	2.275 20.653	2.542 31.949	2.958 58.073	3.490 112.28	4.048 203.48	4.689 368.25	5.452 679.86	6.365 1279.7	7.346 2307.9
0.00380 1/ 263	1.836 8.3041	1.993 11.410	2.144 15.155	2.337 21.219	2.612 32.824	3.039 59.665	3.586 115.36	4.159 209.05	4.817 378.34	5.602 698.49	6.539 1314.8	7.548 2371.2
0.00400 1/ 250	1.883 8.5198	2.045 11.707	2.200 15.548	2.398 21.771	2.680 33.677	3.118 61.215	3.679 118.36	4.267 214.48	4.942 388.17	5.747 716.64	6.709 1348.9	7.744 2432.8
0.00420 1/ 238	1.930 8.7302	2.095 11.996	2.254 15.932	2.457 22.308	2.746 34.509	3.195 62.727	3.770 121.28	4.372 219.78	5.064 397.76	5.889 734.34	6.875 1382.2	7.935 2492.9
0.00440 1/ 227	1.975 8.9357	2.144 12.278	2.307 16.307	2.515 22.833	2.811 35.321	3.270 64.203	3.859 124.14	4.475 224.95	5.184 407.12	6.028 751.62	7.037 1414.8	8.122 2551.5
0.00460 1/ 217	2.020 9.1365	2.193 12.554	2.359 16.674	2.571 23.346	2.874 36.115	3.343 65.646	3.945 126.93	4.576 230.01	5.300 416.27	6.163 768.51	7.195 1446.6	8.304 2608.9
0.00480 1/ 208	2.063 9.3331	2.240 12.824	2.410 17.032	2.627 23.849	2.936 36.891	3.415 67.058	4.030 129.66	4.674 234.96	5.414 425.22	6.296 785.04	7.349 1477.7	8.483 2665.0
0.00500 1/ 200	2.106 9.5255	2.286 13.089	2.459 17.384	2.681 24.340	2.996 37.652	3.486 68.441	4.113 132.33	4.771 239.80	5.526 433.99	6.426 801.23	7.501 1508.2	8.658 2719.9
0.00550 1/ 182	2.208 9.9905	2.398 13.727	2.579 18.232	2.812 25.528	3.143 39.490	3.656 71.781	4.314 138.79	5.004 251.51	5.795 455.17	6.739 840.34	7.867 1581.8	9.080 2852.7
0.00600 1/ 167	2.307 10.435	2.504 14.338	2.694 19.043	2.937 26.664	3.282 41.246	3.818 74.973	4.506 144.96	5.226 262.69	6.053 475.41	7.039 877.71	8.217 1652.1	9.484 2979.5
0.00650 1/ 154	2.401 10.861	2.606 14.923	2.804 19.821	3.057 27.752	3.416 42.930	3.974 78.035	4.690 150.88	5.439 273.42	6.300 494.83	7.327 913.55	8.552 1719.6	9.871 3101.2
0.00700 1/ 143	2.491 11.271	2.705 15.487	2.910 20.569	3.172 28.800	3.545 44.551	4.124 80.981	4.867 156.58	5.645 283.74	6.538 513.51	7.603 948.03	8.875 1784.5	10.24 3218.3
0.00750 1/ 133	2.579 11.666	2.800 16.030	3.012 21.291	3.283 29.811	3.670 46.115	4.269 83.823	5.038 162.07	5.843 293.70	6.768 531.53	7.870 981.31	9.187 1847.1	10.60 3331.2
0.00800 1/ 125	2.663 12.049	2.892 16.556	3.111 21.989	3.391 30.789	3.790 47.627	4.409 86.572	5.203 167.39	6.035 303.33	6.990 548.96	8.128 1013.5	9.488 1907.7	10.95 3440.5
0.00850 1/ 118	2.745 12.420	2.981 17.066	3.207 22.666	3.496 31.736	3.907 49.093	4.545 89.237	5.363 172.54	6.220 312.66	7.205 565.86	8.378 1044.7	9.780 1966.4	11.29 3546.4
0.00900 1/ 111	2.825 12.780	3.067 17.560	3.300 23.323	3.597 32.656	4.020 50.516	4.677 91.824	5.519 177.54	6.401 321.73	7.414 582.26	8.621 1075.0	10.06 2023.4	11.62 3649.2
0.00950 1/ 105	2.902 13.130	3.151 18.042	3.390 23.962	3.695 33.551	4.130 51.901	4.805 94.340	5.670 182.41	6.576 330.55	7.617 598.22	8.857 1104.4	10.34 2078.9	11.93 3749.2
0.01000 1/ 100	2.978 13.471	3.233 18.510	3.478 24.585	3.791 34.423	4.237 53.249	4.930 96.791	5.817 187.14	6.747 339.13	7.815 613.76	9.087 1133.1	10.61 2132.9	12.24 3846.6
0.01100 1/ 91	3.123 14.129	3.391 19.414	3.648 25.785	3.976 36.103	4.444 55.848	5.170 101.52	6.101 196.28	7.076 355.69	8.196 643.72	9.531 1188.4	11.13 2237.0	12.84 4034.3
0.01200 1/ 83	3.262 14.757	3.542 20.277	3.810 26.931	4.153 37.709	4.642 58.331	5.400 106.03	6.373 205.01	7.391 371.50	8.561 672.34	9.955 1241.3	11.62 2336.5	13.41 4213.7
0.01300 1/ 77	3.395 15.360	3.686 21.105	3.966 28.031	4.323 39.248	4.831 60.713	5.621 110.36	6.633 213.38	7.693 386.67	8.910 699.80	10.36 1292.0	12.10 2431.9	13.96 4385.8
0.01400 1/ 71	3.523 15.940	3.825 21.902	4.115 29.089	4.486 40.730	5.014 63.005	5.833 114.53	6.883 221.43	7.983 401.27	9.246 726.21	10.75 1340.7	12.55 2523.7	14.49 4551.3
	40	44	48	55	65	80	100	130	160	200	260	330

θ_{medial} for part-full circular pipes.

$k_s = 60.0$ mm $S = 0.0030$ to 0.014

$k_s = 60.0$ mm
$S = 0.015$ to 0.075

ie hydraulic gradient =
1 in 66 to 1 in 13

Water (or sewage) at 15°C;
full bore conditions.

velocities in ms^{-1}
discharges in m^{3}s^{-1}

A45
continued

Gradient	(Equivalent) Pipe diameters in m											
	2.400	2.700	3.000	3.400	4.000	5.000	6.400	8.000	10.00	12.60	16.00	20.00
0.01500 1/ 67	3.647 16.499	3.960 22.671	4.260 30.110	4.644 42.160	5.190 65.217	6.037 118.55	7.125 229.21	8.263 415.35	9.571 751.70	11.13 1387.8	12.99 2612.2	15.00 4711.1
0.01600 1/ 62	3.767 17.040	4.089 23.414	4.399 31.098	4.796 43.542	5.360 67.356	6.235 122.43	7.359 236.72	8.534 428.97	9.885 776.36	11.49 1433.3	13.42 2697.9	15.49 4865.6
0.01700 1/ 59	3.883 17.565	4.215 24.135	4.535 32.055	4.943 44.883	5.525 69.429	6.427 126.20	7.585 244.01	8.797 442.18	10.19 800.25	11.85 1477.4	13.83 2780.9	15.96 5015.3
0.01800 1/ 56	3.995 18.074	4.337 24.834	4.666 32.984	5.087 46.184	5.685 71.442	6.614 129.86	7.805 251.08	9.052 455.00	10.48 823.45	12.19 1520.2	14.23 2861.6	16.43 5160.7
0.01900 1/ 53	4.105 18.569	4.456 25.515	4.794 33.888	5.226 47.449	5.841 73.399	6.795 133.42	8.019 257.96	9.300 467.46	10.77 846.01	12.53 1561.9	14.62 2940.0	16.88 5302.2
0.02000 1/ 50	4.211 19.052	4.572 26.178	4.919 34.768	5.362 48.682	5.993 75.306	6.971 136.88	8.227 264.66	9.542 479.61	11.05 867.99	12.85 1602.5	15.00 3016.4	17.32 5439.9
0.02200 1/ 45	4.417 19.982	4.795 27.456	5.159 36.465	5.624 51.058	6.285 78.982	7.312 143.57	8.629 277.58	10.01 503.02	11.59 910.36	13.48 1680.7	15.73 3163.6	18.16 5705.4
0.02400 1/ 42	4.613 20.870	5.009 28.677	5.388 38.087	5.874 53.329	6.565 82.494	7.637 149.95	9.012 289.93	10.45 525.39	12.11 950.84	14.08 1755.4	16.43 3304.3	18.97 5959.1
0.02600 1/ 38	4.802 21.722	5.213 29.848	5.608 39.642	6.114 55.506	6.833 85.862	7.949 156.07	9.380 301.76	10.88 546.84	12.60 989.67	14.65 1827.1	17.11 3439.2	19.74 6202.5
0.02800 1/ 36	4.983 22.542	5.410 30.974	5.820 41.139	6.344 57.602	7.091 89.104	8.249 161.96	9.734 313.16	11.29 567.48	13.08 1027.0	15.21 1896.1	17.75 3569.0	20.49 6436.6
0.03000 1/ 33	5.158 23.334	5.600 32.061	6.024 42.583	6.567 59.623	7.340 92.231	8.538 167.65	10.08 324.15	11.69 587.40	13.54 1063.1	15.74 1962.6	18.37 3694.3	21.21 6662.5
0.03200 1/ 31	5.327 24.099	5.783 33.113	6.222 43.979	6.782 61.579	7.580 95.256	8.818 173.15	10.41 334.78	12.07 606.66	13.98 1097.9	16.26 2027.0	18.98 3815.4	21.90 6881.0
0.03400 1/ 29	5.491 24.841	5.961 34.132	6.413 45.333	6.991 63.474	7.814 98.188	9.090 178.48	10.73 345.08	12.44 625.34	14.41 1131.7	16.76 2089.4	19.56 3932.9	22.58 7092.8
0.03600 1/ 28	5.650 25.561	6.134 35.122	6.599 46.647	7.194 65.314	8.040 101.03	9.353 183.65	11.04 355.09	12.80 643.47	14.83 1164.5	17.24 2150.0	20.13 4046.9	23.23 7298.4
0.03800 1/ 26	5.805 26.261	6.302 36.084	6.780 47.925	7.391 67.104	8.260 103.80	9.610 188.68	11.34 364.82	13.15 661.10	15.23 1196.5	17.71 2208.9	20.68 4157.8	23.87 7498.4
0.04000 1/ 25	5.956 26.944	6.466 37.021	6.956 49.170	7.583 68.847	8.475 106.50	9.859 193.59	11.63 374.29	13.49 678.27	15.63 1227.5	18.18 2266.3	21.22 4265.8	24.49 7693.2
0.04200 1/ 24	6.103 27.609	6.626 37.936	7.128 50.385	7.770 70.548	8.684 109.13	10.10 198.37	11.92 383.54	13.83 695.02	16.02 1257.8	18.62 2322.2	21.74 4371.1	25.09 7883.2
0.04400 1/ 23	6.247 28.259	6.782 38.829	7.296 51.570	7.953 72.208	8.889 111.70	10.34 203.03	12.20 392.56	14.15 711.38	16.39 1287.4	19.06 2376.9	22.25 4474.0	25.68 8068.7
0.04600 1/ 22	6.387 28.894	6.934 39.701	7.460 52.729	8.132 73.831	9.088 114.21	10.57 207.60	12.48 401.39	14.47 727.37	16.76 1316.4	19.49 2430.3	22.75 4574.5	26.26 8250.1
0.04800 1/ 21	6.524 29.515	7.083 40.555	7.620 53.864	8.307 75.419	9.284 116.66	10.80 212.06	12.75 410.02	14.78 743.01	17.12 1344.7	19.91 2482.6	23.24 4672.9	26.83 8427.5
0.05000 1/ 20	6.659 30.124	7.229 41.391	7.777 54.974	8.478 76.974	9.475 119.07	11.02 216.44	13.01 418.47	15.09 758.33	17.47 1372.4	20.32 2533.8	23.72 4769.3	27.38 8601.3
0.05500 1/ 18	6.984 31.594	7.582 43.412	8.157 57.658	8.892 80.731	9.938 124.88	11.56 227.00	13.64 438.90	15.82 795.35	18.33 1439.4	21.31 2657.4	24.88 5002.1	28.72 9021.1
0.06000 1/ 17	7.294 32.999	7.919 45.342	8.520 60.221	9.287 84.321	10.38 130.44	12.08 237.09	14.25 458.42	16.53 830.71	19.14 1503.4	22.26 2775.6	25.98 5224.5	29.99 9422.2
0.06500 1/ 15	7.592 34.347	8.243 47.194	8.867 62.680	9.666 87.764	10.80 135.76	12.57 246.77	14.83 477.13	17.20 864.63	19.92 1564.8	23.17 2888.9	27.05 5437.8	31.22 9807.0
0.07000 1/ 14	7.879 35.643	8.554 48.975	9.202 65.047	10.03 91.077	11.21 140.89	13.04 256.09	15.39 495.15	17.85 897.27	20.68 1623.9	24.04 2998.0	28.07 5643.1	32.39 10177
0.07500 1/ 13	8.155 36.894	8.854 50.694	9.525 67.330	10.38 94.274	11.60 145.83	13.50 265.08	15.93 512.52	18.48 928.77	21.40 1680.9	24.89 3103.2	29.05 5841.2	33.53 10534
	40	44	50	55	65	80	110	130	160	210	260	330

θ_{medial} **for part-full circular pipes.**

A46

k_s = 150 mm
S = 0.00010 to 0.00048

ie hydraulic gradient =
1 in 10000 to 1 in 2083

Water (or sewage) at 15°C;
full bore conditions.

velocities in ms^{-1}
discharges in m^{3}s^{-1}

Gradient	(Equivalent) Pipe diameters in m											
	2.400	2.700	3.000	3.400	4.000	5.000	6.400	8.000	10.00	12.60	16.00	20.00
0.00010 1/10000	0.243 1.0999	0.265 1.5192	0.287 2.0266	0.314 2.8518	0.353 4.4385	0.414 8.1307	0.493 15.844	0.575 28.900	0.670 52.618	0.784 97.706	0.920 184.93	1.067 335.12
0.00011 1/ 9091	0.255 1.1536	0.278 1.5933	0.301 2.1255	0.329 2.9910	0.370 4.6551	0.434 8.5276	0.517 16.618	0.603 30.310	0.703 55.187	0.822 102.47	0.965 193.96	1.119 351.48
0.00012 1/ 8333	0.266 1.2049	0.291 1.6642	0.314 2.2201	0.344 3.1241	0.387 4.8622	0.454 8.9068	0.540 17.357	0.630 31.658	0.734 57.641	0.858 107.03	1.008 202.58	1.169 367.11
0.00013 1/ 7692	0.277 1.2541	0.303 1.7322	0.327 2.3107	0.358 3.2516	0.403 5.0607	0.472 9.2706	0.562 18.066	0.656 32.951	0.764 59.995	0.893 111.40	1.049 210.86	1.216 382.10
0.00014 1/ 7143	0.288 1.3015	0.314 1.7976	0.339 2.3980	0.372 3.3744	0.418 5.2518	0.490 9.6206	0.583 18.748	0.680 34.195	0.793 62.260	0.927 115.61	1.088 218.82	1.262 396.53
0.00015 1/ 6667	0.298 1.3472	0.325 1.8607	0.351 2.4822	0.385 3.4929	0.433 5.4362	0.507 9.9583	0.603 19.406	0.704 35.396	0.821 64.445	0.960 119.67	1.126 226.50	1.306 410.44
0.00016 1/ 6250	0.308 1.3914	0.336 1.9217	0.363 2.5636	0.397 3.6074	0.447 5.6145	0.524 10.285	0.623 20.042	0.727 36.556	0.847 66.559	0.991 123.59	1.163 233.92	1.349 423.91
0.00017 1/ 5882	0.317 1.4342	0.346 1.9809	0.374 2.6425	0.410 3.7185	0.461 5.7873	0.540 10.601	0.642 20.659	0.750 37.682	0.874 68.607	1.022 127.39	1.199 241.12	1.391 436.95
0.00018 1/ 5556	0.326 1.4758	0.356 2.0383	0.385 2.7191	0.421 3.8263	0.474 5.9551	0.556 10.909	0.661 21.258	0.771 38.774	0.899 70.596	1.051 131.09	1.234 248.12	1.431 449.62
0.00019 1/ 5263	0.335 1.5162	0.366 2.0942	0.395 2.7936	0.433 3.9312	0.487 6.1183	0.571 11.208	0.679 21.841	0.793 39.837	0.923 72.531	1.080 134.68	1.268 254.91	1.470 461.94
0.00020 1/ 5000	0.344 1.5556	0.375 2.1486	0.405 2.8662	0.444 4.0333	0.500 6.2772	0.586 11.499	0.697 22.408	0.813 40.872	0.947 74.415	1.108 138.18	1.301 261.54	1.509 473.94
0.00022 1/ 4545	0.361 1.6316	0.394 2.2534	0.425 3.0061	0.466 4.2302	0.524 6.5837	0.614 12.060	0.731 23.502	0.853 42.867	0.994 78.048	1.162 144.92	1.364 274.30	1.582 497.08
0.00024 1/ 4167	0.377 1.7041	0.411 2.3537	0.444 3.1398	0.487 4.4183	0.547 6.8765	0.642 12.597	0.763 24.547	0.891 44.773	1.038 81.519	1.214 151.37	1.425 286.50	1.653 519.18
0.00026 1/ 3846	0.392 1.7737	0.428 2.4498	0.462 3.2681	0.507 4.5987	0.570 7.1573	0.668 13.111	0.794 25.550	0.927 46.601	1.080 84.847	1.264 157.55	1.483 298.20	1.720 540.38
0.00028 1/ 3571	0.407 1.8407	0.444 2.5423	0.480 3.3914	0.526 4.7724	0.591 7.4275	0.693 13.606	0.824 26.514	0.962 48.361	1.121 88.050	1.311 163.50	1.539 309.46	1.785 560.78
0.00030 1/ 3333	0.421 1.9053	0.460 2.6315	0.497 3.5105	0.544 4.9399	0.612 7.6882	0.717 14.084	0.853 27.445	0.996 50.058	1.160 91.141	1.357 169.24	1.593 320.32	1.848 580.46
0.00032 1/ 3125	0.435 1.9678	0.475 2.7178	0.513 3.6256	0.562 5.1019	0.632 7.9404	0.741 14.546	0.881 28.345	1.029 51.700	1.199 94.130	1.402 174.79	1.645 330.82	1.908 599.50
0.00034 1/ 2941	0.448 2.0284	0.489 2.8015	0.529 3.7372	0.579 5.2589	0.651 8.1848	0.764 14.993	0.908 29.217	1.060 53.291	1.235 97.027	1.445 180.17	1.696 341.01	1.967 617.95
0.00036 1/ 2778	0.461 2.0872	0.503 2.8827	0.544 3.8456	0.596 5.4114	0.670 8.4221	0.786 15.428	0.935 30.064	1.091 54.836	1.271 99.840	1.487 185.39	1.745 350.89	2.024 635.87
0.00038 1/ 2632	0.474 2.1444	0.517 2.9617	0.559 3.9510	0.612 5.5597	0.689 8.6529	0.807 15.851	0.960 30.888	1.121 56.339	1.306 102.58	1.528 190.47	1.793 360.51	2.079 653.29
0.00040 1/ 2500	0.486 2.2001	0.531 3.0387	0.573 4.0536	0.628 5.7042	0.706 8.8777	0.828 16.263	0.985 31.691	1.150 57.803	1.340 105.24	1.567 195.42	1.840 369.87	2.134 670.27
0.00042 1/ 2381	0.498 2.2544	0.544 3.1137	0.588 4.1538	0.644 5.8450	0.724 9.0969	0.849 16.664	1.009 32.473	1.178 59.230	1.373 107.84	1.606 200.24	1.885 379.01	2.186 686.82
0.00044 1/ 2273	0.510 2.3075	0.557 3.1870	0.601 4.2515	0.659 5.9826	0.741 9.3110	0.869 17.056	1.033 33.238	1.206 60.624	1.405 110.38	1.644 204.96	1.929 387.93	2.238 702.98
0.00046 1/ 2174	0.522 2.3594	0.569 3.2586	0.615 4.3471	0.674 6.1171	0.758 9.5203	0.888 17.440	1.056 33.985	1.233 61.987	1.437 112.86	1.681 209.56	1.973 396.65	2.288 718.78
0.00048 1/ 2083	0.533 2.4101	0.581 3.3287	0.628 4.4406	0.688 6.2487	0.774 9.7251	0.907 17.815	1.079 34.716	1.260 63.320	1.468 115.29	1.717 214.07	2.015 405.18	2.337 734.24
	16	17	19	22	26	32	42	50	65	80	100	130

θ_{medial} for part-full circular pipes.

k_s = 150 mm   S = 0.00010 to 0.00048

$k_s = 150$ mm
$S = 0.00050$ to 0.00280

Water (or sewage) at 15°C;
full bore conditions.

ie hydraulic gradient =
1 in 2000 to 1 in 357

velocities in ms^{-1}
discharges in m^{3}s^{-1}

Gradient	(Equivalent) Pipe diameters in m											
	2.400	2.700	3.000	3.400	4.000	5.000	6.400	8.000	10.00	12.60	16.00	20.00
0.00050 1/ 2000	0.544 2.4598	0.593 3.3974	0.641 4.5322	0.702 6.3775	0.790 9.9256	0.926 18.182	1.101 35.432	1.286 64.626	1.498 117.66	1.752 218.49	2.057 413.53	2.385 749.38
0.00055 1/ 1818	0.570 2.5799	0.622 3.5632	0.672 4.7534	0.737 6.6888	0.828 10.410	0.971 19.070	1.155 37.161	1.348 67.780	1.571 123.41	1.838 229.15	2.157 433.72	2.502 785.96
0.00060 1/ 1667	0.596 2.6946	0.650 3.7217	0.702 4.9648	0.769 6.9863	0.865 10.873	1.014 19.918	1.207 38.814	1.408 70.794	1.641 128.89	1.919 239.34	2.253 453.00	2.613 820.91
0.00065 1/ 1538	0.620 2.8047	0.677 3.8737	0.731 5.1675	0.801 7.2716	0.901 11.317	1.056 20.731	1.256 40.399	1.466 73.685	1.708 134.16	1.998 249.11	2.345 471.50	2.720 854.43
0.00070 1/ 1429	0.643 2.9105	0.702 4.0199	0.759 5.3626	0.831 7.5461	0.935 11.744	1.096 21.514	1.303 41.924	1.521 76.467	1.773 139.22	2.073 258.52	2.434 489.30	2.822 886.68
0.00075 1/ 1333	0.666 3.0127	0.727 4.1610	0.785 5.5508	0.860 7.8110	0.967 12.157	1.134 22.269	1.349 43.395	1.575 79.151	1.835 144.11	2.146 267.59	2.519 506.48	2.921 917.81
0.00080 1/ 1250	0.688 3.1115	0.751 4.2975	0.811 5.7329	0.889 8.0671	0.999 12.555	1.171 22.999	1.393 44.818	1.626 81.747	1.895 148.84	2.216 276.37	2.602 523.09	3.017 947.91
0.00085 1/ 1176	0.709 3.2073	0.774 4.4298	0.836 5.9093	0.916 8.3154	1.030 12.942	1.207 23.707	1.436 46.198	1.676 84.262	1.953 153.42	2.285 284.87	2.682 539.19	3.110 977.08
0.00090 1/ 1111	0.730 3.3003	0.796 4.5582	0.860 6.0807	0.942 8.5565	1.060 13.317	1.242 24.394	1.478 47.537	1.725 86.705	2.010 157.86	2.351 293.13	2.759 554.82	3.200 1005.4
0.00095 1/ 1053	0.750 3.3907	0.818 4.6831	0.884 6.2473	0.968 8.7910	1.089 13.682	1.276 25.063	1.518 48.840	1.772 89.082	2.065 162.19	2.415 301.16	2.835 570.02	3.288 1033.0
0.00100 1/ 1000	0.769 3.4788	0.839 4.8048	0.907 6.4096	0.993 9.0194	1.117 14.037	1.310 25.714	1.558 50.109	1.818 91.396	2.119 166.40	2.478 308.99	2.909 584.83	3.373 1059.8
0.00110 1/ 909	0.807 3.6486	0.880 5.0393	0.951 6.7225	1.042 9.4597	1.172 14.722	1.374 26.969	1.634 52.555	1.907 95.857	2.222 174.53	2.599 324.07	3.051 613.37	3.538 1111.5
0.00120 1/ 833	0.842 3.8109	0.919 5.2634	0.993 7.0214	1.088 9.8803	1.224 15.377	1.435 28.168	1.706 54.892	1.992 100.12	2.321 182.29	2.715 338.48	3.186 640.65	3.695 1160.9
0.00130 1/ 769	0.877 3.9665	0.957 5.4783	1.034 7.3081	1.133 10.284	1.274 16.005	1.493 29.319	1.776 57.133	2.073 104.21	2.416 189.73	2.825 352.30	3.316 666.81	3.846 1208.4
0.00140 1/ 714	0.910 4.1162	0.993 5.6851	1.073 7.5840	1.175 10.672	1.322 16.609	1.550 30.426	1.843 59.290	2.151 108.14	2.507 196.89	2.932 365.60	3.442 691.98	3.992 1254.0
0.00150 1/ 667	0.942 4.2607	1.028 5.8847	1.111 7.8502	1.217 11.047	1.368 17.192	1.604 31.493	1.908 61.371	2.227 111.94	2.595 203.80	3.035 378.43	3.562 716.27	4.132 1298.0
0.00160 1/ 625	0.973 4.4005	1.062 6.0777	1.147 8.1077	1.257 11.409	1.413 17.756	1.657 32.526	1.970 63.384	2.300 115.61	2.680 210.49	3.135 390.85	3.679 739.76	4.267 1340.5
0.00170 1/ 588	1.003 4.5359	1.094 6.2648	1.182 8.3572	1.295 11.760	1.456 18.303	1.708 33.527	2.031 65.334	2.371 119.17	2.762 216.97	3.231 402.87	3.793 762.53	4.398 1381.8
0.00180 1/ 556	1.032 4.6674	1.126 6.4464	1.217 8.5996	1.333 12.101	1.499 18.833	1.757 34.499	2.090 67.229	2.439 122.62	2.843 223.26	3.325 414.55	3.902 784.64	4.526 1421.9
0.00190 1/ 526	1.060 4.7953	1.157 6.6230	1.250 8.8352	1.369 12.433	1.540 19.349	1.805 35.445	2.147 69.071	2.506 125.98	2.920 229.37	3.416 425.91	4.009 806.14	4.650 1460.8
0.00200 1/ 500	1.088 4.9199	1.187 6.7951	1.282 9.0647	1.405 12.756	1.580 19.852	1.852 36.366	2.203 70.865	2.571 129.25	2.996 235.33	3.505 436.98	4.114 827.08	4.771 1498.8
0.00220 1/ 455	1.141 5.1600	1.245 7.1268	1.345 9.5072	1.474 13.378	1.657 20.821	1.942 38.141	2.310 74.324	2.697 135.56	3.143 246.82	3.676 458.31	4.314 867.45	5.004 1571.9
0.00240 1/ 417	1.191 5.3895	1.300 7.4437	1.405 9.9300	1.539 13.973	1.731 21.747	2.029 39.837	2.413 77.629	2.817 141.59	3.282 257.79	3.839 478.69	4.506 906.02	5.226 1641.8
0.00260 1/ 385	1.240 5.6096	1.353 7.7477	1.462 10.335	1.602 14.544	1.801 22.635	2.112 41.463	2.512 80.799	2.932 147.37	3.416 268.32	3.996 498.23	4.690 943.02	5.440 1708.9
0.00280 1/ 357	1.287 5.8214	1.404 8.0402	1.517 10.726	1.662 15.093	1.869 23.489	2.191 43.029	2.606 83.849	3.043 152.94	3.545 278.45	4.147 517.04	4.867 978.62	5.645 1773.4
	16	18	20	22	26	32	42	50	65	85	100	130

θ_{medial} for part-full circular pipes.

$k_s = 150$ mm $S = 0.00050$ to 0.00280

$k_s = 150$ mm
$S = 0.0030$ to 0.014

Water (or sewage) at 15°C;
full bore conditions.

ie hydraulic gradient =
1 in 333 to 1 in 71

velocities in ms^{-1}
discharges in m^{3}s^{-1}

Gradient	(Equivalent) Pipe diameters in m											
	2.400	2.700	3.000	3.400	4.000	5.000	6.400	8.000	10.00	12.60	16.00	20.00
0.00300 1/ 333	1.332 6.0257	1.454 8.3224	1.571 11.102	1.721 15.622	1.935 24.314	2.268 44.539	2.698 86.792	3.149 158.30	3.670 288.22	4.292 535.19	5.038 1013.0	5.843 1835.6
0.00320 1/ 313	1.376 6.2233	1.501 8.5953	1.622 11.466	1.777 16.135	1.998 25.111	2.343 46.000	2.786 89.639	3.253 163.50	3.790 297.68	4.433 552.74	5.203 1046.2	6.035 1895.8
0.00340 1/ 294	1.418 6.4149	1.547 8.8599	1.672 11.819	1.832 16.631	2.060 25.884	2.415 47.415	2.872 92.398	3.353 168.53	3.907 306.84	4.569 569.75	5.363 1078.4	6.220 1954.2
0.00360 1/ 278	1.459 6.6008	1.592 9.1167	1.721 12.162	1.885 17.114	2.120 26.635	2.485 48.790	2.955 95.076	3.450 173.41	4.020 315.73	4.702 586.27	5.519 1109.6	6.401 2010.8
0.00380 1/ 263	1.499 6.7817	1.636 9.3666	1.768 12.495	1.937 17.583	2.178 27.364	2.553 50.127	3.036 97.682	3.545 178.17	4.130 324.39	4.831 602.34	5.670 1140.1	6.576 2065.9
0.00400 1/ 250	1.538 6.9579	1.678 9.6099	1.814 12.820	1.987 18.039	2.234 28.075	2.619 51.429	3.115 100.22	3.637 182.79	4.238 332.81	4.956 617.99	5.817 1169.7	6.747 2119.6
0.00420 1/ 238	1.576 7.1297	1.720 9.8472	1.858 13.136	2.036 18.485	2.289 28.769	2.684 52.699	3.192 102.69	3.726 187.31	4.342 341.03	5.079 633.25	5.961 1198.6	6.914 2172.0
0.00440 1/ 227	1.613 7.2975	1.760 10.079	1.902 13.445	2.084 18.920	2.343 29.446	2.747 53.940	3.267 105.11	3.814 191.72	4.444 349.06	5.198 648.15	6.101 1226.8	7.076 2223.1
0.00460 1/ 217	1.649 7.4615	1.800 10.305	1.945 13.748	2.131 19.345	2.396 30.108	2.809 55.152	3.341 107.47	3.900 196.03	4.544 356.90	5.315 662.72	6.239 1254.3	7.235 2273.0
0.00480 1/ 208	1.685 7.6220	1.839 10.527	1.987 14.043	2.177 19.761	2.447 30.755	2.869 56.338	3.413 109.79	3.984 200.24	4.642 364.58	5.429 676.97	6.373 1281.3	7.391 2321.9
0.00500 1/ 200	1.720 7.7792	1.877 10.744	2.028 14.333	2.221 20.169	2.498 31.389	2.928 57.500	3.483 112.05	4.066 204.37	4.738 372.10	5.541 690.93	6.504 1307.7	7.543 2369.8
0.00550 1/ 182	1.804 8.1589	1.968 11.269	2.127 15.032	2.330 21.153	2.620 32.921	3.071 60.306	3.653 117.52	4.264 214.35	4.969 390.26	5.812 724.65	6.822 1371.6	7.911 2485.5
0.00600 1/ 167	1.884 8.5217	2.056 11.770	2.221 15.701	2.433 22.094	2.736 34.385	3.208 62.988	3.815 122.74	4.454 223.88	5.190 407.61	6.070 756.88	7.125 1432.6	8.263 2596.0
0.00650 1/ 154	1.961 8.8697	2.140 12.250	2.312 16.342	2.533 22.996	2.848 35.789	3.339 65.560	3.971 127.76	4.636 233.02	5.402 424.26	6.318 787.78	7.416 1491.0	8.601 2702.0
0.00700 1/ 143	2.035 9.2045	2.220 12.713	2.399 16.959	2.628 23.864	2.956 37.140	3.465 68.035	4.121 132.58	4.811 241.82	5.606 440.27	6.556 817.52	7.696 1547.3	8.925 2804.0
0.00750 1/ 133	2.106 9.5276	2.298 13.159	2.483 17.554	2.721 24.702	3.059 38.444	3.587 70.423	4.266 137.23	4.980 250.30	5.802 455.72	6.787 846.21	7.966 1601.6	9.239 2902.4
0.00800 1/ 125	2.175 9.8401	2.374 13.591	2.565 18.130	2.810 25.512	3.160 39.705	3.704 72.733	4.406 141.73	5.143 258.51	5.993 470.67	7.009 873.97	8.227 1654.2	9.542 2997.6
0.00850 1/ 118	2.242 10.143	2.447 14.009	2.644 18.688	2.896 26.297	3.257 40.927	3.818 74.971	4.541 146.09	5.301 266.47	6.177 485.16	7.225 900.86	8.480 1705.1	9.835 3089.8
0.00900 1/ 111	2.307 10.437	2.518 14.415	2.720 19.230	2.980 27.059	3.351 42.113	3.929 77.145	4.673 150.33	5.455 274.19	6.356 499.22	7.434 926.98	8.726 1754.5	10.12 3179.4
0.00950 1/ 105	2.370 10.723	2.587 14.810	2.795 19.757	3.062 27.801	3.443 43.267	4.037 79.259	4.801 154.45	5.604 281.71	6.530 512.90	7.638 952.38	8.965 1802.6	10.40 3266.5
0.01000 1/ 100	2.432 11.002	2.654 15.195	2.868 20.270	3.142 28.523	3.533 44.392	4.141 81.318	4.926 158.46	5.750 289.03	6.700 526.23	7.836 977.12	9.198 1849.4	10.67 3351.4
0.01100 1/ 91	2.551 11.539	2.783 15.936	3.008 21.259	3.295 29.915	3.705 46.558	4.344 85.287	5.166 166.20	6.031 303.13	7.027 551.91	8.219 1024.8	9.647 1939.7	11.19 3515.0
0.01200 1/ 83	2.664 12.052	2.907 16.645	3.141 22.205	3.441 31.246	3.870 48.629	4.537 89.079	5.396 173.59	6.299 316.61	7.340 576.45	8.584 1070.4	10.08 2025.9	11.69 3671.3
0.01300 1/ 77	2.773 12.544	3.026 17.325	3.270 23.111	3.582 32.521	4.028 50.614	4.722 92.716	5.616 180.67	6.556 329.54	7.639 599.99	8.935 1114.1	10.49 2108.7	12.16 3821.2
0.01400 1/ 71	2.877 13.017	3.140 17.979	3.393 23.984	3.717 33.749	4.180 52.525	4.900 96.216	5.828 187.49	6.803 341.98	7.928 622.64	9.272 1156.2	10.88 2188.3	12.62 3965.4
	16	18	20	22	26	32	42	55	65	85	110	130

θ_{medial} for part-full circular pipes.

$k_s = 150$ mm $S = 0.0030$ to 0.014

k_s = 150 mm
S = 0.015 to 0.075

Water (or sewage) at 15°C;
full bore conditions.

A46
continued

ie hydraulic gradient =
1 in 66 to 1 in 13

velocities in ms^{-1}
discharges in m^{3}s^{-1}

Gradient	(Equivalent) Pipe diameters in m											
	2.400	2.700	3.000	3.400	4.000	5.000	6.400	8.000	10.00	12.60	16.00	20.00
0.01500 1/ 67	2.978 13.474	3.250 18.610	3.512 24.826	3.848 34.934	4.327 54.368	5.072 99.594	6.033 194.08	7.042 353.98	8.206 644.49	9.598 1196.7	11.27 2265.1	13.07 4104.6
0.01600 1/ 62	3.076 13.916	3.357 19.220	3.627 25.640	3.974 36.079	4.468 56.152	5.239 102.86	6.231 200.44	7.273 365.59	8.475 665.63	9.912 1236.0	11.63 2339.4	13.49 4239.2
0.01700 1/ 59	3.171 14.344	3.460 19.812	3.739 26.429	4.096 37.190	4.606 57.880	5.400 106.03	6.422 206.61	7.497 376.84	8.736 686.12	10.22 1274.0	11.99 2411.4	13.91 4369.7
0.01800 1/ 56	3.263 14.760	3.561 20.386	3.847 27.195	4.215 38.268	4.739 59.558	5.556 109.10	6.609 212.60	7.714 387.77	8.989 706.01	10.51 1311.0	12.34 2481.3	14.31 4496.4
0.01900 1/ 53	3.352 15.165	3.658 20.945	3.953 27.940	4.330 39.317	4.869 61.190	5.709 112.09	6.790 218.43	7.926 398.40	9.236 725.35	10.80 1346.9	12.68 2549.3	14.70 4619.6
0.02000 1/ 50	3.439 15.559	3.753 21.489	4.055 28.666	4.443 40.338	4.996 62.779	5.857 115.00	6.966 224.10	8.132 408.75	9.475 744.20	11.08 1381.9	13.01 2615.5	15.09 4739.6
0.02200 1/ 45	3.607 16.318	3.936 22.538	4.253 30.065	4.660 42.307	5.240 65.844	6.143 120.61	7.306 235.04	8.529 428.70	9.938 780.52	11.62 1449.3	13.64 2743.1	15.82 4970.9
0.02400 1/ 42	3.767 17.044	4.111 23.540	4.443 31.402	4.867 44.188	5.473 68.771	6.416 125.98	7.631 245.49	8.908 447.76	10.38 815.23	12.14 1513.8	14.25 2865.1	16.53 5192.0
0.02600 1/ 38	3.921 17.740	4.279 24.501	4.624 32.685	5.066 45.992	5.696 71.580	6.678 131.12	7.943 255.51	9.272 466.04	10.80 848.52	12.64 1575.6	14.83 2982.1	17.20 5404.0
0.02800 1/ 36	4.069 18.409	4.441 25.426	4.798 33.918	5.257 47.729	5.911 74.282	6.930 136.07	8.242 265.16	9.622 483.63	11.21 880.55	13.11 1635.0	15.39 3094.7	17.85 5608.0
0.03000 1/ 33	4.212 19.055	4.597 26.318	4.967 35.109	5.441 49.404	6.119 76.889	7.173 140.85	8.532 274.47	9.959 500.61	11.60 911.45	13.57 1692.4	15.93 3203.3	18.48 5804.8
0.03200 1/ 31	4.350 19.680	4.747 27.182	5.130 36.260	5.620 51.024	6.319 79.411	7.409 145.47	8.812 283.47	10.29 517.03	11.99 941.35	14.02 1747.9	16.45 3308.4	19.08 5995.2
0.03400 1/ 29	4.484 20.286	4.894 28.018	5.288 37.376	5.793 52.594	6.514 81.855	7.637 149.94	9.083 292.19	10.60 532.94	12.35 970.32	14.45 1801.7	16.96 3410.2	19.67 6179.7
0.03600 1/ 28	4.614 20.874	5.035 28.830	5.441 38.460	5.961 54.119	6.703 84.228	7.858 154.29	9.346 300.66	10.91 548.39	12.71 998.45	14.87 1854.0	17.45 3509.0	20.24 6358.9
0.03800 1/ 26	4.741 21.446	5.173 29.620	5.590 39.514	6.124 55.602	6.886 86.536	8.073 158.52	9.602 308.90	11.21 563.42	13.06 1025.8	15.28 1904.8	17.93 3605.2	20.80 6533.1
0.04000 1/ 25	4.864 22.003	5.308 30.390	5.735 40.540	6.283 57.047	7.065 88.784	8.283 162.64	9.852 316.93	11.50 578.05	13.40 1052.5	15.67 1954.3	18.40 3698.9	21.34 6702.8
0.04200 1/ 24	4.984 22.547	5.439 31.140	5.877 41.541	6.438 58.455	7.240 90.976	8.488 166.65	10.09 324.75	11.78 592.33	13.73 1078.4	16.06 2002.5	18.85 3790.2	21.86 6868.4
0.04400 1/ 23	5.101 23.077	5.567 31.873	6.015 42.519	6.590 59.831	7.410 93.117	8.687 170.57	10.33 332.39	12.06 606.27	14.05 1103.8	16.44 2049.6	19.29 3879.4	22.38 7030.0
0.04600 1/ 22	5.216 23.596	5.692 32.590	6.150 43.475	6.738 61.176	7.577 95.210	8.883 174.41	10.56 339.87	12.33 619.89	14.37 1128.6	16.81 2095.7	19.73 3966.6	22.88 7188.0
0.04800 1/ 21	5.328 24.104	5.814 33.290	6.283 44.410	6.883 62.492	7.740 97.258	9.074 178.16	10.79 347.17	12.60 633.23	14.68 1152.9	17.17 2140.8	20.15 4051.9	23.37 7342.6
0.05000 1/ 20	5.438 24.601	5.934 33.977	6.412 45.325	7.025 63.780	7.899 99.263	9.261 181.83	11.01 354.33	12.86 646.28	14.98 1176.7	17.52 2184.9	20.57 4135.4	23.85 7494.0
0.05500 1/ 18	5.703 25.801	6.224 35.635	6.725 47.538	7.368 66.893	8.285 104.11	9.713 190.71	11.55 371.63	13.48 677.83	15.71 1234.1	18.38 2291.6	21.57 4337.3	25.02 7859.8
0.06000 1/ 17	5.957 26.949	6.501 37.220	7.024 49.652	7.695 69.868	8.653 108.74	10.14 199.19	12.07 388.15	14.08 707.97	16.41 1289.0	19.20 2393.5	22.53 4530.2	26.13 8209.3
0.06500 1/ 15	6.200 28.049	6.766 38.740	7.311 51.679	8.010 72.721	9.006 113.18	10.56 207.32	12.56 404.00	14.66 736.88	17.08 1341.6	19.98 2491.2	23.45 4715.1	27.20 8544.5
0.07000 1/ 14	6.434 29.108	7.022 40.202	7.587 53.630	8.312 75.466	9.346 117.45	10.96 215.15	13.03 419.25	15.21 764.69	17.73 1392.3	20.73 2585.2	24.34 4893.1	28.22 8867.0
0.07500 1/ 13	6.660 30.130	7.268 41.613	7.853 55.512	8.604 78.115	9.674 121.57	11.34 222.70	13.49 433.97	15.75 791.53	18.35 1441.1	21.46 2676.0	25.19 5064.9	29.22 9178.2
	16	18	20	22	26	34	42	55	65	85	110	130

θ_{medial} for part-full circular pipes.

k_s = 150 mm   S = 0.0015 to 0.075

A47

k_s = 300 mm
S = 0.00010 to 0.00048

ie hydraulic gradient =
1 in 10000 to 1 in 2083

Water (or sewage) at 15°C;
full bore conditions.

velocities in ms^{-1}
discharges in m^{3}s^{-1}

Gradient	(Equivalent) Pipe diameters in m											
	2.400	2.700	3.000	3.400	4.000	5.000	6.400	8.000	10.00	12.60	16.00	20.00
0.00010	0.202	0.222	0.241	0.265	0.300	0.355	0.425	0.500	0.586	0.689	0.813	0.948
1/10000	0.9132	1.2685	1.7004	2.4057	3.7687	6.9606	13.675	25.111	45.998	85.908	163.49	297.67
0.00011	0.212	0.232	0.252	0.278	0.315	0.372	0.446	0.524	0.614	0.723	0.853	0.994
1/ 9091	0.9578	1.3304	1.7834	2.5231	3.9527	7.3004	14.343	26.336	48.244	90.101	171.47	312.20
0.00012	0.221	0.243	0.264	0.290	0.329	0.388	0.466	0.547	0.642	0.755	0.891	1.038
1/ 8333	1.0004	1.3896	1.8627	2.6353	4.1285	7.6250	14.981	27.507	50.389	94.108	179.10	326.08
0.00013	0.230	0.253	0.274	0.302	0.342	0.404	0.485	0.570	0.668	0.786	0.927	1.080
1/ 7692	1.0412	1.4463	1.9388	2.7430	4.2971	7.9364	15.593	28.631	52.447	97.951	186.41	339.40
0.00014	0.239	0.262	0.285	0.314	0.355	0.419	0.503	0.591	0.693	0.815	0.962	1.121
1/ 7143	1.0805	1.5009	2.0120	2.8465	4.4593	8.2360	16.181	29.711	54.427	101.65	193.45	352.21
0.00015	0.247	0.271	0.295	0.325	0.367	0.434	0.521	0.612	0.717	0.844	0.996	1.160
1/ 6667	1.1184	1.5536	2.0826	2.9464	4.6158	8.5251	16.749	30.754	56.337	105.22	200.24	364.57
0.00016	0.255	0.280	0.304	0.335	0.379	0.448	0.538	0.632	0.741	0.871	1.029	1.199
1/ 6250	1.1551	1.6046	2.1509	3.0431	4.7672	8.8047	17.298	31.763	58.185	108.67	206.81	376.53
0.00017	0.263	0.289	0.314	0.345	0.391	0.462	0.554	0.651	0.764	0.898	1.060	1.235
1/ 5882	1.1907	1.6540	2.2171	3.1367	4.9139	9.0757	17.831	32.741	59.975	112.01	213.17	388.12
0.00018	0.271	0.297	0.323	0.356	0.402	0.476	0.570	0.670	0.786	0.924	1.091	1.271
1/ 5556	1.2252	1.7019	2.2814	3.2277	5.0564	9.3389	18.348	33.690	61.714	115.26	219.35	399.37
0.00019	0.278	0.305	0.332	0.365	0.413	0.489	0.586	0.689	0.807	0.950	1.121	1.306
1/ 5263	1.2588	1.7486	2.3439	3.3161	5.1950	9.5948	18.851	34.613	63.405	118.42	225.36	410.32
0.00020	0.285	0.313	0.340	0.375	0.424	0.501	0.601	0.706	0.828	0.974	1.150	1.340
1/ 5000	1.2915	1.7940	2.4048	3.4023	5.3299	9.8441	19.340	35.512	65.053	121.49	231.22	420.98
0.00022	0.299	0.329	0.357	0.393	0.445	0.526	0.631	0.741	0.869	1.022	1.206	1.405
1/ 4545	1.3545	1.8816	2.5222	3.5684	5.5901	10.325	20.284	37.246	68.228	127.42	242.50	441.52
0.00024	0.313	0.343	0.373	0.411	0.465	0.549	0.659	0.774	0.907	1.067	1.260	1.468
1/ 4167	1.4148	1.9652	2.6343	3.7271	5.8387	10.784	21.186	38.902	71.262	133.09	253.29	461.16
0.00026	0.326	0.357	0.388	0.427	0.484	0.572	0.685	0.806	0.944	1.111	1.311	1.528
1/ 3846	1.4725	2.0455	2.7419	3.8793	6.0771	11.224	22.052	40.490	74.172	138.52	263.63	479.99
0.00028	0.338	0.371	0.403	0.443	0.502	0.593	0.711	0.836	0.980	1.153	1.361	1.586
1/ 3571	1.5281	2.1227	2.8454	4.0257	6.3065	11.648	22.884	42.019	76.972	143.75	273.58	498.11
0.00030	0.350	0.384	0.417	0.459	0.519	0.614	0.736	0.865	1.014	1.193	1.408	1.641
1/ 3333	1.5818	2.1972	2.9453	4.1670	6.5279	12.057	23.687	43.494	79.673	148.80	283.18	515.59
0.00032	0.361	0.396	0.430	0.474	0.537	0.634	0.760	0.894	1.048	1.232	1.455	1.695
1/ 3125	1.6337	2.2693	3.0419	4.3037	6.7420	12.452	24.464	44.920	82.286	153.68	292.47	532.50
0.00034	0.372	0.409	0.444	0.489	0.553	0.654	0.784	0.921	1.080	1.270	1.499	1.747
1/ 2941	1.6839	2.3391	3.1355	4.4361	6.9495	12.835	25.217	46.303	84.819	158.41	301.47	548.89
0.00036	0.383	0.420	0.456	0.503	0.569	0.673	0.807	0.948	1.111	1.307	1.543	1.798
1/ 2778	1.7328	2.4069	3.2264	4.5648	7.1510	13.207	25.948	47.645	87.278	163.00	310.21	564.80
0.00038	0.394	0.432	0.469	0.517	0.585	0.691	0.829	0.974	1.142	1.343	1.585	1.847
1/ 2632	1.7802	2.4729	3.3148	4.6899	7.3470	13.569	26.659	48.951	89.670	167.47	318.71	580.28
0.00040	0.404	0.443	0.481	0.530	0.600	0.709	0.850	0.999	1.171	1.378	1.626	1.895
1/ 2500	1.8265	2.5371	3.4010	4.8117	7.5378	13.922	27.352	50.223	91.999	171.82	326.99	595.35
0.00042	0.414	0.454	0.493	0.543	0.615	0.727	0.871	1.024	1.200	1.412	1.666	1.942
1/ 2381	1.8716	2.5998	3.4850	4.9305	7.7240	14.266	28.027	51.463	94.271	176.06	335.07	610.06
0.00044	0.423	0.465	0.505	0.556	0.629	0.744	0.892	1.048	1.229	1.445	1.706	1.988
1/ 2273	1.9157	2.6610	3.5670	5.0466	7.9058	14.601	28.687	52.674	96.490	180.21	342.95	624.41
0.00046	0.433	0.475	0.516	0.568	0.643	0.760	0.912	1.071	1.256	1.478	1.744	2.032
1/ 2174	1.9587	2.7208	3.6471	5.1600	8.0834	14.930	29.332	53.858	98.658	184.26	350.66	638.45
0.00048	0.442	0.485	0.527	0.581	0.657	0.777	0.931	1.095	1.283	1.510	1.782	2.076
1/ 2083	2.0008	2.7793	3.7256	5.2710	8.2573	15.251	29.963	55.016	100.78	188.22	358.20	652.18
	8	9	10	11	13	16	22	26	32	42	55	65

θ_{medial} for part-full circular pipes.

k_s = 300 mm   S = 0.00010 to 0.00048

$k_s = 300$ mm
$S = 0.00050$ to 0.00280

Water (or sewage) at 15°C;
full bore conditions.

ie hydraulic gradient =
1 in 2000 to 1 in 357

velocities in ms^{-1}
discharges in m^{3}s^{-1}

Gradient	(Equivalent) Pipe diameters in m											
	2.400	2.700	3.000	3.400	4.000	5.000	6.400	8.000	10.00	12.60	16.00	20.00
0.00050	0.451	0.495	0.538	0.593	0.671	0.793	0.951	1.117	1.310	1.541	1.818	2.119
1/ 2000	2.0421	2.8366	3.8024	5.3797	8.4276	15.565	30.580	56.151	102.86	192.10	365.59	665.63
0.00055	0.473	0.520	0.564	0.621	0.703	0.831	0.997	1.172	1.374	1.616	1.907	2.222
1/ 1818	2.1418	2.9751	3.9880	5.6423	8.8389	16.325	32.073	58.891	107.88	201.48	383.43	698.11
0.00060	0.494	0.543	0.589	0.649	0.735	0.868	1.041	1.224	1.435	1.688	1.992	2.321
1/ 1667	2.2370	3.1074	4.1654	5.8932	9.2320	17.051	33.499	61.510	112.68	210.44	400.48	729.16
0.00065	0.515	0.565	0.613	0.676	0.765	0.904	1.084	1.274	1.493	1.757	2.073	2.416
1/ 1538	2.3284	3.2343	4.3355	6.1338	9.6090	17.747	34.867	64.022	117.28	219.03	416.84	758.93
0.00070	0.534	0.586	0.636	0.701	0.794	0.938	1.125	1.322	1.550	1.823	2.151	2.507
1/ 1429	2.4163	3.3564	4.4991	6.3654	9.9717	18.417	36.183	66.439	121.70	227.30	432.57	787.58
0.00075	0.553	0.607	0.659	0.726	0.821	0.971	1.164	1.368	1.604	1.887	2.227	2.595
1/ 1333	2.5011	3.4742	4.6570	6.5888	10.322	19.063	37.453	68.771	125.98	235.28	447.75	815.22
0.00080	0.571	0.627	0.680	0.750	0.848	1.003	1.202	1.413	1.657	1.949	2.300	2.680
1/ 1250	2.5831	3.5881	4.8098	6.8049	10.660	19.689	38.682	71.026	130.11	242.99	462.44	841.96
0.00085	0.589	0.646	0.701	0.773	0.874	1.034	1.239	1.457	1.708	2.009	2.371	2.763
1/ 1176	2.6626	3.6986	4.9578	7.0143	10.988	20.295	39.872	73.212	134.11	250.47	476.67	867.87
0.00090	0.606	0.665	0.722	0.795	0.900	1.064	1.275	1.499	1.757	2.067	2.440	2.843
1/ 1111	2.7398	3.8058	5.1016	7.2177	11.307	20.883	41.028	75.335	138.00	257.73	490.49	893.03
0.00095	0.622	0.683	0.742	0.817	0.924	1.093	1.310	1.540	1.805	2.124	2.506	2.921
1/ 1053	2.8149	3.9101	5.2414	7.4155	11.617	21.455	42.153	77.399	141.78	264.79	503.93	917.51
0.00100	0.638	0.701	0.761	0.838	0.948	1.121	1.344	1.580	1.852	2.179	2.571	2.996
1/ 1000	2.8880	4.0117	5.3775	7.6081	11.919	22.013	43.248	79.410	145.47	271.67	517.02	941.34
0.00110	0.670	0.735	0.798	0.879	0.995	1.176	1.410	1.657	1.943	2.285	2.697	3.143
1/ 909	3.0290	4.2075	5.6400	7.9795	12.500	23.087	45.359	83.286	152.57	284.93	542.26	987.29
0.00120	0.699	0.768	0.833	0.918	1.039	1.228	1.473	1.731	2.029	2.387	2.817	3.282
1/ 833	3.1637	4.3946	5.8908	8.3343	13.056	24.114	47.375	86.989	159.35	297.60	566.37	1031.2
0.00130	0.728	0.799	0.867	0.955	1.081	1.278	1.533	1.801	2.112	2.484	2.932	3.416
1/ 769	3.2929	4.5741	6.1314	8.6746	13.589	25.098	49.310	90.541	165.86	309.76	589.50	1073.3
0.00140	0.755	0.829	0.900	0.992	1.122	1.327	1.591	1.869	2.191	2.578	3.043	3.545
1/ 714	3.4172	4.7467	6.3628	9.0021	14.102	26.046	51.171	93.959	172.12	321.45	611.75	1113.8
0.00150	0.782	0.858	0.932	1.026	1.162	1.373	1.646	1.935	2.268	2.668	3.149	3.670
1/ 667	3.5371	4.9133	6.5861	9.3181	14.597	26.960	52.968	97.257	178.16	332.73	633.22	1152.9
0.00160	0.808	0.886	0.962	1.060	1.200	1.418	1.700	1.998	2.343	2.756	3.253	3.790
1/ 625	3.6531	5.0745	6.8022	9.6237	15.076	27.844	54.705	100.45	184.00	343.64	653.99	1190.7
0.00170	0.832	0.914	0.992	1.093	1.237	1.462	1.753	2.060	2.415	2.841	3.353	3.907
1/ 588	3.7656	5.2307	7.0115	9.9199	15.540	28.701	56.388	103.54	189.66	354.22	674.12	1227.4
0.00180	0.857	0.940	1.021	1.124	1.272	1.504	1.804	2.120	2.485	2.923	3.450	4.020
1/ 556	3.8747	5.3823	7.2148	10.207	15.991	29.533	58.023	106.54	195.16	364.49	693.66	1262.9
0.00190	0.880	0.966	1.049	1.155	1.307	1.545	1.853	2.178	2.553	3.003	3.545	4.130
1/ 526	3.9809	5.5298	7.4125	10.487	16.429	30.343	59.613	109.46	200.51	374.48	712.67	1297.6
0.00200	0.903	0.991	1.076	1.185	1.341	1.585	1.901	2.234	2.619	3.081	3.637	4.238
1/ 500	4.0843	5.6735	7.6051	10.760	16.856	31.131	61.162	112.30	205.72	384.21	731.18	1331.3
0.00220	0.947	1.039	1.128	1.243	1.407	1.663	1.994	2.343	2.747	3.232	3.814	4.444
1/ 455	4.2837	5.9504	7.9763	11.285	17.678	32.650	64.147	117.78	215.76	402.96	766.87	1396.2
0.00240	0.989	1.085	1.179	1.298	1.469	1.737	2.083	2.447	2.869	3.375	3.984	4.642
1/ 417	4.4742	6.2150	8.3310	11.787	18.464	34.102	66.999	123.02	225.35	420.88	800.97	1458.3
0.00260	1.029	1.130	1.227	1.351	1.529	1.808	2.168	2.547	2.986	3.513	4.146	4.832
1/ 385	4.6569	6.4688	8.6711	12.268	19.218	35.495	69.735	128.05	234.56	438.06	833.68	1517.9
0.00280	1.068	1.172	1.273	1.402	1.587	1.876	2.250	2.644	3.099	3.646	4.303	5.014
1/ 357	4.8327	6.7130	8.9985	12.731	19.944	36.835	72.368	132.88	243.41	454.60	865.15	1575.2
	8	9	10	11	13	17	22	26	34	42	55	65

θ_{medial} for part-full circular pipes.

$k_s = 300$ mm $S = 0.00050$ to 0.00280

$k_s = 300$ mm
$S = 0.0030$ to 0.014

Water (or sewage) at 15°C;
full bore conditions

ie hydraulic gradient =
1 in 333 to 1 in 71

velocities in ms^{-1}
discharges in m^{3}s^{-1}

Gradient	(Equivalent) Pipe diameters in m											
	2.400	2.700	3.000	3.400	4.000	5.000	6.400	8.000	10.00	12.60	16.00	20.00
0.00300 1/ 333	1.106 5.0023	1.214 6.9486	1.318 9.3143	1.451 13.178	1.643 20.644	1.942 38.128	2.329 74.908	2.736 137.54	3.208 251.95	3.774 470.56	4.454 895.52	5.190 1630.5
0.00320 1/ 313	1.142 5.1664	1.253 7.1765	1.361 9.6198	1.499 13.610	1.697 21.321	2.006 39.378	2.405 77.364	2.826 142.05	3.313 260.22	3.898 485.99	4.600 924.89	5.360 1683.9
0.00340 1/ 294	1.177 5.3254	1.292 7.3973	1.403 9.9159	1.545 14.029	1.749 21.977	2.067 40.590	2.479 79.745	2.913 146.43	3.415 268.23	4.018 500.94	4.742 953.35	5.525 1735.8
0.00360 1/ 278	1.211 5.4798	1.329 7.6118	1.443 10.203	1.590 14.436	1.800 22.614	2.127 41.767	2.551 82.057	2.998 150.67	3.514 276.00	4.134 515.47	4.879 980.99	5.685 1786.1
0.00380 1/ 263	1.244 5.6299	1.366 7.8204	1.483 10.483	1.634 14.831	1.849 23.234	2.185 42.911	2.621 84.306	3.080 154.80	3.610 283.57	4.247 529.59	5.013 1007.9	5.841 1835.0
0.00400 1/ 250	1.277 5.7762	1.401 8.0235	1.522 10.755	1.676 15.217	1.897 23.838	2.242 44.026	2.689 86.496	3.160 158.82	3.704 290.93	4.358 543.35	5.143 1034.1	5.993 1882.7
0.00420 1/ 238	1.308 5.9188	1.436 8.2217	1.559 11.021	1.717 15.592	1.944 24.426	2.298 45.113	2.755 88.632	3.238 162.74	3.796 298.12	4.465 556.77	5.270 1059.6	6.141 1929.2
0.00440 1/ 227	1.339 6.0581	1.470 8.4152	1.596 11.280	1.758 15.959	1.990 25.001	2.352 46.175	2.820 90.718	3.314 166.57	3.885 305.13	4.570 569.87	5.394 1084.5	6.285 1974.6
0.00460 1/ 217	1.369 6.1943	1.503 8.6043	1.632 11.534	1.797 16.318	2.034 25.563	2.405 47.213	2.883 92.757	3.388 170.32	3.972 311.99	4.673 582.68	5.515 1108.9	6.427 2019.0
0.00480 1/ 208	1.399 6.3275	1.535 8.7894	1.667 11.782	1.836 16.669	2.078 26.113	2.456 48.228	2.945 94.752	3.461 173.98	4.058 318.70	4.774 595.21	5.634 1132.7	6.565 2062.4
0.00500 1/ 200	1.428 6.4580	1.567 8.9706	1.701 12.025	1.874 17.013	2.121 26.651	2.507 49.223	3.006 96.706	3.533 177.57	4.142 325.27	4.872 607.49	5.750 1156.1	6.700 2104.9
0.00550 1/ 182	1.497 6.7732	1.643 9.4085	1.784 12.612	1.965 17.843	2.224 27.952	2.629 51.625	3.153 101.43	3.705 186.23	4.344 341.15	5.110 637.14	6.031 1212.5	7.027 2207.7
0.00600 1/ 167	1.564 7.0744	1.716 9.8268	1.864 13.173	2.053 18.636	2.323 29.195	2.746 53.921	3.293 105.94	3.870 194.52	4.537 356.32	5.337 665.47	6.299 1266.5	7.340 2305.8
0.00650 1/ 154	1.628 7.3633	1.786 10.228	1.940 13.710	2.136 19.397	2.418 30.387	2.858 56.123	3.427 110.26	4.028 202.46	4.722 370.87	5.555 692.64	6.556 1318.2	7.639 2400.0
0.00700 1/ 143	1.689 7.6412	1.854 10.614	2.013 14.228	2.217 20.130	2.509 31.534	2.966 58.241	3.557 114.42	4.180 210.10	4.900 384.87	5.765 718.79	6.804 1367.9	7.928 2490.6
0.00750 1/ 133	1.748 7.9094	1.919 10.987	2.083 14.727	2.295 20.836	2.597 32.641	3.070 60.285	3.682 118.44	4.327 217.48	5.072 398.38	5.967 744.02	7.042 1415.9	8.206 2578.0
0.00800 1/ 125	1.806 8.1688	1.982 11.347	2.152 15.210	2.370 21.520	2.683 33.712	3.171 62.262	3.802 122.32	4.468 224.61	5.239 411.44	6.163 768.42	7.273 1462.4	8.475 2662.5
0.00850 1/ 118	1.861 8.4202	2.043 11.696	2.218 15.678	2.443 22.182	2.765 34.749	3.269 64.179	3.919 126.09	4.606 231.52	5.400 424.10	6.352 792.06	7.497 1507.4	8.736 2744.5
0.00900 1/ 111	1.915 8.6644	2.102 12.035	2.282 16.133	2.514 22.825	2.845 35.756	3.363 66.039	4.033 129.74	4.739 238.23	5.556 436.40	6.536 815.03	7.714 1551.1	8.989 2824.0
0.00950 1/ 105	1.968 8.9018	2.160 12.365	2.345 16.575	2.583 23.450	2.923 36.736	3.456 67.849	4.144 133.30	4.869 244.76	5.709 448.36	6.716 837.36	7.926 1593.6	9.236 2901.4
0.01000 1/ 100	2.019 9.1330	2.216 12.686	2.406 17.006	2.650 24.060	2.999 37.691	3.545 69.612	4.251 136.76	4.996 251.12	5.857 460.01	6.890 859.12	8.132 1635.0	9.475 2976.8
0.01100 1/ 91	2.117 9.5788	2.324 13.306	2.523 17.836	2.779 25.234	3.146 39.530	3.718 73.009	4.459 143.44	5.240 263.38	6.143 482.46	7.226 901.05	8.529 1714.8	9.938 3122.1
0.01200 1/ 83	2.212 10.005	2.427 13.897	2.635 18.629	2.903 26.356	3.286 41.288	3.884 76.256	4.657 149.82	5.473 275.09	6.416 503.91	7.548 941.11	8.908 1791.0	10.38 3260.9
0.01300 1/ 77	2.302 10.413	2.526 14.465	2.743 19.390	3.021 27.432	3.420 42.974	4.042 79.370	4.847 155.93	5.696 286.32	6.678 524.49	7.856 979.54	9.272 1864.2	10.80 3394.1
0.01400 1/ 71	2.389 10.806	2.622 15.011	2.847 20.121	3.135 28.468	3.549 44.596	4.195 82.366	5.030 161.82	5.911 297.13	6.930 544.29	8.152 1016.5	9.622 1934.5	11.21 3522.2
	8	9	10	11	13	17	22	26	34	42	55	65

θ_{medial} **for part-full circular pipes.**

$k_s = 300$ mm $S = 0.0030$ to 0.014

$k_s = 300$ mm
$S = 0.015$ to 0.075

Water (or sewage) at 15°C; full bore conditions.

ie hydraulic gradient = 1 in 66 to 1 in 13

velocities in ms^{-1}
discharges in m^{3}s^{-1}

Gradient	(Equivalent) Pipe diameters in m											
	2.400	2.700	3.000	3.400	4.000	5.000	6.400	8.000	10.00	12.60	16.00	20.00
0.01500 1/ 67	2.473 11.186	2.714 15.538	2.947 20.828	3.246 29.467	3.673 46.162	4.342 85.257	5.207 167.50	6.119 307.56	7.173 563.39	8.439 1052.2	9.959 2002.4	11.61 3645.8
0.01600 1/ 62	2.554 11.553	2.803 16.047	3.043 21.511	3.352 30.433	3.794 47.675	4.484 88.053	5.377 172.99	6.319 317.64	7.409 581.87	8.715 1086.7	10.29 2068.1	11.99 3765.4
0.01700 1/ 59	2.632 11.908	2.889 16.541	3.137 22.173	3.455 31.370	3.911 49.143	4.623 90.763	5.543 178.32	6.514 327.42	7.637 599.78	8.983 1120.2	10.60 2131.8	12.35 3881.3
0.01800 1/ 56	2.709 12.253	2.973 17.021	3.228 22.816	3.555 32.279	4.024 50.567	4.757 93.394	5.704 183.49	6.703 336.91	7.858 617.16	9.244 1152.6	10.91 2193.6	12.71 3993.8
0.01900 1/ 53	2.783 12.589	3.054 17.487	3.316 23.441	3.653 33.164	4.134 51.953	4.887 95.953	5.860 188.52	6.886 346.14	8.073 634.08	9.497 1184.2	11.21 2253.7	13.06 4103.2
0.02000 1/ 50	2.855 12.916	3.134 17.941	3.402 24.050	3.748 34.026	4.242 53.303	5.014 98.446	6.012 193.41	7.065 355.14	8.283 650.55	9.744 1215.0	11.50 2312.2	13.40 4209.8
0.02200 1/ 45	2.994 13.547	3.287 18.817	3.568 25.224	3.931 35.686	4.449 55.904	5.259 103.25	6.306 202.85	7.410 372.47	8.687 682.30	10.22 1274.3	12.06 2425.1	14.05 4415.3
0.02400 1/ 42	3.128 14.149	3.433 19.654	3.727 26.345	4.105 37.273	4.647 58.390	5.492 107.84	6.586 211.87	7.740 389.03	9.074 712.64	10.67 1330.9	12.60 2532.9	14.68 4611.6
0.02600 1/ 38	3.255 14.727	3.573 20.456	3.879 27.421	4.273 38.795	4.836 60.775	5.717 112.25	6.855 220.52	8.056 404.92	9.444 741.74	11.11 1385.3	13.11 2636.3	15.28 4800.0
0.02800 1/ 36	3.378 15.283	3.708 21.229	4.026 28.456	4.434 40.260	5.019 63.069	5.932 116.48	7.114 228.85	8.360 420.20	9.801 769.74	11.53 1437.6	13.61 2735.9	15.86 4981.1
0.03000 1/ 33	3.497 15.819	3.838 21.974	4.167 29.455	4.590 41.673	5.195 65.282	6.141 120.57	7.363 236.88	8.653 434.95	10.14 796.76	11.93 1488.0	14.08 2831.9	16.41 5156.0
0.03200 1/ 31	3.611 16.338	3.964 22.694	4.304 30.421	4.740 43.039	5.365 67.423	6.342 124.53	7.605 244.65	8.937 449.22	10.48 822.89	12.33 1536.8	14.55 2924.8	16.95 5325.1
0.03400 1/ 29	3.723 16.841	4.086 23.393	4.436 31.357	4.886 44.364	5.531 69.498	6.537 128.36	7.839 252.18	9.212 463.04	10.80 848.21	12.70 1584.1	14.99 3014.8	17.47 5489.0
0.03600 1/ 28	3.831 17.329	4.204 24.071	4.565 32.266	5.028 45.650	5.691 71.513	6.727 132.08	8.066 259.49	9.479 476.47	11.11 872.80	13.07 1630.1	15.43 3102.2	17.98 5648.1
0.03800 1/ 26	3.935 17.804	4.319 24.731	4.690 33.150	5.166 46.901	5.847 73.473	6.911 135.70	8.287 266.60	9.739 489.52	11.42 896.72	13.43 1674.7	15.85 3187.2	18.47 5802.9
0.04000 1/ 25	4.038 18.266	4.432 25.373	4.812 34.012	5.300 48.119	5.999 75.382	7.091 139.22	8.503 273.53	9.992 502.24	11.71 920.01	13.78 1718.2	16.26 3270.0	18.95 5953.6
0.04200 1/ 24	4.137 18.717	4.541 26.000	4.930 34.852	5.431 49.308	6.147 77.243	7.266 142.66	8.713 280.28	10.24 514.64	12.00 942.73	14.12 1760.7	16.67 3350.7	19.42 6100.6
0.04400 1/ 23	4.235 19.158	4.648 26.612	5.047 35.672	5.559 50.468	6.291 79.061	7.437 146.02	8.918 286.88	10.48 526.75	12.29 964.92	14.45 1802.1	17.06 3429.6	19.88 6244.2
0.04600 1/ 22	4.330 19.588	4.752 27.210	5.160 36.473	5.684 51.602	6.433 80.838	7.604 149.30	9.118 293.33	10.71 538.59	12.56 986.61	14.78 1842.6	17.44 3506.7	20.32 6384.5
0.04800 1/ 21	4.423 20.010	4.855 27.795	5.271 37.258	5.806 52.712	6.571 82.577	7.767 152.51	9.314 299.63	10.95 550.18	12.83 1007.8	15.10 1882.2	17.82 3582.1	20.76 6521.9
0.05000 1/ 20	4.514 20.422	4.955 28.368	5.380 38.026	5.926 53.799	6.707 84.279	7.928 155.66	9.506 305.81	11.17 561.52	13.10 1028.6	15.41 1921.0	18.18 3655.9	21.19 6656.3
0.05500 1/ 18	4.735 21.419	5.196 29.753	5.642 39.882	6.215 56.425	7.034 88.393	8.314 163.25	9.970 320.74	11.72 588.93	13.74 1078.8	16.16 2014.8	19.07 3834.4	22.22 6981.2
0.06000 1/ 17	4.945 22.371	5.428 31.076	5.893 41.656	6.491 58.934	7.347 92.323	8.684 170.51	10.41 335.00	12.24 615.12	14.35 1126.8	16.88 2104.4	19.92 4004.9	23.21 7291.7
0.06500 1/ 15	5.147 23.285	5.649 32.344	6.134 43.357	6.756 61.341	7.647 96.093	9.039 177.48	10.84 348.68	12.74 640.23	14.93 1172.8	17.57 2190.3	20.73 4168.4	24.16 7589.4
0.07000 1/ 14	5.341 24.164	5.862 33.565	6.365 44.993	7.011 63.656	7.936 99.721	9.380 184.18	11.25 361.84	13.22 664.40	15.50 1217.1	18.23 2273.0	21.51 4325.8	25.07 7875.9
0.07500 1/ 13	5.529 25.012	6.068 34.744	6.589 46.572	7.257 65.890	8.214 103.22	9.709 190.64	11.64 374.54	13.68 687.72	16.04 1259.8	18.87 2352.8	22.27 4477.6	25.95 8152.3
	8	9	10	11	13	17	22	26	34	42	55	65

θ_{medial} for part-full circular pipes.

$k_s = 300$ mm $S = 0.0015$ to 0.075

k_s = 600 mm
S = 0.00010 to 0.00048

Water (or sewage) at 15°C;
full bore conditions,

ie hydraulic gradient =
1 in 10000 to 1 in 2083

velocities in ms^{-1}
discharges in m^{3}s^{-1}

Gradient	(Equivalent) Pipe diameters in m											
	2.400	2.700	3.000	3.400	4.000	5.000	6.400	8.000	10.00	12.60	16.00	20.00
0.00010			0.194	0.216	0.247	0.295	0.358	0.424	0.501	0.594	0.707	0.828
1/10000			1.3740	1.9595	3.0988	5.7903	11.506	21.321	39.377	74.109	142.05	260.22
0.00011			0.204	0.226	0.259	0.309	0.375	0.445	0.526	0.623	0.741	0.869
1/ 9091			1.4411	2.0551	3.2500	6.0729	12.068	22.361	41.299	77.726	148.99	272.92
0.00012			0.213	0.236	0.270	0.323	0.392	0.465	0.549	0.651	0.774	0.907
1/ 8333			1.5052	2.1465	3.3946	6.3429	12.604	23.356	43.136	81.182	155.61	285.05
0.00013			0.222	0.246	0.281	0.336	0.408	0.484	0.572	0.678	0.806	0.944
1/ 7692			1.5667	2.2342	3.5332	6.6020	13.119	24.309	44.897	84.497	161.96	296.69
0.00014			0.230	0.255	0.292	0.349	0.423	0.502	0.593	0.703	0.836	0.980
1/ 7143			1.6258	2.3185	3.6666	6.8512	13.614	25.227	46.592	87.687	168.08	307.89
0.00015			0.238	0.264	0.302	0.361	0.438	0.519	0.614	0.728	0.865	1.014
1/ 6667			1.6829	2.3999	3.7953	7.0917	14.092	26.112	48.228	90.765	173.98	318.70
0.00016			0.246	0.273	0.312	0.373	0.452	0.537	0.634	0.752	0.894	1.048
1/ 6250			1.7381	2.4786	3.9197	7.3242	14.554	26.969	49.809	93.741	179.68	329.15
0.00017			0.253	0.281	0.322	0.385	0.466	0.553	0.654	0.775	0.921	1.080
1/ 5882			1.7916	2.5549	4.0404	7.5497	15.002	27.799	51.342	96.626	185.21	339.28
0.00018			0.261	0.290	0.331	0.396	0.480	0.569	0.673	0.797	0.948	1.111
1/ 5556			1.8435	2.6290	4.1575	7.7686	15.437	28.605	52.831	99.428	190.58	349.12
0.00019			0.268	0.297	0.340	0.406	0.493	0.585	0.691	0.819	0.974	1.142
1/ 5263			1.8940	2.7010	4.2715	7.9814	15.860	29.389	54.278	102.15	195.81	358.68
0.00020			0.275	0.305	0.349	0.417	0.506	0.600	0.709	0.841	0.999	1.171
1/ 5000			1.9432	2.7712	4.3824	8.1888	16.272	30.152	55.689	104.81	200.89	368.00
0.00022			0.288	0.320	0.366	0.437	0.531	0.629	0.744	0.882	1.048	1.229
1/ 4545			2.0381	2.9064	4.5963	8.5885	17.066	31.624	58.407	109.92	210.70	385.96
0.00024			0.301	0.334	0.382	0.457	0.554	0.657	0.777	0.921	1.095	1.283
1/ 4167			2.1287	3.0357	4.8007	8.9704	17.825	33.030	61.004	114.81	220.07	403.13
0.00026			0.313	0.348	0.398	0.476	0.577	0.684	0.808	0.958	1.139	1.336
1/ 3846			2.2156	3.1596	4.9968	9.3367	18.553	34.379	63.495	119.50	229.05	419.59
0.00028			0.325	0.361	0.413	0.493	0.598	0.710	0.839	0.995	1.182	1.386
1/ 3571			2.2993	3.2789	5.1854	9.6892	19.253	35.677	65.892	124.01	237.70	435.43
0.00030			0.337	0.374	0.427	0.511	0.619	0.735	0.868	1.029	1.224	1.435
1/ 3333			2.3800	3.3940	5.3674	10.029	19.929	36.929	68.205	128.36	246.04	450.71
0.00032			0.348	0.386	0.441	0.528	0.640	0.759	0.897	1.063	1.264	1.482
1/ 3125			2.4580	3.5053	5.5434	10.358	20.583	38.140	70.441	132.57	254.11	465.49
0.00034			0.358	0.398	0.455	0.544	0.660	0.782	0.924	1.096	1.303	1.527
1/ 2941			2.5337	3.6132	5.7140	10.677	21.216	39.314	72.609	136.65	261.93	479.82
0.00036			0.369	0.410	0.468	0.560	0.679	0.805	0.951	1.128	1.341	1.572
1/ 2778			2.6072	3.7180	5.8797	10.987	21.831	40.453	74.714	140.61	269.53	493.73
0.00038			0.379	0.421	0.481	0.575	0.697	0.827	0.977	1.159	1.377	1.615
1/ 2632			2.6786	3.8198	6.0408	11.288	22.430	41.562	76.762	144.47	276.91	507.26
0.00040			0.389	0.432	0.493	0.590	0.715	0.848	1.003	1.189	1.413	1.657
1/ 2500			2.7482	3.9191	6.1978	11.581	23.012	42.642	78.756	148.22	284.11	520.44
0.00042			0.398	0.442	0.505	0.604	0.733	0.869	1.028	1.218	1.448	1.698
1/ 2381			2.8161	4.0159	6.3508	11.867	23.581	43.695	80.701	151.88	291.12	533.29
0.00044			0.408	0.453	0.517	0.619	0.750	0.890	1.052	1.247	1.482	1.737
1/ 2273			2.8823	4.1104	6.5003	12.146	24.136	44.723	82.600	155.45	297.97	545.84
0.00046			0.417	0.463	0.529	0.632	0.767	0.910	1.075	1.275	1.515	1.777
1/ 2174			2.9471	4.2028	6.6464	12.419	24.678	45.728	84.457	158.95	304.67	558.11
0.00048			0.426	0.473	0.540	0.646	0.784	0.929	1.098	1.302	1.548	1.815
1/ 2083			3.0105	4.2931	6.7893	12.686	25.209	46.712	86.273	162.37	311.22	570.11
			5	6	7	8	11	13	17	20	26	34

θ_{medial} for part-full circular pipes.

k_s = 600 mm   S = 0.00010 to 0.00048

k_s = 600 mm
S = 0.00050 to 0.00280

Water (or sewage) at 15°C;
full bore conditions.

ie hydraulic gradient =
1 in 2000 to 1 in 357

velocities in ms^{-1}
discharges in m^{3}s^{-1}

Gradient	(Equivalent) Pipe diameters in m											
	2.400	2.700	3.000	3.400	4.000	5.000	6.400	8.000	10.00	12.60	16.00	20.00
0.00050 1/ 2000			0.435 3.0726	0.483 4.3817	0.551 6.9293	0.659 12.948	0.800 25.729	0.948 47.675	1.121 88.052	1.329 165.71	1.580 317.64	1.852 581.87
0.00055 1/ 1818			0.456 3.2226	0.506 4.5956	0.578 7.2676	0.692 13.580	0.839 26.984	0.995 50.002	1.176 92.350	1.394 173.80	1.657 333.15	1.943 610.27
0.00060 1/ 1667			0.476 3.3659	0.529 4.7999	0.604 7.5907	0.722 14.184	0.876 28.184	1.039 52.225	1.228 96.456	1.456 181.53	1.731 347.96	2.029 637.40
0.00065 1/ 1538			0.496 3.5033	0.550 4.9959	0.629 7.9007	0.752 14.763	0.912 29.335	1.081 54.358	1.278 100.39	1.515 188.94	1.801 362.17	2.112 663.43
0.00070 1/ 1429			0.514 3.6355	0.571 5.1845	0.652 8.1989	0.780 15.320	0.946 30.443	1.122 56.410	1.327 104.18	1.573 196.08	1.869 375.84	2.191 688.47
0.00075 1/ 1333			0.532 3.7632	0.591 5.3665	0.675 8.4867	0.808 15.858	0.980 31.511	1.162 58.390	1.373 107.84	1.628 202.96	1.935 389.03	2.268 712.64
0.00080 1/ 1250			0.550 3.8866	0.610 5.5425	0.698 8.7650	0.834 16.378	1.012 32.545	1.200 60.305	1.418 111.38	1.681 209.61	1.998 401.79	2.343 736.01
0.00085 1/ 1176			0.567 4.0062	0.629 5.7131	0.719 9.0348	0.860 16.882	1.043 33.546	1.237 62.161	1.462 114.81	1.733 216.07	2.060 414.16	2.415 758.66
0.00090 1/ 1111			0.583 4.1223	0.647 5.8787	0.740 9.2967	0.885 17.371	1.073 34.519	1.273 63.963	1.504 118.13	1.783 222.33	2.120 426.16	2.485 780.66
0.00095 1/ 1053			0.599 4.2353	0.665 6.0398	0.760 9.5515	0.909 17.847	1.102 35.465	1.307 65.716	1.545 121.37	1.832 228.42	2.178 437.84	2.553 802.05
0.00100 1/ 1000			0.615 4.3453	0.683 6.1967	0.780 9.7996	0.933 18.311	1.131 36.386	1.341 67.423	1.586 124.52	1.880 234.36	2.234 449.22	2.619 822.88
0.00110 1/ 909			0.645 4.5574	0.716 6.4991	0.818 10.278	0.978 19.205	1.186 38.162	1.407 70.714	1.663 130.60	1.971 245.80	2.343 471.14	2.747 863.05
0.00120 1/ 833			0.673 4.7601	0.748 6.7881	0.854 10.735	1.022 20.059	1.239 39.859	1.469 73.858	1.737 136.41	2.059 256.72	2.447 492.09	2.869 901.42
0.00130 1/ 769			0.701 4.9545	0.778 7.0653	0.889 11.173	1.063 20.878	1.290 41.487	1.529 76.874	1.808 141.98	2.143 267.21	2.547 512.18	2.986 938.23
0.00140 1/ 714			0.727 5.1415	0.808 7.3320	0.923 11.595	1.103 21.666	1.338 43.053	1.587 79.776	1.876 147.34	2.224 277.29	2.644 531.52	3.099 973.65
0.00150 1/ 667			0.753 5.3219	0.836 7.5894	0.955 12.002	1.142 22.426	1.385 44.564	1.643 82.576	1.942 152.51	2.302 287.03	2.736 550.17	3.208 1007.8
0.00160 1/ 625			0.778 5.4965	0.863 7.8383	0.986 12.396	1.180 23.162	1.431 46.025	1.697 85.284	2.006 157.51	2.377 296.44	2.826 568.22	3.313 1040.9
0.00170 1/ 588			0.802 5.6657	0.890 8.0795	1.017 12.777	1.216 23.875	1.475 47.442	1.749 87.909	2.067 162.36	2.451 305.56	2.913 585.71	3.415 1072.9
0.00180 1/ 556			0.825 5.8299	0.916 8.3138	1.046 13.148	1.251 24.567	1.517 48.817	1.800 90.458	2.127 167.07	2.522 314.42	2.998 602.69	3.514 1104.0
0.00190 1/ 526			0.847 5.9897	0.941 8.5416	1.075 13.508	1.285 25.240	1.559 50.155	1.849 92.937	2.185 171.65	2.591 323.04	3.080 619.20	3.610 1134.3
0.00200 1/ 500			0.869 6.1453	0.965 8.7635	1.103 13.859	1.319 25.896	1.600 51.458	1.897 95.351	2.242 176.11	2.658 331.43	3.160 635.29	3.704 1163.7
0.00220 1/ 455			0.912 6.4452	1.012 9.1912	1.157 14.535	1.383 27.160	1.678 53.969	1.990 100.00	2.352 184.70	2.788 347.61	3.314 666.30	3.885 1220.5
0.00240 1/ 417			0.952 6.7318	1.057 9.5999	1.208 15.182	1.445 28.368	1.752 56.369	2.078 104.45	2.456 192.91	2.912 363.06	3.461 695.92	4.058 1274.8
0.00260 1/ 385			0.991 7.0067	1.101 9.9919	1.257 15.802	1.504 29.526	1.824 58.671	2.163 108.72	2.557 200.79	3.031 377.89	3.603 724.34	4.224 1326.9
0.00280 1/ 357			1.029 7.2712	1.142 10.369	1.305 16.398	1.561 30.640	1.893 60.886	2.244 112.82	2.653 208.37	3.145 392.15	3.739 751.68	4.383 1377.0
			5	6	7	8	11	13	17	20	26	34

θ_{medial} for part-full circular pipes.

k_s = 600 mm   S = 0.00050 to 0.00280

k_s = 600 mm
S = 0.0030 to 0.014

ie hydraulic gradient =
1 in 333 to 1 in 71

Water (or sewage) at 15°C;
full bore conditions.

velocities in ms^{-1}
discharges in m^{3}s^{-1}

Gradient	(Equivalent) Pipe diameters in m											
	2.400	2.700	3.000	3.400	4.000	5.000	6.400	8.000	10.00	12.60	16.00	20.00
0.00300 1/ 333			1.065 7.5264	1.182 10.733	1.351 16.974	1.615 31.716	1.959 63.023	2.323 116.78	2.746 215.68	3.255 405.92	3.870 778.07	4.537 1425.3
0.00320 1/ 313			1.100 7.7733	1.221 11.085	1.395 17.530	1.668 32.756	2.023 65.090	2.399 120.61	2.836 222.76	3.362 419.23	3.997 803.58	4.686 1472.0
0.00340 1/ 294			1.134 8.0125	1.259 11.426	1.438 18.070	1.720 33.764	2.086 67.093	2.473 124.32	2.924 229.61	3.466 432.13	4.120 828.31	4.830 1517.3
0.00360 1/ 278			1.166 8.2448	1.295 11.758	1.480 18.594	1.769 34.743	2.146 69.038	2.545 127.93	3.008 236.27	3.566 444.66	4.239 852.33	4.970 1561.3
0.00380 1/ 263			1.198 8.4707	1.330 12.080	1.520 19.103	1.818 35.695	2.205 70.930	2.615 131.43	3.091 242.74	3.664 456.85	4.355 875.68	5.106 1604.1
0.00400 1/ 250			1.229 8.6908	1.365 12.394	1.560 19.599	1.865 36.622	2.262 72.773	2.683 134.85	3.171 249.05	3.759 468.71	4.468 898.43	5.239 1645.8
0.00420 1/ 238			1.260 8.9054	1.399 12.700	1.598 20.083	1.911 37.527	2.318 74.570	2.749 138.18	3.249 255.20	3.852 480.29	4.579 920.62	5.368 1686.4
0.00440 1/ 227			1.290 9.1150	1.432 12.998	1.636 20.556	1.956 38.410	2.373 76.325	2.814 141.43	3.326 261.21	3.943 491.59	4.687 942.29	5.494 1726.1
0.00460 1/ 217			1.318 9.3198	1.464 13.291	1.673 21.018	2.000 39.273	2.426 78.040	2.877 144.61	3.401 267.08	4.031 502.64	4.792 963.46	5.618 1764.9
0.00480 1/ 208			1.347 9.5203	1.495 13.576	1.709 21.470	2.043 40.118	2.478 79.718	2.939 147.72	3.474 272.82	4.118 513.45	4.895 984.18	5.739 1802.9
0.00500 1/ 200			1.375 9.7166	1.526 13.856	1.744 21.913	2.085 40.945	2.529 81.362	2.999 150.76	3.545 278.45	4.203 524.04	4.996 1004.5	5.857 1840.0
0.00550 1/ 182			1.442 10.191	1.601 14.533	1.829 22.982	2.187 42.944	2.653 85.333	3.146 158.12	3.718 292.04	4.408 549.62	5.240 1053.5	6.143 1929.8
0.00600 1/ 167			1.506 10.644	1.672 15.179	1.910 24.004	2.284 44.853	2.771 89.128	3.286 165.15	3.884 305.02	4.604 574.06	5.473 1100.4	6.416 2015.7
0.00650 1/ 154			1.567 11.079	1.740 15.799	1.988 24.985	2.378 46.685	2.884 92.767	3.420 171.90	4.042 317.48	4.792 597.50	5.696 1145.3	6.678 2098.0
0.00700 1/ 143			1.626 11.497	1.806 16.395	2.063 25.928	2.467 48.447	2.993 96.269	3.549 178.39	4.195 329.46	4.973 620.05	5.911 1188.5	6.930 2177.2
0.00750 1/ 133			1.684 11.900	1.869 16.971	2.136 26.838	2.554 50.147	3.098 99.648	3.673 184.65	4.342 341.03	5.147 641.81	6.119 1230.2	7.173 2253.6
0.00800 1/ 125			1.739 12.291	1.930 17.527	2.206 27.718	2.638 51.792	3.199 102.92	3.794 190.70	4.485 352.21	5.316 662.86	6.319 1270.6	7.409 2327.5
0.00850 1/ 118			1.792 12.669	1.990 18.067	2.274 28.571	2.719 53.386	3.298 106.08	3.911 196.57	4.623 363.05	5.480 683.26	6.514 1309.7	7.637 2399.1
0.00900 1/ 111			1.844 13.036	2.048 18.590	2.340 29.399	2.798 54.934	3.393 109.16	4.024 202.27	4.757 373.58	5.639 703.07	6.703 1347.7	7.858 2468.7
0.00950 1/ 105			1.895 13.393	2.104 19.100	2.404 30.205	2.874 56.439	3.486 112.15	4.134 207.81	4.887 383.81	5.793 722.34	6.886 1384.6	8.073 2536.3
0.01000 1/ 100			1.944 13.741	2.158 19.596	2.466 30.990	2.949 57.905	3.577 115.06	4.242 213.21	5.014 393.78	5.944 741.10	7.065 1420.5	8.283 2602.2
0.01100 1/ 91			2.039 14.412	2.264 20.552	2.586 32.502	3.093 60.732	3.751 120.68	4.449 223.62	5.259 413.01	6.234 777.28	7.410 1489.9	8.687 2729.2
0.01200 1/ 83			2.130 15.053	2.364 21.466	2.701 33.947	3.231 63.432	3.918 126.05	4.647 233.56	5.492 431.37	6.511 811.84	7.740 1556.1	9.074 2850.6
0.01300 1/ 77			2.217 15.668	2.461 22.343	2.812 35.334	3.362 66.022	4.078 131.19	4.836 243.10	5.717 448.98	6.777 844.99	8.056 1619.7	9.444 2967.0
0.01400 1/ 71			2.300 16.259	2.554 23.186	2.918 36.667	3.489 68.515	4.232 136.15	5.019 252.28	5.932 465.93	7.033 876.89	8.360 1680.8	9.801 3079.0
			5	6	7	8	11	13	17	20	26	34

θ_{medial} for part-full circular pipes.

k_s = 600 mm   S = 0.0030 to 0.014

| | | | | | | | | | | A48 |
| | | | | | | | | | | continued |

$k_s = 600$ mm

$S = 0.015$ to 0.075

Water (or sewage) at 15°C; full bore conditions.

ie hydraulic gradient = 1 in 66 to 1 in 13

velocities in ms^{-1}
discharges in m^{3}s^{-1}

Gradient	(Equivalent) Pipe diameters in m										
	2.400	2.700	3.000	3.400	4.000	5.000	6.400	8.000	10.00	12.60	16.00	20.00
0.01500 1/ 67			2.381 16.830	2.643 24.000	3.020 37.954	3.612 70.919	4.381 140.92	5.195 261.13	6.141 482.29	7.279 907.66	8.653 1739.8	10.14 3187.0
0.01600 1/ 62			2.459 17.382	2.730 24.787	3.119 39.199	3.730 73.245	4.524 145.55	5.365 269.69	6.342 498.10	7.518 937.43	8.937 1796.9	10.48 3291.5
0.01700 1/ 59			2.535 17.917	2.814 25.550	3.215 40.406	3.845 75.499	4.664 150.03	5.531 277.99	6.537 513.43	7.749 966.28	9.212 1852.2	10.80 3392.9
0.01800 1/ 56			2.608 18.436	2.896 26.291	3.309 41.577	3.957 77.688	4.799 154.37	5.691 286.05	6.727 528.32	7.974 994.30	9.479 1905.9	11.11 3491.2
0.01900 1/ 53			2.680 18.941	2.975 27.011	3.399 42.716	4.065 79.817	4.930 158.60	5.847 293.89	6.911 542.80	8.193 1021.5	9.739 1958.1	11.42 3586.9
0.02000 1/ 50			2.749 19.433	3.052 27.713	3.488 43.826	4.171 81.891	5.058 162.73	5.999 301.53	7.091 556.90	8.405 1048.1	9.992 2009.0	11.71 3680.1
0.02200 1/ 45			2.883 20.382	3.201 29.066	3.658 45.965	4.374 85.888	5.305 170.67	6.291 316.24	7.437 584.08	8.816 1099.2	10.48 2107.0	12.29 3859.7
0.02400 1/ 42			3.012 21.288	3.344 30.358	3.820 48.009	4.569 89.707	5.541 178.26	6.571 330.31	7.767 610.05	9.208 1148.1	10.95 2200.7	12.83 4031.3
0.02600 1/ 38			3.135 22.157	3.480 31.598	3.976 49.969	4.755 93.370	5.767 185.54	6.840 343.79	8.085 634.96	9.584 1195.0	11.39 2290.6	13.36 4195.9
0.02800 1/ 36			3.253 22.994	3.612 32.790	4.127 51.856	4.935 96.894	5.985 192.54	7.098 356.77	8.390 658.93	9.946 1240.1	11.82 2377.0	13.86 4354.3
0.03000 1/ 33			3.367 23.801	3.738 33.941	4.271 53.676	5.108 100.30	6.195 199.30	7.347 369.29	8.684 682.06	10.29 1283.6	12.24 2460.5	14.35 4507.1
0.03200 1/ 31			3.478 24.581	3.861 35.054	4.411 55.436	5.276 103.58	6.398 205.83	7.588 381.41	8.969 704.42	10.63 1325.7	12.64 2541.2	14.82 4655.0
0.03400 1/ 29			3.585 25.338	3.980 36.133	4.547 57.142	5.438 106.77	6.595 212.17	7.821 393.14	9.245 726.10	10.96 1366.5	13.03 2619.4	15.27 4798.2
0.03600 1/ 28			3.689 26.073	4.095 37.181	4.679 58.799	5.596 109.87	6.786 218.32	8.048 404.54	9.513 747.16	11.28 1406.1	13.41 2695.3	15.72 4937.3
0.03800 1/ 26			3.790 26.787	4.207 38.200	4.807 60.410	5.749 112.88	6.972 224.30	8.269 415.63	9.774 767.63	11.59 1444.7	13.77 2769.2	16.15 5072.6
0.04000 1/ 25			3.888 27.483	4.317 39.192	4.932 61.979	5.898 115.81	7.154 230.13	8.483 426.42	10.03 787.57	11.89 1482.2	14.13 2841.1	16.57 5204.4
0.04200 1/ 24			3.984 28.162	4.423 40.160	5.054 63.510	6.044 118.67	7.330 235.81	8.693 436.96	10.28 807.02	12.18 1518.8	14.48 2911.3	16.98 5332.9
0.04400 1/ 23			4.078 28.824	4.527 41.105	5.173 65.005	6.186 121.46	7.503 241.36	8.898 447.24	10.52 826.01	12.47 1554.6	14.82 2979.8	17.37 5458.4
0.04600 1/ 22			4.169 29.472	4.629 42.029	5.289 66.466	6.325 124.19	7.671 246.78	9.097 457.29	10.75 844.58	12.75 1589.5	15.15 3046.7	17.77 5581.1
0.04800 1/ 21			4.259 30.106	4.729 42.933	5.403 67.895	6.461 126.86	7.836 252.09	9.293 467.12	10.98 862.74	13.02 1623.7	15.48 3112.3	18.15 5701.1
0.05000 1/ 20			4.347 30.727	4.826 43.818	5.514 69.295	6.594 129.48	7.998 257.29	9.485 476.76	11.21 880.53	13.29 1657.2	15.80 3176.4	18.52 5818.7
0.05500 1/ 18			4.559 32.227	5.062 45.957	5.783 72.677	6.916 135.80	8.388 269.85	9.948 500.03	11.76 923.51	13.94 1738.0	16.57 3331.5	19.43 6102.7
0.06000 1/ 17			4.762 33.660	5.287 48.000	6.041 75.909	7.224 141.84	8.761 281.85	10.39 522.26	12.28 964.57	14.56 1815.3	17.31 3479.6	20.29 6374.1
0.06500 1/ 15			4.956 35.034	5.503 49.960	6.287 79.009	7.519 147.63	9.119 293.36	10.81 543.59	12.78 1004.0	15.15 1889.5	18.01 3621.7	21.12 6634.3
0.07000 1/ 14			5.143 36.356	5.710 51.846	6.525 81.991	7.803 153.20	9.463 304.43	11.22 564.11	13.27 1041.9	15.73 1960.8	18.69 3758.4	21.91 6884.8
0.07500 1/ 13			5.324 37.633	5.911 53.666	6.754 84.869	8.076 158.58	9.795 315.12	11.62 583.91	13.73 1078.4	16.28 2029.6	19.35 3890.3	22.68 7126.4
			5	6	7	8	11	13	17	20	26	34

θ_{medial} for part-full circular pipes.

$k_s = 600$ mm $S = 0.0015$ to 0.075

B1

Q varies with D to the exponent x as tabulated

Colebrook-White solutions

DV in metres and metres per second

D/ k_s both in metres

DV	0.005	0.01	0.02	0.05	0.10	0.20	0.50	1	2	5	10	20	50	100	200
D/ k_s															
5	2.85	2.845	2.845	2.845	2.845	2.845	2.845	2.84	2.84	2.84	2.84	2.84	2.84	2.84	2.84
10	2.79	2.785	2.78	2.78	2.78	2.775	2.775	2.775	2.775	2.775	2.775	2.775	2.775	2.775	2.775
20	2.755	2.745	2.74	2.735	2.735	2.735	2.735	2.73	2.73	2.73	2.73	2.73	2.73	2.73	2.73
50	2.74	2.72	2.705	2.70	2.695	2.695	2.69	2.69	2.69	2.69	2.69	2.69	2.69	2.69	2.69
100	2.74	2.71	2.695	2.68	2.675	2.67	2.67	2.67	2.67	2.67	2.67	2.67	2.67	2.67	2.67
200	2.745	2.715	2.69	2.67	2.66	2.655	2.655	2.655	2.65	2.65	2.65	2.65	2.65	2.65	2.65
500	2.75	2.72	2.695	2.67	2.655	2.645	2.64	2.635	2.635	2.635	2.635	2.635	2.635	2.635	2.635
10^3	2.755	2.73	2.705	2.675	2.655	2.645	2.63	2.625	2.625	2.625	2.62	2.62	2.62	2.62	2.62
2×10^3	2.755	2.73	2.705	2.68	2.66	2.645	2.63	2.62	2.615	2.615	2.615	2.615	2.61	2.61	2.61
5×10^3	2.755	2.73	2.71	2.685	2.67	2.655	2.635	2.62	2.615	2.61	2.605	2.605	2.60	2.60	2.60
10^4	2.755	2.73	2.71	2.69	2.675	2.66	2.64	2.625	2.615	2.605	2.60	2.60	2.595	2.595	2.595
2×10^4	2.755	2.735	2.71	2.69	2.675	2.66	2.645	2.63	2.62	2.605	2.60	2.595	2.59	2.59	2.59
5×10^4	2.755	2.735	2.715	2.69	2.675	2.665	2.65	2.635	2.625	2.61	2.60	2.595	2.585	2.585	2.585
10^5	2.755	2.735	2.715	2.69	2.675	2.665	2.65	2.64	2.63	2.615	2.605	2.595	2.585	2.585	2.58
2×10^5	2.755	2.735	2.715	2.69	2.675	2.665	2.65	2.64	2.63	2.62	2.61	2.60	2.585	2.585	2.58
5×10^5	2.755	2.735	2.715	2.69	2.675	2.665	2.65	2.64	2.635	2.62	2.615	2.605	2.595	2.585	2.58
10^6	2.755	2.735	2.715	2.69	2.68	2.665	2.65	2.64	2.635	2.625	2.615	2.61	2.60	2.59	2.58
D/ k_s															
R	4.38 $\times10^3$	8.76 $\times10^3$	1.75 $\times10^4$	4.38 $\times10^4$	8.76 $\times10^4$	1.75 $\times10^5$	4.38 $\times10^5$	8.76 $\times10^5$	1.75 $\times10^6$	4.38 $\times10^6$	8.76 $\times10^6$	1.75 $\times10^7$	4.38 $\times10^7$	8.76 $\times10^7$	1.75 $\times10^8$

Equivalent Reynolds number for water at 15°C

$$Q_R = Q_T \left[\frac{D_G}{D_T} \right]^x \left[\frac{S_G}{S_T} \right]^{1/y} \left[\frac{(k_s)_T}{(k_s)_G} \right]^u \tag{8}$$

For V_R, Q_T is replaced by V_T and exponent x is replaced by (x-2)

$$D_R = D_T \left[\frac{Q_G}{Q_T} \right]^{1/x} \left[\frac{S_T}{S_G} \right]^{1/z} \left[\frac{(k_s)_G}{(k_s)_T} \right]^w \tag{10}$$

(For Manning solutions, x = 2.665)

Colebrook-White solutions

DV in metres and metres per second

D/ k_s both in metres

DV	0.005	0.01	0.02	0.05	0.10	0.20	0.50	1	2	5	10	20	50	100	200
D/ k_s															
5	1.98	1.99	1.995	2.00	2.00	2.00	2.00	2.00	2.00	2.00	2.00	2.00	2.00	2.00	2.00
10	1.965	1.98	1.99	1.995	2.00	2.00	2.00	2.00	2.00	2.00	2.00	2.00	2.00	2.00	2.00
20	1.935	1.965	1.98	1.995	1.995	2.00	2.00	2.00	2.00	2.00	2.00	2.00	2.00	2.00	2.00
50	1.88	1.93	1.96	1.985	1.99	1.995	2.00	2.00	2.00	2.00	2.00	2.00	2.00	2.00	2.00
100	1.83	1.885	1.93	1.965	1.985	1.99	1.995	2.00	2.00	2.00	2.00	2.00	2.00	2.00	2.00
200	1.785	1.835	1.89	1.94	1.965	1.985	1.995	1.995	2.00	2.00	2.00	2.00	2.00	2.00	2.00
500	1.745	1.785	1.835	1.895	1.935	1.96	1.985	1.99	1.995	2.00	2.00	2.00	2.00	2.00	2.00
10^3	1.725	1.76	1.80	1.855	1.90	1.935	1.97	1.985	1.99	1.995	2.00	2.00	2.00	2.00	2.00
2×10^3	1.715	1.745	1.78	1.82	1.86	1.90	1.945	1.97	1.985	1.995	1.995	2.00	2.00	2.00	2.00
5×10^3	1.71	1.74	1.76	1.795	1.825	1.86	1.905	1.94	1.965	1.985	1.99	1.995	2.00	2.00	2.00
10^4	1.71	1.735	1.755	1.785	1.81	1.835	1.875	1.91	1.94	1.97	1.985	1.99	1.995	2.00	2.00
2×10^4	1.71	1.735	1.755	1.78	1.80	1.82	1.85	1.88	1.91	1.95	1.97	1.985	1.995	1.995	2.00
5×10^4	1.71	1.73	1.75	1.775	1.795	1.81	1.83	1.85	1.875	1.915	1.94	1.965	1.985	1.99	1.995
10^5	1.71	1.73	1.75	1.775	1.79	1.805	1.825	1.84	1.86	1.89	1.915	1.945	1.97	1.985	1.99
2×10^5	1.71	1.73	1.75	1.775	1.79	1.80	1.82	1.835	1.85	1.87	1.895	1.92	1.95	1.97	1.985
5×10^5	1.71	1.73	1.75	1.775	1.79	1.80	1.82	1.83	1.84	1.855	1.875	1.89	1.92	1.945	1.965
10^6	1.71	1.73	1.75	1.775	1.79	1.80	1.815	1.83	1.84	1.85	1.865	1.88	1.90	1.925	1.945
D/ k_s															
R	4.38×10^3	8.76×10^3	1.75×10^4	4.38×10^4	8.76×10^4	1.75×10^5	4.38×10^5	8.76×10^5	1.75×10^6	4.38×10^6	8.76×10^6	1.75×10^7	4.38×10^7	8.76×10^7	1.75×10^8

Equivalent Reynolds number for water at 15°C

$$Q_R = Q_T \left[\frac{D_G}{D_T} \right]^x \left[\frac{S_G}{S_T} \right]^{1/y} \left[\frac{(k_s)_T}{(k_s)_G} \right]^u \qquad (8)$$

For V_R , Q_T is replaced by V_T and exponent x is replaced by (x-2)

$$S_R = S_T \left[\frac{Q_G}{Q_T} \right]^y \left[\frac{D_T}{D_G} \right]^z \left[\frac{(k_s)_G}{(k_s)_T} \right]^v \qquad (9)$$

(For Manning solutions, y = 2.00)

B3   S varies inversely with D to the exponent z as tabulated

Colebrook-White solutions

DV in metres and metres per second

D/ k_s both in metres

DV	0.005	0.01	0.02	0.05	0.10	0.20	0.50	1	2	5	10	20	50	100	200
D/ k_s															
5	5.65	5.665	5.675	5.68	5.685	5.69	5.685	5.685	5.685	5.685	5.685	5.685	5.685	5.685	5.685
10	5.485	5.52	5.535	5.545	5.55	5.55	5.555	5.555	5.555	5.555	5.555	5.555	5.555	5.555	5.555
20	5.34	5.40	5.43	5.45	5.455	5.46	5.465	5.465	5.465	5.465	5.465	5.465	5.465	5.465	5.465
50	5.15	5.24	5.305	5.35	5.365	5.375	5.38	5.38	5.38	5.385	5.385	5.385	5.385	5.385	5.385
100	5.005	5.11	5.20	5.275	5.305	5.32	5.33	5.335	5.335	5.335	5.34	5.34	5.34	5.34	5.34
200	4.89	4.99	5.085	5.19	5.24	5.27	5.29	5.295	5.30	5.30	5.30	5.30	5.305	5.305	5.305
500	4.79	4.86	4.945	5.06	5.135	5.19	5.23	5.25	5.255	5.26	5.265	5.265	5.265	5.265	5.265
10^3	4.75	4.805	4.86	4.96	5.04	5.115	5.18	5.21	5.225	5.235	5.24	5.24	5.245	5.245	5.245
2×10^3	4.73	4.77	4.815	4.885	4.955	5.03	5.115	5.16	5.19	5.21	5.215	5.22	5.225	5.225	5.225
5×10^3	4.715	4.745	4.775	4.825	4.87	4.93	5.015	5.08	5.13	5.17	5.185	5.195	5.20	5.20	5.205
10^4	4.71	4.74	4.765	4.80	4.835	4.875	4.945	5.01	5.065	5.13	5.155	5.175	5.185	5.185	5.19
2×10^4	4.71	4.735	4.76	4.79	4.81	4.84	4.895	4.945	5.005	5.075	5.12	5.145	5.165	5.17	5.175
5×10^4	4.71	4.735	4.755	4.78	4.80	4.82	4.85	4.885	4.925	4.995	5.05	5.095	5.13	5.15	5.155
10^5	4.71	4.73	4.755	4.775	4.795	4.81	4.835	4.855	4.885	4.94	4.99	5.045	5.095	5.125	5.14
2×10^5	4.71	4.73	4.75	4.775	4.79	4.805	4.825	4.845	4.865	4.90	4.94	4.99	5.05	5.09	5.115
5×10^5	4.71	4.73	4.75	4.775	4.79	4.805	4.82	4.835	4.845	4.87	4.895	4.93	4.985	5.03	5.07
10^6	4.71	4.73	4.75	4.775	4.79	4.805	4.815	4.83	4.84	4.86	4.875	4.90	4.94	4.98	5.025
D/ k_s															
R	4.38×10^3	8.76×10^3	1.75×10^4	4.38×10^4	8.76×10^4	1.75×10^5	4.38×10^5	8.76×10^5	1.75×10^6	4.38×10^6	8.76×10^6	1.75×10^7	4.38×10^7	8.76×10^7	1.75×10^8

Equivalent Reynolds number for water at 15°C

$$S_R \ = \ S_T \left[\frac{Q_G}{Q_T} \right]^y \left[\frac{D_T}{D_G} \right]^z \left[\frac{(k_s)_G}{(k_s)_T} \right]^v \tag{9}$$

$$D_R \ = \ D_T \left[\frac{Q_G}{Q_T} \right]^{1/x} \left[\frac{S_T}{S_G} \right]^{1/z} \left[\frac{(k_s)_G}{(k_s)_T} \right]^w \tag{10}$$

(For Manning solutions, z = 5.33)

Q varies inversely with k_s to the exponent u as tabulated

Colebrook-White solutions

DV in metres and metres per second

D/ k_s both in metres

DV	0.005	0.01	0.02	0.05	0.10	0.20	0.50	1	2	5	10	20	50	100	200
D/ k_s															
5	0.335	0.34	0.34	0.34	0.345	0.345	0.345	0.345	0.345	0.345	0.345	0.345	0.345	0.345	0.345
10	0.265	0.27	0.275	0.275	0.275	0.275	0.275	0.275	0.275	0.275	0.275	0.275	0.275	0.275	0.275
20	0.21	0.22	0.225	0.23	0.23	0.23	0.23	0.23	0.23	0.23	0.23	0.23	0.23	0.23	0.23
50	0.145	0.16	0.175	0.185	0.19	0.19	0.19	0.19	0.19	0.19	0.19	0.19	0.19	0.19	0.19
100	0.095	0.12	0.14	0.155	0.16	0.165	0.17	0.17	0.17	0.17	0.17	0.17	0.17	0.17	0.17
200	0.06	0.08	0.105	0.125	0.14	0.145	0.15	0.15	0.15	0.15	0.15	0.15	0.15	0.15	0.15
500	0.03	0.04	0.06	0.085	0.105	0.115	0.125	0.13	0.13	0.13	0.135	0.135	0.135	0.135	0.135
10^3	0.015	0.025	0.035	0.06	0.075	0.095	0.11	0.115	0.12	0.12	0.12	0.12	0.12	0.12	0.12
2×10^3	0.01	0.01	0.02	0.035	0.05	0.07	0.09	0.10	0.105	0.11	0.11	0.11	0.11	0.11	0.11
5×10^3	0.005	0.005	0.01	0.015	0.025	0.04	0.06	0.075	0.085	0.095	0.10	0.10	0.10	0.10	0.10
10^4	0.00	0.005	0.005	0.01	0.015	0.02	0.04	0.055	0.065	0.08	0.085	0.09	0.095	0.095	0.095
2×10^4	0.00	0.00	0.00	0.005	0.005	0.01	0.025	0.035	0.05	0.065	0.075	0.08	0.085	0.09	0.09
5×10^4	0.00	0.00	0.00	0.00	0.005	0.005	0.01	0.025	0.025	0.045	0.055	0.065	0.075	0.08	0.08
10^5	0.00	0.00	0.00	0.00	0.00	0.005	0.005	0.01	0.015	0.025	0.04	0.05	0.065	0.07	0.075
2×10^5	0.00	0.00	0.00	0.00	0.00	0.00	0.005	0.005	0.01	0.015	0.025	0.035	0.05	0.06	0.065
5×10^5	0.00	0.00	0.00	0.00	0.00	0.00	0.00	0.00	0.005	0.005	0.01	0.02	0.035	0.045	0.055
10^6	0.00	0.00	0.00	0.00	0.00	0.00	0.00	0.00	0.00	0.005	0.005	0.01	0.02	0.03	0.04
D/ k_s															
R	4.38×10^3	8.76×10^3	1.75×10^4	4.38×10^4	8.76×10^4	1.75×10^5	4.38×10^5	8.76×10^5	1.75×10^6	4.38×10^6	8.76×10^6	1.75×10^7	4.38×10^7	8.76×10^7	1.75×10^8

Equivalent Reynolds number for water at 15°C

$$Q_R = Q_T \left[\frac{D_G}{D_T}\right]^x \left[\frac{S_G}{S_T}\right]^{1/y} \left[\frac{(k_s)_T}{(k_s)_G}\right]^u \tag{8}$$

For V_R, Q_T is replaced by V_T and exponent x is replaced by (x-2)

Note that if the change in value of k_s is sufficient to modify the value of D/ k_s as leading to the selection of the appropriate line on the table, this shift should influence accordingly the estimate of medial value of exponent.

B5

S varies with k_s to the exponent v as tabulated

Colebrook-White solutions

DV in metres and metres per second

D/k_s both in metres

DV	0.005	0.01	0.02	0.05	0.10	0.20	0.50	1	2	5	10	20	50	100	200
D/k_s															
5	0.67	0.675	0.68	0.685	0.685	0.685	0.685	0.685	0.685	0.685	0.685	0.685	0.685	0.685	0.685
10	0.52	0.535	0.545	0.55	0.55	0.555	0.555	0.555	0.555	0.555	0.555	0.555	0.555	0.555	0.555
20	0.405	0.43	0.45	0.46	0.46	0.465	0.465	0.465	0.465	0.465	0.465	0.465	0.465	0.465	0.465
50	0.27	0.31	0.345	0.365	0.375	0.38	0.38	0.385	0.385	0.385	0.385	0.385	0.385	0.385	0.385
100	0.18	0.225	0.27	0.305	0.32	0.33	0.335	0.335	0.335	0.34	0.34	0.34	0.34	0.34	0.34
200	0.105	0.15	0.195	0.245	0.27	0.285	0.30	0.30	0.30	0.30	0.30	0.305	0.305	0.305	0.305
500	0.05	0.075	0.11	0.165	0.20	0.23	0.25	0.255	0.26	0.265	0.265	0.265	0.265	0.265	0.265
10^3	0.025	0.04	0.065	0.105	0.145	0.18	0.21	0.225	0.235	0.24	0.24	0.24	0.245	0.245	0.245
2×10^3	0.015	0.02	0.035	0.065	0.095	0.13	0.17	0.195	0.21	0.215	0.22	0.225	0.225	0.225	0.225
5×10^3	0.005	0.01	0.015	0.03	0.045	0.07	0.11	0.145	0.165	0.19	0.195	0.20	0.20	0.205	0.205
10^4	0.005	0.005	0.005	0.015	0.025	0.04	0.07	0.10	0.13	0.16	0.175	0.18	0.185	0.19	0.19
2×10^4	0.00	0.00	0.005	0.01	0.015	0.02	0.04	0.065	0.09	0.13	0.15	0.16	0.17	0.175	0.175
5×10^4	0.00	0.00	0.00	0.005	0.005	0.01	0.02	0.03	0.05	0.08	0.11	0.13	0.15	0.155	0.16
10^5	0.00	0.00	0.00	0.00	0.005	0.005	0.01	0.015	0.03	0.05	0.075	0.10	0.125	0.14	0.145
2×10^5	0.00	0.00	0.00	0.00	0.00	0.00	0.005	0.01	0.015	0.03	0.045	0.07	0.10	0.12	0.13
5×10^5	0.00	0.00	0.00	0.00	0.00	0.00	0.00	0.005	0.005	0.015	0.02	0.035	0.065	0.085	0.105
10^6	0.00	0.00	0.00	0.00	0.00	0.00	0.00	0.00	0.005	0.005	0.01	0.02	0.04	0.06	0.08
D/k_s															
R	4.38 $\times 10^3$	8.76 $\times 10^3$	1.75 $\times 10^4$	4.38 $\times 10^4$	8.76 $\times 10^4$	1.75 $\times 10^5$	4.38 $\times 10^5$	8.76 $\times 10^5$	1.75 $\times 10^6$	4.38 $\times 10^6$	8.76 $\times 10^6$	1.75 $\times 10^7$	4.38 $\times 10^7$	8.76 $\times 10^7$	1.75 $\times 10^8$

Equivalent Reynolds number for water at 15°C

$$S_R = S_T \left[\frac{Q_G}{Q_T} \right]^y \left[\frac{D_T}{D_G} \right]^z \left[\frac{(k_s)_G}{(k_s)_T} \right]^v \qquad (9)$$

Note that if the change in value of k_s is sufficient to modify the value of D/k_s as leading to the selection of the appropriate line on the table, this shift should influence accordingly the estimate of medial value of exponent.

Colebrook-White solutions

DV in metres and metres per second

D/ k_s both in metres

DV	0.005	0.01	0.02	0.05	0.10	0.20	0.50	1	2	5	10	20	50	100	200
D/ k_s															
5	0.12	0.12	0.12	0.12	0.12	0.12	0.12	0.12	0.12	0.12	0.12	0.12	0.12	0.12	0.12
10	0.095	0.095	0.10	0.10	0.10	0.10	0.10	0.10	0.10	0.10	0.10	0.10	0.10	0.10	0.10
20	0.075	0.08	0.08	0.085	0.085	0.085	0.085	0.085	0.085	0.085	0.085	0.085	0.085	0.085	0.085
50	0.05	0.06	0.065	0.07	0.07	0.07	0.07	0.07	0.07	0.07	0.07	0.07	0.07	0.07	0.07
100	0.035	0.045	0.05	0.06	0.06	0.06	0.065	0.065	0.065	0.065	0.065	0.065	0.065	0.065	0.065
200	0.02	0.03	0.04	0.045	0.05	0.055	0.055	0.055	0.055	0.055	0.055	0.055	0.055	0.055	0.055
500	0.01	0.015	0.02	0.03	0.04	0.045	0.05	0.05	0.05	0.05	0.05	0.05	0.05	0.05	0.05
10^3	0.005	0.01	0.015	0.02	0.03	0.035	0.04	0.045	0.045	0.045	0.045	0.045	0.045	0.045	0.045
2×10^3	0.005	0.005	0.005	0.015	0.02	0.025	0.035	0.035	0.04	0.04	0.04	0.045	0.045	0.045	0.045
5×10^3	0.00	0.00	0.005	0.005	0.01	0.015	0.02	0.03	0.035	0.04	0.04	0.04	0.04	0.04	0.04
10^4	0.00	0.00	0.00	0.005	0.005	0.01	0.015	0.02	0.025	0.03	0.035	0.035	0.035	0.035	0.035
2×10^4	0.00	0.00	0.00	0.00	0.005	0.005	0.01	0.015	0.02	0.025	0.03	0.03	0.035	0.035	0.035
5×10^4	0.00	0.00	0.00	0.00	0.00	0.00	0.005	0.005	0.01	0.015	0.02	0.025	0.03	0.03	0.03
10^5	0.00	0.00	0.00	0.00	0.00	0.00	0.00	0.005	0.005	0.01	0.015	0.02	0.025	0.025	0.03
2×10^5	0.00	0.00	0.00	0.00	0.00	0.00	0.00	0.00	0.005	0.005	0.01	0.015	0.02	0.025	0.025
5×10^5	0.00	0.00	0.00	0.00	0.00	0.00	0.00	0.00	0.00	0.005	0.005	0.005	0.015	0.015	0.02
10^6	0.00	0.00	0.00	0.00	0.00	0.00	0.00	0.00	0.00	0.00	0.00	0.005	0.01	0.01	0.015
D/ k_s															
R	4.38×10^3	8.76×10^3	1.75×10^4	4.38×10^4	8.76×10^4	1.75×10^5	4.38×10^5	8.76×10^5	1.75×10^6	4.38×10^6	8.76×10^6	1.75×10^7	4.38×10^7	8.76×10^7	1.75×10^8

Equivalent Reynolds number for water at 15°C

$$D_R = D_T \left[\frac{Q_G}{Q_T} \right]^{1/x} \left[\frac{S_T}{S_G} \right]^{1/z} \left[\frac{(k_s)_G}{(k_s)_T} \right]^{w} \qquad (10)$$

Note that if the change in value of k_s is sufficient to modify the value of D/ k_s as leading to the selection of the appropriate line on the table, this shift should influence accordingly the estimate of medial value of exponent.

C1

Circular pipe

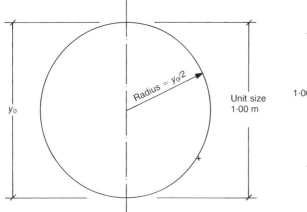

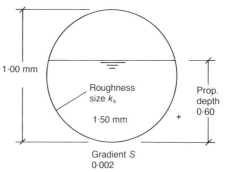

Medial case
for discharge ratio
for Colebrook-White
solutions
$\Theta_{ep} = 300$

Prop dpth	U.equ. dia. $D_{ep(u)}$	Equiv. disch. factor	Unit sect. area	Unit wetted perim.	Unit surf. brdth	Unit mean depth	Discharge ratios $Q/Q_{0.60}$		U.crit. disch. Q_{uc}
Y	(m)	J	A_u (m²)	P_u (m)	B_{us} (m)	y_{um} (m)	medial	Mann'g	(m³s⁻¹)
0.02	0.0528	0.5849	0.0037	0.2838	0.2800	0.0134	0.0010	0.0010	0.0014
0.04	0.1047	0.8165	0.0105	0.4027	0.3919	0.0269	0.0045	0.0044	0.0054
0.06	0.1555	0.9869	0.0192	0.4949	0.4750	0.0405	0.0109	0.0105	0.0121
0.08	0.2053	1.1246	0.0294	0.5735	0.5426	0.0542	0.0201	0.0194	0.0215
0.10	0.2541	1.2404	0.0409	0.6435	0.6000	0.0681	0.0321	0.0311	0.0334
0.11	0.2781	1.2921	0.0470	0.6761	0.6258	0.0751	0.0392	0.0380	0.0403
0.12	0.3018	1.3403	0.0534	0.7075	0.6499	0.0821	0.0470	0.0455	0.0479
0.13	0.3253	1.3853	0.0600	0.7377	0.6726	0.0892	0.0555	0.0538	0.0561
0.14	0.3485	1.4276	0.0668	0.7670	0.6940	0.0963	0.0647	0.0627	0.0649
0.15	0.3715	1.4673	0.0739	0.7954	0.7141	0.1034	0.0745	0.0724	0.0744
0.16	0.3942	1.5047	0.0811	0.8230	0.7332	0.1106	0.0850	0.0826	0.0845
0.17	0.4166	1.5399	0.0885	0.8500	0.7513	0.1178	0.0962	0.0936	0.0952
0.18	0.4388	1.5732	0.0961	0.8763	0.7684	0.1251	0.1081	0.1052	0.1065
0.19	0.4607	1.6045	0.1039	0.9021	0.7846	0.1324	0.1205	0.1175	0.1184
0.20	0.4824	1.6342	0.1118	0.9273	0.8000	0.1398	0.1336	0.1303	0.1309
0.21	0.5037	1.6622	0.1199	0.9521	0.8146	0.1472	0.1473	0.1439	0.1440
0.22	0.5248	1.6886	0.1281	0.9764	0.8285	0.1546	0.1616	0.1580	0.1578
0.23	0.5457	1.7136	0.1365	1.0004	0.8417	0.1621	0.1765	0.1727	0.1721
0.24	0.5662	1.7372	0.1449	1.0239	0.8542	0.1697	0.1920	0.1880	0.1870
0.25	0.5865	1.7595	0.1535	1.0472	0.8660	0.1773	0.2081	0.2039	0.2025
0.26	0.6065	1.7805	0.1623	1.0701	0.8773	0.1850	0.2247	0.2203	0.2185
0.27	0.6262	1.8003	0.1711	1.0928	0.8879	0.1927	0.2418	0.2373	0.2352
0.28	0.6457	1.8189	0.1800	1.1152	0.8980	0.2005	0.2594	0.2549	0.2524
0.29	0.6649	1.8365	0.1890	1.1374	0.9075	0.2083	0.2776	0.2729	0.2702
0.30	0.6838	1.8529	0.1982	1.1593	0.9165	0.2162	0.2962	0.2915	0.2886
0.31	0.7024	1.8684	0.2074	1.1810	0.9250	0.2242	0.3153	0.3105	0.3075
0.32	0.7207	1.8828	0.2167	1.2025	0.9330	0.2322	0.3349	0.3301	0.3270
0.33	0.7387	1.8963	0.2260	1.2239	0.9404	0.2404	0.3550	0.3501	0.3470
0.34	0.7565	1.9088	0.2355	1.2451	0.9474	0.2485	0.3754	0.3705	0.3676
0.35	0.7740	1.9204	0.2450	1.2661	0.9539	0.2568	0.3963	0.3914	0.3888
0.36	0.7911	1.9311	0.2546	1.2870	0.9600	0.2652	0.4175	0.4127	0.4105
0.37	0.8080	1.9410	0.2642	1.3078	0.9656	0.2736	0.4392	0.4343	0.4327
0.38	0.8246	1.9501	0.2739	1.3284	0.9708	0.2821	0.4612	0.4564	0.4555
0.39	0.8409	1.9583	0.2836	1.3490	0.9755	0.2907	0.4835	0.4788	0.4788
0.40	0.8569	1.9658	0.2934	1.3694	0.9798	0.2994	0.5062	0.5016	0.5027
0.41	0.8726	1.9724	0.3032	1.3898	0.9837	0.3082	0.5292	0.5247	0.5271
0.42	0.8880	1.9783	0.3130	1.4101	0.9871	0.3171	0.5525	0.5481	0.5521
0.43	0.9031	1.9835	0.3229	1.4303	0.9902	0.3261	0.5761	0.5718	0.5775
0.44	0.9179	1.9879	0.3328	1.4505	0.9928	0.3353	0.5999	0.5958	0.6035
0.45	0.9323	1.9916	0.3428	1.4706	0.9950	0.3445	0.6239	0.6200	0.6301

Prop dpth	U.equ. dia. $D_{ep(u)}$	Equiv. disch. factor	Unit sect. area	Unit wetted perim.	Unit surf. brdth	Unit mean depth	Discharge ratios $Q/Q_{0.60}$		U.crit. disch. Q_{uc}
Y	(m)	J	A_u (m²)	P_u (m)	B_{us} (m)	y_{um} (m)	medial	Mann'g	(m³s⁻¹)
0.46	0.9465	1.9947	0.3527	1.4907	0.9968	0.3539	0.6482	0.6444	0.6571
0.47	0.9604	1.9970	0.3627	1.5108	0.9982	0.3634	0.6727	0.6691	0.6847
0.48	0.9739	1.9986	0.3727	1.5308	0.9992	0.3730	0.6973	0.6940	0.7128
0.49	0.9871	1.9996	0.3827	1.5508	0.9998	0.3828	0.7222	0.7190	0.7415
0.50	1.0000	2.0000	0.3927	1.5708	1.0000	0.3927	0.7471	0.7442	0.7706
0.51	1.0126	1.9996	0.4027	1.5908	0.9998	0.4028	0.7722	0.7696	0.8003
0.52	1.0248	1.9987	0.4127	1.6108	0.9992	0.4130	0.7974	0.7950	0.8306
0.53	1.0367	1.9971	0.4227	1.6308	0.9982	0.4234	0.8227	0.8205	0.8613
0.54	1.0483	1.9949	0.4327	1.6509	0.9968	0.4340	0.8480	0.8461	0.8926
0.55	1.0595	1.9920	0.4426	1.6710	0.9950	0.4448	0.8734	0.8718	0.9245
0.56	1.0704	1.9886	0.4526	1.6911	0.9928	0.4558	0.8987	0.8975	0.9568
0.57	1.0810	1.9845	0.4625	1.7113	0.9902	0.4671	0.9241	0.9232	0.9898
0.58	1.0912	1.9798	0.4724	1.7315	0.9871	0.4785	0.9495	0.9488	1.0232
0.59	1.1011	1.9746	0.4822	1.7518	0.9837	0.4902	0.9748	0.9744	1.0573
0.60	1.1106	1.9687	0.4920	1.7722	0.9798	0.5022	1.0000	1.0000	1.0919
0.61	1.1197	1.9623	0.5018	1.7926	0.9755	0.5144	1.0252	1.0255	1.1271
0.62	1.1285	1.9553	0.5115	1.8132	0.9708	0.5269	1.0502	1.0508	1.1628
0.63	1.1369	1.9476	0.5212	1.8338	0.9656	0.5398	1.0751	1.0760	1.1992
0.64	1.1449	1.9394	0.5308	1.8546	0.9600	0.5530	1.0998	1.1010	1.2362
0.65	1.1526	1.9306	0.5404	1.8755	0.9539	0.5665	1.1243	1.1259	1.2738
0.66	1.1599	1.9213	0.5499	1.8965	0.9474	0.5804	1.1486	1.1505	1.3120
0.67	1.1667	1.9113	0.5594	1.9177	0.9404	0.5948	1.1727	1.1749	1.3510
0.68	1.1732	1.9007	0.5687	1.9391	0.9330	0.6096	1.1965	1.1990	1.3906
0.69	1.1793	1.8896	0.5780	1.9606	0.9250	0.6249	1.2200	1.2227	1.4309
0.70	1.1849	1.8779	0.5872	1.9823	0.9165	0.6407	1.2431	1.2462	1.4720
0.71	1.1902	1.8655	0.5964	2.0042	0.9075	0.6571	1.2659	1.2693	1.5139
0.72	1.1950	1.8526	0.6054	2.0264	0.8980	0.6741	1.2884	1.2920	1.5566
0.73	1.1994	1.8390	0.6143	2.0488	0.8879	0.6919	1.3104	1.3142	1.6001
0.74	1.2033	1.8249	0.6231	2.0715	0.8773	0.7103	1.3319	1.3360	1.6446
0.75	1.2067	1.8101	0.6319	2.0944	0.8660	0.7296	1.3530	1.3573	1.6901
0.76	1.2097	1.7947	0.6405	2.1176	0.8542	0.7498	1.3735	1.3780	1.7367
0.77	1.2123	1.7786	0.6489	2.1412	0.8417	0.7710	1.3935	1.3982	1.7844
0.78	1.2143	1.7618	0.6573	2.1652	0.8285	0.7933	1.4130	1.4178	1.8334
0.79	1.2158	1.7444	0.6655	2.1895	0.8146	0.8169	1.4317	1.4367	1.8837
0.80	1.2168	1.7263	0.6736	2.2143	0.8000	0.8420	1.4498	1.4549	1.9355
0.81	1.2172	1.7074	0.6815	2.2395	0.7846	0.8686	1.4672	1.4724	1.9890
0.82	1.2171	1.6879	0.6893	2.2653	0.7684	0.8970	1.4838	1.4891	2.0443
0.83	1.2164	1.6675	0.6969	2.2916	0.7513	0.9276	1.4996	1.5049	2.1018
0.84	1.2150	1.6463	0.7043	2.3186	0.7332	0.9605	1.5146	1.5198	2.1616
0.85	1.2131	1.6243	0.7115	2.3462	0.7141	0.9963	1.5286	1.5338	2.2241
0.86	1.2104	1.6013	0.7186	2.3746	0.6940	1.0354	1.5416	1.5467	2.2897
0.87	1.2071	1.5775	0.7254	2.4039	0.6726	1.0785	1.5535	1.5585	2.3591
0.88	1.2029	1.5525	0.7320	2.4341	0.6499	1.1263	1.5643	1.5691	2.4328
0.89	1.1980	1.5265	0.7384	2.4655	0.6258	1.1800	1.5739	1.5784	2.5118
0.90	1.1921	1.4992	0.7445	2.4981	0.6000	1.2409	1.5821	1.5864	2.5972
0.91	1.1853	1.4706	0.7504	2.5322	0.5724	1.3110	1.5889	1.5928	2.6906
0.92	1.1775	1.4404	0.7560	2.5681	0.5426	1.3933	1.5940	1.5975	2.7943
0.93	1.1684	1.4085	0.7612	2.6061	0.5103	1.4917	1.5973	1.6004	2.9115
0.94	1.1579	1.3744	0.7662	2.6467	0.4750	1.6131	1.5986	1.6011	3.0472
0.95	1.1458	1.3378	0.7707	2.6906	0.4359	1.7681	1.5975	1.5994	3.2093
0.96	1.1316	1.2980	0.7749	2.7389	0.3919	1.9771	1.5936	1.5947	3.4119
0.97	1.1148	1.2537	0.7785	2.7934	0.3412	2.2819	1.5861	1.5863	3.6829
0.98	1.0941	1.2027	0.7816	2.8578	0.2800	2.7916	1.5738	1.5728	4.0898
0.99	1.0663	1.1389	0.7841	2.9413	0.1990	3.9401	1.5533	1.5509	4.8738
1.00	1.0000	1.0000	0.7854	3.1416	0.0000	-	1.4942	1.4884	-

C1(a)   Proportional discharges in part-full circular pipes

Prop. depth	Coefficient θ for full pipe $= [\{ k_s / D \} + \{ 1 / (3600\, D\, S^{1/3}) \}]^{-1}$ (Water at 15°C - k_s and D in m)											
	5	10	20	50	100	200	500	1000	2000	5000	10000	20000
0.02	0.0000	0.0002	0.0004	0.0005	0.0006	0.0007	0.0007	0.0008	0.0008	0.0008	0.0008	0.0008
0.04	0.0011	0.0017	0.0021	0.0025	0.0028	0.0030	0.0032	0.0033	0.0034	0.0035	0.0036	0.0036
0.06	0.0036	0.0048	0.0056	0.0064	0.0068	0.0071	0.0075	0.0077	0.0079	0.0081	0.0082	0.0083
0.08	0.0080	0.0097	0.0109	0.0121	0.0127	0.0132	0.0137	0.0141	0.0144	0.0147	0.0149	0.0150
0.10	0.0142	0.0165	0.0182	0.0196	0.0205	0.0212	0.0219	0.0224	0.0228	0.0232	0.0234	0.0236
0.12	0.0224	0.0253	0.0273	0.0291	0.0302	0.0311	0.0320	0.0325	0.0330	0.0335	0.0338	0.0340
0.14	0.0325	0.0360	0.0383	0.0405	0.0418	0.0428	0.0439	0.0446	0.0451	0.0457	0.0461	0.0463
0.16	0.0446	0.0486	0.0513	0.0538	0.0552	0.0564	0.0576	0.0584	0.0590	0.0597	0.0601	0.0604
0.18	0.0587	0.0631	0.0661	0.0689	0.0704	0.0717	0.0731	0.0739	0.0747	0.0754	0.0759	0.0762
0.20	0.0748	0.0795	0.0827	0.0857	0.0874	0.0888	0.0902	0.0912	0.0920	0.0928	0.0933	0.0936
0.22	0.0927	0.0977	0.1011	0.1042	0.1060	0.1075	0.1090	0.1100	0.1109	0.1117	0.1122	0.1126
0.24	0.1124	0.1176	0.1211	0.1244	0.1263	0.1278	0.1294	0.1304	0.1313	0.1322	0.1327	0.1331
0.26	0.1340	0.1392	0.1428	0.1462	0.1481	0.1496	0.1513	0.1525	0.1532	0.1541	0.1546	0.1550
0.28	0.1572	0.1625	0.1661	0.1694	0.1713	0.1729	0.1745	0.1756	0.1765	0.1774	0.1779	0.1783
0.30	0.1821	0.1873	0.1908	0.1941	0.1960	0.1975	0.1992	0.2002	0.2011	0.2020	0.2025	0.2029
0.32	0.2085	0.2135	0.2170	0.2202	0.2220	0.2234	0.2250	0.2260	0.2269	0.2278	0.2283	0.2286
0.34	0.2364	0.2412	0.2444	0.2475	0.2492	0.2506	0.2521	0.2530	0.2538	0.2547	0.2551	0.2555
0.36	0.2657	0.2701	0.2731	0.2759	0.2775	0.2788	0.2802	0.2811	0.2818	0.2826	0.2831	0.2834
0.38	0.2962	0.3002	0.3029	0.3055	0.3069	0.3081	0.3093	0.3101	0.3108	0.3115	0.3119	0.3122
0.40	0.3279	0.3314	0.3338	0.3360	0.3373	0.3383	0.3394	0.3401	0.3407	0.3413	0.3417	0.3419
0.42	0.3607	0.3636	0.3656	0.3674	0.3685	0.3693	0.3703	0.3708	0.3713	0.3719	0.3722	0.3724
0.44	0.3944	0.3967	0.3982	0.3997	0.4005	0.4011	0.4019	0.4023	0.4027	0.4031	0.4033	0.4035
0.46	0.4289	0.4305	0.4316	0.4326	0.4331	0.4336	0.4341	0.4344	0.4347	0.4349	0.4351	0.4352
0.48	0.4642	0.4650	0.4655	0.4661	0.4663	0.4666	0.4668	0.4670	0.4671	0.4673	0.4674	0.4674
0.50	0.5000	0.5000	0.5000	0.5000	0.5000	0.5000	0.5000	0.5000	0.5000	0.5000	0.5000	0.5000
0.52	0.5363	0.5355	0.5349	0.5343	0.5340	0.5338	0.5335	0.5333	0.5332	0.5330	0.5330	0.5329
0.54	0.5729	0.5712	0.5700	0.5689	0.5683	0.5678	0.5672	0.5669	0.5666	0.5663	0.5661	0.5660
0.56	0.6097	0.6071	0.6053	0.6036	0.6026	0.6019	0.6010	0.6005	0.6000	0.5996	0.5993	0.5991
0.58	0.6466	0.6430	0.6406	0.6383	0.6370	0.6360	0.6348	0.6341	0.6335	0.6329	0.6325	0.6322
0.60	0.6834	0.6788	0.6757	0.6729	0.6712	0.6699	0.6685	0.6676	0.6668	0.6660	0.6655	0.6652
0.62	0.7199	0.7144	0.7107	0.7072	0.7052	0.7036	0.7019	0.7008	0.6999	0.6989	0.6984	0.6979
0.64	0.7560	0.7496	0.7452	0.7411	0.7388	0.7370	0.7349	0.7337	0.7326	0.7314	0.7308	0.7303
0.66	0.7915	0.7842	0.7792	0.7745	0.7719	0.7698	0.7675	0.7660	0.7648	0.7635	0.7627	0.7622
0.68	0.8263	0.8181	0.8125	0.8073	0.8043	0.8019	0.7993	0.7977	0.7963	0.7948	0.7940	0.7934
0.70	0.8601	0.8511	0.8449	0.8391	0.8359	0.8333	0.8304	0.8286	0.8271	0.8255	0.8245	0.8239
0.72	0.8928	0.8830	0.8763	0.8700	0.8665	0.8636	0.8605	0.8586	0.8569	0.8552	0.8542	0.8534
0.74	0.9242	0.9137	0.9065	0.8998	0.8960	0.8929	0.8896	0.8875	0.8857	0.8838	0.8827	0.8819
0.76	0.9540	0.9429	0.9352	0.9281	0.9241	0.9208	0.9173	0.9151	0.9132	0.9112	0.9101	0.9092
0.78	0.9821	0.9704	0.9624	0.9549	0.9507	0.9473	0.9436	0.9413	0.9393	0.9372	0.9360	0.9351
0.80	1.0081	0.9960	0.9877	0.9800	0.9756	0.9720	0.9682	0.9658	0.9637	0.9616	0.9603	0.9594

Proportional discharges in part-full circular pipes

Prop. depth	Coefficient θ for full pipe $= [\{ k_s / D \} + \{ 1 / (3600\, D\, S^{1/3}) \}]^{-1}$ (Water at 15°C - k_s and D in m)											
	5	10	20	50	100	200	500	1000	2000	5000	10000	20000
0.82	1.0318	1.0194	1.0109	1.0030	0.9985	0.9948	0.9909	0.9884	0.9863	0.9841	0.9828	0.9819
0.84	1.0529	1.0403	1.0317	1.0237	1.0191	1.0154	1.0115	1.0090	1.0068	1.0046	1.0033	1.0023
0.86	1.0709	1.0583	1.0497	1.0417	1.0372	1.0335	1.0295	1.0270	0.0249	1.0227	1.0214	1.0204
0.88	1.0854	1.0731	1.0646	1.0568	1.0523	1.0487	1.0448	1.0423	1.0403	1.0380	1.0368	1.0359
0.90	1.0959	1.0840	1.0759	1.0683	1.0640	1.0605	1.0568	1.0544	1.0524	1.0503	1.0491	1.0482
0.92	1.1015	1.0904	1.0828	1.0757	1.0716	1.0684	1.0648	1.0626	1.0607	1.0587	1.0576	1.0568
0.94	1.1012	1.0912	1.0843	1.0779	1.0742	1.0713	0.0681	1.0661	1.0644	1.0626	1.0616	1.0608
0.96	1.0929	1.0845	1.0787	1.0733	1.0702	1.0677	1.0650	1.0634	1.0619	1.0604	1.0595	1.0589
0.98	1.0723	1.0662	1.0620	1.0581	1.0559	1.0541	1.0522	1.0510	1.0499	1.0488	1.0482	1.0478
1.00	1.0000	1.0000	1.0000	1.0000	1.0000	1.0000	1.0000	1.0000	1.0000	1.0000	1.0000	1.0000

Corrections to assessed proportional depths for circular pipes, as based on shift of θ ratio.

Prop. Depth	θ/θ_{medial} at 0.60 proportional depth								
	0.05	0.10	0.20	0.50	1	2	5	10	20
0.05	+ 0.0075	+ 0.005	+ 0.003	+ 0.001	--	− 0.001	− 0.0015	− 0.002	− 0.0025
0.10	+ 0.0085	+ 0.0055	+ 0.0035	+ 0.0015	--	− 0.0015	− 0.002	− 0.003	− 0.0035
0.15	+ 0.009	+ 0.006	+ 0.004	+ 0.0015	--	− 0.0015	− 0.0025	− 0.0035	− 0.004
0.20	+ 0.009	+ 0.006	+ 0.004	+ 0.0015	--	− 0.0015	− 0.0025	− 0.0035	− 0.004
0.25	+ 0.0085	+ 0.006	+ 0.0035	+ 0.0015	--	− 0.0015	− 0.0025	− 0.003	− 0.004
0.30	+ 0.008	+ 0.005	+ 0.003	+ 0.001	--	− 0.001	− 0.0025	− 0.003	− 0.0035
0.35	+ 0.007	+ 0.0045	+ 0.0025	+ 0.001	--	− 0.001	− 0.002	− 0.0025	− 0.003
0.40	+ 0.0055	+ 0.004	+ 0.0025	+ 0.001	--	− 0.001	− 0.0015	− 0.002	− 0.0025
0.45	+ 0.0045	+ 0.003	+ 0.002	+ 0.0005	--	− 0.0005	− 0.0015	− 0.0015	− 0.002
0.50	+ 0.003	+ 0.002	+ 0.0015	+ 0.0005	--	− 0.0005	− 0.001	− 0.001	− 0.0015
0.55	+ 0.0015	+ 0.001	+ 0.0005	--	--	--	− 0.0005	− 0.0005	− 0.001
0.60	--	--	--	--	--	--	--	--	--
0.65	− 0.0015	− 0.001	− 0.0005	--	--	--	+ 0.0005	+ 0.001	+ 0.001
0.70	− 0.003	− 0.002	− 0.0015	− 0.0005	--	+ 0.0005	+ 0.001	+ 0.0015	+ 0.0015
0.75	− 0.005	− 0.0035	− 0.0025	− 0.001	--	+ 0.001	+ 0.0015	+ 0.002	+ 0.002
0.80	− 0.007	− 0.005	− 0.003	− 0.001	--	+ 0.001	+ 0.002	+ 0.0025	+ 0.003
0.85	− 0.009	− 0.006	− 0.004	− 0.0015	--	+ 0.0015	+ 0.0025	+ 0.0035	+ 0.004
0.90	− 0.0135	− 0.009	− 0.006	− 0.0025	--	+ 0.002	+ 0.004	+ 0.005	+ 0.0065

Note : These values derive from θ_{medial} for 1.50 m diameter pipe at 0.60 proportional depth with gradient 0.002 and roughness size k_s of 1.50 mm.

C2

Form 1 egg-shape (3:2 old type)

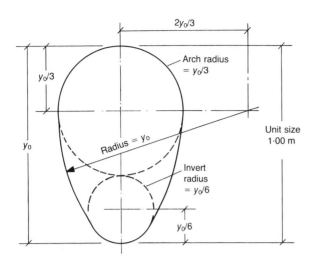

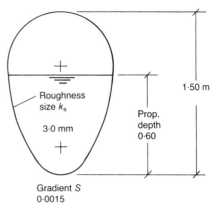

Unit size 1·00 m

Medial case for discharge ratio for Colebrook-White solutions $\Theta_{ep} = 220$

1·50 m

Prop. depth 0·60

Roughness size k_s 3·0 mm

Gradient S 0·0015

Prop dpth	U.equ. dia. $D_{ep(u)}$	Equiv. disch. factor	Unit sect. area	Unit wetted perim.	Unit surf. brdth	Unit mean depth	Discharge ratios $Q/Q_{0.60}$		U.crit. disch. Q_{uc}
Y	(m)	J	A_u (m²)	P_u (m)	B_{us} (m)	y_{um} (m)	medial	Mann'g	(m³s⁻¹)
0.02	0.0518	0.9869	0.0021	0.1650	0.1583	0.0135	0.0012	0.0012	0.0008
0.04	0.1006	1.3403	0.0059	0.2358	0.2166	0.0274	0.0051	0.0051	0.0031
0.06	0.1463	1.5732	0.0107	0.2921	0.2561	0.0417	0.0120	0.0118	0.0068
0.08	0.1883	1.7287	0.0161	0.3422	0.2863	0.0563	0.0215	0.0211	0.0120
0.10	0.2262	1.8167	0.0221	0.3912	0.3146	0.0703	0.0333	0.0328	0.0184
0.11	0.2440	1.8454	0.0253	0.4154	0.3281	0.0772	0.0402	0.0395	0.0220
0.12	0.2612	1.8674	0.0287	0.4393	0.3414	0.0840	0.0476	0.0467	0.0260
0.13	0.2778	1.8843	0.0322	0.4631	0.3543	0.0908	0.0556	0.0546	0.0303
0.14	0.2939	1.8971	0.0358	0.4868	0.3668	0.0975	0.0641	0.0631	0.0350
0.15	0.3097	1.9068	0.0395	0.5102	0.3790	0.1042	0.0733	0.0721	0.0399
0.16	0.3250	1.9140	0.0433	0.5335	0.3910	0.1109	0.0831	0.0817	0.0452
0.17	0.3400	1.9193	0.0473	0.5566	0.4025	0.1175	0.0934	0.0919	0.0508
0.18	0.3547	1.9229	0.0514	0.5796	0.4138	0.1242	0.1043	0.1027	0.0567
0.19	0.3692	1.9252	0.0556	0.6024	0.4248	0.1309	0.1158	0.1141	0.0630
0.20	0.3833	1.9264	0.0599	0.6251	0.4355	0.1375	0.1279	0.1260	0.0696
0.21	0.3972	1.9266	0.0643	0.6476	0.4459	0.1442	0.1406	0.1386	0.0765
0.22	0.4108	1.9261	0.0688	0.6700	0.4561	0.1509	0.1538	0.1517	0.0837
0.23	0.4242	1.9249	0.0734	0.6923	0.4659	0.1576	0.1676	0.1653	0.0913
0.24	0.4374	1.9232	0.0781	0.7145	0.4755	0.1643	0.1819	0.1796	0.0992
0.25	0.4504	1.9209	0.0829	0.7365	0.4848	0.1711	0.1968	0.1943	0.1074
0.26	0.4632	1.9183	0.0878	0.7585	0.4938	0.1778	0.2122	0.2097	0.1160
0.27	0.4757	1.9152	0.0928	0.7803	0.5026	0.1847	0.2282	0.2256	0.1249
0.28	0.4881	1.9118	0.0979	0.8021	0.5111	0.1915	0.2447	0.2420	0.1341
0.29	0.5003	1.9081	0.1030	0.8237	0.5194	0.1984	0.2618	0.2590	0.1437
0.30	0.5123	1.9042	0.1083	0.8453	0.5274	0.2053	0.2793	0.2765	0.1536
0.31	0.5242	1.8999	0.1136	0.8667	0.5351	0.2122	0.2974	0.2945	0.1639
0.32	0.5358	1.8955	0.1190	0.8881	0.5426	0.2192	0.3160	0.3130	0.1744
0.33	0.5473	1.8908	0.1244	0.9094	0.5499	0.2263	0.3351	0.3320	0.1853
0.34	0.5586	1.8860	0.1300	0.9306	0.5569	0.2333	0.3547	0.3516	0.1966
0.35	0.5698	1.8809	0.1356	0.9517	0.5637	0.2405	0.3747	0.3716	0.2082
0.36	0.5808	1.8757	0.1412	0.9727	0.5703	0.2477	0.3952	0.3921	0.2201
0.37	0.5916	1.8703	0.1470	0.9937	0.5766	0.2549	0.4162	0.4131	0.2324
0.38	0.6023	1.8648	0.1528	1.0146	0.5827	0.2622	0.4376	0.4345	0.2449
0.39	0.6128	1.8591	0.1586	1.0355	0.5886	0.2695	0.4595	0.4564	0.2579
0.40	0.6231	1.8533	0.1645	1.0562	0.5942	0.2769	0.4818	0.4787	0.2711
0.41	0.6333	1.8473	0.1705	1.0770	0.5997	0.2843	0.5045	0.5015	0.2847
0.42	0.6433	1.8413	0.1765	1.0976	0.6049	0.2919	0.5276	0.5247	0.2986
0.43	0.6532	1.8350	0.1826	1.1182	0.6098	0.2994	0.5511	0.5483	0.3129
0.44	0.6629	1.8287	0.1887	1.1388	0.6146	0.3071	0.5750	0.5723	0.3275
0.45	0.6725	1.8223	0.1949	1.1593	0.6192	0.3148	0.5993	0.5966	0.3424

Form 1 egg-shape (3:2 old type)

Prop dpth Y	U.equ. dia. $D_{ep(u)}$ (m)	Equiv. disch. factor J	Unit sect. area A_u (m²)	Unit wetted perim. P_u (m)	Unit surf. brdth B_{us} (m)	Unit mean depth y_{um} (m)	Discharge ratios $Q/Q_{0.60}$ medial	Mann'g	U.crit. disch. Q_{uc} (m³s⁻¹)
0.46	0.6819	1.8157	0.2011	1.1798	0.6235	0.3226	0.6240	0.6214	0.3577
0.47	0.6911	1.8090	0.2074	1.2002	0.6276	0.3304	0.6490	0.6465	0.3733
0.48	0.7002	1.8022	0.2137	1.2206	0.6315	0.3383	0.6743	0.6719	0.3892
0.49	0.7091	1.7953	0.2200	1.2409	0.6352	0.3463	0.7000	0.6977	0.4054
0.50	0.7179	1.7883	0.2264	1.2612	0.6387	0.3544	0.7259	0.7239	0.4220
0.51	0.7266	1.7812	0.2328	1.2815	0.6420	0.3626	0.7522	0.7503	0.4389
0.52	0.7350	1.7740	0.2392	1.3017	0.6450	0.3708	0.7788	0.7770	0.4562
0.53	0.7434	1.7666	0.2457	1.3219	0.6479	0.3792	0.8056	0.8040	0.4737
0.54	0.7515	1.7592	0.2522	1.3421	0.6506	0.3876	0.8327	0.8313	0.4916
0.55	0.7596	1.7517	0.2587	1.3622	0.6530	0.3961	0.8601	0.8589	0.5098
0.56	0.7674	1.7441	0.2652	1.3824	0.6553	0.4048	0.8876	0.8867	0.5284
0.57	0.7752	1.7364	0.2718	1.4025	0.6573	0.4135	0.9154	0.9147	0.5473
0.58	0.7827	1.7285	0.2784	1.4225	0.6591	0.4223	0.9434	0.9429	0.5665
0.59	0.7901	1.7206	0.2850	1.4426	0.6608	0.4313	0.9716	0.9714	0.5860
0.60	0.7974	1.7126	0.2916	1.4627	0.6622	0.4403	1.0000	1.0000	0.6059
0.61	0.8045	1.7046	0.2982	1.4827	0.6635	0.4495	1.0286	1.0288	0.6261
0.62	0.8115	1.6964	0.3048	1.5027	0.6645	0.4588	1.0573	1.0578	0.6466
0.63	0.8183	1.6881	0.3115	1.5228	0.6653	0.4682	1.0861	1.0869	0.6675
0.64	0.8249	1.6797	0.3182	1.5428	0.6660	0.4777	1.1150	1.1161	0.6886
0.65	0.8314	1.6713	0.3248	1.5628	0.6664	0.4874	1.1441	1.1454	0.7102
0.66	0.8377	1.6627	0.3315	1.5828	0.6666	0.4973	1.1732	1.1749	0.7320
0.67	0.8439	1.6541	0.3381	1.6028	0.6666	0.5072	1.2025	1.2044	0.7542
0.68	0.8499	1.6454	0.3448	1.6228	0.6661	0.5176	1.2317	1.2340	0.7769
0.69	0.8558	1.6365	0.3515	1.6428	0.6650	0.5285	1.2610	1.2635	0.8001
0.70	0.8614	1.6274	0.3581	1.6629	0.6633	0.5399	1.2902	1.2931	0.8240
0.71	0.8669	1.6181	0.3647	1.6830	0.6610	0.5518	1.3194	1.3225	0.8484
0.72	0.8721	1.6085	0.3713	1.7032	0.6581	0.5643	1.3484	1.3518	0.8735
0.73	0.8770	1.5985	0.3779	1.7235	0.6545	0.5774	1.3771	1.3809	0.8992
0.74	0.8817	1.5883	0.3844	1.7440	0.6503	0.5911	1.4057	1.4098	0.9255
0.75	0.8861	1.5776	0.3909	1.7646	0.6455	0.6056	1.4339	1.4383	0.9526
0.76	0.8902	1.5665	0.3973	1.7853	0.6400	0.6208	1.4618	1.4665	0.9804
0.77	0.8940	1.5549	0.4037	1.8062	0.6338	0.6369	1.4893	1.4942	1.0089
0.78	0.8975	1.5429	0.4100	1.8274	0.6270	0.6540	1.5162	1.5214	1.0383
0.79	0.9006	1.5303	0.4162	1.8488	0.6194	0.6720	1.5427	1.5481	1.0685
0.80	0.9033	1.5171	0.4224	1.8704	0.6110	0.6913	1.5685	1.5742	1.0998
0.81	0.9056	1.5033	0.4284	1.8924	0.6019	0.7118	1.5936	1.5995	1.1320
0.82	0.9075	1.4889	0.4344	1.9148	0.5919	0.7339	1.6180	1.6241	1.1654
0.83	0.9090	1.4738	0.4403	1.9375	0.5811	0.7576	1.6415	1.6478	1.2001
0.84	0.9100	1.4580	0.4460	1.9607	0.5694	0.7833	1.6641	1.6705	1.2362
0.85	0.9105	1.4414	0.4517	1.9843	0.5568	0.8112	1.6857	1.6922	1.2740
0.86	0.9104	1.4240	0.4572	2.0086	0.5431	0.8418	1.7062	1.7128	1.3136
0.87	0.9098	1.4056	0.4625	2.0335	0.5283	0.8756	1.7255	1.7321	1.3553
0.88	0.9086	1.3863	0.4677	2.0591	0.5122	0.9131	1.7434	1.7500	1.3996
0.89	0.9067	1.3658	0.4728	2.0856	0.4949	0.9553	1.7599	1.7664	1.4470
0.90	0.9042	1.3442	0.4776	2.1130	0.4761	1.0032	1.7747	1.7812	1.4981
0.91	0.9008	1.3213	0.4823	2.1416	0.4556	1.0585	1.7877	1.7941	1.5539
0.92	0.8965	1.2969	0.4867	2.1716	0.4333	1.1234	1.7988	1.8049	1.6155
0.93	0.8913	1.2708	0.4909	2.2033	0.4087	1.2011	1.8076	1.8134	1.6849
0.94	0.8849	1.2427	0.4949	2.2370	0.3816	1.2970	1.8139	1.8193	1.7650
0.95	0.8772	1.2122	0.4986	2.2734	0.3512	1.4197	1.8172	1.8221	1.8603
0.96	0.8678	1.1785	0.5019	2.3133	0.3166	1.5851	1.8169	1.8213	1.9788
0.97	0.8563	1.1408	0.5049	2.3583	0.2764	1.8266	1.8122	1.8159	2.1368
0.98	0.8418	1.0967	0.5074	2.4112	0.2274	2.2309	1.8014	1.8042	2.3733
0.99	0.8217	1.0411	0.5094	2.4796	0.1621	3.1429	1.7808	1.7823	2.8279
1.00	0.7725	0.9181	0.5105	2.6433	0.0000	-	1.7155	1.7140	-

C3

Form 1 egg-shape with 5% lining

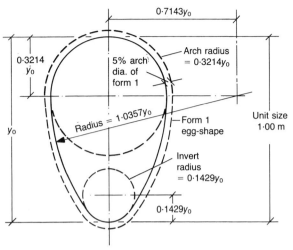

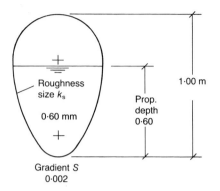

Prop dpth	U.equ. dia. $D_{ep(u)}$	Equiv. disch. factor	Unit sect. area	Unit wetted perim.	Unit surf. brdth	Unit mean depth	Discharge ratios $Q/Q_{0.60}$		U.crit. disch. Q_{uc}
Y	(m)	J	A_u (m²)	P_u (m)	B_{us} (m)	y_{um} (m)	medial	Mann'g	(m³s⁻¹)
0.04	0.0996	1.4276	0.0055	0.2191	0.1983	0.0275	0.0053	0.0051	0.0028
0.08	0.1834	1.7915	0.0147	0.3215	0.2619	0.0563	0.0213	0.0206	0.0109
0.12	0.2518	1.8923	0.0263	0.4180	0.3158	0.0833	0.0469	0.0453	0.0238
0.16	0.3122	1.9167	0.0399	0.5117	0.3645	0.1095	0.0818	0.0794	0.0414
0.20	0.3676	1.9152	0.0554	0.6029	0.4085	0.1356	0.1260	0.1229	0.0639
0.24	0.4193	1.9028	0.0725	0.6922	0.4480	0.1619	0.1794	0.1756	0.0914
0.28	0.4678	1.8851	0.0912	0.7797	0.4833	0.1887	0.2418	0.2375	0.1240
0.32	0.5137	1.8643	0.1112	0.8656	0.5148	0.2160	0.3128	0.3081	0.1618
0.36	0.5570	1.8413	0.1323	0.9503	0.5424	0.2439	0.3919	0.3871	0.2047
0.40	0.5978	1.8166	0.1545	1.0338	0.5665	0.2727	0.4785	0.4738	0.2527
0.42	0.6174	1.8037	0.1660	1.0752	0.5773	0.2875	0.5244	0.5199	0.2786
0.44	0.6363	1.7904	0.1776	1.1164	0.5872	0.3025	0.5720	0.5677	0.3059
0.46	0.6546	1.7768	0.1894	1.1575	0.5962	0.3177	0.6211	0.6172	0.3344
0.48	0.6724	1.7629	0.2014	1.1983	0.6044	0.3333	0.6717	0.6681	0.3642
0.50	0.6896	1.7486	0.2136	1.2390	0.6118	0.3491	0.7236	0.7205	0.3952
0.52	0.7062	1.7340	0.2259	1.2795	0.6184	0.3653	0.7768	0.7742	0.4276
0.54	0.7223	1.7190	0.2383	1.3199	0.6242	0.3818	0.8311	0.8291	0.4612
0.56	0.7377	1.7038	0.2509	1.3602	0.6292	0.3987	0.8865	0.8851	0.4961
0.58	0.7526	1.6882	0.2635	1.4005	0.6335	0.4160	0.9429	0.9421	0.5322
0.60	0.7669	1.6724	0.2762	1.4406	0.6369	0.4337	1.0000	1.0000	0.5696
0.62	0.7806	1.6562	0.2890	1.4807	0.6395	0.4518	1.0579	1.0587	0.6083
0.64	0.7938	1.6397	0.3018	1.5207	0.6414	0.4705	1.1164	1.1180	0.6482
0.66	0.8063	1.6230	0.3146	1.5608	0.6425	0.4897	1.1753	1.1778	0.6894
0.68	0.8183	1.6059	0.3275	1.6008	0.6429	0.5094	1.2347	1.2380	0.7319
0.70	0.8297	1.5885	0.3403	1.6408	0.6414	0.5306	1.2942	1.2985	0.7763
0.72	0.8403	1.5703	0.3531	1.6810	0.6375	0.5539	1.3535	1.3587	0.8230
0.74	0.8500	1.5510	0.3658	1.7215	0.6310	0.5797	1.4122	1.4184	0.8722
0.76	0.8586	1.5304	0.3783	1.7626	0.6219	0.6084	1.4699	1.4769	0.9241
0.78	0.8661	1.5080	0.3907	1.8043	0.6100	0.6404	1.5259	1.5338	0.9790
0.80	0.8722	1.4835	0.4027	1.8469	0.5952	0.6766	1.5799	1.5886	1.0374
0.82	0.8768	1.4567	0.4145	1.8908	0.5773	0.7179	1.6313	1.6406	1.0997
0.84	0.8797	1.4273	0.4258	1.9362	0.5559	0.7659	1.6793	1.6892	1.1669
0.86	0.8806	1.3947	0.4367	1.9835	0.5307	0.8229	1.7234	1.7336	1.2404
0.88	0.8793	1.3586	0.4470	2.0333	0.5010	0.8922	1.7625	1.7728	1.3222
0.90	0.8755	1.3182	0.4567	2.0865	0.4660	0.9800	1.7958	1.8060	1.4157
0.92	0.8686	1.2726	0.4656	2.1442	0.4244	1.0970	1.8218	1.8315	1.5271
0.94	0.8577	1.2201	0.4736	2.2085	0.3740	1.2662	1.8387	1.8475	1.6688
0.96	0.8416	1.1578	0.4805	2.2836	0.3106	1.5470	1.8434	1.8507	1.8714
0.98	0.8166	1.0781	0.4859	2.3797	0.2232	2.1765	1.8295	1.8343	2.2446
1.00	0.7498	0.9033	0.4888	2.6077	0.0000	-	1.7452	1.7435	-

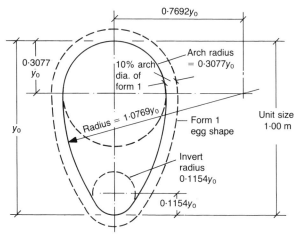

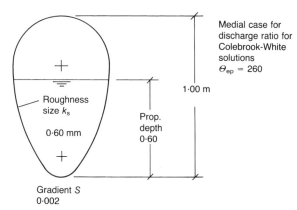

Prop dpth	U.equ. dia. $D_{ep(u)}$	Equiv. disch. factor	Unit sect. area	Unit wetted perim.	Unit surf. brdth	Unit mean depth	Discharge ratios $Q/Q_{0.60}$		U.crit. disch. Q_{uc}
Y	(m)	J	A_u (m²)	P_u (m)	B_{us} (m)	y_{um} (m)	medial	Mann'g	(m³s⁻¹)
0.04	0.0979	1.5512	0.0048	0.1982	0.1747	0.0278	0.0051	0.0049	0.0025
0.08	0.1754	1.8524	0.0130	0.2974	0.2334	0.0559	0.0202	0.0195	0.0097
0.12	0.2385	1.9054	0.0234	0.3932	0.2861	0.0820	0.0445	0.0431	0.0210
0.16	0.2949	1.9045	0.0359	0.4864	0.3339	0.1074	0.0780	0.0759	0.0368
0.20	0.3470	1.8882	0.0501	0.5774	0.3772	0.1328	0.1211	0.1181	0.0572
0.24	0.3960	1.8668	0.0660	0.6664	0.4162	0.1585	0.1735	0.1699	0.0823
0.28	0.4422	1.8432	0.0833	0.7538	0.4513	0.1847	0.2352	0.2310	0.1122
0.32	0.4860	1.8184	0.1020	0.8397	0.4826	0.2114	0.3057	0.3011	0.1469
0.36	0.5275	1.7927	0.1219	0.9244	0.5103	0.2389	0.3846	0.3800	0.1866
0.40	0.5667	1.7663	0.1428	1.0080	0.5345	0.2672	0.4715	0.4669	0.2311
0.42	0.5855	1.7528	0.1536	1.0494	0.5454	0.2816	0.5177	0.5133	0.2553
0.44	0.6037	1.7390	0.1646	1.0906	0.5554	0.2964	0.5656	0.5614	0.2806
0.46	0.6214	1.7250	0.1758	1.1317	0.5647	0.3114	0.6152	0.6113	0.3072
0.48	0.6386	1.7109	0.1872	1.1726	0.5731	0.3266	0.6663	0.6628	0.3350
0.50	0.6552	1.6964	0.1987	1.2133	0.5808	0.3422	0.7189	0.7158	0.3641
0.52	0.6713	1.6818	0.2104	1.2539	0.5876	0.3581	0.7728	0.7702	0.3943
0.54	0.6868	1.6669	0.2222	1.2944	0.5937	0.3743	0.8280	0.8260	0.4258
0.56	0.7018	1.6518	0.2342	1.3347	0.5991	0.3909	0.8844	0.8829	0.4585
0.58	0.7162	1.6364	0.2462	1.3750	0.6036	0.4078	0.9417	0.9410	0.4924
0.60	0.7301	1.6208	0.2583	1.4152	0.6075	0.4252	1.0000	1.0000	0.5275
0.62	0.7435	1.6049	0.2705	1.4553	0.6105	0.4430	1.0591	1.0599	0.5638
0.64	0.7563	1.5888	0.2827	1.4953	0.6128	0.4613	1.1189	1.1205	0.6013
0.66	0.7685	1.5725	0.2950	1.5354	0.6144	0.4801	1.1793	1.1818	0.6401
0.68	0.7802	1.5559	0.3073	1.5754	0.6152	0.4995	1.2401	1.2435	0.6801
0.70	0.7914	1.5391	0.3196	1.6154	0.6152	0.5195	1.3013	1.3056	0.7214
0.72	0.8019	1.5218	0.3319	1.6555	0.6129	0.5415	1.3625	1.3678	0.7648
0.74	0.8117	1.5037	0.3441	1.6958	0.6079	0.5660	1.4233	1.4296	0.8107
0.76	0.8205	1.4843	0.3562	1.7365	0.6003	0.5933	1.4833	1.4905	0.8592
0.78	0.8282	1.4634	0.3681	1.7778	0.5899	0.6240	1.5418	1.5499	0.9106
0.80	0.8346	1.4406	0.3798	1.8200	0.5765	0.6588	1.5984	1.6073	0.9653
0.82	0.8396	1.4156	0.3911	1.8633	0.5599	0.6986	1.6524	1.6621	1.0237
0.84	0.8430	1.3879	0.4021	1.9081	0.5399	0.7449	1.7031	1.7134	1.0869
0.86	0.8445	1.3573	0.4127	1.9547	0.5160	0.7999	1.7498	1.7605	1.1558
0.88	0.8439	1.3231	0.4227	2.0037	0.4876	0.8669	1.7916	1.8025	1.2326
0.90	0.8408	1.2847	0.4322	2.0560	0.4540	0.9518	1.8274	1.8382	1.3204
0.92	0.8347	1.2412	0.4409	2.1127	0.4139	1.0651	1.8557	1.8661	1.4248
0.94	0.8248	1.1909	0.4487	2.1758	0.3651	1.2289	1.8746	1.8841	1.5576
0.96	0.8098	1.1310	0.4554	2.2493	0.3034	1.5008	1.8809	1.8890	1.7470
0.98	0.7863	1.0540	0.4607	2.3435	0.2182	2.1107	1.8679	1.8736	2.0958
1.00	0.7225	0.8843	0.4636	2.5666	0.0000	-	1.7830	1.7821	-

C5

Form 2 egg-shape (3:2 new type)

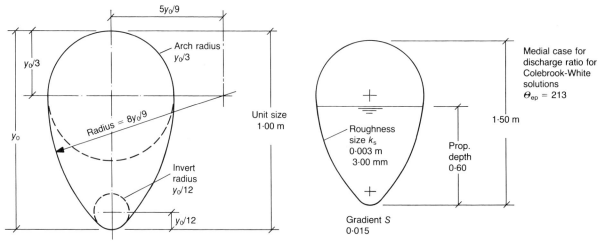

Prop dpth Y	U.equ. dia. $D_{ep(u)}$ (m)	Equiv. disch. factor J	Unit sect. area A_u (m²)	Unit wetted perim. P_u (m)	Unit surf. brdth B_{us} (m)	Unit mean depth y_{um} (m)	Discharge ratios $Q/Q_{0.60}$ medial	Mann'g	U.crit. disch. Q_{uc} (m³s⁻¹)
0.02	0.0503	1.3403	0.0015	0.1179	0.1083	0.0137	0.0008	0.0009	0.0005
0.04	0.0928	1.6611	0.0041	0.1755	0.1497	0.0272	0.0036	0.0036	0.0021
0.06	0.1291	1.7550	0.0075	0.2310	0.1882	0.0396	0.0083	0.0082	0.0046
0.08	0.1626	1.7923	0.0116	0.2850	0.2245	0.0516	0.0150	0.0148	0.0082
0.10	0.1946	1.8106	0.0164	0.3376	0.2586	0.0635	0.0240	0.0237	0.0130
0.11	0.2101	1.8165	0.0191	0.3634	0.2749	0.0694	0.0294	0.0290	0.0157
0.12	0.2254	1.8210	0.0219	0.3889	0.2907	0.0754	0.0354	0.0348	0.0188
0.13	0.2405	1.8246	0.0249	0.4141	0.3061	0.0813	0.0420	0.0413	0.0222
0.14	0.2554	1.8275	0.0280	0.4391	0.3210	0.0873	0.0493	0.0484	0.0259
0.15	0.2701	1.8298	0.0313	0.4638	0.3355	0.0933	0.0571	0.0562	0.0300
0.16	0.2847	1.8317	0.0347	0.4882	0.3496	0.0994	0.0656	0.0645	0.0343
0.17	0.2990	1.8331	0.0383	0.5124	0.3633	0.1055	0.0747	0.0735	0.0390
0.18	0.3132	1.8343	0.0420	0.5365	0.3765	0.1116	0.0845	0.0832	0.0439
0.19	0.3273	1.8351	0.0458	0.5602	0.3894	0.1177	0.0949	0.0935	0.0492
0.20	0.3412	1.8357	0.0498	0.5838	0.4020	0.1239	0.1060	0.1044	0.0549
0.21	0.3549	1.8359	0.0539	0.6073	0.4141	0.1301	0.1177	0.1159	0.0609
0.22	0.3685	1.8360	0.0581	0.6305	0.4259	0.1364	0.1300	0.1281	0.0672
0.23	0.3819	1.8358	0.0624	0.6535	0.4374	0.1427	0.1430	0.1410	0.0738
0.24	0.3952	1.8353	0.0668	0.6764	0.4485	0.1490	0.1566	0.1545	0.0808
0.25	0.4083	1.8347	0.0714	0.6991	0.4593	0.1554	0.1708	0.1686	0.0881
0.26	0.4213	1.8338	0.0760	0.7217	0.4697	0.1618	0.1857	0.1834	0.0957
0.27	0.4341	1.8327	0.0808	0.7441	0.4798	0.1683	0.2012	0.1988	0.1037
0.28	0.4468	1.8314	0.0856	0.7664	0.4897	0.1748	0.2173	0.2148	0.1121
0.29	0.4593	1.8299	0.0905	0.7885	0.4992	0.1814	0.2341	0.2314	0.1208
0.30	0.4717	1.8282	0.0956	0.8105	0.5084	0.1880	0.2514	0.2487	0.1298
0.31	0.4839	1.8263	0.1007	0.8324	0.5173	0.1947	0.2693	0.2665	0.1392
0.32	0.4960	1.8242	0.1059	0.8542	0.5259	0.2014	0.2878	0.2850	0.1489
0.33	0.5080	1.8219	0.1112	0.8759	0.5342	0.2082	0.3069	0.3040	0.1589
0.34	0.5198	1.8194	0.1166	0.8974	0.5423	0.2150	0.3266	0.3236	0.1693
0.35	0.5314	1.8167	0.1221	0.9189	0.5500	0.2219	0.3469	0.3438	0.1801
0.36	0.5429	1.8139	0.1276	0.9402	0.5575	0.2289	0.3676	0.3646	0.1912
0.37	0.5542	1.8108	0.1332	0.9615	0.5647	0.2359	0.3890	0.3859	0.2026
0.38	0.5654	1.8076	0.1389	0.9827	0.5717	0.2430	0.4108	0.4077	0.2144
0.39	0.5765	1.8041	0.1447	1.0038	0.5784	0.2501	0.4332	0.4301	0.2266
0.40	0.5873	1.8005	0.1505	1.0248	0.5848	0.2573	0.4561	0.4530	0.2390
0.41	0.5981	1.7967	0.1564	1.0457	0.5909	0.2646	0.4795	0.4765	0.2518
0.42	0.6087	1.7928	0.1623	1.0666	0.5968	0.2719	0.5034	0.5004	0.2650
0.43	0.6191	1.7886	0.1683	1.0873	0.6025	0.2793	0.5277	0.5248	0.2785
0.44	0.6294	1.7843	0.1743	1.1081	0.6079	0.2868	0.5525	0.5496	0.2924
0.45	0.6395	1.7798	0.1804	1.1287	0.6130	0.2943	0.5778	0.5750	0.3066

Form 2 egg-shape (3:2 new type)

Prop dpth Y	U.equ. dia. $D_{ep(u)}$ (m)	Equiv. disch. factor J	Unit sect. area A_u (m²)	Unit wetted perim. P_u (m)	Unit surf. brdth B_{us} (m)	Unit mean depth y_{um} (m)	Discharge ratios $Q/Q_{0.60}$ medial	Mann'g	U.crit. disch. Q_{uc} (m³s⁻¹)
0.46	0.6494	1.7752	0.1866	1.1493	0.6179	0.3020	0.6034	0.6007	0.3211
0.47	0.6592	1.7704	0.1928	1.1698	0.6226	0.3097	0.6295	0.6270	0.3360
0.48	0.6689	1.7654	0.1991	1.1903	0.6270	0.3175	0.6560	0.6536	0.3512
0.49	0.6784	1.7602	0.2053	1.2107	0.6312	0.3253	0.6829	0.6806	0.3668
0.50	0.6877	1.7549	0.2117	1.2311	0.6351	0.3333	0.7102	0.7080	0.3827
0.51	0.6969	1.7495	0.2180	1.2515	0.6388	0.3413	0.7378	0.7358	0.3989
0.52	0.7059	1.7438	0.2244	1.2718	0.6423	0.3494	0.7658	0.7639	0.4155
0.53	0.7148	1.7380	0.2309	1.2920	0.6455	0.3577	0.7941	0.7924	0.4324
0.54	0.7235	1.7321	0.2374	1.3123	0.6485	0.3660	0.8227	0.8212	0.4497
0.55	0.7321	1.7260	0.2439	1.3324	0.6513	0.3744	0.8516	0.8503	0.4673
0.56	0.7405	1.7198	0.2504	1.3526	0.6538	0.3830	0.8808	0.8798	0.4852
0.57	0.7487	1.7134	0.2569	1.3727	0.6561	0.3916	0.9102	0.9094	0.5035
0.58	0.7567	1.7068	0.2635	1.3928	0.6582	0.4003	0.9399	0.9394	0.5221
0.59	0.7646	1.7001	0.2701	1.4129	0.6600	0.4092	0.9699	0.9696	0.5411
0.60	0.7724	1.6933	0.2767	1.4330	0.6617	0.4182	1.0000	1.0000	0.5604
0.61	0.7800	1.6863	0.2833	1.4530	0.6631	0.4273	1.0304	1.0306	0.5800
0.62	0.7874	1.6792	0.2900	1.4731	0.6642	0.4366	1.0609	1.0614	0.6000
0.63	0.7946	1.6719	0.2966	1.4931	0.6652	0.4459	1.0916	1.0924	0.6203
0.64	0.8017	1.6645	0.3033	1.5131	0.6659	0.4555	1.1224	1.1236	0.6409
0.65	0.8086	1.6570	0.3099	1.5331	0.6664	0.4651	1.1534	1.1548	0.6619
0.66	0.8154	1.6493	0.3166	1.5531	0.6666	0.4749	1.1845	1.1862	0.6832
0.67	0.8220	1.6415	0.3233	1.5731	0.6666	0.4849	1.2156	1.2177	0.7049
0.68	0.8284	1.6335	0.3299	1.5931	0.6661	0.4953	1.2469	1.2493	0.7271
0.69	0.8346	1.6253	0.3366	1.6132	0.6650	0.5061	1.2781	1.2809	0.7499
0.70	0.8406	1.6169	0.3432	1.6332	0.6633	0.5174	1.3093	1.3124	0.7732
0.71	0.8464	1.6082	0.3498	1.6534	0.6610	0.5293	1.3404	1.3438	0.7970
0.72	0.8519	1.5992	0.3564	1.6736	0.6581	0.5416	1.3714	1.3751	0.8215
0.73	0.8572	1.5898	0.3630	1.6939	0.6545	0.5546	1.4021	1.4063	0.8466
0.74	0.8622	1.5800	0.3695	1.7143	0.6503	0.5682	1.4326	1.4371	0.8723
0.75	0.8669	1.5698	0.3760	1.7349	0.6455	0.5825	1.4628	1.4676	0.8987
0.76	0.8713	1.5591	0.3824	1.7556	0.6400	0.5976	1.4927	1.4978	0.9258
0.77	0.8754	1.5480	0.3888	1.7766	0.6338	0.6134	1.5220	1.5275	0.9536
0.78	0.8791	1.5363	0.3951	1.7977	0.6270	0.6302	1.5509	1.5566	0.9823
0.79	0.8825	1.5240	0.4013	1.8191	0.6194	0.6480	1.5792	1.5852	1.0117
0.80	0.8855	1.5112	0.4075	1.8408	0.6110	0.6669	1.6068	1.6131	1.0421
0.81	0.8881	1.4977	0.4136	1.8628	0.6019	0.6871	1.6338	1.6403	1.0735
0.82	0.8902	1.4835	0.4195	1.8851	0.5919	0.7087	1.6599	1.6667	1.1060
0.83	0.8919	1.4686	0.4254	1.9078	0.5811	0.7320	1.6852	1.6921	1.1398
0.84	0.8931	1.4530	0.4312	1.9310	0.5694	0.7571	1.7094	1.7166	1.1748
0.85	0.8938	1.4365	0.4368	1.9547	0.5568	0.7845	1.7326	1.7399	1.2115
0.86	0.8940	1.4192	0.4423	1.9789	0.5431	0.8144	1.7547	1.7620	1.2499
0.87	0.8936	1.4009	0.4476	2.0038	0.5283	0.8474	1.7754	1.7828	1.2904
0.88	0.8925	1.3816	0.4528	2.0294	0.5122	0.8840	1.7947	1.8022	1.3334
0.89	0.8909	1.3613	0.4579	2.0559	0.4949	0.9252	1.8125	1.8199	1.3792
0.90	0.8884	1.3397	0.4627	2.0834	0.4761	0.9719	1.8286	1.8359	1.4286
0.91	0.8852	1.3167	0.4674	2.1120	0.4556	1.0258	1.8427	1.8499	1.4825
0.92	0.8811	1.2923	0.4718	2.1420	0.4333	1.0890	1.8548	1.8617	1.5420
0.93	0.8760	1.2661	0.4761	2.1737	0.4087	1.1647	1.8644	1.8711	1.6089
0.94	0.8698	1.2379	0.4800	2.2074	0.3816	1.2580	1.8714	1.8777	1.6860
0.95	0.8623	1.2073	0.4837	2.2438	0.3512	1.3773	1.8752	1.8811	1.7776
0.96	0.8530	1.1735	0.4870	2.2837	0.3166	1.5380	1.8753	1.8806	1.8914
0.97	0.8417	1.1355	0.4900	2.3286	0.2764	1.7727	1.8707	1.8752	2.0430
0.98	0.8272	1.0912	0.4925	2.3815	0.2274	2.1654	1.8597	1.8632	2.2696
0.99	0.8073	1.0353	0.4945	2.4499	0.1621	3.0511	1.8384	1.8406	2.7049
1.00	0.7584	0.9116	0.4956	2.6136	0.0000	-	1.7703	1.7693	-

C6

Form 2 egg-shape with 3% lining

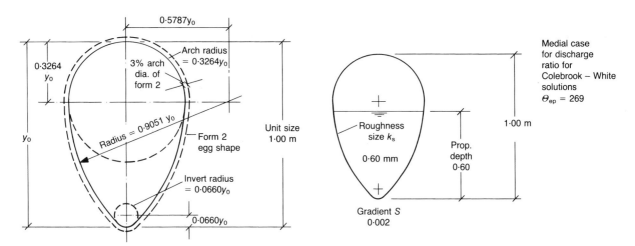

Prop dpth	U.equ. dia. $D_{ep(u)}$	Equiv. disch. factor	Unit sect. area	Unit wetted perim.	Unit surf. brdth	Unit mean depth	Discharge ratios $Q/Q_{0.60}$		U.crit. disch. Q_{uc}
Y	(m)	J	A_u (m²)	P_u (m)	B_{us} (m)	y_{um} (m)	medial	Mann'g	(m³s⁻¹)
0.04	0.0888	1.7174	0.0036	0.1625	0.1352	0.0267	0.0034	0.0033	0.0018
0.08	0.1551	1.7956	0.0105	0.2714	0.2091	0.0503	0.0143	0.0137	0.0074
0.12	0.2158	1.8086	0.0202	0.3749	0.2747	0.0736	0.0341	0.0329	0.0172
0.16	0.2735	1.8127	0.0324	0.4740	0.3331	0.0973	0.0637	0.0617	0.0316
0.20	0.3287	1.8133	0.0468	0.5694	0.3852	0.1215	0.1036	0.1008	0.0511
0.24	0.3816	1.8111	0.0631	0.6619	0.4315	0.1463	0.1538	0.1502	0.0756
0.28	0.4322	1.8060	0.0812	0.7518	0.4726	0.1719	0.2143	0.2100	0.1055
0.32	0.4806	1.7980	0.1009	0.8397	0.5089	0.1982	0.2846	0.2799	0.1407
0.36	0.5267	1.7872	0.1219	0.9258	0.5406	0.2255	0.3644	0.3595	0.1812
0.40	0.5704	1.7736	0.1441	1.0103	0.5681	0.2536	0.4530	0.4482	0.2272
0.42	0.5914	1.7658	0.1556	1.0522	0.5803	0.2681	0.5004	0.4957	0.2522
0.44	0.6118	1.7574	0.1673	1.0937	0.5914	0.2828	0.5497	0.5452	0.2786
0.46	0.6316	1.7483	0.1792	1.1350	0.6016	0.2979	0.6008	0.5967	0.3063
0.48	0.6508	1.7385	0.1913	1.1760	0.6109	0.3132	0.6537	0.6499	0.3354
0.50	0.6694	1.7281	0.2036	1.2169	0.6192	0.3289	0.7081	0.7048	0.3657
0.52	0.6874	1.7172	0.2161	1.2576	0.6265	0.3449	0.7640	0.7612	0.3975
0.54	0.7047	1.7056	0.2287	1.2981	0.6329	0.3613	0.8213	0.8191	0.4305
0.56	0.7215	1.6934	0.2414	1.3384	0.6385	0.3781	0.8798	0.8783	0.4649
0.58	0.7376	1.6807	0.2542	1.3787	0.6431	0.3953	0.9394	0.9386	0.5006
0.60	0.7531	1.6674	0.2671	1.4189	0.6468	0.4130	1.0000	1.0000	0.5376
0.62	0.7679	1.6535	0.2801	1.4590	0.6496	0.4312	1.0614	1.0623	0.5760
0.64	0.7821	1.6391	0.2931	1.4990	0.6515	0.4499	1.1235	1.1253	0.6157
0.66	0.7957	1.6242	0.3062	1.5390	0.6526	0.4691	1.1862	1.1889	0.6567
0.68	0.8086	1.6087	0.3192	1.5790	0.6527	0.4891	1.2493	1.2530	0.6991
0.70	0.8208	1.5926	0.3322	1.6191	0.6506	0.5106	1.3125	1.3172	0.7435
0.72	0.8322	1.5755	0.3452	1.6594	0.6462	0.5343	1.3755	1.3813	0.7902
0.74	0.8425	1.5570	0.3581	1.7000	0.6391	0.5603	1.4378	1.4446	0.8393
0.76	0.8518	1.5369	0.3708	1.7411	0.6295	0.5890	1.4989	1.5067	0.8911
0.78	0.8598	1.5148	0.3832	1.7830	0.6171	0.6210	1.5583	1.5671	0.9457
0.80	0.8663	1.4906	0.3954	1.8258	0.6018	0.6570	1.6155	1.6252	1.0037
0.82	0.8713	1.4638	0.4073	1.8699	0.5834	0.6981	1.6699	1.6803	1.0657
0.84	0.8745	1.4342	0.4187	1.9155	0.5616	0.7456	1.7208	1.7318	1.1323
0.86	0.8756	1.4013	0.4297	1.9630	0.5359	0.8019	1.7674	1.7788	1.2051
0.88	0.8746	1.3647	0.4402	2.0131	0.5057	0.8704	1.8089	1.8204	1.2859
0.90	0.8708	1.3238	0.4499	2.0666	0.4702	0.9568	1.8441	1.8555	1.3782
0.92	0.8640	1.2774	0.4589	2.1247	0.4281	1.0719	1.8717	1.8827	1.4879
0.94	0.8531	1.2241	0.4670	2.1895	0.3772	1.2381	1.8898	1.8997	1.6272
0.96	0.8369	1.1607	0.4739	2.2651	0.3131	1.5135	1.8950	1.9034	1.8258
0.98	0.8118	1.0797	0.4794	2.3619	0.2250	2.1305	1.8807	1.8865	2.1911
1.00	0.7445	0.9025	0.4824	2.5917	0.0000	-	1.7929	1.7920	-

Form 2 egg-shape with 6% lining

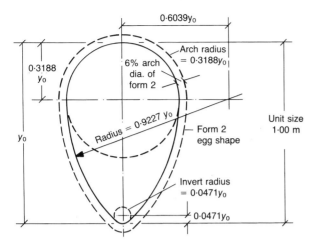

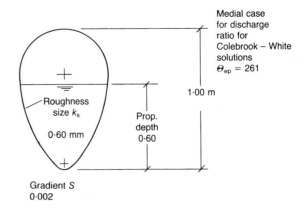

Unit size 1·00 m

Roughness size k_s
0·60 mm

Gradient S
0·002

Medial case for discharge ratio for Colebrook – White solutions $\Theta_{ep} = 261$

Prop. depth 0·60

1·00 m

Prop dpth	U.equ. dia. $D_{ep(u)}$	Equiv. disch. factor	Unit sect. area	Unit wetted perim.	Unit surf. brdth	Unit mean depth	Discharge ratios $Q/Q_{0.60}$		U.crit. disch. Q_{uc}
Y	(m)	J	A_u (m²)	P_u (m)	B_{us} (m)	y_{um} (m)	medial	Mann'g	(m³s⁻¹)
0.04	0.0830	1.7583	0.0031	0.1484	0.1194	0.0258	0.0029	0.0028	0.0015
0.08	0.1456	1.7827	0.0093	0.2566	0.1923	0.0486	0.0129	0.0124	0.0064
0.12	0.2041	1.7833	0.0184	0.3596	0.2572	0.0714	0.0316	0.0305	0.0154
0.16	0.2602	1.7831	0.0298	0.4584	0.3151	0.0946	0.0602	0.0583	0.0287
0.20	0.3141	1.7820	0.0435	0.5537	0.3669	0.1185	0.0991	0.0964	0.0469
0.24	0.3659	1.7789	0.0591	0.6461	0.4131	0.1431	0.1485	0.1451	0.0700
0.28	0.4155	1.7735	0.0765	0.7361	0.4541	0.1684	0.2084	0.2043	0.0982
0.32	0.4630	1.7654	0.0954	0.8239	0.4904	0.1945	0.2785	0.2739	0.1317
0.36	0.5083	1.7547	0.1156	0.9100	0.5223	0.2214	0.3583	0.3534	0.1704
0.40	0.5513	1.7413	0.1371	0.9947	0.5499	0.2493	0.4471	0.4423	0.2144
0.42	0.5720	1.7336	0.1482	1.0365	0.5622	0.2636	0.4948	0.4901	0.2383
0.44	0.5921	1.7253	0.1596	1.0781	0.5735	0.2782	0.5444	0.5399	0.2636
0.46	0.6116	1.7164	0.1712	1.1194	0.5839	0.2931	0.5959	0.5918	0.2902
0.48	0.6305	1.7068	0.1829	1.1605	0.5933	0.3083	0.6493	0.6455	0.3181
0.50	0.6488	1.6967	0.1949	1.2014	0.6018	0.3238	0.7042	0.7009	0.3473
0.52	0.6666	1.6859	0.2070	1.2421	0.6093	0.3397	0.7608	0.7580	0.3778
0.54	0.6837	1.6746	0.2192	1.2826	0.6160	0.3559	0.8188	0.8165	0.4096
0.56	0.7003	1.6628	0.2316	1.3231	0.6217	0.3726	0.8781	0.8765	0.4427
0.58	0.7162	1.6503	0.2441	1.3633	0.6266	0.3896	0.9385	0.9377	0.4771
0.60	0.7315	1.6373	0.2567	1.4035	0.6305	0.4071	1.0000	1.0000	0.5129
0.62	0.7462	1.6238	0.2693	1.4437	0.6336	0.4251	1.0624	1.0633	0.5499
0.64	0.7603	1.6098	0.2820	1.4837	0.6358	0.4435	1.1256	1.1273	0.5882
0.66	0.7738	1.5952	0.2948	1.5238	0.6372	0.4626	1.1894	1.1921	0.6278
0.68	0.7866	1.5802	0.3075	1.5638	0.6377	0.4822	1.2536	1.2574	0.6687
0.70	0.7987	1.5646	0.3202	1.6038	0.6366	0.5031	1.3182	1.3230	0.7113
0.72	0.8101	1.5481	0.3329	1.6439	0.6329	0.5260	1.3826	1.3885	0.7562
0.74	0.8206	1.5304	0.3455	1.6844	0.6267	0.5514	1.4464	1.4534	0.8035
0.76	0.8299	1.5111	0.3580	1.7254	0.6179	0.5794	1.5092	1.5172	0.8534
0.78	0.8381	1.4900	0.3702	1.7671	0.6063	0.6107	1.5704	1.5793	0.9061
0.80	0.8449	1.4667	0.3822	1.8096	0.5917	0.6460	1.6294	1.6392	0.9620
0.82	0.8501	1.4410	0.3939	1.8534	0.5740	0.6862	1.6856	1.6962	1.0218
0.84	0.8536	1.4124	0.4052	1.8986	0.5529	0.7328	1.7383	1.7496	1.0861
0.86	0.8551	1.3806	0.4160	1.9458	0.5279	0.7880	1.7867	1.7984	1.1564
0.88	0.8544	1.3452	0.4263	1.9955	0.4985	0.8551	1.8299	1.8418	1.2344
0.90	0.8512	1.3053	0.4359	2.0485	0.4638	0.9399	1.8667	1.8786	1.3234
0.92	0.8448	1.2601	0.4448	2.1060	0.4224	1.0528	1.8958	1.9072	1.4291
0.94	0.8345	1.2080	0.4527	2.1701	0.3723	1.2159	1.9151	1.9256	1.5633
0.96	0.8189	1.1459	0.4596	2.2449	0.3092	1.4861	1.9213	1.9303	1.7545
0.98	0.7945	1.0664	0.4649	2.3407	0.2223	2.0916	1.9075	1.9140	2.1057
1.00	0.7289	0.8918	0.4679	2.5677	0.0000	-	1.8190	1.8187	-

C8

4:3 egg-shape (WRC)

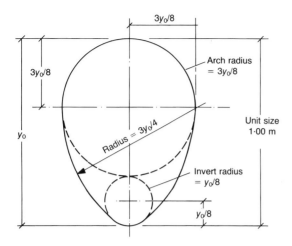

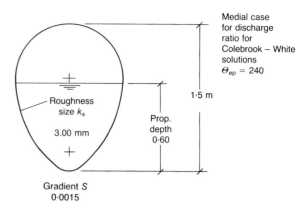

Prop dpth	U.equ. dia. $D_{ep(u)}$	Equiv. disch. factor	Unit sect. area	Unit wetted perim.	Unit surf. brdth	Unit mean depth	Discharge ratios $Q/Q_{0.60}$		U.crit. disch. Q_{uc}
Y	(m)	J	A_u (m²)	P_u (m)	B_{us} (m)	y_{um} (m)	medial	Mann'g	(m³s⁻¹)
0.02	0.0513	1.1246	0.0018	0.1434	0.1356	0.0136	0.0008	0.0009	0.0007
0.04	0.0972	1.4551	0.0051	0.2098	0.1887	0.0270	0.0037	0.0036	0.0026
0.06	0.1375	1.5874	0.0094	0.2721	0.2365	0.0396	0.0085	0.0084	0.0058
0.08	0.1753	1.6603	0.0145	0.3316	0.2805	0.0518	0.0157	0.0154	0.0104
0.10	0.2115	1.7093	0.0206	0.3887	0.3212	0.0640	0.0251	0.0247	0.0163
0.11	0.2292	1.7287	0.0239	0.4165	0.3405	0.0701	0.0308	0.0302	0.0198
0.12	0.2466	1.7460	0.0274	0.4438	0.3590	0.0762	0.0371	0.0364	0.0237
0.13	0.2638	1.7613	0.0310	0.4706	0.3769	0.0824	0.0440	0.0432	0.0279
0.14	0.2808	1.7752	0.0349	0.4970	0.3942	0.0885	0.0516	0.0506	0.0325
0.15	0.2977	1.7878	0.0389	0.5230	0.4108	0.0947	0.0598	0.0587	0.0375
0.16	0.3143	1.7993	0.0431	0.5487	0.4269	0.1010	0.0687	0.0674	0.0429
0.17	0.3307	1.8098	0.0475	0.5740	0.4424	0.1073	0.0782	0.0767	0.0487
0.18	0.3469	1.8195	0.0520	0.5990	0.4574	0.1136	0.0883	0.0867	0.0548
0.19	0.3630	1.8284	0.0566	0.6237	0.4719	0.1199	0.0991	0.0974	0.0614
0.20	0.3789	1.8365	0.0614	0.6481	0.4859	0.1263	0.1106	0.1087	0.0683
0.21	0.3946	1.8440	0.0663	0.6723	0.4994	0.1328	0.1227	0.1206	0.0757
0.22	0.4101	1.8508	0.0714	0.6962	0.5125	0.1393	0.1354	0.1332	0.0834
0.23	0.4255	1.8571	0.0766	0.7198	0.5251	0.1458	0.1488	0.1464	0.0916
0.24	0.4407	1.8627	0.0819	0.7432	0.5373	0.1524	0.1628	0.1603	0.1001
0.25	0.4557	1.8679	0.0873	0.7664	0.5490	0.1590	0.1774	0.1748	0.1090
0.26	0.4705	1.8725	0.0929	0.7894	0.5604	0.1657	0.1926	0.1899	0.1184
0.27	0.4852	1.8767	0.0985	0.8122	0.5713	0.1724	0.2085	0.2056	0.1281
0.28	0.4997	1.8803	0.1043	0.8348	0.5819	0.1792	0.2249	0.2220	0.1383
0.29	0.5140	1.8836	0.1102	0.8573	0.5921	0.1861	0.2420	0.2389	0.1488
0.30	0.5281	1.8864	0.1161	0.8795	0.6019	0.1929	0.2596	0.2565	0.1597
0.31	0.5421	1.8887	0.1222	0.9017	0.6113	0.1999	0.2778	0.2746	0.1711
0.32	0.5559	1.8907	0.1284	0.9236	0.6204	0.2069	0.2966	0.2933	0.1828
0.33	0.5695	1.8922	0.1346	0.9454	0.6291	0.2140	0.3160	0.3126	0.1950
0.34	0.5829	1.8934	0.1409	0.9671	0.6375	0.2211	0.3358	0.3324	0.2075
0.35	0.5961	1.8942	0.1473	0.9887	0.6455	0.2283	0.3563	0.3528	0.2205
0.36	0.6092	1.8946	0.1538	1.0101	0.6532	0.2355	0.3772	0.3737	0.2338
0.37	0.6221	1.8947	0.1604	1.0314	0.6606	0.2428	0.3986	0.3952	0.2475
0.38	0.6348	1.8944	0.1671	1.0527	0.6677	0.2502	0.4206	0.4171	0.2617
0.39	0.6473	1.8938	0.1738	1.0738	0.6745	0.2576	0.4430	0.4395	0.2762
0.40	0.6596	1.8929	0.1805	1.0948	0.6809	0.2651	0.4659	0.4625	0.2911
0.41	0.6718	1.8916	0.1874	1.1157	0.6870	0.2727	0.4893	0.4859	0.3064
0.42	0.6838	1.8900	0.1943	1.1365	0.6929	0.2804	0.5131	0.5097	0.3222
0.43	0.6956	1.8881	0.2012	1.1573	0.6984	0.2881	0.5373	0.5340	0.3383
0.44	0.7072	1.8859	0.2082	1.1780	0.7037	0.2960	0.5619	0.5588	0.3548
0.45	0.7186	1.8834	0.2153	1.1986	0.7086	0.3039	0.5870	0.5839	0.3717

Prop dpth	U.equ. dia. $D_{ep(u)}$	Equiv. disch. factor	Unit sect. area	Unit wetted perim.	Unit surf. brdth	Unit mean depth	Discharge ratios $Q/Q_{0.60}$		U.crit. disch. Q_{uc}
Y	(m)	J	A_u (m^2)	P_u (m)	B_{us} (m)	y_{um} (m)	medial	Mann'g	(m^{3}s^{-1})
0.46	0.7298	1.8806	0.2224	1.2191	0.7132	0.3118	0.6124	0.6095	0.3890
0.47	0.7408	1.8775	0.2296	1.2396	0.7176	0.3199	0.6382	0.6354	0.4066
0.48	0.7517	1.8741	0.2368	1.2600	0.7217	0.3281	0.6644	0.6617	0.4247
0.49	0.7623	1.8705	0.2440	1.2803	0.7255	0.3363	0.6909	0.6883	0.4431
0.50	0.7728	1.8665	0.2513	1.3006	0.7290	0.3447	0.7177	0.7153	0.4620
0.51	0.7831	1.8623	0.2586	1.3209	0.7323	0.3531	0.7448	0.7426	0.4812
0.52	0.7931	1.8579	0.2659	1.3411	0.7352	0.3617	0.7722	0.7702	0.5008
0.53	0.8030	1.8532	0.2733	1.3613	0.7379	0.3704	0.7999	0.7981	0.5208
0.54	0.8127	1.8482	0.2807	1.3815	0.7403	0.3791	0.8279	0.8263	0.5412
0.55	0.8222	1.8429	0.2881	1.4016	0.7425	0.3880	0.8561	0.8547	0.5620
0.56	0.8315	1.8374	0.2955	1.4217	0.7444	0.3970	0.8845	0.8834	0.5831
0.57	0.8406	1.8317	0.3030	1.4417	0.7460	0.4062	0.9131	0.9123	0.6047
0.58	0.8495	1.8257	0.3104	1.4618	0.7473	0.4154	0.9419	0.9413	0.6266
0.59	0.8582	1.8195	0.3179	1.4818	0.7484	0.4248	0.9709	0.9706	0.6489
0.60	0.8667	1.8130	0.3254	1.5018	0.7492	0.4344	1.0000	1.0000	0.6716
0.61	0.8750	1.8063	0.3329	1.5218	0.7497	0.4441	1.0293	1.0296	0.6947
0.62	0.8831	1.7994	0.3404	1.5418	0.7500	0.4539	1.0587	1.0592	0.7182
0.63	0.8910	1.7923	0.3479	1.5618	0.7499	0.4639	1.0881	1.0890	0.7421
0.64	0.8987	1.7849	0.3554	1.5818	0.7494	0.4743	1.1177	1.1189	0.7665
0.65	0.9062	1.7772	0.3629	1.6019	0.7483	0.4849	1.1473	1.1488	0.7914
0.66	0.9134	1.7692	0.3704	1.6219	0.7467	0.4960	1.1768	1.1787	0.8168
0.67	0.9204	1.7609	0.3778	1.6420	0.7446	0.5074	1.2063	1.2085	0.8428
0.68	0.9271	1.7522	0.3853	1.6622	0.7419	0.5193	1.2358	1.2383	0.8694
0.69	0.9335	1.7431	0.3927	1.6825	0.7386	0.5316	1.2651	1.2679	0.8965
0.70	0.9397	1.7336	0.4000	1.7028	0.7348	0.5444	1.2942	1.2973	0.9243
0.71	0.9455	1.7237	0.4074	1.7233	0.7305	0.5577	1.3231	1.3266	0.9526
0.72	0.9511	1.7132	0.4146	1.7439	0.7255	0.5715	1.3517	1.3555	0.9816
0.73	0.9563	1.7024	0.4219	1.7647	0.7200	0.5859	1.3801	1.3842	1.0112
0.74	0.9611	1.6910	0.4290	1.7856	0.7139	0.6010	1.4081	1.4125	1.0416
0.75	0.9656	1.6790	0.4361	1.8067	0.7071	0.6168	1.4356	1.4404	1.0727
0.76	0.9697	1.6665	0.4432	1.8280	0.6997	0.6334	1.4627	1.4678	1.1045
0.77	0.9735	1.6535	0.4501	1.8496	0.6917	0.6508	1.4893	1.4946	1.1372
0.78	0.9768	1.6398	0.4570	1.8714	0.6829	0.6692	1.5154	1.5209	1.1707
0.79	0.9797	1.6255	0.4638	1.8935	0.6735	0.6886	1.5408	1.5466	1.2052
0.80	0.9822	1.6105	0.4705	1.9160	0.6633	0.7093	1.5655	1.5715	1.2408
0.81	0.9842	1.5949	0.4771	1.9388	0.6524	0.7313	1.5895	1.5957	1.2775
0.82	0.9858	1.5785	0.4835	1.9620	0.6406	0.7548	1.6126	1.6190	1.3155
0.83	0.9868	1.5613	0.4899	1.9856	0.6280	0.7800	1.6349	1.6414	1.3548
0.84	0.9873	1.5434	0.4961	2.0097	0.6145	0.8073	1.6561	1.6628	1.3958
0.85	0.9873	1.5246	0.5022	2.0344	0.6000	0.8369	1.6764	1.6831	1.4386
0.86	0.9867	1.5048	0.5081	2.0598	0.5845	0.8693	1.6955	1.7022	1.4834
0.87	0.9854	1.4842	0.5138	2.0858	0.5678	0.9050	1.7133	1.7200	1.5307
0.88	0.9835	1.4624	0.5194	2.1126	0.5499	0.9446	1.7298	1.7365	1.5809
0.89	0.9808	1.4396	0.5248	2.1404	0.5307	0.9890	1.7448	1.7514	1.6345
0.90	0.9774	1.4154	0.5300	2.1692	0.5099	1.0395	1.7581	1.7646	1.6923
0.91	0.9731	1.3900	0.5350	2.1993	0.4874	1.0976	1.7697	1.7760	1.7553
0.92	0.9678	1.3629	0.5398	2.2309	0.4630	1.1657	1.7793	1.7853	1.8251
0.93	0.9615	1.3341	0.5443	2.2642	0.4363	1.2473	1.7867	1.7924	1.9036
0.94	0.9540	1.3032	0.5485	2.2998	0.4069	1.3479	1.7916	1.7968	1.9941
0.95	0.9450	1.2697	0.5524	2.3382	0.3742	1.4764	1.7936	1.7983	2.1019
0.96	0.9343	1.2330	0.5560	2.3803	0.3370	1.6495	1.7921	1.7961	2.2361
0.97	0.9212	1.1919	0.5591	2.4279	0.2939	1.9022	1.7862	1.7894	2.4149
0.98	0.9047	1.1443	0.5618	2.4839	0.2417	2.3248	1.7743	1.7766	2.6825
0.99	0.8824	1.0844	0.5639	2.5563	0.1720	3.2776	1.7528	1.7536	3.1970
1.00	0.8279	0.9528	0.5651	2.7299	0.0000	-	1.6865	1.6842	-

C9

4:3 egg-shape with 3% lining

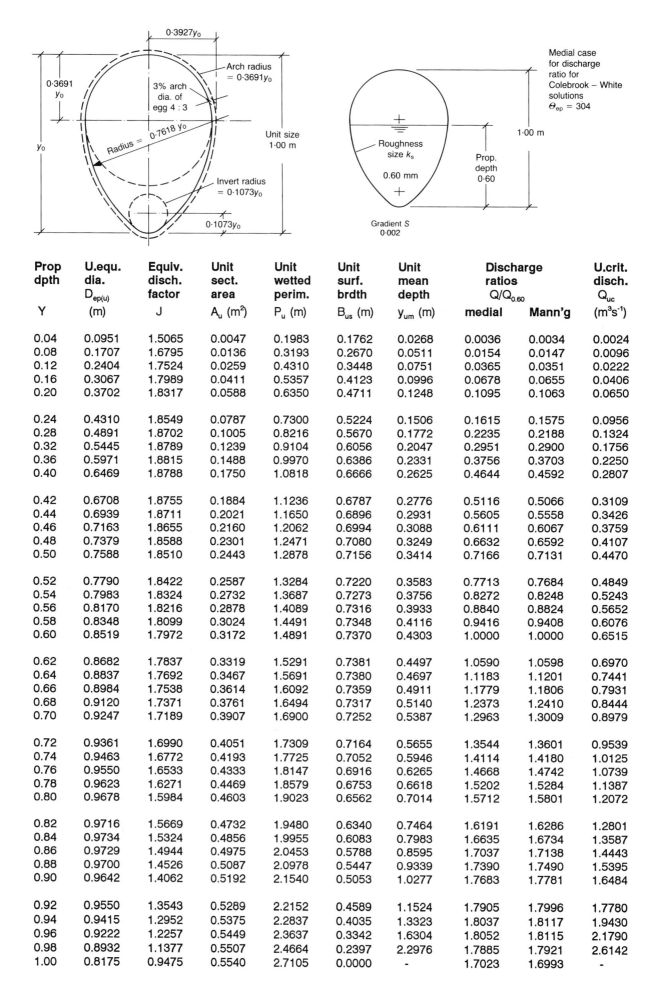

Prop dpth	U.equ. dia. $D_{ep(u)}$	Equiv. disch. factor	Unit sect. area	Unit wetted perim.	Unit surf. brdth	Unit mean depth	Discharge ratios $Q/Q_{0.60}$		U.crit. disch. Q_{uc}
Y	(m)	J	A_u (m²)	P_u (m)	B_{us} (m)	y_{um} (m)	medial	Mann'g	(m³s⁻¹)
0.04	0.0951	1.5065	0.0047	0.1983	0.1762	0.0268	0.0036	0.0034	0.0024
0.08	0.1707	1.6795	0.0136	0.3193	0.2670	0.0511	0.0154	0.0147	0.0096
0.12	0.2404	1.7524	0.0259	0.4310	0.3448	0.0751	0.0365	0.0351	0.0222
0.16	0.3067	1.7989	0.0411	0.5357	0.4123	0.0996	0.0678	0.0655	0.0406
0.20	0.3702	1.8317	0.0588	0.6350	0.4711	0.1248	0.1095	0.1063	0.0650
0.24	0.4310	1.8549	0.0787	0.7300	0.5224	0.1506	0.1615	0.1575	0.0956
0.28	0.4891	1.8702	0.1005	0.8216	0.5670	0.1772	0.2235	0.2188	0.1324
0.32	0.5445	1.8789	0.1239	0.9104	0.6056	0.2047	0.2951	0.2900	0.1756
0.36	0.5971	1.8815	0.1488	0.9970	0.6386	0.2331	0.3756	0.3703	0.2250
0.40	0.6469	1.8788	0.1750	1.0818	0.6666	0.2625	0.4644	0.4592	0.2807
0.42	0.6708	1.8755	0.1884	1.1236	0.6787	0.2776	0.5116	0.5066	0.3109
0.44	0.6939	1.8711	0.2021	1.1650	0.6896	0.2931	0.5605	0.5558	0.3426
0.46	0.7163	1.8655	0.2160	1.2062	0.6994	0.3088	0.6111	0.6067	0.3759
0.48	0.7379	1.8588	0.2301	1.2471	0.7080	0.3249	0.6632	0.6592	0.4107
0.50	0.7588	1.8510	0.2443	1.2878	0.7156	0.3414	0.7166	0.7131	0.4470
0.52	0.7790	1.8422	0.2587	1.3284	0.7220	0.3583	0.7713	0.7684	0.4849
0.54	0.7983	1.8324	0.2732	1.3687	0.7273	0.3756	0.8272	0.8248	0.5243
0.56	0.8170	1.8216	0.2878	1.4089	0.7316	0.3933	0.8840	0.8824	0.5652
0.58	0.8348	1.8099	0.3024	1.4491	0.7348	0.4116	0.9416	0.9408	0.6076
0.60	0.8519	1.7972	0.3172	1.4891	0.7370	0.4303	1.0000	1.0000	0.6515
0.62	0.8682	1.7837	0.3319	1.5291	0.7381	0.4497	1.0590	1.0598	0.6970
0.64	0.8837	1.7692	0.3467	1.5691	0.7380	0.4697	1.1183	1.1201	0.7441
0.66	0.8984	1.7538	0.3614	1.6092	0.7359	0.4911	1.1779	1.1806	0.7931
0.68	0.9120	1.7371	0.3761	1.6494	0.7317	0.5140	1.2373	1.2410	0.8444
0.70	0.9247	1.7189	0.3907	1.6900	0.7252	0.5387	1.2963	1.3009	0.8979
0.72	0.9361	1.6990	0.4051	1.7309	0.7164	0.5655	1.3544	1.3601	0.9539
0.74	0.9463	1.6772	0.4193	1.7725	0.7052	0.5946	1.4114	1.4180	1.0125
0.76	0.9550	1.6533	0.4333	1.8147	0.6916	0.6265	1.4668	1.4742	1.0739
0.78	0.9623	1.6271	0.4469	1.8579	0.6753	0.6618	1.5202	1.5284	1.1387
0.80	0.9678	1.5984	0.4603	1.9023	0.6562	0.7014	1.5712	1.5801	1.2072
0.82	0.9716	1.5669	0.4732	1.9480	0.6340	0.7464	1.6191	1.6286	1.2801
0.84	0.9734	1.5324	0.4856	1.9955	0.6083	0.7983	1.6635	1.6734	1.3587
0.86	0.9729	1.4944	0.4975	2.0453	0.5788	0.8595	1.7037	1.7138	1.4443
0.88	0.9700	1.4526	0.5087	2.0978	0.5447	0.9339	1.7390	1.7490	1.5395
0.90	0.9642	1.4062	0.5192	2.1540	0.5053	1.0277	1.7683	1.7781	1.6484
0.92	0.9550	1.3543	0.5289	2.2152	0.4589	1.1524	1.7905	1.7996	1.7780
0.94	0.9415	1.2952	0.5375	2.2837	0.4035	1.3323	1.8037	1.8117	1.9430
0.96	0.9222	1.2257	0.5449	2.3637	0.3342	1.6304	1.8052	1.8115	2.1790
0.98	0.8932	1.1377	0.5507	2.4664	0.2397	2.2976	1.7885	1.7921	2.6142
1.00	0.8175	0.9475	0.5540	2.7105	0.0000	-	1.7023	1.6993	-

4:3 egg-shape with 6% lining

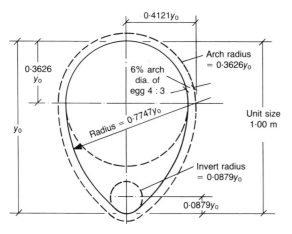

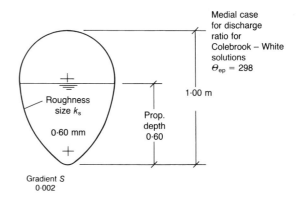

Prop dpth Y	U.equ. dia. $D_{ep(u)}$ (m)	Equiv. disch. factor J	Unit sect. area A_u (m²)	Unit wetted perim. P_u (m)	Unit surf. brdth B_{us} (m)	Unit mean depth y_{um} (m)	Discharge ratios $Q/Q_{0.60}$ medial	Mann'g	U.crit. disch. Q_{uc} (m³s⁻¹)
0.04	0.0922	1.5597	0.0043	0.1856	0.1624	0.0263	0.0033	0.0032	0.0022
0.08	0.1650	1.6954	0.0126	0.3058	0.2520	0.0500	0.0145	0.0139	0.0088
0.12	0.2329	1.7549	0.0243	0.4169	0.3291	0.0737	0.0349	0.0336	0.0206
0.16	0.2978	1.7946	0.0388	0.5213	0.3962	0.0980	0.0655	0.0633	0.0380
0.20	0.3601	1.8232	0.0559	0.6205	0.4548	0.1228	0.1066	0.1035	0.0613
0.24	0.4198	1.8434	0.0751	0.7155	0.5059	0.1484	0.1580	0.1541	0.0906
0.28	0.4770	1.8567	0.0962	0.8071	0.5506	0.1748	0.2197	0.2151	0.1260
0.32	0.5315	1.8637	0.1191	0.8960	0.5893	0.2020	0.2910	0.2860	0.1676
0.36	0.5834	1.8652	0.1433	0.9826	0.6226	0.2302	0.3715	0.3663	0.2153
0.40	0.6325	1.8615	0.1688	1.0675	0.6508	0.2594	0.4604	0.4553	0.2692
0.42	0.6560	1.8579	0.1819	1.1093	0.6630	0.2744	0.5078	0.5028	0.2985
0.44	0.6789	1.8532	0.1953	1.1508	0.6742	0.2897	0.5570	0.5522	0.3292
0.46	0.7010	1.8473	0.2089	1.1920	0.6841	0.3053	0.6078	0.6034	0.3615
0.48	0.7224	1.8405	0.2227	1.2330	0.6930	0.3213	0.6602	0.6562	0.3953
0.50	0.7430	1.8326	0.2366	1.2738	0.7007	0.3377	0.7140	0.7105	0.4306
0.52	0.7630	1.8237	0.2507	1.3143	0.7074	0.3544	0.7691	0.7662	0.4673
0.54	0.7822	1.8138	0.2649	1.3547	0.7130	0.3715	0.8254	0.8231	0.5056
0.56	0.8006	1.8030	0.2792	1.3950	0.7175	0.3891	0.8828	0.8812	0.5454
0.58	0.8183	1.7913	0.2936	1.4351	0.7210	0.4072	0.9410	0.9402	0.5867
0.60	0.8353	1.7787	0.3080	1.4752	0.7235	0.4258	1.0000	1.0000	0.6294
0.62	0.8514	1.7653	0.3225	1.5152	0.7249	0.4449	1.0596	1.0605	0.6737
0.64	0.8668	1.7510	0.3370	1.5552	0.7253	0.4647	1.1197	1.1215	0.7195
0.66	0.8814	1.7358	0.3515	1.5952	0.7239	0.4856	1.1801	1.1828	0.7671
0.68	0.8951	1.7194	0.3660	1.6354	0.7202	0.5081	1.2404	1.2442	0.8169
0.70	0.9078	1.7017	0.3803	1.6758	0.7144	0.5324	1.3004	1.3051	0.8690
0.72	0.9193	1.6823	0.3945	1.7167	0.7062	0.5587	1.3596	1.3653	0.9234
0.74	0.9295	1.6610	0.4085	1.7581	0.6956	0.5873	1.4177	1.4243	0.9805
0.76	0.9384	1.6377	0.4223	1.8001	0.6825	0.6188	1.4742	1.4818	1.0403
0.78	0.9458	1.6121	0.4358	1.8431	0.6668	0.6536	1.5288	1.5372	1.1034
0.80	0.9516	1.5841	0.4490	1.8872	0.6482	0.6926	1.5809	1.5900	1.1702
0.82	0.9556	1.5533	0.4617	1.9327	0.6266	0.7369	1.6300	1.6397	1.2413
0.84	0.9576	1.5194	0.4740	1.9800	0.6015	0.7881	1.6756	1.6857	1.3178
0.86	0.9575	1.4822	0.4858	2.0294	0.5725	0.8485	1.7169	1.7273	1.4013
0.88	0.9549	1.4411	0.4969	2.0816	0.5390	0.9219	1.7533	1.7636	1.4940
0.90	0.9494	1.3954	0.5073	2.1374	0.5001	1.0144	1.7836	1.7937	1.6000
0.92	0.9405	1.3442	0.5169	2.1981	0.4544	1.1374	1.8067	1.8161	1.7262
0.94	0.9275	1.2858	0.5254	2.2660	0.3996	1.3149	1.8207	1.8290	1.8867
0.96	0.9086	1.2171	0.5327	2.3453	0.3311	1.6089	1.8226	1.8293	2.1162
0.98	0.8802	1.1299	0.5385	2.4472	0.2375	2.2670	1.8062	1.8103	2.5391
1.00	0.8057	0.9412	0.5417	2.6892	0.0000	-	1.7195	1.7168	-

C11

U-shaped (free surface)

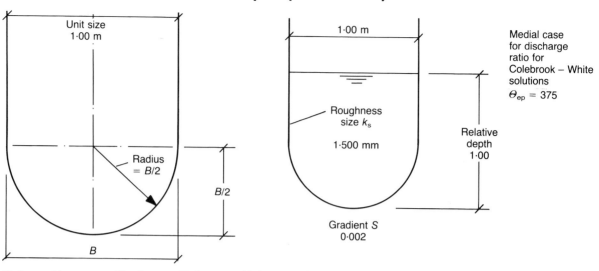

Rel. dpth	U.equ. dia. $D_{ep(u)}$	Equiv. disch. factor	Unit sect. area	Unit wetted perim.	Unit surf. brdth	Unit mean depth	Discharge ratios $Q/Q_{1.00}$		U.crit. disch. Q_{uc}
Y	(m)	J	A_u (m²)	P_u (m)	B_{us} (m)	y_{um} (m)	medial	Mann'g	(m³s⁻¹)
0.02	0.0528	0.5849	0.0037	0.2838	0.2800	0.0134	0.0005	0.0005	0.0014
0.04	0.1047	0.8165	0.0105	0.4027	0.3919	0.0269	0.0022	0.0021	0.0054
0.06	0.1555	0.9869	0.0192	0.4949	0.4750	0.0405	0.0052	0.0050	0.0121
0.08	0.2053	1.1246	0.0294	0.5735	0.5426	0.0542	0.0096	0.0092	0.0215
0.10	0.2541	1.2404	0.0409	0.6435	0.6000	0.0681	0.0154	0.0148	0.0334
0.12	0.3018	1.3403	0.0534	0.7075	0.6499	0.0821	0.0225	0.0216	0.0479
0.14	0.3485	1.4276	0.0668	0.7670	0.6940	0.0963	0.0310	0.0298	0.0649
0.16	0.3942	1.5047	0.0811	0.8230	0.7332	0.1106	0.0407	0.0392	0.0845
0.18	0.4388	1.5732	0.0961	0.8763	0.7684	0.1251	0.0518	0.0500	0.1065
0.20	0.4824	1.6342	0.1118	0.9273	0.8000	0.1398	0.0640	0.0619	0.1309
0.22	0.5248	1.6886	0.1281	0.9764	0.8285	0.1546	0.0774	0.0750	0.1578
0.24	0.5662	1.7372	0.1449	1.0239	0.8542	0.1697	0.0920	0.0893	0.1870
0.26	0.6065	1.7805	0.1623	1.0701	0.8773	0.1850	0.1076	0.1046	0.2185
0.28	0.6457	1.8189	0.1800	1.1152	0.8980	0.2005	0.1243	0.1210	0.2524
0.30	0.6838	1.8529	0.1982	1.1593	0.9165	0.2162	0.1419	0.1384	0.2886
0.32	0.7207	1.8828	0.2167	1.2025	0.9330	0.2322	0.1604	0.1567	0.3270
0.34	0.7565	1.9088	0.2355	1.2451	0.9474	0.2485	0.1798	0.1759	0.3676
0.36	0.7911	1.9311	0.2546	1.2870	0.9600	0.2652	0.2000	0.1959	0.4105
0.38	0.8246	1.9501	0.2739	1.3284	0.9708	0.2821	0.2209	0.2167	0.4555
0.40	0.8569	1.9658	0.2934	1.3694	0.9798	0.2994	0.2425	0.2382	0.5027
0.42	0.8880	1.9783	0.3130	1.4101	0.9871	0.3171	0.2647	0.2602	0.5521
0.44	0.9179	1.9879	0.3328	1.4505	0.9928	0.3353	0.2874	0.2829	0.6035
0.46	0.9465	1.9947	0.3527	1.4907	0.9968	0.3539	0.3105	0.3060	0.6571
0.48	0.9739	1.9986	0.3727	1.5308	0.9992	0.3730	0.3341	0.3295	0.7128
0.50	1.0000	2.0000	0.3927	1.5708	1.0000	0.3927	0.3579	0.3534	0.7706
0.52	1.0248	1.9987	0.4127	1.6108	1.0000	0.4127	0.3820	0.3775	0.8303
0.54	1.0485	1.9953	0.4327	1.6508	1.0000	0.4327	0.4063	0.4018	0.8913
0.56	1.0710	1.9899	0.4527	1.6908	1.0000	0.4527	0.4308	0.4264	0.9538
0.58	1.0924	1.9829	0.4727	1.7308	1.0000	0.4727	0.4555	0.4512	1.0177
0.60	1.1129	1.9744	0.4927	1.7708	1.0000	0.4927	0.4803	0.4761	1.0830
0.62	1.1325	1.9648	0.5127	1.8108	1.0000	0.5127	0.5054	0.5013	1.1496
0.64	1.1513	1.9542	0.5327	1.8508	1.0000	0.5327	0.5305	0.5265	1.2175
0.66	1.1692	1.9427	0.5527	1.8908	1.0000	0.5527	0.5558	0.5520	1.2867
0.68	1.1865	1.9304	0.5727	1.9308	1.0000	0.5727	0.5812	0.5776	1.3572
0.70	1.2030	1.9176	0.5927	1.9708	1.0000	0.5927	0.6068	0.6033	1.4289
0.72	1.2188	1.9042	0.6127	2.0108	1.0000	0.6127	0.6324	0.6291	1.5019
0.74	1.2341	1.8904	0.6327	2.0508	1.0000	0.6327	0.6582	0.6550	1.5760
0.76	1.2487	1.8762	0.6527	2.0908	1.0000	0.6527	0.6840	0.6811	1.6513
0.78	1.2628	1.8618	0.6727	2.1308	1.0000	0.6727	0.7100	0.7072	1.7278
0.80	1.2764	1.8472	0.6927	2.1708	1.0000	0.6927	0.7360	0.7334	1.8054

Rel. dpth	U.equ. dia. $D_{ep(u)}$	Equiv. disch. factor	Unit sect. area	Unit wetted perim.	Unit surf. brdth	Unit mean depth	Discharge ratios $Q/Q_{1.00}$		U.crit. disch. Q_{uc}
Y	(m)	J	A_u (m²)	P_u (m)	B_{us} (m)	y_{um} (m)	medial	Mann'g	(m³s⁻¹)
0.82	1.2895	1.8323	0.7127	2.2108	1.0000	0.7127	0.7621	0.7598	1.8842
0.84	1.3021	1.8174	0.7327	2.2508	1.0000	0.7327	0.7883	0.7862	1.9640
0.86	1.3143	1.8024	0.7527	2.2908	1.0000	0.7527	0.8145	0.8127	2.0450
0.88	1.3261	1.7873	0.7727	2.3308	1.0000	0.7727	0.8409	0.8392	2.1270
0.90	1.3374	1.7722	0.7927	2.3708	1.0000	0.7927	0.8672	0.8659	2.2102
0.92	1.3484	1.7571	0.8127	2.4108	1.0000	0.8127	0.8937	0.8926	2.2943
0.94	1.3591	1.7421	0.8327	2.4508	1.0000	0.8327	0.9202	0.9193	2.3795
0.96	1.3694	1.7271	0.8527	2.4908	1.0000	0.8527	0.9468	0.9462	2.4658
0.98	1.3793	1.7122	0.8727	2.5308	1.0000	0.8727	0.9734	0.9731	2.5530
1.00	1.3890	1.6973	0.8927	2.5708	1.0000	0.8927	1.0000	1.0000	2.6413
1.02	1.3983	1.6826	0.9127	2.6108	1.0000	0.9127	1.0267	1.0270	2.7306
1.04	1.4074	1.6680	0.9327	2.6508	1.0000	0.9327	1.0535	1.0540	2.8208
1.06	1.4162	1.6535	0.9527	2.6908	1.0000	0.9527	1.0803	1.0811	2.9120
1.08	1.4248	1.6391	0.9727	2.7308	1.0000	0.9727	1.1071	1.1083	3.0042
1.10	1.4331	1.6248	0.9927	2.7708	1.0000	0.9927	1.1340	1.1354	3.0973
1.12	1.4412	1.6107	1.0127	2.8108	1.0000	1.0127	1.1609	1.1627	3.1914
1.14	1.4490	1.5968	1.0327	2.8508	1.0000	1.0327	1.1878	1.1899	3.2864
1.16	1.4566	1.5830	1.0527	2.8908	1.0000	1.0527	1.2148	1.2172	3.3823
1.18	1.4640	1.5693	1.0727	2.9308	1.0000	1.0727	1.2418	1.2445	3.4792
1.20	1.4713	1.5558	1.0927	2.9708	1.0000	1.0927	1.2689	1.2719	3.5769
1.24	1.4851	1.5293	1.1327	3.0508	1.0000	1.1327	1.3230	1.3267	3.7751
1.28	1.4983	1.5034	1.1727	3.1308	1.0000	1.1727	1.3773	1.3817	3.9769
1.32	1.5108	1.4782	1.2127	3.2108	1.0000	1.2127	1.4317	1.4368	4.1821
1.36	1.5227	1.4536	1.2527	3.2908	1.0000	1.2527	1.4862	1.4919	4.3907
1.40	1.5340	1.4297	1.2927	3.3708	1.0000	1.2927	1.5407	1.5472	4.6026
1.44	1.5448	1.4063	1.3327	3.4508	1.0000	1.3327	1.5954	1.6025	4.8179
1.48	1.5551	1.3837	1.3727	3.5308	1.0000	1.3727	1.6501	1.6580	5.0364
1.52	1.5650	1.3616	1.4127	3.6108	1.0000	1.4127	1.7049	1.7135	5.2582
1.56	1.5744	1.3401	1.4527	3.6908	1.0000	1.4527	1.7597	1.7691	5.4831
1.60	1.5834	1.3192	1.4927	3.7708	1.0000	1.4927	1.8146	1.8247	5.7111
1.64	1.5921	1.2988	1.5327	3.8508	1.0000	1.5327	1.8696	1.8805	5.9422
1.68	1.6004	1.2790	1.5727	3.9308	1.0000	1.5727	1.9246	1.9362	6.1763
1.72	1.6084	1.2598	1.6127	4.0108	1.0000	1.6127	1.9797	1.9921	6.4134
1.76	1.6160	1.2410	1.6527	4.0908	1.0000	1.6527	2.0348	2.0480	6.6535
1.80	1.6234	1.2228	1.6927	4.1708	1.0000	1.6927	2.0899	2.1039	6.8965
1.84	1.6305	1.2050	1.7327	4.2508	1.0000	1.7327	2.1451	2.1599	7.1424
1.88	1.6373	1.1877	1.7727	4.3308	1.0000	1.7727	2.2004	2.2159	7.3912
1.92	1.6439	1.1708	1.8127	4.4108	1.0000	1.8127	2.2557	2.2720	7.6427
1.96	1.6502	1.1544	1.8527	4.4908	1.0000	1.8527	2.3110	2.3281	7.8971
2.00	1.6563	1.1384	1.8927	4.5708	1.0000	1.8927	2.3663	2.3842	8.1542
2.05	1.6637	1.1190	1.9427	4.6708	1.0000	1.9427	2.4355	2.4544	8.4795
2.10	1.6707	1.1002	1.9927	4.7708	1.0000	1.9927	2.5048	2.5247	8.8089
2.15	1.6775	1.0819	2.0427	4.8708	1.0000	2.0427	2.5741	2.5950	9.1425
2.20	1.6840	1.0643	2.0927	4.9708	1.0000	2.0927	2.6434	2.6654	9.4803
2.25	1.6902	1.0471	2.1427	5.0708	1.0000	2.1427	2.7128	2.7358	9.8220
2.30	1.6962	1.0305	2.1927	5.1708	1.0000	2.1927	2.7822	2.8063	10.168
2.35	1.7020	1.0144	2.2427	5.2708	1.0000	2.2427	2.8517	2.8768	10.518
2.40	1.7075	0.9988	2.2927	5.3708	1.0000	2.2927	2.9212	2.9473	10.871
2.45	1.7129	0.9836	2.3427	5.4708	1.0000	2.3427	2.9907	3.0178	11.229
2.50	1.7180	0.9688	2.3927	5.5708	1.0000	2.3927	3.0602	3.0884	11.590
2.55	1.7230	0.9545	2.4427	5.6708	1.0000	2.4427	3.1298	3.1591	11.955
2.60	1.7278	0.9406	2.4927	5.7708	1.0000	2.4927	3.1994	3.2297	12.324
2.65	1.7324	0.9270	2.5427	5.8708	1.0000	2.5427	3.2690	3.3004	12.697
2.70	1.7369	0.9139	2.5927	5.9708	1.0000	2.5927	3.3386	3.3711	13.073
2.75	1.7413	0.9011	2.6427	6.0708	1.0000	2.6427	3.4083	3.4418	13.453

C12 U-shaped capped (running full - modified table)

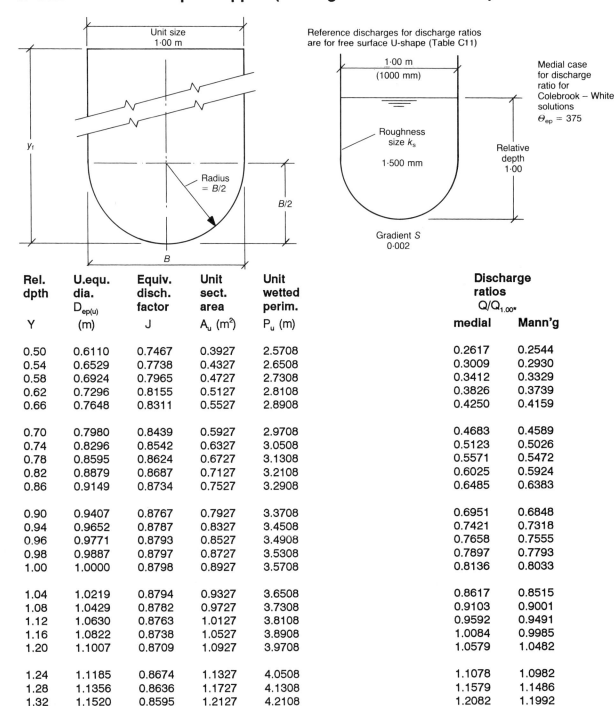

Unit size 1·00 m

Reference discharges for discharge ratios are for free surface U-shape (Table C11)

1·00 m (1000 mm)

Medial case for discharge ratio for Colebrook – White solutions $\Theta_{ep} = 375$

Roughness size k_s 1·500 mm

Relative depth 1·00

Radius = B/2

B/2

Gradient S 0·002

y_f

B

Rel. dpth Y	U.equ. dia. $D_{ep(u)}$ (m)	Equiv. disch. factor J	Unit sect. area A_u (m²)	Unit wetted perim. P_u (m)	Discharge ratios $Q/Q_{1.00*}$ medial	Mann'g
0.50	0.6110	0.7467	0.3927	2.5708	0.2617	0.2544
0.54	0.6529	0.7738	0.4327	2.6508	0.3009	0.2930
0.58	0.6924	0.7965	0.4727	2.7308	0.3412	0.3329
0.62	0.7296	0.8155	0.5127	2.8108	0.3826	0.3739
0.66	0.7648	0.8311	0.5527	2.8908	0.4250	0.4159
0.70	0.7980	0.8439	0.5927	2.9708	0.4683	0.4589
0.74	0.8296	0.8542	0.6327	3.0508	0.5123	0.5026
0.78	0.8595	0.8624	0.6727	3.1308	0.5571	0.5472
0.82	0.8879	0.8687	0.7127	3.2108	0.6025	0.5924
0.86	0.9149	0.8734	0.7527	3.2908	0.6485	0.6383
0.90	0.9407	0.8767	0.7927	3.3708	0.6951	0.6848
0.94	0.9652	0.8787	0.8327	3.4508	0.7421	0.7318
0.96	0.9771	0.8793	0.8527	3.4908	0.7658	0.7555
0.98	0.9887	0.8797	0.8727	3.5308	0.7897	0.7793
1.00	1.0000	0.8798	0.8927	3.5708	0.8136	0.8033
1.04	1.0219	0.8794	0.9327	3.6508	0.8617	0.8515
1.08	1.0429	0.8782	0.9727	3.7308	0.9103	0.9001
1.12	1.0630	0.8763	1.0127	3.8108	0.9592	0.9491
1.16	1.0822	0.8738	1.0527	3.8908	1.0084	0.9985
1.20	1.1007	0.8709	1.0927	3.9708	1.0579	1.0482
1.24	1.1185	0.8674	1.1327	4.0508	1.1078	1.0982
1.28	1.1356	0.8636	1.1727	4.1308	1.1579	1.1486
1.32	1.1520	0.8595	1.2127	4.2108	1.2082	1.1992
1.36	1.1678	0.8550	1.2527	4.2908	1.2588	1.2500
1.40	1.1830	0.8503	1.2927	4.3708	1.3096	1.3011
1.44	1.1977	0.8454	1.3327	4.4508	1.3606	1.3525
1.48	1.2119	0.8403	1.3727	4.5308	1.4118	1.4040
1.52	1.2256	0.8350	1.4127	4.6108	1.4632	1.4558
1.56	1.2388	0.8296	1.4527	4.6908	1.5148	1.5078
1.60	1.2515	0.8241	1.4927	4.7708	1.5666	1.5599
1.64	1.2639	0.8185	1.5327	4.8508	1.6185	1.6122
1.68	1.2758	0.8129	1.5727	4.9308	1.6705	1.6647
1.72	1.2874	0.8071	1.6127	5.0108	1.7227	1.7173
1.76	1.2986	0.8013	1.6527	5.0908	1.7750	1.7701
1.80	1.3094	0.7955	1.6927	5.1708	1.8275	1.8230
1.84	1.3200	0.7897	1.7327	5.2508	1.8801	1.8761
1.88	1.3302	0.7839	1.7727	5.3308	1.9328	1.9293
1.92	1.3401	0.7780	1.8127	5.4108	1.9856	1.9826
1.96	1.3497	0.7722	1.8527	5.4908	2.0385	2.0360
2.00	1.3590	0.7664	1.8927	5.5708	2.0915	2.0896

Oval (running full - modified table)

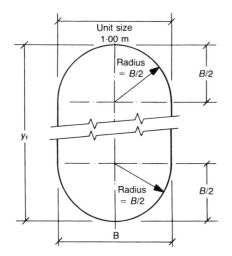

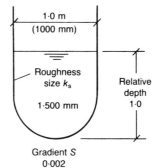

Reference discharges for discharge ratios are for free surface U-shape (Table C11)

Medial case for discharge ratio for Colebrook – White solutions $Q_{ep} = 375$

Rel. dpth	U.equ. dia. $D_{ep(u)}$	Equiv. disch. factor	Unit sect. area	Unit wetted perim.	Discharge ratios $Q/Q_{1.00*}$	
Y	(m)	J	A_u (m²)	P_u (m)	medial	Mann'g
1.04	1.0248	0.9994	0.8254	3.2216	0.7640	0.7550
1.08	1.0485	0.9976	0.8654	3.3016	0.8126	0.8037
1.12	1.0710	0.9949	0.9054	3.3816	0.8616	0.8528
1.16	1.0924	0.9914	0.9454	3.4616	0.9110	0.9024
1.20	1.1129	0.9872	0.9854	3.5416	0.9607	0.9523
1.24	1.1325	0.9824	1.0254	3.6216	1.0107	1.0025
1.28	1.1513	0.9771	1.0654	3.7016	1.0610	1.0531
1.32	1.1692	0.9713	1.1054	3.7816	1.1116	1.1040
1.36	1.1865	0.9652	1.1454	3.8616	1.1625	1.1551
1.40	1.2030	0.9588	1.1854	3.9416	1.2136	1.2065
1.44	1.2188	0.9521	1.2254	4.0216	1.2649	1.2582
1.48	1.2341	0.9452	1.2654	4.1016	1.3164	1.3100
1.52	1.2487	0.9381	1.3054	4.1816	1.3681	1.3621
1.56	1.2628	0.9309	1.3454	4.2616	1.4199	1.4144
1.60	1.2764	0.9236	1.3854	4.3416	1.4720	1.4669
1.64	1.2895	0.9162	1.4254	4.4216	1.5242	1.5195
1.68	1.3021	0.9087	1.4654	4.5016	1.5766	1.5724
1.72	1.3143	0.9012	1.5054	4.5816	1.6291	1.6253
1.76	1.3261	0.8937	1.5454	4.6616	1.6817	1.6785
1.80	1.3374	0.8861	1.5854	4.7416	1.7345	1.7317
1.84	1.3484	0.8786	1.6254	4.8216	1.7874	1.7852
1.88	1.3591	0.8711	1.6654	4.9016	1.8404	1.8387
1.92	1.3694	0.8636	1.7054	4.9816	1.8935	1.8923
1.96	1.3793	0.8561	1.7454	5.0616	1.9467	1.9461
2.00	1.3890	0.8487	1.7854	5.1416	2.0000	2.0000
2.04	1.3983	0.8413	1.8254	5.2216	2.0535	2.0540
2.08	1.4074	0.8340	1.8654	5.3016	2.1070	2.1081
2.12	1.4162	0.8267	1.9054	5.3816	2.1605	2.1622
2.16	1.4248	0.8195	1.9454	5.4616	2.2142	2.2165
2.20	1.4331	0.8124	1.9854	5.5416	2.2680	2.2709
2.24	1.4412	0.8054	2.0254	5.6216	2.3218	2.3253
2.28	1.4490	0.7984	2.0654	5.7016	2.3757	2.3798
2.32	1.4566	0.7915	2.1054	5.7816	2.4296	2.4344
2.36	1.4640	0.7847	2.1454	5.8616	2.4836	2.4891
2.40	1.4713	0.7779	2.1854	5.9416	2.5377	2.5438
2.44	1.4783	0.7712	2.2254	6.0216	2.5919	2.5986
2.48	1.4851	0.7646	2.2654	6.1016	2.6461	2.6535
2.52	1.4918	0.7581	2.3054	6.1816	2.7003	2.7084
2.56	1.4983	0.7517	2.3454	6.2616	2.7546	2.7634
2.60	1.5046	0.7454	2.3854	6.3416	2.8090	2.8184

C14

Rectangular (free surface)

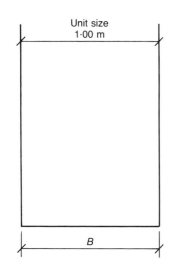

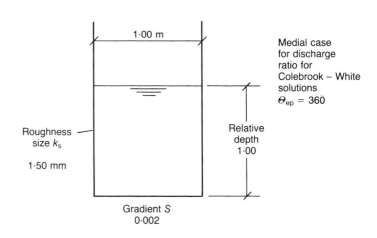

Medial case for discharge ratio for Colebrook – White solutions $\Theta_{ep} = 360$

Rel. dpth	U.equ. dia. $D_{ep(u)}$	Equiv. disch. factor	Unit sect. area	Unit wetted perim.	Unit surf. brdth	Unit mean depth	Discharge ratios $Q/Q_{1.00}$		U.crit. disch. Q_{uc}
Y	(m)	J	A_u (m²)	P_u (m)	B_{us} (m)	y_{um} (m)	medial	Mann'g	(m³s⁻¹)
0.02	0.0769	0.2324	0.0200	1.0400	1.0000	0.0200	0.0031	0.0030	0.0089
0.04	0.1481	0.4309	0.0400	1.0800	1.0000	0.0400	0.0096	0.0092	0.0251
0.06	0.2143	0.6011	0.0600	1.1200	1.0000	0.0600	0.0185	0.0177	0.0460
0.08	0.2759	0.7471	0.0800	1.1600	1.0000	0.0800	0.0291	0.0280	0.0709
0.10	0.3333	0.8726	0.1000	1.2000	1.0000	0.1000	0.0412	0.0397	0.0990
0.12	0.3871	0.9807	0.1200	1.2400	1.0000	0.1200	0.0545	0.0526	0.1302
0.14	0.4375	1.0738	0.1400	1.2800	1.0000	0.1400	0.0689	0.0666	0.1640
0.16	0.4848	1.1539	0.1600	1.3200	1.0000	0.1600	0.0842	0.0815	0.2004
0.18	0.5294	1.2229	0.1800	1.3600	1.0000	0.1800	0.1002	0.0972	0.2391
0.20	0.5714	1.2823	0.2000	1.4000	1.0000	0.2000	0.1169	0.1137	0.2801
0.22	0.6111	1.3332	0.2200	1.4400	1.0000	0.2200	0.1343	0.1308	0.3231
0.24	0.6486	1.3769	0.2400	1.4800	1.0000	0.2400	0.1522	0.1485	0.3682
0.26	0.6842	1.4141	0.2600	1.5200	1.0000	0.2600	0.1706	0.1667	0.4152
0.28	0.7179	1.4458	0.2800	1.5600	1.0000	0.2800	0.1894	0.1853	0.4640
0.30	0.7500	1.4726	0.3000	1.6000	1.0000	0.3000	0.2087	0.2044	0.5146
0.32	0.7805	1.4951	0.3200	1.6400	1.0000	0.3200	0.2283	0.2239	0.5669
0.34	0.8095	1.5138	0.3400	1.6800	1.0000	0.3400	0.2483	0.2438	0.6208
0.36	0.8372	1.5291	0.3600	1.7200	1.0000	0.3600	0.2685	0.2640	0.6764
0.38	0.8636	1.5416	0.3800	1.7600	1.0000	0.3800	0.2891	0.2845	0.7336
0.40	0.8889	1.5514	0.4000	1.8000	1.0000	0.4000	0.3099	0.3053	0.7922
0.42	0.9130	1.5589	0.4200	1.8400	1.0000	0.4200	0.3310	0.3263	0.8524
0.44	0.9362	1.5644	0.4400	1.8800	1.0000	0.4400	0.3523	0.3476	0.9140
0.46	0.9583	1.5680	0.4600	1.9200	1.0000	0.4600	0.3738	0.3691	0.9770
0.48	0.9796	1.5701	0.4800	1.9600	1.0000	0.4800	0.3955	0.3908	1.0414
0.50	1.0000	1.5708	0.5000	2.0000	1.0000	0.5000	0.4173	0.4127	1.1072
0.52	1.0196	1.5702	0.5200	2.0400	1.0000	0.5200	0.4394	0.4348	1.1743
0.54	1.0385	1.5684	0.5400	2.0800	1.0000	0.5400	0.4616	0.4571	1.2427
0.56	1.0566	1.5657	0.5600	2.1200	1.0000	0.5600	0.4839	0.4796	1.3123
0.58	1.0741	1.5621	0.5800	2.1600	1.0000	0.5800	0.5064	0.5021	1.3833
0.60	1.0909	1.5578	0.6000	2.2000	1.0000	0.6000	0.5290	0.5249	1.4554
0.62	1.1071	1.5527	0.6200	2.2400	1.0000	0.6200	0.5518	0.5477	1.5288
0.64	1.1228	1.5471	0.6400	2.2800	1.0000	0.6400	0.5746	0.5707	1.6034
0.66	1.1379	1.5409	0.6600	2.3200	1.0000	0.6600	0.5976	0.5938	1.6791
0.68	1.1525	1.5342	0.6800	2.3600	1.0000	0.6800	0.6206	0.6171	1.7560
0.70	1.1667	1.5271	0.7000	2.4000	1.0000	0.7000	0.6438	0.6404	1.8340
0.72	1.1803	1.5197	0.7200	2.4400	1.0000	0.7200	0.6671	0.6638	1.9132
0.74	1.1935	1.5119	0.7400	2.4800	1.0000	0.7400	0.6904	0.6873	1.9935
0.76	1.2063	1.5039	0.7600	2.5200	1.0000	0.7600	0.7138	0.7109	2.0748
0.78	1.2187	1.4956	0.7800	2.5600	1.0000	0.7800	0.7373	0.7346	2.1573
0.80	1.2308	1.4871	0.8000	2.6000	1.0000	0.8000	0.7609	0.7584	2.2408

Rel. dpth	U.equ. dia. $D_{ep(u)}$	Equiv. disch. factor	Unit sect. area	Unit wetted perim.	Unit surf. brdth	Unit mean depth	Discharge ratios $Q/Q_{1.00}$		U.crit. disch. Q_{uc}
Y	(m)	J	A_u (m²)	P_u (m)	B_{us} (m)	y_{um} (m)	medial	Mann'g	(m³s⁻¹)
0.82	1.2424	1.4784	0.8200	2.6400	1.0000	0.8200	0.7845	0.7823	2.3253
0.84	1.2537	1.4696	0.8400	2.6800	1.0000	0.8400	0.8082	0.8062	2.4109
0.86	1.2647	1.4607	0.8600	2.7200	1.0000	0.8600	0.8320	0.8302	2.4975
0.88	1.2754	1.4517	0.8800	2.7600	1.0000	0.8800	0.8559	0.8543	2.5851
0.90	1.2857	1.4425	0.9000	2.8000	1.0000	0.9000	0.8798	0.8784	2.6738
0.92	1.2958	1.4333	0.9200	2.8400	1.0000	0.9200	0.9037	0.9026	2.7634
0.94	1.3056	1.4241	0.9400	2.8800	1.0000	0.9400	0.9277	0.9269	2.8540
0.96	1.3151	1.4148	0.9600	2.9200	1.0000	0.9600	0.9518	0.9512	2.9456
0.98	1.3243	1.4055	0.9800	2.9600	1.0000	0.9800	0.9759	0.9756	3.0381
1.00	1.3333	1.3962	1.0000	3.0000	1.0000	1.0000	1.0000	1.0000	3.1316
1.02	1.3421	1.3869	1.0200	3.0400	1.0000	1.0200	1.0242	1.0245	3.2260
1.04	1.3506	1.3776	1.0400	3.0800	1.0000	1.0400	1.0485	1.0490	3.3213
1.06	1.3590	1.3683	1.0600	3.1200	1.0000	1.0600	1.0727	1.0735	3.4176
1.08	1.3671	1.3591	1.0800	3.1600	1.0000	1.0800	1.0971	1.0982	3.5148
1.10	1.3750	1.3499	1.1000	3.2000	1.0000	1.1000	1.1214	1.1228	3.6128
1.12	1.3827	1.3407	1.1200	3.2400	1.0000	1.1200	1.1458	1.1475	3.7118
1.14	1.3902	1.3315	1.1400	3.2800	1.0000	1.1400	1.1702	1.1722	3.8117
1.16	1.3976	1.3225	1.1600	3.3200	1.0000	1.1600	1.1947	1.1970	3.9124
1.18	1.4048	1.3134	1.1800	3.3600	1.0000	1.1800	1.2192	1.2218	4.0141
1.20	1.4118	1.3044	1.2000	3.4000	1.0000	1.2000	1.2437	1.2466	4.1165
1.24	1.4253	1.2867	1.2400	3.4800	1.0000	1.2400	1.2928	1.2964	4.3241
1.28	1.4382	1.2691	1.2800	3.5600	1.0000	1.2800	1.3421	1.3463	4.5350
1.32	1.4505	1.2519	1.3200	3.6400	1.0000	1.3200	1.3914	1.3963	4.7492
1.36	1.4624	1.2350	1.3600	3.7200	1.0000	1.3600	1.4409	1.4464	4.9667
1.40	1.4737	1.2183	1.4000	3.8000	1.0000	1.4000	1.4904	1.4966	5.1874
1.44	1.4845	1.2020	1.4400	3.8800	1.0000	1.4400	1.5400	1.5469	5.4113
1.48	1.4949	1.1860	1.4800	3.9600	1.0000	1.4800	1.5897	1.5973	5.6384
1.52	1.5050	1.1703	1.5200	4.0400	1.0000	1.5200	1.6395	1.6478	5.8685
1.56	1.5146	1.1549	1.5600	4.1200	1.0000	1.5600	1.6893	1.6983	6.1016
1.60	1.5238	1.1398	1.6000	4.2000	1.0000	1.6000	1.7392	1.7490	6.3378
1.64	1.5327	1.1250	1.6400	4.2800	1.0000	1.6400	1.7892	1.7997	6.5770
1.68	1.5413	1.1105	1.6800	4.3600	1.0000	1.6800	1.8392	1.8504	6.8191
1.72	1.5495	1.0964	1.7200	4.4400	1.0000	1.7200	1.8893	1.9013	7.0640
1.76	1.5575	1.0825	1.7600	4.5200	1.0000	1.7600	1.9394	1.9521	7.3119
1.80	1.5652	1.0689	1.8000	4.6000	1.0000	1.8000	1.9896	2.0031	7.5626
1.84	1.5726	1.0557	1.8400	4.6800	1.0000	1.8400	2.0398	2.0541	7.8160
1.88	1.5798	1.0427	1.8800	4.7600	1.0000	1.8800	2.0901	2.1051	8.0723
1.92	1.5868	1.0299	1.9200	4.8400	1.0000	1.9200	2.1404	2.1562	8.3313
1.96	1.5935	1.0175	1.9600	4.9200	1.0000	1.9600	2.1908	2.2073	8.5930
2.00	1.6000	1.0053	2.0000	5.0000	1.0000	2.0000	2.2412	2.2585	8.8574
2.05	1.6078	0.9904	2.0500	5.1000	1.0000	2.0500	2.3042	2.3225	9.1916
2.10	1.6154	0.9759	2.1000	5.2000	1.0000	2.1000	2.3673	2.3866	9.5299
2.15	1.6226	0.9618	2.1500	5.3000	1.0000	2.1500	2.4304	2.4507	9.8723
2.20	1.6296	0.9481	2.2000	5.4000	1.0000	2.2000	2.4936	2.5149	10.219
2.25	1.6364	0.9347	2.2500	5.5000	1.0000	2.2500	2.5568	2.5791	10.569
2.30	1.6429	0.9216	2.3000	5.6000	1.0000	2.3000	2.6201	2.6434	10.923
2.35	1.6491	0.9089	2.3500	5.7000	1.0000	2.3500	2.6834	2.7078	11.281
2.40	1.6552	0.8965	2.4000	5.8000	1.0000	2.4000	2.7468	2.7721	11.643
2.45	1.6610	0.8844	2.4500	5.9000	1.0000	2.4500	2.8101	2.8365	12.009
2.50	1.6667	0.8726	2.5000	6.0000	1.0000	2.5000	2.8736	2.9010	12.379
2.55	1.6721	0.8612	2.5500	6.1000	1.0000	2.5500	2.9370	2.9655	12.752
2.60	1.6774	0.8499	2.6000	6.2000	1.0000	2.6000	3.0005	3.0300	13.129
2.65	1.6825	0.8390	2.6500	6.3000	1.0000	2.6500	3.0640	3.0945	13.509
2.70	1.6875	0.8283	2.7000	6.4000	1.0000	2.7000	3.1275	3.1591	13.893
2.75	1.6923	0.8179	2.7500	6.5000	1.0000	2.7500	3.1911	3.2237	14.281

C15 Rectangular capped (running full - modified table)

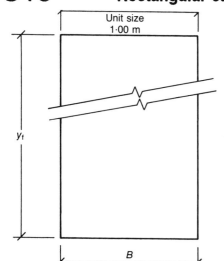

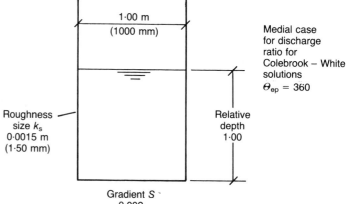

Unit size
1·00 m

Reference discharges for discharge ratios
are for free surface rectangular channel (Table C14)

1·00 m
(1000 mm)

Medial case
for discharge
ratio for
Colebrook – White
solutions
$\Theta_{ep} = 360$

Roughness
size k_s
0·0015 m
(1·50 mm)

Relative
depth
1·00

Gradient S
0·002

Rel. dpth	U.equ. dia. $D_{ep(u)}$	Equiv. disch. factor	Unit sect. area	Unit wetted perim.	Discharge ratios $Q/Q_{1.00*}$	
Y	(m)	J	A_u (m²)	P_u (m)	medial	Mann'g
0.02	0.0392	0.0604	0.0200	2.0400	0.0019	0.0019
0.04	0.0769	0.1162	0.0400	2.0800	0.0061	0.0060
0.06	0.1132	0.1678	0.0600	2.1200	0.0120	0.0116
0.08	0.1481	0.2155	0.0800	2.1600	0.0192	0.0185
0.10	0.1818	0.2596	0.1000	2.2000	0.0276	0.0265
0.12	0.2143	0.3005	0.1200	2.2400	0.0369	0.0355
0.14	0.2456	0.3384	0.1400	2.2800	0.0472	0.0453
0.16	0.2759	0.3735	0.1600	2.3200	0.0582	0.0560
0.18	0.3051	0.4061	0.1800	2.3600	0.0700	0.0673
0.20	0.3333	0.4363	0.2000	2.4000	0.0825	0.0794
0.22	0.3607	0.4643	0.2200	2.4400	0.0955	0.0920
0.24	0.3871	0.4904	0.2400	2.4800	0.1091	0.1052
0.26	0.4127	0.5145	0.2600	2.5200	0.1232	0.1190
0.28	0.4375	0.5369	0.2800	2.5600	0.1378	0.1332
0.30	0.4615	0.5577	0.3000	2.6000	0.1528	0.1479
0.32	0.4848	0.5770	0.3200	2.6400	0.1683	0.1630
0.34	0.5075	0.5949	0.3400	2.6800	0.1842	0.1786
0.36	0.5294	0.6115	0.3600	2.7200	0.2004	0.1945
0.38	0.5507	0.6269	0.3800	2.7600	0.2170	0.2108
0.40	0.5714	0.6411	0.4000	2.8000	0.2339	0.2274
0.42	0.5915	0.6544	0.4200	2.8400	0.2511	0.2443
0.44	0.6111	0.6666	0.4400	2.8800	0.2686	0.2616
0.46	0.6301	0.6779	0.4600	2.9200	0.2863	0.2791
0.48	0.6486	0.6884	0.4800	2.9600	0.3044	0.2969
0.50	0.6667	0.6981	0.5000	3.0000	0.3226	0.3150
0.52	0.6842	0.7071	0.5200	3.0400	0.3412	0.3333
0.54	0.7013	0.7153	0.5400	3.0800	0.3599	0.3519
0.56	0.7179	0.7229	0.5600	3.1200	0.3789	0.3706
0.58	0.7342	0.7299	0.5800	3.1600	0.3980	0.3896
0.60	0.7500	0.7363	0.6000	3.2000	0.4174	0.4089
0.62	0.7654	0.7422	0.6200	3.2400	0.4369	0.4283
0.64	0.7805	0.7475	0.6400	3.2800	0.4566	0.4478
0.66	0.7952	0.7524	0.6600	3.3200	0.4765	0.4676
0.68	0.8095	0.7569	0.6800	3.3600	0.4965	0.4876
0.70	0.8235	0.7609	0.7000	3.4000	0.5167	0.5077
0.72	0.8372	0.7646	0.7200	3.4400	0.5371	0.5280
0.74	0.8506	0.7678	0.7400	3.4800	0.5576	0.5484
0.76	0.8636	0.7708	0.7600	3.5200	0.5782	0.5690
0.78	0.8764	0.7734	0.7800	3.5600	0.5990	0.5897
0.80	0.8889	0.7757	0.8000	3.6000	0.6198	0.6105

Rel. dpth	U.equ. dia. $D_{ep(u)}$	Equiv. disch. factor	Unit sect. area	Unit wetted perim.	Discharge ratios $Q/Q_{1.00*}$	
Y	(m)	J	A_u (m²)	P_u (m)	medial	Mann'g
0.82	0.9011	0.7777	0.8200	3.6400	0.6408	0.6315
0.84	0.9130	0.7794	0.8400	3.6800	0.6620	0.6526
0.86	0.9247	0.7809	0.8600	3.7200	0.6832	0.6738
0.88	0.9362	0.7822	0.8800	3.7600	0.7045	0.6952
0.90	0.9474	0.7832	0.9000	3.8000	0.7260	0.7166
0.92	0.9583	0.7840	0.9200	3.8400	0.7475	0.7382
0.94	0.9691	0.7846	0.9400	3.8800	0.7692	0.7599
0.96	0.9796	0.7851	0.9600	3.9200	0.7909	0.7816
0.98	0.9899	0.7853	0.9800	3.9600	0.8128	0.8035
1.00	1.0000	0.7854	1.0000	4.0000	0.8347	0.8255
1.02	1.0099	0.7853	1.0200	4.0400	0.8567	0.8475
1.04	1.0196	0.7851	1.0400	4.0800	0.8788	0.8697
1.06	1.0291	0.7847	1.0600	4.1200	0.9009	0.8919
1.08	1.0385	0.7842	1.0800	4.1600	0.9231	0.9142
1.10	1.0476	0.7836	1.1000	4.2000	0.9455	0.9366
1.12	1.0566	0.7829	1.1200	4.2400	0.9678	0.9591
1.14	1.0654	0.7820	1.1400	4.2800	0.9903	0.9817
1.16	1.0741	0.7811	1.1600	4.3200	1.0128	1.0043
1.18	1.0826	0.7800	1.1800	4.3600	1.0354	1.0270
1.20	1.0909	0.7789	1.2000	4.4000	1.0580	1.0497
1.24	1.1071	0.7764	1.2400	4.4800	1.1035	1.0955
1.28	1.1228	0.7735	1.2800	4.5600	1.1492	1.1414
1.32	1.1379	0.7704	1.3200	4.6400	1.1951	1.1877
1.36	1.1525	0.7671	1.3600	4.7200	1.2413	1.2341
1.40	1.1667	0.7636	1.4000	4.8000	1.2876	1.2808
1.44	1.1803	0.7598	1.4400	4.8800	1.3341	1.3276
1.48	1.1935	0.7560	1.4800	4.9600	1.3808	1.3747
1.52	1.2063	0.7519	1.5200	5.0400	1.4276	1.4219
1.56	1.2187	0.7478	1.5600	5.1200	1.4746	1.4693
1.60	1.2308	0.7436	1.6000	5.2000	1.5218	1.5169
1.64	1.2424	0.7392	1.6400	5.2800	1.5691	1.5646
1.68	1.2537	0.7348	1.6800	5.3600	1.6165	1.6125
1.72	1.2647	0.7303	1.7200	5.4400	1.6641	1.6605
1.76	1.2754	0.7258	1.7600	5.5200	1.7117	1.7086
1.80	1.2857	0.7213	1.8000	5.6000	1.7595	1.7569
1.84	1.2958	0.7167	1.8400	5.6800	1.8074	1.8053
1.88	1.3056	0.7121	1.8800	5.7600	1.8554	1.8538
1.92	1.3151	0.7074	1.9200	5.8400	1.9035	1.9024
1.96	1.3243	0.7028	1.9600	5.9200	1.9517	1.9512
2.00	1.3333	0.6981	2.0000	6.0000	2.0000	2.0000
2.05	1.3443	0.6923	2.0500	6.1000	2.0605	2.0612
2.10	1.3548	0.6865	2.1000	6.2000	2.1212	2.1225
2.15	1.3651	0.6807	2.1500	6.3000	2.1819	2.1840
2.20	1.3750	0.6749	2.2000	6.4000	2.2428	2.2456
2.25	1.3846	0.6692	2.2500	6.5000	2.3038	2.3073
2.30	1.3939	0.6635	2.3000	6.6000	2.3649	2.3692
2.35	1.4030	0.6578	2.3500	6.7000	2.4261	2.4311
2.40	1.4118	0.6522	2.4000	6.8000	2.4874	2.4932
2.45	1.4203	0.6466	2.4500	6.9000	2.5488	2.5554
2.50	1.4286	0.6411	2.5000	7.0000	2.6103	2.6177
2.55	1.4366	0.6357	2.5500	7.1000	2.6718	2.6800
2.60	1.4444	0.6302	2.6000	7.2000	2.7335	2.7425
2.65	1.4521	0.6249	2.6500	7.3000	2.7952	2.8051
2.70	1.4595	0.6196	2.7000	7.4000	2.8570	2.8677
2.75	1.4667	0.6143	2.7500	7.5000	2.9189	2.9304

C16 Arch culvert - radius 0.5 breadth (running full - modified table)

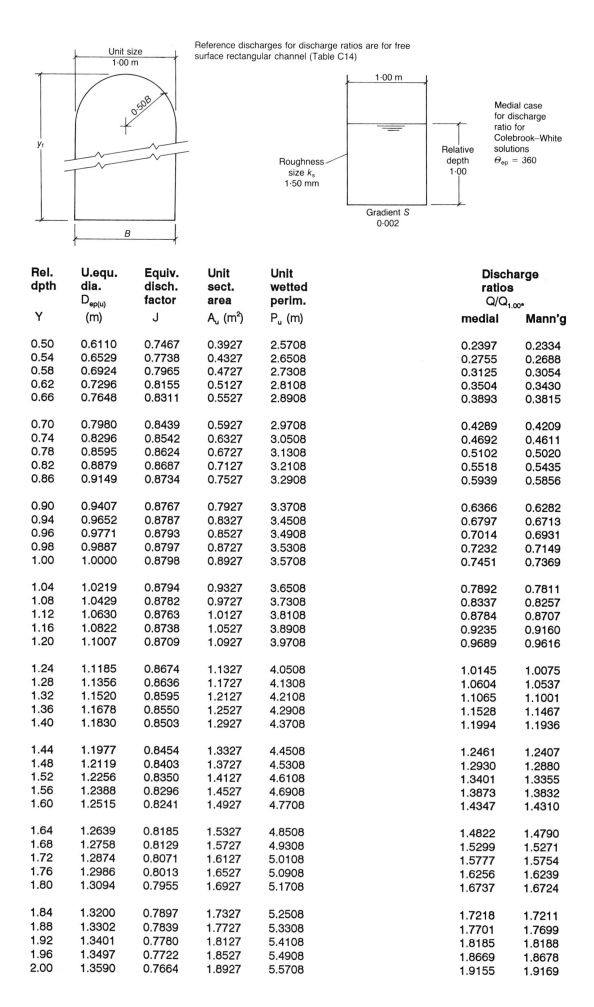

Reference discharges for discharge ratios are for free surface rectangular channel (Table C14)

Unit size 1.00 m

0.50B

y_f

B

1.00 m

Roughness size k_s 1.50 mm

Relative depth 1.00

Gradient S 0.002

Medial case for discharge ratio for Colebrook–White solutions $\Theta_{ep} = 360$

Rel. dpth Y	U.equ. dia. $D_{ep(u)}$ (m)	Equiv. disch. factor J	Unit sect. area A_u (m²)	Unit wetted perim. P_u (m)	Discharge ratios $Q/Q_{1.00*}$ medial	Mann'g
0.50	0.6110	0.7467	0.3927	2.5708	0.2397	0.2334
0.54	0.6529	0.7738	0.4327	2.6508	0.2755	0.2688
0.58	0.6924	0.7965	0.4727	2.7308	0.3125	0.3054
0.62	0.7296	0.8155	0.5127	2.8108	0.3504	0.3430
0.66	0.7648	0.8311	0.5527	2.8908	0.3893	0.3815
0.70	0.7980	0.8439	0.5927	2.9708	0.4289	0.4209
0.74	0.8296	0.8542	0.6327	3.0508	0.4692	0.4611
0.78	0.8595	0.8624	0.6727	3.1308	0.5102	0.5020
0.82	0.8879	0.8687	0.7127	3.2108	0.5518	0.5435
0.86	0.9149	0.8734	0.7527	3.2908	0.5939	0.5856
0.90	0.9407	0.8767	0.7927	3.3708	0.6366	0.6282
0.94	0.9652	0.8787	0.8327	3.4508	0.6797	0.6713
0.96	0.9771	0.8793	0.8527	3.4908	0.7014	0.6931
0.98	0.9887	0.8797	0.8727	3.5308	0.7232	0.7149
1.00	1.0000	0.8798	0.8927	3.5708	0.7451	0.7369
1.04	1.0219	0.8794	0.9327	3.6508	0.7892	0.7811
1.08	1.0429	0.8782	0.9727	3.7308	0.8337	0.8257
1.12	1.0630	0.8763	1.0127	3.8108	0.8784	0.8707
1.16	1.0822	0.8738	1.0527	3.8908	0.9235	0.9160
1.20	1.1007	0.8709	1.0927	3.9708	0.9689	0.9616
1.24	1.1185	0.8674	1.1327	4.0508	1.0145	1.0075
1.28	1.1356	0.8636	1.1727	4.1308	1.0604	1.0537
1.32	1.1520	0.8595	1.2127	4.2108	1.1065	1.1001
1.36	1.1678	0.8550	1.2527	4.2908	1.1528	1.1467
1.40	1.1830	0.8503	1.2927	4.3708	1.1994	1.1936
1.44	1.1977	0.8454	1.3327	4.4508	1.2461	1.2407
1.48	1.2119	0.8403	1.3727	4.5308	1.2930	1.2880
1.52	1.2256	0.8350	1.4127	4.6108	1.3401	1.3355
1.56	1.2388	0.8296	1.4527	4.6908	1.3873	1.3832
1.60	1.2515	0.8241	1.4927	4.7708	1.4347	1.4310
1.64	1.2639	0.8185	1.5327	4.8508	1.4822	1.4790
1.68	1.2758	0.8129	1.5727	4.9308	1.5299	1.5271
1.72	1.2874	0.8071	1.6127	5.0108	1.5777	1.5754
1.76	1.2986	0.8013	1.6527	5.0908	1.6256	1.6239
1.80	1.3094	0.7955	1.6927	5.1708	1.6737	1.6724
1.84	1.3200	0.7897	1.7327	5.2508	1.7218	1.7211
1.88	1.3302	0.7839	1.7727	5.3308	1.7701	1.7699
1.92	1.3401	0.7780	1.8127	5.4108	1.8185	1.8188
1.96	1.3497	0.7722	1.8527	5.4908	1.8669	1.8678
2.00	1.3590	0.7664	1.8927	5.5708	1.9155	1.9169

Arch culvert - radius 0.75 breadth (running full - modified table) C17

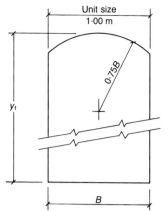

Unit size
1·00 m

0.75B

y_f

B

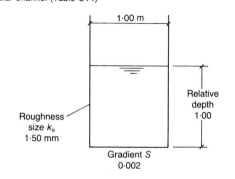

Reference discharges for discharge ratios are for free surface rectangular channel (Table C14)

1·00 m

Relative depth 1·00

Roughness size k_s 1·50 mm

Gradient S 0·002

Medial case for discharge ratio for Colebrook–White solutions $\Theta_{ep} = 360$

Rel. dpth	U.equ. dia. $D_{ep(u)}$	Equiv. disch. factor	Unit sect. area	Unit wetted perim.	Discharge ratios $Q/Q_{1.00*}$	
Y	(m)	J	A_u (m²)	P_u (m)	medial	Mann'g
0.50	0.6488	0.7514	0.4400	2.7126	0.2790	0.2722
0.54	0.6875	0.7734	0.4800	2.7926	0.3159	0.3086
0.58	0.7240	0.7918	0.5200	2.8726	0.3537	0.3461
0.62	0.7586	0.8072	0.5600	2.9526	0.3924	0.3845
0.66	0.7914	0.8198	0.6000	3.0326	0.4318	0.4237
0.70	0.8224	0.8301	0.6400	3.1126	0.4720	0.4637
0.74	0.8519	0.8383	0.6800	3.1926	0.5129	0.5044
0.78	0.8800	0.8447	0.7200	3.2726	0.5543	0.5458
0.82	0.9067	0.8496	0.7600	3.3526	0.5963	0.5877
0.86	0.9322	0.8532	0.8000	3.4326	0.6388	0.6302
0.90	0.9565	0.8555	0.8400	3.5126	0.6817	0.6731
0.94	0.9798	0.8567	0.8800	3.5926	0.7251	0.7166
0.96	0.9910	0.8570	0.9000	3.6326	0.7469	0.7385
0.98	1.0020	0.8571	0.9200	3.6726	0.7688	0.7604
1.00	1.0127	0.8570	0.9400	3.7126	0.7909	0.7825
1.04	1.0336	0.8561	0.9800	3.7926	0.8352	0.8270
1.08	1.0535	0.8546	1.0200	3.8726	0.8798	0.8718
1.12	1.0727	0.8526	1.0600	3.9526	0.9247	0.9169
1.16	1.0911	0.8500	1.1000	4.0326	0.9700	0.9623
1.20	1.1088	0.8470	1.1400	4.1126	1.0154	1.0081
1.24	1.1258	0.8435	1.1800	4.1926	1.0612	1.0541
1.28	1.1421	0.8398	1.2200	4.2726	1.1072	1.1004
1.32	1.1579	0.8357	1.2600	4.3526	1.1533	1.1469
1.36	1.1731	0.8314	1.3000	4.4326	1.1997	1.1936
1.40	1.1878	0.8269	1.3400	4.5126	1.2463	1.2406
1.44	1.2019	0.8221	1.3800	4.5926	1.2931	1.2877
1.48	1.2156	0.8173	1.4200	4.6726	1.3401	1.3351
1.52	1.2288	0.8122	1.4600	4.7526	1.3872	1.3826
1.56	1.2415	0.8071	1.5000	4.8326	1.4345	1.4303
1.60	1.2539	0.8018	1.5400	4.9126	1.4819	1.4782
1.64	1.2659	0.7965	1.5800	4.9926	1.5295	1.5262
1.68	1.2774	0.7911	1.6200	5.0726	1.5771	1.5744
1.72	1.2886	0.7857	1.6600	5.1526	1.6250	1.6227
1.76	1.2995	0.7802	1.7000	5.2326	1.6729	1.6711
1.80	1.3101	0.7747	1.7400	5.3126	1.7210	1.7197
1.84	1.3203	0.7692	1.7800	5.3926	1.7691	1.7684
1.88	1.3302	0.7636	1.8200	5.4726	1.8174	1.8172
1.92	1.3399	0.7581	1.8600	5.5526	1.8657	1.8661
1.96	1.3493	0.7525	1.9000	5.6326	1.9142	1.9151
2.00	1.3584	0.7470	1.9400	5.7126	1.9627	1.9642

C18

Box culvert (9% splays) (free surface)

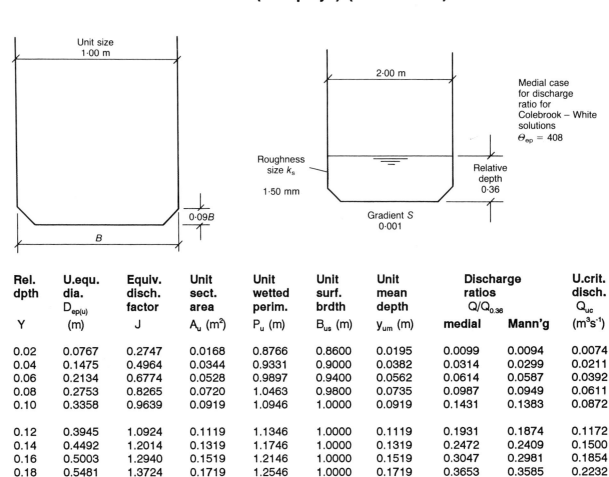

Unit size 1·00 m

0·09B

B

2·00 m

Medial case for discharge ratio for Colebrook − White solutions $\Theta_{ep} = 408$

Roughness size k_s 1·50 mm

Relative depth 0·36

Gradient S 0·001

Rel. dpth Y	U.equ. dia. $D_{ep(u)}$ (m)	Equiv. disch. factor J	Unit sect. area A_u (m²)	Unit wetted perim. P_u (m)	Unit surf. brdth B_{us} (m)	Unit mean depth y_{um} (m)	Discharge ratios $Q/Q_{0.36}$ medial	Mann'g	U.crit. disch. Q_{uc} (m³s⁻¹)
0.02	0.0767	0.2747	0.0168	0.8766	0.8600	0.0195	0.0099	0.0094	0.0074
0.04	0.1475	0.4964	0.0344	0.9331	0.9000	0.0382	0.0314	0.0299	0.0211
0.06	0.2134	0.6774	0.0528	0.9897	0.9400	0.0562	0.0614	0.0587	0.0392
0.08	0.2753	0.8265	0.0720	1.0463	0.9800	0.0735	0.0987	0.0949	0.0611
0.10	0.3358	0.9639	0.0919	1.0946	1.0000	0.0919	0.1431	0.1383	0.0872
0.12	0.3945	1.0924	0.1119	1.1346	1.0000	0.1119	0.1931	0.1874	0.1172
0.14	0.4492	1.2014	0.1319	1.1746	1.0000	0.1319	0.2472	0.2409	0.1500
0.16	0.5003	1.2940	0.1519	1.2146	1.0000	0.1519	0.3047	0.2981	0.1854
0.18	0.5481	1.3724	0.1719	1.2546	1.0000	0.1719	0.3653	0.3585	0.2232
0.20	0.5929	1.4389	0.1919	1.2946	1.0000	0.1919	0.4285	0.4217	0.2633
0.22	0.6351	1.4950	0.2119	1.3346	1.0000	0.2119	0.4940	0.4875	0.3055
0.24	0.6748	1.5423	0.2319	1.3746	1.0000	0.2319	0.5616	0.5556	0.3497
0.26	0.7123	1.5819	0.2519	1.4146	1.0000	0.2519	0.6311	0.6256	0.3959
0.28	0.7477	1.6149	0.2719	1.4546	1.0000	0.2719	0.7021	0.6975	0.4440
0.30	0.7812	1.6421	0.2919	1.4946	1.0000	0.2919	0.7747	0.7710	0.4939
0.32	0.8130	1.6644	0.3119	1.5346	1.0000	0.3119	0.8486	0.8460	0.5455
0.34	0.8432	1.6822	0.3319	1.5746	1.0000	0.3319	0.9238	0.9224	0.5988
0.36	0.8718	1.6963	0.3519	1.6146	1.0000	0.3519	1.0000	1.0000	0.6537
0.38	0.8991	1.7071	0.3719	1.6546	1.0000	0.3719	1.0773	1.0788	0.7102
0.40	0.9251	1.7150	0.3919	1.6946	1.0000	0.3919	1.1555	1.1586	0.7683
0.42	0.9499	1.7203	0.4119	1.7346	1.0000	0.4119	1.2346	1.2394	0.8278
0.44	0.9735	1.7235	0.4319	1.7746	1.0000	0.4319	1.3145	1.3210	0.8889
0.46	0.9962	1.7246	0.4519	1.8146	1.0000	0.4519	1.3951	1.4035	0.9513
0.48	1.0178	1.7241	0.4719	1.8546	1.0000	0.4719	1.4764	1.4868	1.0152
0.50	1.0386	1.7221	0.4919	1.8946	1.0000	0.4919	1.5583	1.5708	1.0804
0.52	1.0584	1.7188	0.5119	1.9346	1.0000	0.5119	1.6408	1.6555	1.1469
0.54	1.0775	1.7143	0.5319	1.9746	1.0000	0.5319	1.7239	1.7408	1.2148
0.56	1.0958	1.7088	0.5519	2.0146	1.0000	0.5519	1.8075	1.8266	1.2840
0.58	1.1134	1.7025	0.5719	2.0546	1.0000	0.5719	1.8915	1.9130	1.3544
0.60	1.1304	1.6954	0.5919	2.0946	1.0000	0.5919	1.9760	2.0000	1.4260
0.64	1.1624	1.6792	0.6319	2.1746	1.0000	0.6319	2.1463	2.1752	1.5730
0.68	1.1921	1.6610	0.6719	2.2546	1.0000	0.6719	2.3180	2.3522	1.7247
0.72	1.2198	1.6414	0.7119	2.3346	1.0000	0.7119	2.4911	2.5307	1.8810
0.76	1.2456	1.6206	0.7519	2.4146	1.0000	0.7519	2.6653	2.7105	2.0417
0.80	1.2698	1.5991	0.7919	2.4946	1.0000	0.7919	2.8406	2.8915	2.2068
0.84	1.2925	1.5771	0.8319	2.5746	1.0000	0.8319	3.0168	3.0736	2.3761
0.88	1.3138	1.5548	0.8719	2.6546	1.0000	0.8719	3.1939	3.2568	2.5495
0.92	1.3339	1.5324	0.9119	2.7346	1.0000	0.9119	3.3718	3.4408	2.7270
0.96	1.3528	1.5100	0.9519	2.8146	1.0000	0.9519	3.5504	3.6256	2.9084
1.00	1.3707	1.4877	0.9919	2.8946	1.0000	0.9919	3.7296	3.8112	3.0936

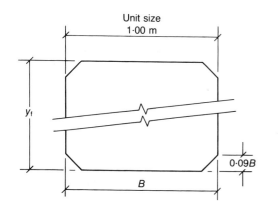

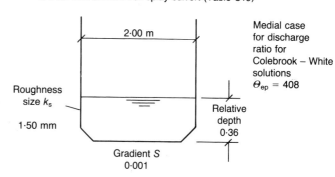

Unit size
1·00 m

Reference discharges for discharge ratios
are for free surface 9% splay culvert (Table C18)

2·00 m

Medial case
for discharge
ratio for
Colebrook – White
solutions
$\Theta_{ep} = 408$

Roughness
size k_s

1·50 mm

Relative
depth
0·36

Gradient S
0·001

0·09B

B

Rel. dpth Y	U.equ. dia. $D_{ep(u)}$ (m)	Equiv. disch. factor J	Unit sect. area A_u (m²)	Unit wetted perim. P_u (m)	Discharge ratios $Q/Q_{0.36*}$ medial	Mann'g
0.33	0.5125	0.6574	0.3138	2.4491	0.6392	0.6258
0.35	0.5364	0.6770	0.3338	2.4891	0.6998	0.6862
0.36	0.5481	0.6862	0.3438	2.5091	0.7306	0.7170
0.38	0.5709	0.7035	0.3638	2.5491	0.7932	0.7796
0.40	0.5929	0.7195	0.3838	2.5891	0.8570	0.8435
0.42	0.6144	0.7341	0.4038	2.6291	0.9220	0.9087
0.44	0.6351	0.7475	0.4238	2.6691	0.9881	0.9750
0.46	0.6553	0.7599	0.4438	2.7091	1.0552	1.0426
0.48	0.6748	0.7712	0.4638	2.7491	1.1232	1.1111
0.50	0.6938	0.7815	0.4838	2.7891	1.1922	1.1807
0.52	0.7123	0.7910	0.5038	2.8291	1.2621	1.2512
0.54	0.7303	0.7996	0.5238	2.8691	1.3328	1.3227
0.56	0.7477	0.8075	0.5438	2.9091	1.4043	1.3950
0.58	0.7647	0.8146	0.5638	2.9491	1.4765	1.4681
0.60	0.7812	0.8211	0.5838	2.9891	1.5494	1.5420
0.62	0.7973	0.8269	0.6038	3.0291	1.6230	1.6166
0.64	0.8130	0.8322	0.6238	3.0691	1.6972	1.6920
0.66	0.8283	0.8369	0.6438	3.1091	1.7721	1.7681
0.68	0.8432	0.8411	0.6638	3.1491	1.8475	1.8448
0.70	0.8577	0.8449	0.6838	3.1891	1.9235	1.9221
0.71	0.8648	0.8466	0.6938	3.2091	1.9617	1.9610
0.73	0.8788	0.8497	0.7138	3.2491	2.0385	2.0392
0.75	0.8924	0.8524	0.7338	3.2891	2.1158	2.1179
0.76	0.8991	0.8536	0.7438	3.3091	2.1546	2.1575
0.77	0.9057	0.8547	0.7538	3.3291	2.1936	2.1972
0.78	0.9122	0.8557	0.7638	3.3491	2.2326	2.2371
0.80	0.9251	0.8575	0.7838	3.3891	2.3111	2.3172
0.82	0.9376	0.8590	0.8038	3.4291	2.3899	2.3977
0.84	0.9499	0.8602	0.8238	3.4691	2.4692	2.4787
0.85	0.9559	0.8607	0.8338	3.4891	2.5090	2.5194
0.86	0.9618	0.8611	0.8438	3.5091	2.5489	2.5602
0.87	0.9677	0.8614	0.8538	3.5291	2.5889	2.6011
0.88	0.9735	0.8617	0.8638	3.5491	2.6290	2.6421
0.89	0.9793	0.8620	0.8738	3.5691	2.6691	2.6832
0.90	0.9850	0.8621	0.8838	3.5891	2.7094	2.7244
0.91	0.9906	0.8623	0.8938	3.6091	2.7498	2.7657
0.92	0.9962	0.8623	0.9038	3.6291	2.7902	2.8071
0.94	1.0071	0.8623	0.9238	3.6691	2.8713	2.8902
0.95	1.0125	0.8622	0.9338	3.6891	2.9120	2.9319
1.00	1.0386	0.8611	0.9838	3.7891	3.1166	3.1416

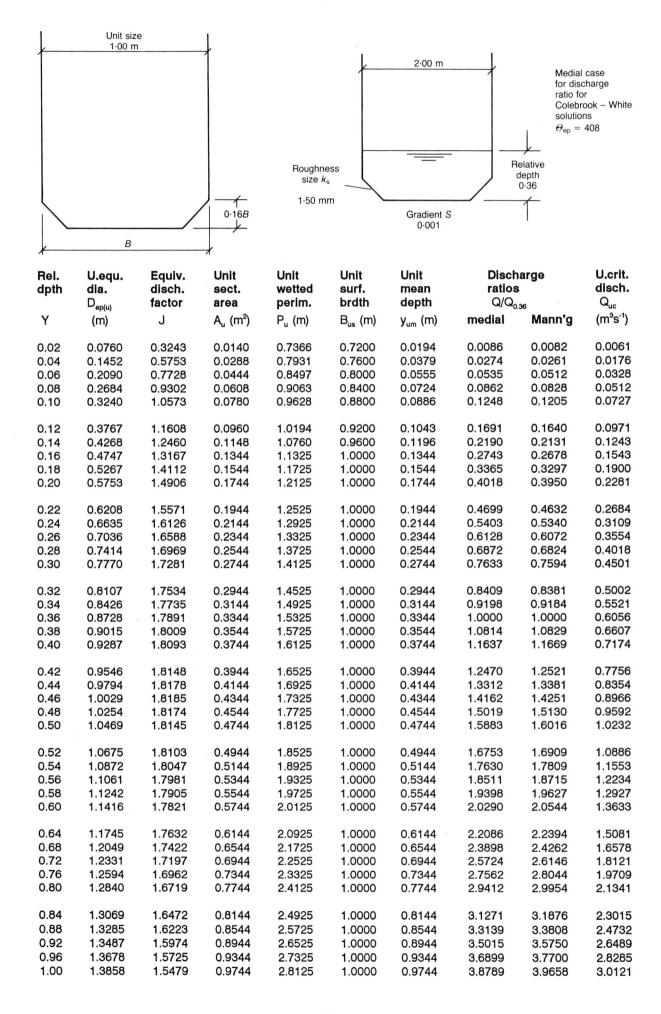

Unit size 1·00 m

0·16B

B

2·00 m

Roughness size k_s

1·50 mm

Gradient S 0·001

Relative depth 0·36

Medial case for discharge ratio for Colebrook – White solutions $\Theta_{ep} = 408$

Rel. dpth	U.equ. dia. $D_{ep(u)}$	Equiv. disch. factor	Unit sect. area	Unit wetted perim.	Unit surf. brdth	Unit mean depth	Discharge ratios $Q/Q_{0.36}$		U.crit. disch. Q_{uc}
Y	(m)	J	A_u (m²)	P_u (m)	B_{us} (m)	y_{um} (m)	medial	Mann'g	(m³s⁻¹)
0.02	0.0760	0.3243	0.0140	0.7366	0.7200	0.0194	0.0086	0.0082	0.0061
0.04	0.1452	0.5753	0.0288	0.7931	0.7600	0.0379	0.0274	0.0261	0.0176
0.06	0.2090	0.7728	0.0444	0.8497	0.8000	0.0555	0.0535	0.0512	0.0328
0.08	0.2684	0.9302	0.0608	0.9063	0.8400	0.0724	0.0862	0.0828	0.0512
0.10	0.3240	1.0573	0.0780	0.9628	0.8800	0.0886	0.1248	0.1205	0.0727
0.12	0.3767	1.1608	0.0960	1.0194	0.9200	0.1043	0.1691	0.1640	0.0971
0.14	0.4268	1.2460	0.1148	1.0760	0.9600	0.1196	0.2190	0.2131	0.1243
0.16	0.4747	1.3167	0.1344	1.1325	1.0000	0.1344	0.2743	0.2678	0.1543
0.18	0.5267	1.4112	0.1544	1.1725	1.0000	0.1544	0.3365	0.3297	0.1900
0.20	0.5753	1.4906	0.1744	1.2125	1.0000	0.1744	0.4018	0.3950	0.2281
0.22	0.6208	1.5571	0.1944	1.2525	1.0000	0.1944	0.4699	0.4632	0.2684
0.24	0.6635	1.6126	0.2144	1.2925	1.0000	0.2144	0.5403	0.5340	0.3109
0.26	0.7036	1.6588	0.2344	1.3325	1.0000	0.2344	0.6128	0.6072	0.3554
0.28	0.7414	1.6969	0.2544	1.3725	1.0000	0.2544	0.6872	0.6824	0.4018
0.30	0.7770	1.7281	0.2744	1.4125	1.0000	0.2744	0.7633	0.7594	0.4501
0.32	0.8107	1.7534	0.2944	1.4525	1.0000	0.2944	0.8409	0.8381	0.5002
0.34	0.8426	1.7735	0.3144	1.4925	1.0000	0.3144	0.9198	0.9184	0.5521
0.36	0.8728	1.7891	0.3344	1.5325	1.0000	0.3344	1.0000	1.0000	0.6056
0.38	0.9015	1.8009	0.3544	1.5725	1.0000	0.3544	1.0814	1.0829	0.6607
0.40	0.9287	1.8093	0.3744	1.6125	1.0000	0.3744	1.1637	1.1669	0.7174
0.42	0.9546	1.8148	0.3944	1.6525	1.0000	0.3944	1.2470	1.2521	0.7756
0.44	0.9794	1.8178	0.4144	1.6925	1.0000	0.4144	1.3312	1.3381	0.8354
0.46	1.0029	1.8185	0.4344	1.7325	1.0000	0.4344	1.4162	1.4251	0.8966
0.48	1.0254	1.8174	0.4544	1.7725	1.0000	0.4544	1.5019	1.5130	0.9592
0.50	1.0469	1.8145	0.4744	1.8125	1.0000	0.4744	1.5883	1.6016	1.0232
0.52	1.0675	1.8103	0.4944	1.8525	1.0000	0.4944	1.6753	1.6909	1.0886
0.54	1.0872	1.8047	0.5144	1.8925	1.0000	0.5144	1.7630	1.7809	1.1553
0.56	1.1061	1.7981	0.5344	1.9325	1.0000	0.5344	1.8511	1.8715	1.2234
0.58	1.1242	1.7905	0.5544	1.9725	1.0000	0.5544	1.9398	1.9627	1.2927
0.60	1.1416	1.7821	0.5744	2.0125	1.0000	0.5744	2.0290	2.0544	1.3633
0.64	1.1745	1.7632	0.6144	2.0925	1.0000	0.6144	2.2086	2.2394	1.5081
0.68	1.2049	1.7422	0.6544	2.1725	1.0000	0.6544	2.3898	2.4262	1.6578
0.72	1.2331	1.7197	0.6944	2.2525	1.0000	0.6944	2.5724	2.6146	1.8121
0.76	1.2594	1.6962	0.7344	2.3325	1.0000	0.7344	2.7562	2.8044	1.9709
0.80	1.2840	1.6719	0.7744	2.4125	1.0000	0.7744	2.9412	2.9954	2.1341
0.84	1.3069	1.6472	0.8144	2.4925	1.0000	0.8144	3.1271	3.1876	2.3015
0.88	1.3285	1.6223	0.8544	2.5725	1.0000	0.8544	3.3139	3.3808	2.4732
0.92	1.3487	1.5974	0.8944	2.6525	1.0000	0.8944	3.5015	3.5750	2.6489
0.96	1.3678	1.5725	0.9344	2.7325	1.0000	0.9344	3.6899	3.7700	2.8285
1.00	1.3858	1.5479	0.9744	2.8125	1.0000	0.9744	3.8789	3.9658	3.0121

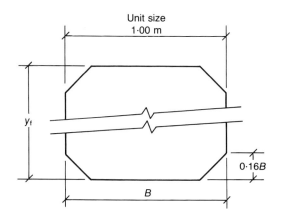

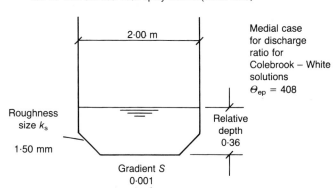

Rel. dpth Y	U.equ. dia. $D_{ep(u)}$ (m)	Equiv. disch. factor J	Unit sect. area A_u (m²)	Unit wetted perim. P_u (m)	Discharge ratios $Q/Q_{0.36*}$ medial	Mann'g
0.33	0.4880	0.6709	0.2788	2.2851	0.5790	0.5659
0.35	0.5140	0.6945	0.2988	2.3251	0.6413	0.6278
0.36	0.5267	0.7056	0.3088	2.3451	0.6730	0.6595
0.38	0.5514	0.7263	0.3288	2.3851	0.7376	0.7240
0.40	0.5753	0.7453	0.3488	2.4251	0.8036	0.7900
0.42	0.5984	0.7626	0.3688	2.4651	0.8710	0.8576
0.44	0.6208	0.7785	0.3888	2.5051	0.9397	0.9265
0.46	0.6425	0.7931	0.4088	2.5451	1.0096	0.9967
0.48	0.6635	0.8063	0.4288	2.5851	1.0805	1.0681
0.50	0.6839	0.8184	0.4488	2.6251	1.1526	1.1407
0.52	0.7036	0.8294	0.4688	2.6651	1.2256	1.2143
0.54	0.7228	0.8394	0.4888	2.7051	1.2995	1.2890
0.56	0.7414	0.8485	0.5088	2.7451	1.3744	1.3647
0.58	0.7595	0.8567	0.5288	2.7851	1.4500	1.4413
0.60	0.7770	0.8641	0.5488	2.8251	1.5265	1.5188
0.62	0.7941	0.8707	0.5688	2.8651	1.6038	1.5971
0.64	0.8107	0.8767	0.5888	2.9051	1.6817	1.6762
0.66	0.8269	0.8820	0.6088	2.9451	1.7604	1.7561
0.68	0.8426	0.8867	0.6288	2.9851	1.8396	1.8367
0.70	0.8579	0.8909	0.6488	3.0251	1.9195	1.9180
0.71	0.8654	0.8928	0.6588	3.0451	1.9597	1.9589
0.73	0.8801	0.8962	0.6788	3.0851	2.0405	2.0412
0.75	0.8944	0.8991	0.6988	3.1251	2.1218	2.1241
0.76	0.9015	0.9004	0.7088	3.1451	2.1627	2.1658
0.77	0.9084	0.9016	0.7188	3.1651	2.2037	2.2076
0.78	0.9153	0.9027	0.7288	3.1851	2.2448	2.2496
0.80	0.9287	0.9046	0.7488	3.2251	2.3274	2.3339
0.82	0.9418	0.9062	0.7688	3.2651	2.4105	2.4187
0.84	0.9546	0.9074	0.7888	3.3051	2.4941	2.5041
0.85	0.9609	0.9079	0.7988	3.3251	2.5360	2.5470
0.86	0.9671	0.9083	0.8088	3.3451	2.5780	2.5900
0.87	0.9733	0.9086	0.8188	3.3651	2.6202	2.6331
0.88	0.9794	0.9089	0.8288	3.3851	2.6624	2.6763
0.89	0.9853	0.9091	0.8388	3.4051	2.7048	2.7196
0.90	0.9913	0.9092	0.8488	3.4251	2.7472	2.7631
0.91	0.9971	0.9093	0.8588	3.4451	2.7897	2.8066
0.92	1.0029	0.9093	0.8688	3.4651	2.8324	2.8503
0.94	1.0143	0.9091	0.8888	3.5051	2.9179	2.9379
0.95	1.0199	0.9089	0.8988	3.5251	2.9608	2.9819
1.00	1.0469	0.9073	0.9488	3.6251	3.1766	3.2031

C22   Box culvert upright (23% splays) (free surface)

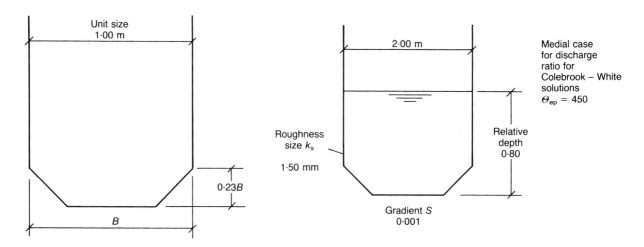

Unit size 1·00 m

2·00 m

Medial case for discharge ratio for Colebrook – White solutions $\Theta_{ep} = 450$

Roughness size k_s

1·50 mm

Relative depth 0·80

0·23B

B

Gradient S 0·001

Rel. dpth	U.equ. dia. $D_{ep(u)}$	Equiv. disch. factor	Unit sect. area	Unit wetted perim.	Unit surf. brdth	Unit mean depth	Discharge ratios $Q/Q_{0.80}$		U.crit. disch. Q_{uc}
Y	(m)	J	A_u (m²)	P_u (m)	B_{us} (m)	y_{um} (m)	medial	Mann'g	(m³s⁻¹)
0.04	0.1421	0.6834	0.0232	0.6531	0.6200	0.0374	0.0076	0.0072	0.0141
0.08	0.2589	1.0615	0.0496	0.7663	0.7000	0.0709	0.0240	0.0229	0.0413
0.12	0.3602	1.2869	0.0792	0.8794	0.7800	0.1015	0.0475	0.0455	0.0790
0.16	0.4514	1.4286	0.1120	0.9925	0.8600	0.1302	0.0777	0.0747	0.1266
0.20	0.5354	1.5212	0.1480	1.1057	0.9400	0.1574	0.1145	0.1107	0.1839
0.24	0.6182	1.6044	0.1871	1.2105	1.0000	0.1871	0.1585	0.1540	0.2534
0.28	0.7039	1.7135	0.2271	1.2905	1.0000	0.2271	0.2089	0.2038	0.3389
0.32	0.7795	1.7869	0.2671	1.3705	1.0000	0.2671	0.2620	0.2566	0.4323
0.36	0.8469	1.8341	0.3071	1.4505	1.0000	0.3071	0.3173	0.3117	0.5329
0.40	0.9071	1.8619	0.3471	1.5305	1.0000	0.3471	0.3744	0.3689	0.6404
0.44	0.9614	1.8753	0.3871	1.6105	1.0000	0.3871	0.4330	0.4276	0.7542
0.48	1.0106	1.8779	0.4271	1.6905	1.0000	0.4271	0.4929	0.4878	0.8741
0.52	1.0553	1.8724	0.4671	1.7705	1.0000	0.4671	0.5538	0.5491	0.9997
0.56	1.0961	1.8608	0.5071	1.8505	1.0000	0.5071	0.6157	0.6114	1.1308
0.60	1.1336	1.8446	0.5471	1.9305	1.0000	0.5471	0.6783	0.6745	1.2672
0.64	1.1680	1.8251	0.5871	2.0105	1.0000	0.5871	0.7416	0.7384	1.4087
0.68	1.1999	1.8031	0.6271	2.0905	1.0000	0.6271	0.8055	0.8030	1.5551
0.72	1.2294	1.7793	0.6671	2.1705	1.0000	0.6671	0.8699	0.8682	1.7063
0.76	1.2568	1.7543	0.7071	2.2505	1.0000	0.7071	0.9347	0.9339	1.8620
0.80	1.2823	1.7285	0.7471	2.3305	1.0000	0.7471	1.0000	1.0000	2.0222
0.84	1.3061	1.7022	0.7871	2.4105	1.0000	0.7871	1.0657	1.0665	2.1868
0.88	1.3284	1.6756	0.8271	2.4905	1.0000	0.8271	1.1316	1.1335	2.3556
0.92	1.3493	1.6490	0.8671	2.5705	1.0000	0.8671	1.1979	1.2007	2.5285
0.96	1.3689	1.6225	0.9071	2.6505	1.0000	0.9071	1.2644	1.2683	2.7055
1.00	1.3874	1.5962	0.9471	2.7305	1.0000	0.9471	1.3312	1.3361	2.8864
1.08	1.4213	1.5447	1.0271	2.8905	1.0000	1.0271	1.4654	1.4725	3.2597
1.16	1.4517	1.4950	1.1071	3.0505	1.0000	1.1071	1.6004	1.6097	3.6479
1.24	1.4790	1.4472	1.1871	3.2105	1.0000	1.1871	1.7359	1.7476	4.0503
1.32	1.5037	1.4016	1.2671	3.3705	1.0000	1.2671	1.8720	1.8861	4.4666
1.40	1.5262	1.3581	1.3471	3.5305	1.0000	1.3471	2.0085	2.0251	4.8962
1.48	1.5468	1.3167	1.4271	3.6905	1.0000	1.4271	2.1455	2.1646	5.3388
1.56	1.5656	1.2773	1.5071	3.8505	1.0000	1.5071	2.2827	2.3044	5.7939
1.64	1.5829	1.2399	1.5871	4.0105	1.0000	1.5871	2.4203	2.4446	6.2613
1.72	1.5989	1.2044	1.6671	4.1705	1.0000	1.6671	2.5581	2.5851	6.7407
1.80	1.6137	1.1707	1.7471	4.3305	1.0000	1.7471	2.6962	2.7259	7.2316
1.88	1.6275	1.1386	1.8271	4.4905	1.0000	1.8271	2.8345	2.8669	7.7340
1.96	1.6403	1.1081	1.9071	4.6505	1.0000	1.9071	2.9730	3.0081	8.2475
2.04	1.6523	1.0790	1.9871	4.8105	1.0000	1.9871	3.1116	3.1495	8.7718
2.12	1.6635	1.0514	2.0671	4.9705	1.0000	2.0671	3.2504	3.2911	9.3068
2.20	1.6740	1.0250	2.1471	5.1305	1.0000	2.1471	3.3893	3.4328	9.8523

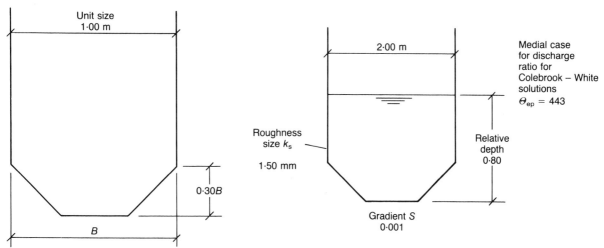

Rel. dpth	U.equ. dia. $D_{ep(u)}$	Equiv. disch. factor	Unit sect. area	Unit wetted perim.	Unit surf. brdth	Unit mean depth	Discharge ratios $Q/Q_{0.80}$		U.crit. disch. Q_{uc}
Y	(m)	J	A_u (m²)	P_u (m)	B_{us} (m)	y_{um} (m)	medial	Mann'g	(m³s⁻¹)
0.04	0.1372	0.8399	0.0176	0.5131	0.4800	0.0367	0.0059	0.0056	0.0106
0.08	0.2453	1.2303	0.0384	0.6263	0.5600	0.0686	0.0191	0.0181	0.0315
0.12	0.3376	1.4342	0.0624	0.7394	0.6400	0.0975	0.0381	0.0365	0.0610
0.16	0.4204	1.5491	0.0896	0.8525	0.7200	0.1244	0.0631	0.0606	0.0990
0.20	0.4971	1.6170	0.1200	0.9657	0.8000	0.1500	0.0940	0.0908	0.1455
0.24	0.5695	1.6584	0.1536	1.0788	0.8800	0.1745	0.1312	0.1272	0.2010
0.28	0.6389	1.6840	0.1904	1.1920	0.9600	0.1983	0.1750	0.1703	0.2655
0.32	0.7140	1.7408	0.2300	1.2885	1.0000	0.2300	0.2267	0.2215	0.3454
0.36	0.7892	1.8116	0.2700	1.3685	1.0000	0.2700	0.2835	0.2779	0.4393
0.40	0.8560	1.8566	0.3100	1.4485	1.0000	0.3100	0.3425	0.3369	0.5405
0.44	0.9159	1.8824	0.3500	1.5285	1.0000	0.3500	0.4035	0.3979	0.6484
0.48	0.9698	1.8941	0.3900	1.6085	1.0000	0.3900	0.4659	0.4606	0.7627
0.52	1.0186	1.8952	0.4300	1.6885	1.0000	0.4300	0.5297	0.5247	0.8830
0.56	1.0630	1.8883	0.4700	1.7685	1.0000	0.4700	0.5946	0.5901	1.0090
0.60	1.1036	1.8755	0.5100	1.8485	1.0000	0.5100	0.6605	0.6565	1.1406
0.64	1.1408	1.8583	0.5500	1.9285	1.0000	0.5500	0.7271	0.7238	1.2773
0.68	1.1750	1.8378	0.5900	2.0085	1.0000	0.5900	0.7945	0.7919	1.4192
0.72	1.2066	1.8149	0.6300	2.0885	1.0000	0.6300	0.8625	0.8607	1.5659
0.76	1.2359	1.7904	0.6700	2.1685	1.0000	0.6700	0.9310	0.9301	1.7174
0.80	1.2630	1.7647	0.7100	2.2485	1.0000	0.7100	1.0000	1.0000	1.8735
0.84	1.2884	1.7382	0.7500	2.3285	1.0000	0.7500	1.0695	1.0704	2.0340
0.88	1.3120	1.7113	0.7900	2.4085	1.0000	0.7900	1.1393	1.1412	2.1989
0.92	1.3341	1.6842	0.8300	2.4885	1.0000	0.8300	1.2094	1.2125	2.3680
0.96	1.3549	1.6571	0.8700	2.5685	1.0000	0.8700	1.2799	1.2840	2.5412
1.00	1.3743	1.6302	0.9100	2.6485	1.0000	0.9100	1.3507	1.3559	2.7185
1.08	1.4100	1.5772	0.9900	2.8085	1.0000	0.9900	1.4929	1.5005	3.0847
1.16	1.4418	1.5258	1.0700	2.9685	1.0000	1.0700	1.6360	1.6461	3.4661
1.24	1.4703	1.4764	1.1500	3.1285	1.0000	1.1500	1.7797	1.7924	3.8620
1.32	1.4961	1.4292	1.2300	3.2885	1.0000	1.2300	1.9241	1.9394	4.2719
1.40	1.5195	1.3842	1.3100	3.4485	1.0000	1.3100	2.0690	2.0870	4.6953
1.48	1.5408	1.3414	1.3900	3.6085	1.0000	1.3900	2.2143	2.2351	5.1320
1.56	1.5603	1.3007	1.4700	3.7685	1.0000	1.4700	2.3600	2.3837	5.5813
1.64	1.5782	1.2620	1.5500	3.9285	1.0000	1.5500	2.5061	2.5326	6.0431
1.72	1.5947	1.2253	1.6300	4.0885	1.0000	1.6300	2.6524	2.6819	6.5169
1.80	1.6100	1.1905	1.7100	4.2485	1.0000	1.7100	2.7990	2.8314	7.0025
1.88	1.6241	1.1574	1.7900	4.4085	1.0000	1.7900	2.9458	2.9812	7.4996
1.96	1.6373	1.1259	1.8700	4.5685	1.0000	1.8700	3.0928	3.1313	8.0080
2.04	1.6496	1.0959	1.9500	4.7285	1.0000	1.9500	3.2400	3.2815	8.5273
2.12	1.6610	1.0674	2.0300	4.8885	1.0000	2.0300	3.3874	3.4320	9.0574
2.20	1.6718	1.0403	2.1100	5.0485	1.0000	2.1100	3.5349	3.5826	9.5981

C24

Standard Horseshoe

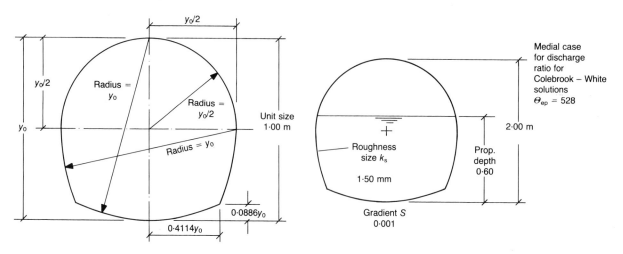

Prop dpth	U.equ. dia. $D_{ep(u)}$	Equiv. disch. factor	Unit sect. area	Unit wetted perim.	Unit surf. brdth	Unit mean depth	Discharge ratios $Q/Q_{0.60}$		U.crit. disch. Q_{uc}
Y	(m)	J	A_u (m²)	P_u (m)	B_{us} (m)	y_{um} (m)	medial	Mann'g	(m³s⁻¹)
0.02	0.0531	0.4162	0.0053	0.4007	0.3980	0.0134	0.0014	0.0013	0.0019
0.04	0.1057	0.5849	0.0150	0.5676	0.5600	0.0268	0.0061	0.0058	0.0077
0.06	0.1578	0.7117	0.0275	0.6963	0.6823	0.0402	0.0146	0.0138	0.0173
0.08	0.2093	0.8165	0.0422	0.8054	0.7838	0.0538	0.0270	0.0256	0.0306
0.10	0.2680	0.9644	0.0585	0.8731	0.8330	0.0702	0.0441	0.0418	0.0485
0.11	0.2989	1.0494	0.0669	0.8949	0.8416	0.0795	0.0540	0.0514	0.0590
0.12	0.3288	1.1269	0.0753	0.9165	0.8500	0.0886	0.0647	0.0617	0.0702
0.13	0.3576	1.1976	0.0839	0.9381	0.8581	0.0977	0.0760	0.0727	0.0821
0.14	0.3856	1.2622	0.0925	0.9596	0.8659	0.1068	0.0879	0.0843	0.0947
0.15	0.4126	1.3214	0.1012	0.9810	0.8735	0.1158	0.1004	0.0965	0.1079
0.16	0.4388	1.3755	0.1100	1.0023	0.8809	0.1248	0.1135	0.1092	0.1217
0.17	0.4643	1.4251	0.1188	1.0235	0.8880	0.1338	0.1271	0.1225	0.1361
0.18	0.4890	1.4706	0.1277	1.0447	0.8948	0.1427	0.1412	0.1364	0.1511
0.19	0.5131	1.5124	0.1367	1.0657	0.9015	0.1516	0.1558	0.1507	0.1667
0.20	0.5365	1.5508	0.1457	1.0867	0.9079	0.1605	0.1708	0.1655	0.1829
0.21	0.5592	1.5861	0.1549	1.1077	0.9141	0.1694	0.1863	0.1808	0.1996
0.22	0.5814	1.6184	0.1640	1.1285	0.9200	0.1783	0.2023	0.1965	0.2169
0.23	0.6030	1.6482	0.1733	1.1493	0.9257	0.1872	0.2186	0.2127	0.2347
0.24	0.6240	1.6755	0.1825	1.1701	0.9312	0.1960	0.2353	0.2293	0.2531
0.25	0.6446	1.7005	0.1919	1.1908	0.9365	0.2049	0.2524	0.2463	0.2720
0.26	0.6646	1.7235	0.2013	1.2114	0.9415	0.2138	0.2699	0.2636	0.2914
0.27	0.6841	1.7446	0.2107	1.2320	0.9464	0.2227	0.2878	0.2814	0.3114
0.28	0.7032	1.7639	0.2202	1.2525	0.9510	0.2315	0.3059	0.2995	0.3318
0.29	0.7219	1.7815	0.2297	1.2730	0.9554	0.2405	0.3244	0.3180	0.3528
0.30	0.7401	1.7976	0.2393	1.2934	0.9596	0.2494	0.3433	0.3368	0.3742
0.31	0.7579	1.8122	0.2489	1.3138	0.9636	0.2583	0.3624	0.3559	0.3962
0.32	0.7753	1.8255	0.2586	1.3342	0.9673	0.2673	0.3818	0.3753	0.4187
0.33	0.7923	1.8375	0.2683	1.3545	0.9709	0.2763	0.4015	0.3951	0.4416
0.34	0.8089	1.8484	0.2780	1.3747	0.9742	0.2853	0.4215	0.4151	0.4650
0.35	0.8251	1.8581	0.2878	1.3950	0.9774	0.2944	0.4418	0.4354	0.4889
0.36	0.8410	1.8669	0.2975	1.4152	0.9803	0.3035	0.4622	0.4560	0.5133
0.37	0.8565	1.8746	0.3074	1.4354	0.9830	0.3127	0.4830	0.4768	0.5382
0.38	0.8717	1.8814	0.3172	1.4555	0.9855	0.3219	0.5039	0.4979	0.5635
0.39	0.8866	1.8874	0.3271	1.4757	0.9879	0.3311	0.5251	0.5192	0.5893
0.40	0.9011	1.8925	0.3370	1.4958	0.9900	0.3404	0.5464	0.5407	0.6156
0.41	0.9153	1.8969	0.3469	1.5159	0.9919	0.3497	0.5680	0.5624	0.6424
0.42	0.9292	1.9005	0.3568	1.5360	0.9936	0.3591	0.5898	0.5844	0.6696
0.43	0.9428	1.9034	0.3667	1.5560	0.9951	0.3685	0.6117	0.6065	0.6972
0.44	0.9561	1.9057	0.3767	1.5761	0.9964	0.3781	0.6338	0.6288	0.7253
0.45	0.9690	1.9073	0.3867	1.5961	0.9975	0.3876	0.6560	0.6513	0.7539

Standard Horseshoe

Prop dpth	U.equ. dia. $D_{ep(u)}$	Equiv. disch. factor	Unit sect. area	Unit wetted perim.	Unit surf. brdth	Unit mean depth	Discharge ratios $Q/Q_{0.60}$		U.crit. disch. Q_{uc}
Y	(m)	J	A_u (m²)	P_u (m)	B_{us} (m)	y_{um} (m)	medial	Mann'g	(m³s⁻¹)
0.46	0.9817	1.9084	0.3966	1.6161	0.9984	0.3973	0.6784	0.6739	0.7829
0.47	0.9941	1.9089	0.4066	1.6361	0.9991	0.4070	0.7009	0.6967	0.8124
0.48	1.0063	1.9088	0.4166	1.6561	0.9996	0.4168	0.7236	0.7196	0.8423
0.49	1.0181	1.9082	0.4266	1.6761	0.9999	0.4267	0.7463	0.7426	0.8727
0.50	1.0297	1.9072	0.4366	1.6961	1.0000	0.4366	0.7692	0.7658	0.9035
0.51	1.0410	1.9057	0.4466	1.7161	0.9998	0.4467	0.7922	0.7891	0.9348
0.52	1.0520	1.9036	0.4566	1.7361	0.9992	0.4570	0.8152	0.8124	0.9666
0.53	1.0628	1.9012	0.4666	1.7562	0.9982	0.4674	0.8383	0.8358	0.9990
0.54	1.0733	1.8982	0.4766	1.7762	0.9968	0.4781	0.8614	0.8593	1.0320
0.55	1.0834	1.8948	0.4865	1.7963	0.9950	0.4890	0.8846	0.8828	1.0654
0.56	1.0933	1.8909	0.4965	1.8164	0.9928	0.5001	0.9077	0.9063	1.0995
0.57	1.1029	1.8865	0.5064	1.8366	0.9902	0.5114	0.9309	0.9298	1.1341
0.58	1.1122	1.8817	0.5163	1.8568	0.9871	0.5230	0.9540	0.9532	1.1692
0.59	1.1212	1.8764	0.5261	1.8771	0.9837	0.5349	0.9770	0.9767	1.2050
0.60	1.1298	1.8706	0.5360	1.8975	0.9798	0.5470	1.0000	1.0000	1.2413
0.61	1.1382	1.8643	0.5457	1.9179	0.9755	0.5594	1.0229	1.0232	1.2782
0.62	1.1462	1.8575	0.5555	1.9385	0.9708	0.5722	1.0457	1.0464	1.3158
0.63	1.1539	1.8502	0.5651	1.9591	0.9656	0.5853	1.0683	1.0694	1.3539
0.64	1.1612	1.8425	0.5748	1.9799	0.9600	0.5987	1.0908	1.0922	1.3927
0.65	1.1682	1.8342	0.5843	2.0008	0.9539	0.6126	1.1131	1.1148	1.4322
0.66	1.1749	1.8255	0.5939	2.0219	0.9474	0.6268	1.1351	1.1373	1.4723
0.67	1.1812	1.8162	0.6033	2.0430	0.9404	0.6415	1.1570	1.1595	1.5132
0.68	1.1871	1.8065	0.6127	2.0644	0.9330	0.6567	1.1786	1.1814	1.5547
0.69	1.1927	1.7962	0.6219	2.0859	0.9250	0.6724	1.1999	1.2031	1.5971
0.70	1.1978	1.7854	0.6312	2.1076	0.9165	0.6886	1.2210	1.2244	1.6402
0.71	1.2026	1.7741	0.6403	2.1296	0.9075	0.7055	1.2417	1.2454	1.6842
0.72	1.2070	1.7623	0.6493	2.1517	0.8980	0.7231	1.2620	1.2661	1.7290
0.73	1.2110	1.7499	0.6582	2.1741	0.8879	0.7413	1.2820	1.2863	1.7748
0.74	1.2146	1.7370	0.6671	2.1968	0.8773	0.7604	1.3016	1.3061	1.8216
0.75	1.2178	1.7235	0.6758	2.2197	0.8660	0.7803	1.3207	1.3255	1.8694
0.76	1.2205	1.7094	0.6844	2.2430	0.8542	0.8012	1.3393	1.3444	1.9184
0.77	1.2227	1.6948	0.6929	2.2666	0.8417	0.8232	1.3575	1.3627	1.9686
0.78	1.2245	1.6795	0.7012	2.2905	0.8285	0.8464	1.3751	1.3805	2.0202
0.79	1.2259	1.6636	0.7094	2.3149	0.8146	0.8709	1.3921	1.3977	2.0732
0.80	1.2267	1.6471	0.7175	2.3396	0.8000	0.8969	1.4085	1.4142	2.1279
0.81	1.2270	1.6300	0.7254	2.3649	0.7846	0.9246	1.4243	1.4301	2.1844
0.82	1.2268	1.6121	0.7332	2.3906	0.7684	0.9542	1.4394	1.4452	2.2428
0.83	1.2260	1.5935	0.7408	2.4169	0.7513	0.9861	1.4538	1.4595	2.3036
0.84	1.2246	1.5742	0.7482	2.4439	0.7332	1.0205	1.4673	1.4731	2.3669
0.85	1.2226	1.5541	0.7554	2.4715	0.7141	1.0578	1.4800	1.4857	2.4332
0.86	1.2200	1.5331	0.7625	2.4999	0.6940	1.0987	1.4918	1.4974	2.5029
0.87	1.2167	1.5113	0.7693	2.5292	0.6726	1.1438	1.5027	1.5081	2.5766
0.88	1.2127	1.4885	0.7759	2.5594	0.6499	1.1939	1.5125	1.5177	2.6550
0.89	1.2078	1.4646	0.7823	2.5908	0.6258	1.2501	1.5212	1.5261	2.7392
0.90	1.2022	1.4396	0.7884	2.6234	0.6000	1.3141	1.5287	1.5333	2.8304
0.91	1.1956	1.4133	0.7943	2.6575	0.5724	1.3878	1.5348	1.5390	2.9303
0.92	1.1879	1.3856	0.7999	2.6934	0.5426	1.4742	1.5395	1.5432	3.0414
0.93	1.1791	1.3562	0.8052	2.7314	0.5103	1.5778	1.5425	1.5457	3.1672
0.94	1.1690	1.3248	0.8101	2.7720	0.4750	1.7055	1.5437	1.5462	3.3130
0.95	1.1572	1.2910	0.8146	2.8159	0.4359	1.8689	1.5427	1.5444	3.4876
0.96	1.1435	1.2542	0.8188	2.8642	0.3919	2.0892	1.5391	1.5400	3.7061
0.97	1.1271	1.2132	0.8225	2.9188	0.3412	2.4107	1.5323	1.5321	3.9989
0.98	1.1070	1.1658	0.8256	2.9831	0.2800	2.9485	1.5211	1.5196	4.4393
0.99	1.0800	1.1064	0.8280	3.0666	0.1990	4.1608	1.5024	1.4992	5.2890
1.00	1.0154	0.9764	0.8293	3.2669	0.0000	-	1.4485	1.4411	-

C25　Metcalf and Eddy (M and E) Horseshoe

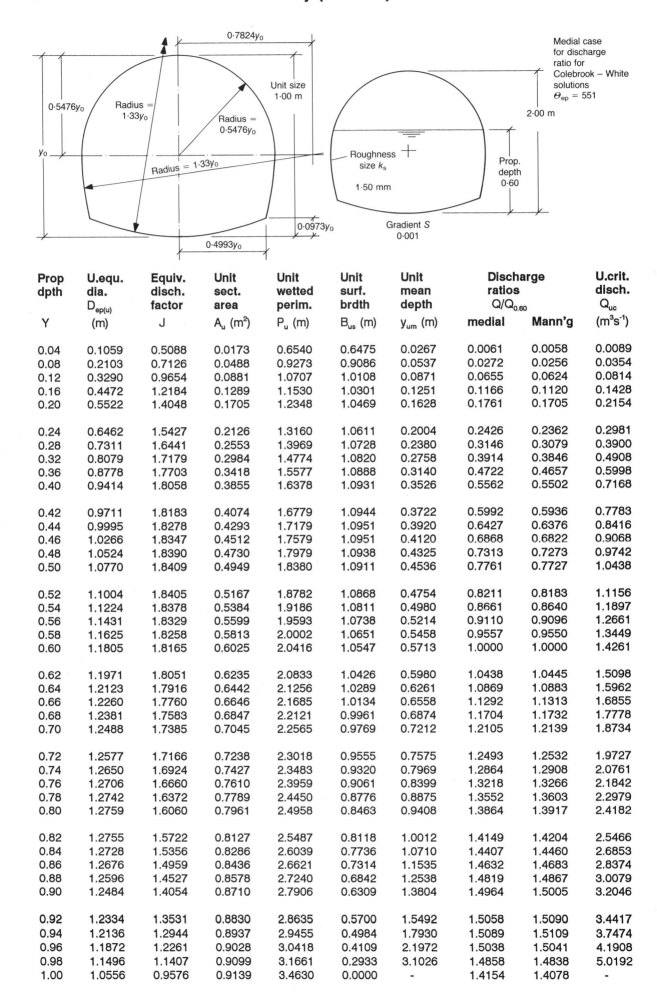

Prop dpth	U.equ. dia. $D_{ep(u)}$	Equiv. disch. factor	Unit sect. area	Unit wetted perim.	Unit surf. brdth	Unit mean depth	Discharge ratios $Q/Q_{0.60}$		U.crit. disch. Q_{uc}
Y	(m)	J	A_u (m²)	P_u (m)	B_{us} (m)	y_{um} (m)	medial	Mann'g	(m³s⁻¹)
0.04	0.1059	0.5088	0.0173	0.6540	0.6475	0.0267	0.0061	0.0058	0.0089
0.08	0.2103	0.7126	0.0488	0.9273	0.9086	0.0537	0.0272	0.0256	0.0354
0.12	0.3290	0.9654	0.0881	1.0707	1.0108	0.0871	0.0655	0.0624	0.0814
0.16	0.4472	1.2184	0.1289	1.1530	1.0301	0.1251	0.1166	0.1120	0.1428
0.20	0.5522	1.4048	0.1705	1.2348	1.0469	0.1628	0.1761	0.1705	0.2154
0.24	0.6462	1.5427	0.2126	1.3160	1.0611	0.2004	0.2426	0.2362	0.2981
0.28	0.7311	1.6441	0.2553	1.3969	1.0728	0.2380	0.3146	0.3079	0.3900
0.32	0.8079	1.7179	0.2984	1.4774	1.0820	0.2758	0.3914	0.3846	0.4908
0.36	0.8778	1.7703	0.3418	1.5577	1.0888	0.3140	0.4722	0.4657	0.5998
0.40	0.9414	1.8058	0.3855	1.6378	1.0931	0.3526	0.5562	0.5502	0.7168
0.42	0.9711	1.8183	0.4074	1.6779	1.0944	0.3722	0.5992	0.5936	0.7783
0.44	0.9995	1.8278	0.4293	1.7179	1.0951	0.3920	0.6427	0.6376	0.8416
0.46	1.0266	1.8347	0.4512	1.7579	1.0951	0.4120	0.6868	0.6822	0.9068
0.48	1.0524	1.8390	0.4730	1.7979	1.0938	0.4325	0.7313	0.7273	0.9742
0.50	1.0770	1.8409	0.4949	1.8380	1.0911	0.4536	0.7761	0.7727	1.0438
0.52	1.1004	1.8405	0.5167	1.8782	1.0868	0.4754	0.8211	0.8183	1.1156
0.54	1.1224	1.8378	0.5384	1.9186	1.0811	0.4980	0.8661	0.8640	1.1897
0.56	1.1431	1.8329	0.5599	1.9593	1.0738	0.5214	0.9110	0.9096	1.2661
0.58	1.1625	1.8258	0.5813	2.0002	1.0651	0.5458	0.9557	0.9550	1.3449
0.60	1.1805	1.8165	0.6025	2.0416	1.0547	0.5713	1.0000	1.0000	1.4261
0.62	1.1971	1.8051	0.6235	2.0833	1.0426	0.5980	1.0438	1.0445	1.5098
0.64	1.2123	1.7916	0.6442	2.1256	1.0289	0.6261	1.0869	1.0883	1.5962
0.66	1.2260	1.7760	0.6646	2.1685	1.0134	0.6558	1.1292	1.1313	1.6855
0.68	1.2381	1.7583	0.6847	2.2121	0.9961	0.6874	1.1704	1.1732	1.7778
0.70	1.2488	1.7385	0.7045	2.2565	0.9769	0.7212	1.2105	1.2139	1.8734
0.72	1.2577	1.7166	0.7238	2.3018	0.9555	0.7575	1.2493	1.2532	1.9727
0.74	1.2650	1.6924	0.7427	2.3483	0.9320	0.7969	1.2864	1.2908	2.0761
0.76	1.2706	1.6660	0.7610	2.3959	0.9061	0.8399	1.3218	1.3266	2.1842
0.78	1.2742	1.6372	0.7789	2.4450	0.8776	0.8875	1.3552	1.3603	2.2979
0.80	1.2759	1.6060	0.7961	2.4958	0.8463	0.9408	1.3864	1.3917	2.4182
0.82	1.2755	1.5722	0.8127	2.5487	0.8118	1.0012	1.4149	1.4204	2.5466
0.84	1.2728	1.5356	0.8286	2.6039	0.7736	1.0710	1.4407	1.4460	2.6853
0.86	1.2676	1.4959	0.8436	2.6621	0.7314	1.1535	1.4632	1.4683	2.8374
0.88	1.2596	1.4527	0.8578	2.7240	0.6842	1.2538	1.4819	1.4867	3.0079
0.90	1.2484	1.4054	0.8710	2.7906	0.6309	1.3804	1.4964	1.5005	3.2046
0.92	1.2334	1.3531	0.8830	2.8635	0.5700	1.5492	1.5058	1.5090	3.4417
0.94	1.2136	1.2944	0.8937	2.9455	0.4984	1.7930	1.5089	1.5109	3.7474
0.96	1.1872	1.2261	0.9028	3.0418	0.4109	2.1972	1.5038	1.5041	4.1908
0.98	1.1496	1.1407	0.9099	3.1661	0.2933	3.1026	1.4858	1.4838	5.0192
1.00	1.0556	0.9576	0.9139	3.4630	0.0000	-	1.4154	1.4078	-

Babbitt and Baumann (B and B) Horseshoe C26

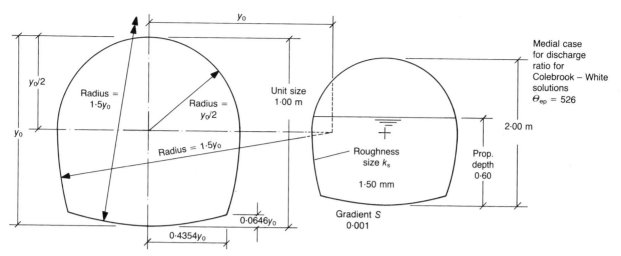

Prop dpth Y	U.equ. dia. $D_{ep(u)}$ (m)	Equiv. disch. factor J	Unit sect. area A_u (m²)	Unit wetted perim. P_u (m)	Unit surf. brdth B_{us} (m)	Unit mean depth y_{um} (m)	Discharge ratios $Q/Q_{0.60}$ medial	Mann'g	U.crit. disch. Q_{uc} (m³s⁻¹)
0.04	0.1060	0.4796	0.0184	0.6944	0.6882	0.0267	0.0073	0.0069	0.0094
0.08	0.2235	0.7666	0.0512	0.9157	0.8800	0.0581	0.0332	0.0314	0.0386
0.12	0.3477	1.0936	0.0868	0.9987	0.9021	0.0962	0.0749	0.0716	0.0843
0.16	0.4562	1.3255	0.1233	1.0811	0.9219	0.1337	0.1265	0.1219	0.1412
0.20	0.5521	1.4913	0.1605	1.1630	0.9394	0.1709	0.1859	0.1803	0.2078
0.24	0.6378	1.6100	0.1984	1.2445	0.9546	0.2079	0.2515	0.2453	0.2833
0.28	0.7148	1.6941	0.2369	1.3255	0.9676	0.2448	0.3225	0.3160	0.3670
0.32	0.7845	1.7525	0.2758	1.4062	0.9783	0.2819	0.3980	0.3915	0.4585
0.36	0.8478	1.7915	0.3151	1.4867	0.9869	0.3193	0.4773	0.4711	0.5576
0.40	0.9055	1.8154	0.3547	1.5669	0.9933	0.3571	0.5597	0.5541	0.6638
0.42	0.9324	1.8228	0.3746	1.6070	0.9957	0.3762	0.6020	0.5967	0.7195
0.44	0.9582	1.8276	0.3945	1.6471	0.9976	0.3955	0.6448	0.6399	0.7770
0.46	0.9828	1.8300	0.4145	1.6871	0.9989	0.4149	0.6882	0.6838	0.8362
0.48	1.0063	1.8304	0.4345	1.7271	0.9997	0.4346	0.7321	0.7282	0.8970
0.50	1.0288	1.8290	0.4545	1.7671	1.0000	0.4545	0.7763	0.7730	0.9595
0.52	1.0503	1.8258	0.4745	1.8071	0.9992	0.4749	0.8209	0.8182	1.0239
0.54	1.0707	1.8210	0.4944	1.8472	0.9968	0.4960	0.8657	0.8636	1.0905
0.56	1.0901	1.8144	0.5143	1.8874	0.9928	0.5181	0.9106	0.9092	1.1594
0.58	1.1083	1.8061	0.5341	1.9278	0.9871	0.5411	0.9554	0.9547	1.2305
0.60	1.1254	1.7961	0.5538	1.9685	0.9798	0.5652	1.0000	1.0000	1.3039
0.62	1.1413	1.7842	0.5733	2.0095	0.9708	0.5906	1.0442	1.0449	1.3798
0.64	1.1559	1.7705	0.5926	2.0509	0.9600	0.6173	1.0879	1.0893	1.4582
0.66	1.1692	1.7550	0.6117	2.0928	0.9474	0.6457	1.1309	1.1330	1.5393
0.68	1.1811	1.7376	0.6305	2.1354	0.9330	0.6758	1.1730	1.1758	1.6232
0.70	1.1916	1.7183	0.6490	2.1786	0.9165	0.7081	1.2141	1.2174	1.7103
0.72	1.2007	1.6970	0.6672	2.2227	0.8980	0.7430	1.2539	1.2578	1.8009
0.74	1.2081	1.6736	0.6849	2.2677	0.8773	0.7807	1.2922	1.2966	1.8952
0.76	1.2139	1.6481	0.7022	2.3139	0.8542	0.8221	1.3288	1.3337	1.9940
0.78	1.2180	1.6203	0.7191	2.3615	0.8285	0.8679	1.3634	1.3687	2.0979
0.80	1.2202	1.5902	0.7354	2.4106	0.8000	0.9192	1.3959	1.4014	2.2079
0.82	1.2204	1.5575	0.7511	2.4616	0.7684	0.9775	1.4258	1.4315	2.3253
0.84	1.2185	1.5221	0.7661	2.5149	0.7332	1.0448	1.4529	1.4585	2.4522
0.86	1.2141	1.4836	0.7804	2.5709	0.6940	1.1245	1.4767	1.4822	2.5914
0.88	1.2071	1.4417	0.7938	2.6304	0.6499	1.2214	1.4968	1.5019	2.7473
0.90	1.1970	1.3957	0.8063	2.6944	0.6000	1.3439	1.5125	1.5171	2.9271
0.92	1.1833	1.3447	0.8178	2.7644	0.5426	1.5071	1.5231	1.5268	3.1438
0.94	1.1649	1.2873	0.8280	2.8430	0.4750	1.7432	1.5273	1.5298	3.4232
0.96	1.1402	1.2203	0.8367	2.9352	0.3919	2.1348	1.5230	1.5239	3.8281
0.98	1.1047	1.1363	0.8434	3.0541	0.2800	3.0123	1.5056	1.5042	4.5842
1.00	1.0152	0.9555	0.8472	3.3379	0.0000	-	1.4352	1.4282	-

M and E Semi-eliptical

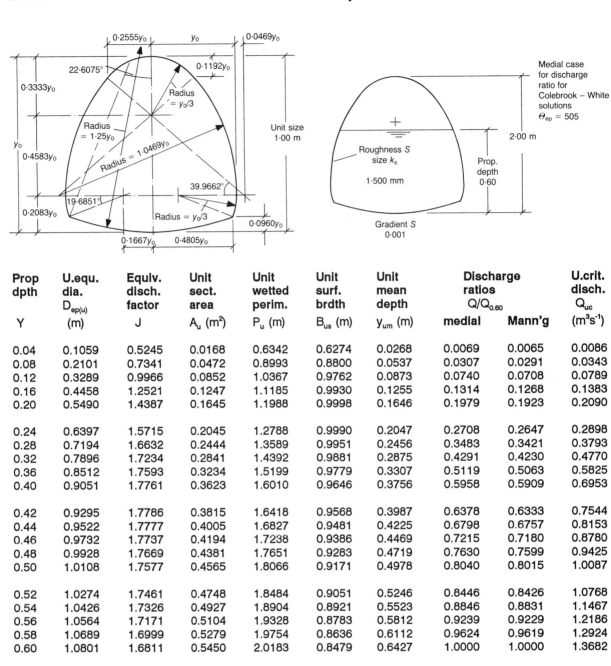

Prop dpth	U.equ. dia. $D_{ep(u)}$	Equiv. disch. factor	Unit sect. area	Unit wetted perim.	Unit surf. brdth	Unit mean depth	Discharge ratios $Q/Q_{0.60}$		U.crit. disch. Q_{uc}
Y	(m)	J	A_u (m²)	P_u (m)	B_{us} (m)	y_{um} (m)	medial	Mann'g	(m³s⁻¹)
0.04	0.1059	0.5245	0.0168	0.6342	0.6274	0.0268	0.0069	0.0065	0.0086
0.08	0.2101	0.7341	0.0472	0.8993	0.8800	0.0537	0.0307	0.0291	0.0343
0.12	0.3289	0.9966	0.0852	1.0367	0.9762	0.0873	0.0740	0.0708	0.0789
0.16	0.4458	1.2521	0.1247	1.1185	0.9930	0.1255	0.1314	0.1268	0.1383
0.20	0.5490	1.4387	0.1645	1.1988	0.9998	0.1646	0.1979	0.1923	0.2090
0.24	0.6397	1.5715	0.2045	1.2788	0.9990	0.2047	0.2708	0.2647	0.2898
0.28	0.7194	1.6632	0.2444	1.3589	0.9951	0.2456	0.3483	0.3421	0.3793
0.32	0.7896	1.7234	0.2841	1.4392	0.9881	0.2875	0.4291	0.4230	0.4770
0.36	0.8512	1.7593	0.3234	1.5199	0.9779	0.3307	0.5119	0.5063	0.5825
0.40	0.9051	1.7761	0.3623	1.6010	0.9646	0.3756	0.5958	0.5909	0.6953
0.42	0.9295	1.7786	0.3815	1.6418	0.9568	0.3987	0.6378	0.6333	0.7544
0.44	0.9522	1.7777	0.4005	1.6827	0.9481	0.4225	0.6798	0.6757	0.8153
0.46	0.9732	1.7737	0.4194	1.7238	0.9386	0.4469	0.7215	0.7180	0.8780
0.48	0.9928	1.7669	0.4381	1.7651	0.9283	0.4719	0.7630	0.7599	0.9425
0.50	1.0108	1.7577	0.4565	1.8066	0.9171	0.4978	0.8040	0.8015	1.0087
0.52	1.0274	1.7461	0.4748	1.8484	0.9051	0.5246	0.8446	0.8426	1.0768
0.54	1.0426	1.7326	0.4927	1.8904	0.8921	0.5523	0.8846	0.8831	1.1467
0.56	1.0564	1.7171	0.5104	1.9328	0.8783	0.5812	0.9239	0.9229	1.2186
0.58	1.0689	1.6999	0.5279	1.9754	0.8636	0.6112	0.9624	0.9619	1.2924
0.60	1.0801	1.6811	0.5450	2.0183	0.8479	0.6427	1.0000	1.0000	1.3682
0.62	1.0899	1.6608	0.5618	2.0617	0.8313	0.6758	1.0367	1.0371	1.4462
0.64	1.0986	1.6392	0.5782	2.1054	0.8137	0.7106	1.0723	1.0731	1.5264
0.66	1.1060	1.6164	0.5943	2.1495	0.7951	0.7475	1.1067	1.1079	1.6091
0.68	1.1121	1.5924	0.6100	2.1941	0.7755	0.7867	1.1398	1.1414	1.6944
0.70	1.1171	1.5673	0.6253	2.2391	0.7547	0.8286	1.1717	1.1735	1.7825
0.72	1.1209	1.5412	0.6402	2.2847	0.7329	0.8736	1.2020	1.2041	1.8738
0.74	1.1234	1.5142	0.6546	2.3308	0.7099	0.9222	1.2309	1.2332	1.9686
0.76	1.1248	1.4863	0.6686	2.3776	0.6857	0.9751	1.2581	1.2605	2.0675
0.78	1.1251	1.4575	0.6821	2.4250	0.6603	1.0330	1.2836	1.2860	2.1709
0.80	1.1241	1.4279	0.6950	2.4731	0.6335	1.0970	1.3072	1.3097	2.2796
0.82	1.1220	1.3976	0.7074	2.5220	0.6054	1.1684	1.3290	1.3314	2.3945
0.84	1.1187	1.3665	0.7192	2.5717	0.5759	1.2488	1.3487	1.3509	2.5169
0.86	1.1142	1.3348	0.7304	2.6223	0.5449	1.3405	1.3663	1.3683	2.6483
0.88	1.1085	1.3023	0.7410	2.6739	0.5123	1.4465	1.3817	1.3834	2.7909
0.90	1.1011	1.2680	0.7509	2.7279	0.4761	1.5772	1.3944	1.3956	2.9531
0.92	1.0910	1.2300	0.7600	2.7865	0.4333	1.7540	1.4033	1.4039	3.1520
0.94	1.0774	1.1868	0.7682	2.8519	0.3816	2.0131	1.4074	1.4072	3.4131
0.96	1.0589	1.1361	0.7752	2.9282	0.3166	2.4480	1.4051	1.4037	3.7981
0.98	1.0319	1.0713	0.7807	3.0260	0.2274	3.4323	1.3926	1.3896	4.5291
1.00	0.9622	0.9277	0.7837	3.2581	0.0000	-	1.3387	1.3314	-

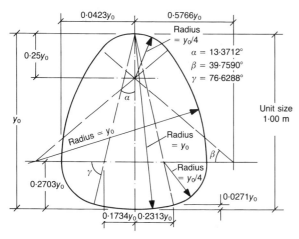

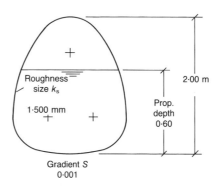

Medial case
for discharge
ratio for
Colebrook – White
solutions
$\Theta_{ep} = 473$

Prop dpth	U.equ. dia. $D_{ep(u)}$	Equiv. disch. factor	Unit sect. area	Unit wetted perim.	Unit surf. brdth	Unit mean depth	Discharge ratios $Q/Q_{0.60}$		U.crit. disch. Q_{uc}
Y	(m)	J	A_u (m²)	P_u (m)	B_{us} (m)	y_{um} (m)	medial	Mann'g	(m³s⁻¹)
0.04	0.1084	0.6196	0.0149	0.5497	0.5413	0.0275	0.0078	0.0073	0.0077
0.08	0.2245	1.0036	0.0394	0.7028	0.6711	0.0588	0.0332	0.0316	0.0299
0.12	0.3341	1.2914	0.0679	0.8129	0.7464	0.0910	0.0739	0.0709	0.0641
0.16	0.4358	1.5097	0.0988	0.9069	0.7956	0.1242	0.1273	0.1231	0.1091
0.20	0.5290	1.6739	0.1313	0.9929	0.8267	0.1588	0.1913	0.1862	0.1639
0.24	0.6132	1.7927	0.1648	1.0746	0.8432	0.1954	0.2634	0.2578	0.2281
0.28	0.6879	1.8713	0.1986	1.1548	0.8468	0.2345	0.3412	0.3355	0.3012
0.32	0.7529	1.9154	0.2324	1.2348	0.8444	0.2753	0.4226	0.4171	0.3819
0.36	0.8094	1.9336	0.2661	1.3150	0.8388	0.3172	0.5062	0.5011	0.4693
0.40	0.8584	1.9324	0.2995	1.3955	0.8300	0.3608	0.5909	0.5865	0.5634
0.42	0.8804	1.9261	0.3160	1.4359	0.8244	0.3834	0.6334	0.6294	0.6128
0.44	0.9007	1.9165	0.3325	1.4764	0.8179	0.4065	0.6759	0.6723	0.6638
0.46	0.9195	1.9040	0.3487	1.5171	0.8106	0.4302	0.7182	0.7150	0.7163
0.48	0.9368	1.8891	0.3649	1.5579	0.8024	0.4547	0.7602	0.7574	0.7705
0.50	0.9527	1.8719	0.3808	1.5989	0.7934	0.4800	0.8017	0.7995	0.8263
0.52	0.9673	1.8527	0.3966	1.6401	0.7835	0.5062	0.8428	0.8410	0.8836
0.54	0.9805	1.8317	0.4122	1.6816	0.7728	0.5334	0.8833	0.8820	0.9426
0.56	0.9924	1.8091	0.4275	1.7232	0.7611	0.5617	0.9231	0.9222	1.0034
0.58	1.0030	1.7851	0.4426	1.7651	0.7486	0.5913	0.9620	0.9616	1.0658
0.60	1.0124	1.7598	0.4575	1.8074	0.7351	0.6223	1.0000	1.0000	1.1301
0.62	1.0206	1.7332	0.4720	1.8499	0.7206	0.6550	1.0370	1.0374	1.1963
0.64	1.0276	1.7056	0.4863	1.8928	0.7052	0.6895	1.0729	1.0736	1.2645
0.66	1.0335	1.6770	0.5002	1.9360	0.6888	0.7262	1.1076	1.1086	1.3349
0.68	1.0382	1.6475	0.5138	1.9796	0.6714	0.7653	1.1409	1.1422	1.4076
0.70	1.0418	1.6172	0.5271	2.0237	0.6529	0.8073	1.1728	1.1743	1.4830
0.72	1.0442	1.5861	0.5399	2.0682	0.6333	0.8526	1.2031	1.2049	1.5612
0.74	1.0455	1.5543	0.5524	2.1133	0.6126	0.9017	1.2319	1.2337	1.6426
0.76	1.0458	1.5218	0.5644	2.1589	0.5907	0.9555	1.2589	1.2608	1.7277
0.78	1.0449	1.4886	0.5760	2.2051	0.5676	1.0148	1.2841	1.2859	1.8170
0.80	1.0429	1.4549	0.5871	2.2519	0.5433	1.0807	1.3073	1.3091	1.9113
0.82	1.0398	1.4206	0.5977	2.2994	0.5177	1.1547	1.3285	1.3301	2.0114
0.84	1.0356	1.3858	0.6078	2.3477	0.4906	1.2388	1.3475	1.3489	2.1185
0.86	1.0303	1.3504	0.6173	2.3968	0.4622	1.3357	1.3643	1.3654	2.2343
0.88	1.0239	1.3146	0.6263	2.4467	0.4322	1.4491	1.3787	1.3794	2.3609
0.90	1.0163	1.2783	0.6346	2.4977	0.4006	1.5841	1.3906	1.3909	2.5013
0.92	1.0074	1.2410	0.6423	2.5502	0.3666	1.7520	1.3998	1.3995	2.6624
0.94	0.9958	1.1995	0.6492	2.6080	0.3250	1.9979	1.4047	1.4036	2.8737
0.96	0.9798	1.1506	0.6552	2.6750	0.2713	2.4152	1.4035	1.4014	3.1887
0.98	0.9563	1.0883	0.6599	2.7604	0.1960	3.3678	1.3925	1.3888	3.7926
1.00	0.8948	0.9492	0.6626	2.9617	0.0000	-	1.3416	1.3340	-

C29

M and E Basket-handle

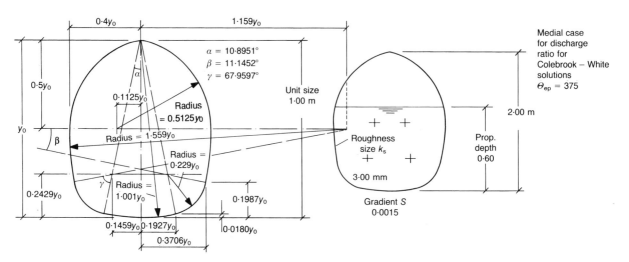

$\alpha = 10.8951°$
$\beta = 11.1452°$
$\gamma = 67.9597°$

Unit size 1.00 m

Medial case for discharge ratio for Colebrook – White solutions $\Theta_{ep} = 375$

2.00 m

Prop. depth 0.60

Roughness size k_s
3.00 mm

Gradient S
0.0015

Prop dpth Y	U.equ. dia. $D_{ep(u)}$ (m)	Equiv. disch. factor J	Unit sect. area A_u (m²)	Unit wetted perim. P_u (m)	Unit surf. brdth B_{us} (m)	Unit mean depth y_{um} (m)	Discharge ratios $Q/Q_{0.60}$ medial	Mann'g	U.crit. disch. Q_{uc} (m³s⁻¹)
0.04	0.1125	0.6872	0.0145	0.5143	0.5041	0.0287	0.0080	0.0078	0.0077
0.08	0.2278	1.1003	0.0371	0.6505	0.6137	0.0604	0.0329	0.0318	0.0285
0.12	0.3344	1.3940	0.0630	0.7535	0.6783	0.0929	0.0720	0.0699	0.0601
0.16	0.4316	1.6078	0.0910	0.8433	0.7188	0.1266	0.1227	0.1196	0.1014
0.20	0.5191	1.7597	0.1203	0.9267	0.7417	0.1621	0.1826	0.1788	0.1516
0.24	0.5961	1.8579	0.1502	1.0080	0.7563	0.1986	0.2492	0.2450	0.2097
0.28	0.6639	1.9152	0.1807	1.0890	0.7688	0.2351	0.3211	0.3166	0.2744
0.32	0.7240	1.9445	0.2117	1.1696	0.7791	0.2717	0.3974	0.3930	0.3456
0.36	0.7777	1.9544	0.2430	1.2501	0.7874	0.3087	0.4774	0.4732	0.4228
0.40	0.8259	1.9503	0.2747	1.3303	0.7936	0.3461	0.5605	0.5566	0.5060
0.42	0.8481	1.9443	0.2906	1.3704	0.7959	0.3651	0.6030	0.5993	0.5498
0.44	0.8692	1.9361	0.3065	1.4104	0.7977	0.3842	0.6460	0.6427	0.5950
0.46	0.8893	1.9261	0.3225	1.4504	0.7990	0.4036	0.6895	0.6865	0.6415
0.48	0.9083	1.9145	0.3385	1.4904	0.7997	0.4232	0.7334	0.7308	0.6895
0.50	0.9264	1.9016	0.3545	1.5305	0.8000	0.4431	0.7777	0.7755	0.7388
0.52	0.9435	1.8874	0.3704	1.5705	0.7992	0.4635	0.8223	0.8204	0.7898
0.54	0.9597	1.8720	0.3864	1.6105	0.7969	0.4849	0.8669	0.8655	0.8426
0.56	0.9749	1.8553	0.4023	1.6507	0.7930	0.5074	0.9116	0.9106	0.8974
0.58	0.9890	1.8372	0.4181	1.6911	0.7874	0.5310	0.9560	0.9555	0.9541
0.60	1.0020	1.8177	0.4338	1.7317	0.7803	0.5559	1.0000	1.0000	1.0129
0.62	1.0139	1.7967	0.4493	1.7727	0.7715	0.5824	1.0435	1.0439	1.0738
0.64	1.0245	1.7743	0.4646	1.8141	0.7610	0.6106	1.0862	1.0871	1.1370
0.66	1.0340	1.7503	0.4797	1.8559	0.7488	0.6407	1.1280	1.1294	1.2026
0.68	1.0422	1.7247	0.4946	1.8983	0.7347	0.6732	1.1687	1.1704	1.2708
0.70	1.0490	1.6975	0.5091	1.9414	0.7187	0.7084	1.2080	1.2101	1.3419
0.72	1.0544	1.6686	0.5233	1.9852	0.7008	0.7468	1.2457	1.2481	1.4162
0.74	1.0584	1.6379	0.5371	2.0300	0.6807	0.7891	1.2816	1.2843	1.4943
0.76	1.0608	1.6055	0.5505	2.0758	0.6583	0.8363	1.3155	1.3183	1.5766
0.78	1.0617	1.5711	0.5635	2.1229	0.6335	0.8894	1.3470	1.3500	1.6641
0.80	1.0608	1.5347	0.5759	2.1714	0.6060	0.9502	1.3760	1.3789	1.7578
0.82	1.0581	1.4962	0.5877	2.2217	0.5756	1.0209	1.4020	1.4048	1.8595
0.84	1.0534	1.4553	0.5989	2.2740	0.5420	1.1050	1.4247	1.4273	1.9713
0.86	1.0466	1.4119	0.6093	2.3287	0.5045	1.2077	1.4437	1.4461	2.0970
0.88	1.0375	1.3657	0.6190	2.3866	0.4628	1.3376	1.4586	1.4605	2.2420
0.90	1.0257	1.3161	0.6278	2.4483	0.4158	1.5099	1.4687	1.4700	2.4158
0.92	1.0109	1.2627	0.6356	2.5150	0.3624	1.7539	1.4734	1.4739	2.6360
0.94	0.9924	1.2043	0.6422	2.5887	0.3006	2.1368	1.4716	1.4711	2.9399
0.96	0.9692	1.1393	0.6475	2.6725	0.2269	2.8537	1.4618	1.4600	3.4256
0.98	0.9392	1.0638	0.6512	2.7735	0.1342	4.8524	1.4412	1.4377	4.4921
1.00	0.8960	0.9661	0.6527	2.9137	0.0000	-	1.4022	1.3964	-

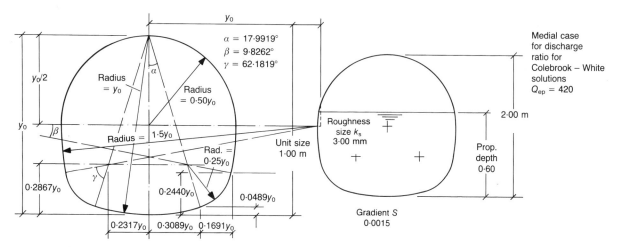

Prop dpth	U.equ. dia. $D_{ep(u)}$	Equiv. disch. factor	Unit sect. area	Unit wetted perim.	Unit surf. brdth	Unit mean depth	Discharge ratios $Q/Q_{0.60}$		U.crit. disch. Q_{uc}
Y	(m)	J	A_u (m²)	P_u (m)	B_{us} (m)	y_{um} (m)	medial	Mann'g	(m³s⁻¹)
0.04	0.1057	0.5849	0.0150	0.5676	0.5600	0.0268	0.0059	0.0057	0.0077
0.08	0.2165	0.8834	0.0417	0.7698	0.7446	0.0559	0.0266	0.0256	0.0309
0.12	0.3294	1.1608	0.0734	0.8915	0.8360	0.0878	0.0618	0.0596	0.0681
0.16	0.4365	1.3840	0.1081	0.9907	0.8944	0.1209	0.1092	0.1059	0.1177
0.20	0.5362	1.5607	0.1447	1.0794	0.9323	0.1552	0.1668	0.1627	0.1785
0.24	0.6279	1.6968	0.1825	1.1625	0.9545	0.1912	0.2326	0.2279	0.2499
0.28	0.7107	1.7953	0.2209	1.2436	0.9676	0.2283	0.3048	0.2997	0.3306
0.32	0.7849	1.8620	0.2599	1.3243	0.9783	0.2656	0.3818	0.3766	0.4194
0.36	0.8519	1.9051	0.2992	1.4048	0.9869	0.3031	0.4629	0.4579	0.5158
0.40	0.9125	1.9305	0.3388	1.4850	0.9933	0.3411	0.5474	0.5428	0.6196
0.42	0.9407	1.9378	0.3587	1.5251	0.9957	0.3602	0.5908	0.5865	0.6741
0.44	0.9676	1.9422	0.3786	1.5651	0.9976	0.3795	0.6348	0.6308	0.7304
0.46	0.9932	1.9439	0.3986	1.6052	0.9989	0.3990	0.6794	0.6758	0.7884
0.48	1.0177	1.9433	0.4186	1.6452	0.9997	0.4187	0.7245	0.7212	0.8481
0.50	1.0410	1.9406	0.4386	1.6852	1.0000	0.4386	0.7699	0.7672	0.9095
0.52	1.0632	1.9361	0.4586	1.7252	0.9992	0.4589	0.8158	0.8135	0.9728
0.54	1.0843	1.9297	0.4785	1.7653	0.9968	0.4801	0.8619	0.8602	1.0383
0.56	1.1043	1.9214	0.4984	1.8055	0.9928	0.5020	0.9081	0.9069	1.1059
0.58	1.1230	1.9112	0.5182	1.8459	0.9871	0.5250	0.9542	0.9535	1.1758
0.60	1.1405	1.8992	0.5379	1.8865	0.9798	0.5490	1.0000	1.0000	1.2481
0.62	1.1567	1.8852	0.5574	1.9275	0.9708	0.5742	1.0455	1.0461	1.3227
0.64	1.1716	1.8693	0.5767	1.9690	0.9600	0.6007	1.0904	1.0916	1.3998
0.66	1.1851	1.8514	0.5958	2.0109	0.9474	0.6289	1.1346	1.1363	1.4796
0.68	1.1972	1.8316	0.6146	2.0534	0.9330	0.6588	1.1779	1.1802	1.5621
0.70	1.2078	1.8097	0.6331	2.0967	0.9165	0.6908	1.2201	1.2229	1.6478
0.72	1.2168	1.7857	0.6512	2.1408	0.8980	0.7252	1.2609	1.2642	1.7368
0.74	1.2243	1.7595	0.6690	2.1858	0.8773	0.7626	1.3002	1.3039	1.8295
0.76	1.2300	1.7311	0.6863	2.2320	0.8542	0.8035	1.3378	1.3418	1.9265
0.78	1.2338	1.7004	0.7032	2.2796	0.8285	0.8487	1.3733	1.3776	2.0286
0.80	1.2358	1.6672	0.7194	2.3287	0.8000	0.8993	1.4065	1.4110	2.1365
0.82	1.2357	1.6313	0.7351	2.3797	0.7684	0.9567	1.4371	1.4417	2.2518
0.84	1.2333	1.5925	0.7502	2.4329	0.7332	1.0231	1.4647	1.4693	2.3761
0.86	1.2285	1.5506	0.7644	2.4890	0.6940	1.1015	1.4889	1.4934	2.5124
0.88	1.2209	1.5050	0.7779	2.5485	0.6499	1.1969	1.5093	1.5134	2.6650
0.90	1.2102	1.4553	0.7904	2.6125	0.6000	1.3173	1.5251	1.5287	2.8408
0.92	1.1957	1.4003	0.8018	2.6825	0.5426	1.4778	1.5355	1.5384	3.0525
0.94	1.1764	1.3385	0.8120	2.7610	0.4750	1.7096	1.5393	1.5412	3.3249
0.96	1.1506	1.2668	0.8207	2.8533	0.3919	2.0941	1.5343	1.5348	3.7193
0.98	1.1137	1.1771	0.8275	2.9722	0.2800	2.9554	1.5157	1.5142	4.4550
1.00	1.0212	0.9853	0.8313	3.2560	0.0000	-	1.4419	1.4357	-

C31

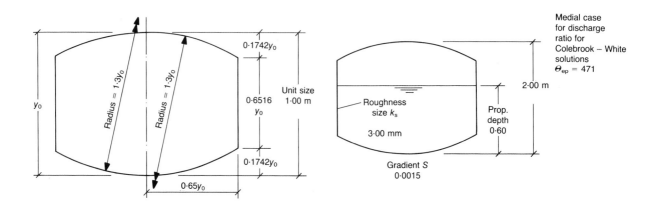

Prop dpth	U.equ. dia. $D_{ep(u)}$	Equiv. disch. factor	Unit sect. area	Unit wetted perim.	Unit surf. brdth	Unit mean depth	Discharge ratios $Q/Q_{0.60}$		U.crit. disch. Q_{uc}
Y	(m)	J	A_u (m²)	P_u (m)	B_{us} (m)	y_{um} (m)	medial	Mann'g	(m³s⁻¹)
0.04	0.1059	0.5145	0.0171	0.6466	0.6400	0.0267	0.0048	0.0046	0.0088
0.08	0.2103	0.7204	0.0482	0.9169	0.8980	0.0537	0.0214	0.0205	0.0350
0.12	0.3131	0.8735	0.0881	1.1259	1.0911	0.0808	0.0508	0.0488	0.0784
0.16	0.4143	0.9985	0.1350	1.3036	1.2496	0.1081	0.0935	0.0902	0.1390
0.20	0.5284	1.1749	0.1867	1.4130	1.3000	0.1436	0.1511	0.1467	0.2215
0.24	0.6394	1.3455	0.2387	1.4930	1.3000	0.1836	0.2183	0.2129	0.3203
0.28	0.7391	1.4762	0.2907	1.5730	1.3000	0.2236	0.2915	0.2856	0.4304
0.32	0.8292	1.5759	0.3427	1.6530	1.3000	0.2636	0.3696	0.3636	0.5509
0.36	0.9109	1.6513	0.3947	1.7330	1.3000	0.3036	0.4517	0.4458	0.6810
0.40	0.9855	1.7076	0.4467	1.8130	1.3000	0.3436	0.5372	0.5317	0.8199
0.42	1.0203	1.7298	0.4727	1.8530	1.3000	0.3636	0.5810	0.5758	0.8925
0.44	1.0537	1.7487	0.4987	1.8930	1.3000	0.3836	0.6255	0.6207	0.9672
0.46	1.0857	1.7645	0.5247	1.9330	1.3000	0.4036	0.6706	0.6662	1.0438
0.48	1.1164	1.7776	0.5507	1.9730	1.3000	0.4236	0.7163	0.7124	1.1224
0.50	1.1459	1.7883	0.5767	2.0130	1.3000	0.4436	0.7624	0.7591	1.2028
0.52	1.1742	1.7968	0.6027	2.0530	1.3000	0.4636	0.8091	0.8063	1.2850
0.54	1.2015	1.8033	0.6287	2.0930	1.3000	0.4836	0.8562	0.8541	1.3691
0.56	1.2277	1.8081	0.6547	2.1330	1.3000	0.5036	0.9038	0.9023	1.4549
0.58	1.2530	1.8114	0.6807	2.1730	1.3000	0.5236	0.9517	0.9509	1.5424
0.60	1.2773	1.8132	0.7067	2.2130	1.3000	0.5436	1.0000	1.0000	1.6316
0.62	1.3008	1.8138	0.7327	2.2530	1.3000	0.5636	1.0487	1.0495	1.7225
0.64	1.3234	1.8132	0.7587	2.2930	1.3000	0.5836	1.0977	1.0993	1.8150
0.66	1.3453	1.8116	0.7847	2.3330	1.3000	0.6036	1.1470	1.1495	1.9091
0.68	1.3665	1.8090	0.8107	2.3730	1.3000	0.6236	1.1966	1.2000	2.0047
0.70	1.3869	1.8056	0.8367	2.4130	1.3000	0.6436	1.2465	1.2508	2.1020
0.72	1.4067	1.8015	0.8627	2.4530	1.3000	0.6636	1.2966	1.3019	2.2007
0.74	1.4259	1.7968	0.8887	2.4930	1.3000	0.6836	1.3470	1.3533	2.3009
0.76	1.4444	1.7914	0.9147	2.5330	1.3000	0.7036	1.3976	1.4049	2.4026
0.78	1.4624	1.7855	0.9407	2.5730	1.3000	0.7236	1.4484	1.4568	2.5058
0.80	1.4798	1.7791	0.9667	2.6130	1.3000	0.7436	1.4995	1.5089	2.6104
0.82	1.4967	1.7722	0.9927	2.6530	1.3000	0.7636	1.5507	1.5613	2.7164
0.84	1.4962	1.7265	1.0183	2.7225	1.2496	0.8149	1.5905	1.6012	2.8787
0.86	1.4850	1.6612	1.0426	2.8083	1.1737	0.8883	1.6207	1.6312	3.0770
0.88	1.4692	1.5915	1.0652	2.9001	1.0911	0.9763	1.6450	1.6548	3.2961
0.90	1.4484	1.5169	1.0861	2.9996	1.0000	1.0861	1.6625	1.6714	3.5448
0.92	1.4218	1.4366	1.1052	3.1092	0.8980	1.2307	1.6721	1.6797	3.8393
0.94	1.3881	1.3488	1.1220	3.2330	0.7808	1.4370	1.6724	1.6782	4.2118
0.96	1.3449	1.2502	1.1362	3.3794	0.6400	1.7754	1.6605	1.6641	4.7410
0.98	1.2857	1.1316	1.1473	3.5694	0.4543	2.5253	1.6302	1.6306	5.7094
1.00	1.1459	0.8941	1.1533	4.0260	0.0000	-	1.5249	1.5181	-

Medial semi-circular

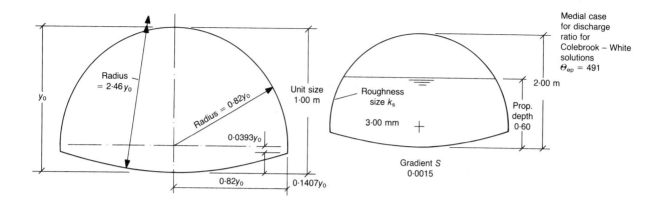

Prop dpth	U.equ. dia. $D_{ep(u)}$	Equiv. disch. factor	Unit sect. area	Unit wetted perim.	Unit surf. brdth	Unit mean depth	Discharge ratios $Q/Q_{0.60}$		U.crit. disch. Q_{uc}
Y	(m)	J	A_u (m²)	P_u (m)	B_{us} (m)	y_{um} (m)	medial	Mann'g	(m³s⁻¹)
0.04	0.1063	0.3757	0.0236	0.8884	0.8836	0.0267	0.0052	0.0050	0.0121
0.08	0.2117	0.5286	0.0666	1.2582	1.2445	0.0535	0.0233	0.0223	0.0482
0.12	0.3163	0.6441	0.1220	1.5431	1.5179	0.0804	0.0557	0.0534	0.1084
0.16	0.4358	0.8004	0.1864	1.7106	1.6400	0.1137	0.1047	0.1010	0.1968
0.20	0.5629	0.9876	0.2520	1.7906	1.6395	0.1537	0.1668	0.1619	0.3094
0.24	0.6789	1.1401	0.3175	1.8707	1.6356	0.1941	0.2369	0.2311	0.4381
0.28	0.7847	1.2635	0.3828	1.9511	1.6278	0.2352	0.3131	0.3069	0.5813
0.32	0.8812	1.3624	0.4477	2.0320	1.6159	0.2770	0.3941	0.3878	0.7379
0.36	0.9690	1.4402	0.5120	2.1136	1.6000	0.3200	0.4785	0.4725	0.9070
0.40	1.0484	1.4998	0.5756	2.1961	1.5799	0.3643	0.5653	0.5598	1.0880
0.42	1.0852	1.5235	0.6071	2.2377	1.5682	0.3871	0.6092	0.6042	1.1829
0.44	1.1200	1.5434	0.6383	2.2798	1.5554	0.4104	0.6534	0.6488	1.2806
0.46	1.1529	1.5598	0.6693	2.3221	1.5414	0.4342	0.6977	0.6935	1.3811
0.48	1.1840	1.5728	0.7000	2.3649	1.5263	0.4586	0.7419	0.7383	1.4845
0.50	1.2132	1.5827	0.7303	2.4081	1.5100	0.4837	0.7860	0.7829	1.5906
0.52	1.2405	1.5895	0.7604	2.4518	1.4924	0.5095	0.8298	0.8273	1.6997
0.54	1.2661	1.5935	0.7900	2.4960	1.4735	0.5362	0.8732	0.8713	1.8116
0.56	1.2898	1.5947	0.8193	2.5408	1.4533	0.5638	0.9161	0.9149	1.9264
0.58	1.3118	1.5934	0.8482	2.5863	1.4316	0.5924	0.9585	0.9578	2.0444
0.60	1.3319	1.5894	0.8766	2.6325	1.4085	0.6223	1.0000	1.0000	2.1655
0.62	1.3502	1.5831	0.9045	2.6795	1.3839	0.6536	1.0407	1.0413	2.2899
0.64	1.3668	1.5743	0.9319	2.7273	1.3576	0.6864	1.0805	1.0816	2.4178
0.66	1.3815	1.5633	0.9588	2.7762	1.3297	0.7211	1.1191	1.1208	2.5496
0.68	1.3943	1.5499	0.9851	2.8261	1.2998	0.7578	1.1564	1.1586	2.6855
0.70	1.4052	1.5344	1.0108	2.8771	1.2681	0.7971	1.1923	1.1950	2.8259
0.72	1.4143	1.5166	1.0358	2.9296	1.2342	0.8393	1.2268	1.2299	2.9715
0.74	1.4213	1.4966	1.0601	2.9835	1.1980	0.8849	1.2595	1.2629	3.1229
0.76	1.4263	1.4743	1.0837	3.0392	1.1593	0.9348	1.2903	1.2940	3.2811
0.78	1.4292	1.4498	1.1065	3.0968	1.1179	0.9898	1.3191	1.3230	3.4473
0.80	1.4299	1.4230	1.1284	3.1566	1.0733	1.0513	1.3456	1.3497	3.6231
0.82	1.4282	1.3937	1.1494	3.2191	1.0253	1.1210	1.3696	1.3737	3.8110
0.84	1.4240	1.3619	1.1694	3.2848	0.9732	1.2015	1.3909	1.3948	4.0140
0.86	1.4171	1.3272	1.1883	3.3542	0.9165	1.2965	1.4091	1.4128	4.2371
0.88	1.4071	1.2894	1.2060	3.4283	0.8542	1.4119	1.4238	1.4271	4.4875
0.90	1.3937	1.2480	1.2224	3.5083	0.7849	1.5575	1.4346	1.4374	4.7773
0.92	1.3762	1.2022	1.2373	3.5963	0.7065	1.7512	1.4408	1.4427	5.1277
0.94	1.3536	1.1507	1.2506	3.6955	0.6158	2.0308	1.4412	1.4422	5.5810
0.96	1.3239	1.0910	1.2618	3.8124	0.5060	2.4939	1.4342	1.4338	6.2403
0.98	1.2822	1.0162	1.2706	3.9638	0.3600	3.5294	1.4155	1.4132	7.4750
1.00	1.1791	0.8561	1.2754	4.3267	0.0000	-	1.3482	1.3415	-

C33 1.30 Pipe arch (corrugated sheet metal)

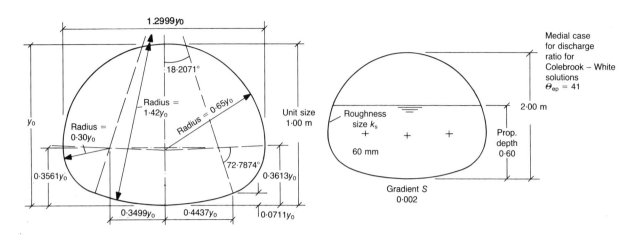

Prop dpth Y	U.equ. dia. $D_{ep(u)}$ (m)	Equiv. disch. factor J	Unit sect. area A_u (m²)	Unit wetted perim. P_u (m)	Unit surf. brdth B_{us} (m)	Unit mean depth y_{um} (m)	Discharge ratios $Q/Q_{0.60}$ medial	Mann'g	U.crit. disch. Q_{uc} (m³s⁻¹)
0.04	0.1060	0.4927	0.0179	0.6757	0.6693	0.0267	0.0038	0.0049	0.0092
0.08	0.2115	0.6972	0.0504	0.9531	0.9347	0.0539	0.0191	0.0220	0.0366
0.12	0.3268	0.9243	0.0908	1.1108	1.0701	0.0848	0.0485	0.0529	0.0828
0.16	0.4413	1.1299	0.1354	1.2269	1.1540	0.1173	0.0909	0.0964	0.1452
0.20	0.5513	1.3060	0.1827	1.3260	1.2123	0.1507	0.1448	0.1510	0.2222
0.24	0.6557	1.4548	0.2321	1.4159	1.2531	0.1852	0.2088	0.2152	0.3128
0.28	0.7539	1.5785	0.2828	1.5005	1.2803	0.2209	0.2815	0.2879	0.4163
0.32	0.8454	1.6789	0.3344	1.5820	1.2955	0.2581	0.3614	0.3673	0.5320
0.36	0.9297	1.7571	0.3863	1.6621	1.2998	0.2972	0.4467	0.4521	0.6595
0.40	1.0062	1.8143	0.4382	1.7422	1.2961	0.3381	0.5360	0.5407	0.7980
0.42	1.0416	1.8358	0.4641	1.7824	1.2924	0.3591	0.5818	0.5860	0.8710
0.44	1.0752	1.8531	0.4899	1.8227	1.2875	0.3805	0.6280	0.6318	0.9465
0.46	1.1070	1.8664	0.5156	1.8632	1.2812	0.4024	0.6746	0.6780	1.0243
0.48	1.1370	1.8761	0.5412	1.9039	1.2737	0.4249	0.7215	0.7244	1.1047
0.50	1.1653	1.8822	0.5666	1.9449	1.2649	0.4479	0.7685	0.7709	1.1874
0.52	1.1918	1.8851	0.5918	1.9861	1.2548	0.4716	0.8154	0.8173	1.2726
0.54	1.2166	1.8848	0.6167	2.0278	1.2432	0.4961	0.8622	0.8636	1.3603
0.56	1.2397	1.8816	0.6415	2.0698	1.2303	0.5214	0.9087	0.9096	1.4506
0.58	1.2611	1.8755	0.6659	2.1123	1.2159	0.5477	0.9547	0.9551	1.5434
0.60	1.2807	1.8667	0.6901	2.1554	1.2000	0.5751	1.0000	1.0000	1.6389
0.62	1.2986	1.8553	0.7139	2.1990	1.1825	0.6037	1.0446	1.0442	1.7372
0.64	1.3148	1.8412	0.7374	2.2433	1.1634	0.6338	1.0883	1.0874	1.8384
0.66	1.3292	1.8247	0.7605	2.2884	1.1426	0.6655	1.1309	1.1296	1.9428
0.68	1.3418	1.8058	0.7831	2.3344	1.1200	0.6992	1.1722	1.1706	2.0506
0.70	1.3526	1.7844	0.8053	2.3813	1.0954	0.7351	1.2121	1.2101	2.1620
0.72	1.3615	1.7606	0.8269	2.4294	1.0688	0.7736	1.2504	1.2481	2.2776
0.74	1.3684	1.7344	0.8480	2.4787	1.0400	0.8154	1.2868	1.2843	2.3979
0.76	1.3734	1.7057	0.8685	2.5294	1.0088	0.8609	1.3212	1.3185	2.5235
0.78	1.3762	1.6746	0.8883	2.5819	0.9749	0.9112	1.3534	1.3505	2.6555
0.80	1.3769	1.6408	0.9075	2.6362	0.9381	0.9674	1.3830	1.3800	2.7950
0.82	1.3752	1.6044	0.9258	2.6929	0.8980	1.0310	1.4098	1.4068	2.9439
0.84	1.3710	1.5650	0.9434	2.7522	0.8542	1.1044	1.4334	1.4305	3.1046
0.86	1.3641	1.5225	0.9600	2.8148	0.8060	1.1911	1.4535	1.4508	3.2808
0.88	1.3542	1.4764	0.9756	2.8816	0.7526	1.2963	1.4697	1.4672	3.4782
0.90	1.3408	1.4262	0.9900	2.9535	0.6928	1.4290	1.4812	1.4791	3.7061
0.92	1.3233	1.3710	1.0032	3.0324	0.6248	1.6056	1.4873	1.4858	3.9808
0.94	1.3007	1.3092	1.0149	3.1212	0.5455	1.8605	1.4867	1.4859	4.3352
0.96	1.2709	1.2378	1.0249	3.2257	0.4490	2.2827	1.4773	1.4776	4.8493
0.98	1.2291	1.1489	1.0327	3.3609	0.3200	3.2272	1.4540	1.4559	5.8095
1.00	1.1259	0.9600	1.0370	3.6842	0.0000	-	1.3730	1.3789	-

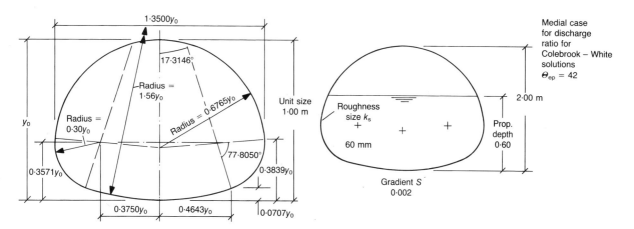

Prop dpth	U.equ. dia. $D_{ep(u)}$	Equiv. disch. factor	Unit sect. area	Unit wetted perim.	Unit surf. brdth	Unit mean depth	Discharge ratios $Q/Q_{0.60}$		U.crit. disch. Q_{uc}
Y	(m)	J	A_u (m²)	P_u (m)	B_{us} (m)	y_{um} (m)	medial	Mann'g	(m³s⁻¹)
0.04	0.1060	0.4704	0.0188	0.7081	0.7020	0.0267	0.0038	0.0049	0.0096
0.08	0.2120	0.6675	0.0529	0.9975	0.9799	0.0539	0.0192	0.0220	0.0384
0.12	0.3287	0.8923	0.0951	1.1572	1.1176	0.0851	0.0487	0.0531	0.0869
0.16	0.4446	1.0964	0.1416	1.2739	1.2023	0.1178	0.0912	0.0967	0.1522
0.20	0.5561	1.2722	0.1909	1.3733	1.2612	0.1514	0.1452	0.1514	0.2327
0.24	0.6622	1.4215	0.2423	1.4634	1.3024	0.1860	0.2094	0.2158	0.3272
0.28	0.7621	1.5465	0.2949	1.5481	1.3298	0.2218	0.2822	0.2885	0.4350
0.32	0.8554	1.6490	0.3485	1.6296	1.3454	0.2590	0.3622	0.3682	0.5554
0.36	0.9415	1.7298	0.4024	1.7098	1.3500	0.2981	0.4479	0.4532	0.6881
0.40	1.0197	1.7896	0.4563	1.7901	1.3443	0.3395	0.5375	0.5420	0.8326
0.42	1.0559	1.8123	0.4832	1.8304	1.3392	0.3608	0.5832	0.5874	0.9089
0.44	1.0902	1.8306	0.5099	1.8709	1.3328	0.3826	0.6295	0.6332	0.9877
0.46	1.1226	1.8448	0.5365	1.9116	1.3252	0.4048	0.6761	0.6794	1.0690
0.48	1.1531	1.8553	0.5629	1.9526	1.3163	0.4276	0.7229	0.7257	1.1527
0.50	1.1819	1.8622	0.5891	1.9939	1.3061	0.4510	0.7698	0.7721	1.2390
0.52	1.2088	1.8657	0.6151	2.0355	1.2947	0.4751	0.8166	0.8184	1.3278
0.54	1.2340	1.8660	0.6409	2.0775	1.2818	0.5000	0.8631	0.8645	1.4192
0.56	1.2574	1.8633	0.6664	2.1199	1.2676	0.5257	0.9093	0.9102	1.5131
0.58	1.2790	1.8577	0.6916	2.1629	1.2520	0.5524	0.9550	0.9554	1.6097
0.60	1.2989	1.8494	0.7165	2.2064	1.2348	0.5802	1.0000	1.0000	1.7091
0.62	1.3170	1.8383	0.7410	2.2506	1.2161	0.6093	1.0442	1.0438	1.8113
0.64	1.3333	1.8247	0.7651	2.2954	1.1958	0.6398	1.0875	1.0866	1.9165
0.66	1.3477	1.8085	0.7888	2.3411	1.1737	0.6720	1.1297	1.1284	2.0250
0.68	1.3604	1.7899	0.8120	2.3877	1.1499	0.7062	1.1705	1.1689	2.1370
0.70	1.3712	1.7688	0.8348	2.4353	1.1241	0.7426	1.2099	1.2080	2.2528
0.72	1.3800	1.7453	0.8570	2.4840	1.0962	0.7818	1.2477	1.2454	2.3729
0.74	1.3869	1.7193	0.8786	2.5341	1.0662	0.8241	1.2836	1.2811	2.4978
0.76	1.3917	1.6909	0.8996	2.5856	1.0337	0.8703	1.3174	1.3148	2.6282
0.78	1.3945	1.6601	0.9200	2.6389	0.9985	0.9213	1.3490	1.3462	2.7652
0.80	1.3950	1.6266	0.9395	2.6941	0.9604	0.9783	1.3781	1.3752	2.9101
0.82	1.3931	1.5904	0.9583	2.7517	0.9190	1.0428	1.4044	1.4015	3.0647
0.84	1.3887	1.5514	0.9763	2.8121	0.8738	1.1173	1.4275	1.4247	3.2316
0.86	1.3816	1.5092	0.9933	2.8758	0.8242	1.2052	1.4472	1.4445	3.4147
0.88	1.3713	1.4635	1.0092	2.9437	0.7693	1.3118	1.4629	1.4605	3.6198
0.90	1.3576	1.4137	1.0240	3.0170	0.7080	1.4464	1.4740	1.4720	3.8566
0.92	1.3398	1.3589	1.0375	3.0974	0.6382	1.6255	1.4797	1.4783	4.1422
0.94	1.3168	1.2977	1.0495	3.1879	0.5571	1.8839	1.4788	1.4782	4.5108
0.96	1.2866	1.2269	1.0596	3.2944	0.4583	2.3119	1.4692	1.4696	5.0455
0.98	1.2442	1.1388	1.0676	3.4322	0.3266	3.2692	1.4460	1.4479	6.0447
1.00	1.1398	0.9518	1.0719	3.7620	0.0000	-	1.3654	1.3713	-

C35

1.40 Pipe arch (corrugated sheet metal)

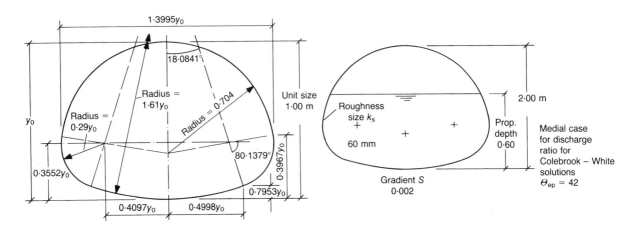

Prop dpth	U.equ. dia.	Equiv. disch. factor	Unit sect. area	Unit wetted perim.	Unit surf. brdth	Unit mean depth	Discharge ratios $Q/Q_{0.60}$		U.crit. disch.
Y	$D_{ep(u)}$ (m)	J	A_u (m²)	P_u (m)	B_{us} (m)	y_{um} (m)	medial	Mann'g	Q_{uc} (m³s⁻¹)
0.04	0.1060	0.4632	0.0191	0.7193	0.7133	0.0267	0.0037	0.0048	0.0098
0.08	0.2109	0.6499	0.0537	1.0193	1.0024	0.0536	0.0186	0.0214	0.0390
0.12	0.3257	0.8557	0.0973	1.1957	1.1588	0.0840	0.0475	0.0519	0.0884
0.16	0.4426	1.0565	0.1456	1.3160	1.2484	0.1166	0.0897	0.0952	0.1557
0.20	0.5558	1.2323	0.1968	1.4168	1.3094	0.1503	0.1436	0.1498	0.2390
0.24	0.6637	1.3833	0.2501	1.5074	1.3518	0.1850	0.2078	0.2142	0.3369
0.28	0.7657	1.5109	0.3048	1.5922	1.3796	0.2209	0.2808	0.2871	0.4486
0.32	0.8611	1.6163	0.3603	1.6737	1.3952	0.2583	0.3611	0.3671	0.5735
0.36	0.9493	1.7004	0.4163	1.7539	1.3994	0.2975	0.4472	0.4525	0.7109
0.40	1.0296	1.7634	0.4721	1.8343	1.3926	0.3390	0.5372	0.5418	0.8609
0.42	1.0666	1.7873	0.4999	1.8748	1.3860	0.3607	0.5832	0.5874	0.9402
0.44	1.1017	1.8067	0.5276	1.9155	1.3782	0.3828	0.6297	0.6334	1.0222
0.46	1.1347	1.8220	0.5550	1.9565	1.3693	0.4054	0.6764	0.6796	1.1066
0.48	1.1659	1.8334	0.5823	1.9978	1.3591	0.4285	0.7233	0.7260	1.1937
0.50	1.1952	1.8411	0.6094	2.0394	1.3476	0.4522	0.7702	0.7725	1.2833
0.52	1.2227	1.8454	0.6362	2.0814	1.3348	0.4766	0.8169	0.8188	1.3755
0.54	1.2483	1.8464	0.6628	2.1238	1.3207	0.5018	0.8635	0.8648	1.4703
0.56	1.2720	1.8443	0.6890	2.1667	1.3053	0.5279	0.9096	0.9105	1.5678
0.58	1.2940	1.8393	0.7150	2.2102	1.2883	0.5550	0.9551	0.9556	1.6680
0.60	1.3141	1.8314	0.7406	2.2542	1.2700	0.5831	1.0000	1.0000	1.7710
0.62	1.3324	1.8208	0.7658	2.2989	1.2500	0.6126	1.0441	1.0436	1.8769
0.64	1.3489	1.8076	0.7906	2.3443	1.2285	0.6435	1.0871	1.0862	1.9860
0.66	1.3635	1.7918	0.8149	2.3906	1.2052	0.6762	1.1290	1.1278	2.0984
0.68	1.3762	1.7735	0.8388	2.4378	1.1801	0.7107	1.1696	1.1680	2.2144
0.70	1.3871	1.7527	0.8621	2.4861	1.1531	0.7476	1.2087	1.2068	2.3343
0.72	1.3959	1.7295	0.8849	2.5356	1.1240	0.7873	1.2461	1.2439	2.4586
0.74	1.4028	1.7039	0.9070	2.5864	1.0927	0.8301	1.2817	1.2793	2.5879
0.76	1.4076	1.6758	0.9286	2.6387	1.0589	0.8769	1.3152	1.3126	2.7230
0.78	1.4102	1.6452	0.9494	2.6928	1.0225	0.9285	1.3465	1.3437	2.8648
0.80	1.4106	1.6120	0.9694	2.7490	0.9831	0.9861	1.3752	1.3724	3.0147
0.82	1.4086	1.5762	0.9887	2.8075	0.9403	1.0514	1.4011	1.3983	3.1747
0.84	1.4040	1.5374	1.0070	2.8689	0.8937	1.1268	1.4239	1.4211	3.3475
0.86	1.3967	1.4956	1.0244	2.9338	0.8427	1.2157	1.4431	1.4406	3.5370
0.88	1.3862	1.4502	1.0407	3.0029	0.7863	1.3236	1.4585	1.4562	3.7493
0.90	1.3723	1.4008	1.0558	3.0775	0.7233	1.4596	1.4693	1.4674	3.9945
0.92	1.3541	1.3465	1.0696	3.1594	0.6519	1.6407	1.4747	1.4734	4.2903
0.94	1.3308	1.2857	1.0818	3.2517	0.5688	1.9019	1.4737	1.4731	4.6720
0.96	1.3001	1.2155	1.0922	3.3603	0.4678	2.3345	1.4639	1.4643	5.2259
0.98	1.2572	1.1282	1.1003	3.5008	0.3332	3.3019	1.4406	1.4425	6.2612
1.00	1.1516	0.9428	1.1048	3.8372	0.0000	-	1.3603	1.3661	-

1.45 Pipe arch (corrugated sheet metal)

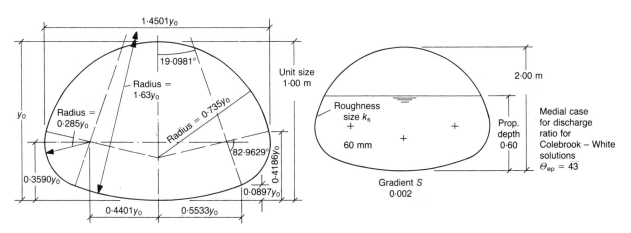

Prop dpth	U.equ. dia. $D_{ep(u)}$	Equiv. disch. factor	Unit sect. area	Unit wetted perim.	Unit surf. brdth	Unit mean depth	Discharge ratios $Q/Q_{0.60}$		U.crit. disch. Q_{uc}
Y	(m)	J	A_u (m²)	P_u (m)	B_{us} (m)	y_{um} (m)	medial	Mann'g	(m³s⁻¹)
0.04	0.1061	0.4604	0.0192	0.7237	0.7178	0.0267	0.0036	0.0047	0.0098
0.08	0.2109	0.6460	0.0541	1.0256	1.0088	0.0536	0.0180	0.0208	0.0392
0.12	0.3217	0.8252	0.0985	1.2249	1.1906	0.0828	0.0458	0.0501	0.0888
0.16	0.4388	1.0201	0.1482	1.3513	1.2881	0.1151	0.0873	0.0927	0.1575
0.20	0.5532	1.1947	0.2012	1.4546	1.3531	0.1487	0.1407	0.1468	0.2429
0.24	0.6628	1.3464	0.2562	1.5464	1.3980	0.1833	0.2046	0.2110	0.3435
0.28	0.7667	1.4761	0.3128	1.6318	1.4278	0.2191	0.2775	0.2839	0.4585
0.32	0.8643	1.5844	0.3703	1.7137	1.4448	0.2563	0.3580	0.3640	0.5870
0.36	0.9548	1.6721	0.4282	1.7939	1.4501	0.2953	0.4445	0.4498	0.7287
0.40	1.0375	1.7391	0.4861	1.8742	1.4442	0.3366	0.5352	0.5397	0.8833
0.42	1.0757	1.7648	0.5150	1.9149	1.4369	0.3584	0.5815	0.5857	0.9654
0.44	1.1117	1.7856	0.5436	1.9559	1.4277	0.3808	0.6283	0.6320	1.0504
0.46	1.1457	1.8021	0.5721	1.9973	1.4173	0.4036	0.6753	0.6785	1.1381
0.48	1.1777	1.8145	0.6003	2.0389	1.4057	0.4270	0.7224	0.7252	1.2285
0.50	1.2077	1.8232	0.6283	2.0809	1.3928	0.4511	0.7696	0.7719	1.3214
0.52	1.2358	1.8283	0.6560	2.1234	1.3787	0.4758	0.8165	0.8184	1.4170
0.54	1.2619	1.8301	0.6834	2.1663	1.3632	0.5013	0.8632	0.8646	1.5153
0.56	1.2862	1.8286	0.7105	2.2097	1.3464	0.5277	0.9094	0.9103	1.6164
0.58	1.3086	1.8242	0.7373	2.2536	1.3282	0.5551	0.9551	0.9555	1.7202
0.60	1.3291	1.8168	0.7636	2.2982	1.3084	0.5836	1.0000	1.0000	1.8269
0.62	1.3477	1.8066	0.7896	2.3435	1.2872	0.6134	1.0441	1.0436	1.9366
0.64	1.3644	1.7938	0.8151	2.3896	1.2643	0.6447	1.0871	1.0862	2.0496
0.66	1.3792	1.7783	0.8402	2.4366	1.2397	0.6777	1.1289	1.1277	2.1659
0.68	1.3921	1.7603	0.8647	2.4845	1.2133	0.7127	1.1694	1.1679	2.2860
0.70	1.4030	1.7397	0.8887	2.5335	1.1849	0.7500	1.2084	1.2065	2.4101
0.72	1.4120	1.7167	0.9121	2.5838	1.1545	0.7900	1.2457	1.2435	2.5387
0.74	1.4189	1.6913	0.9348	2.6355	1.1218	0.8333	1.2811	1.2787	2.6724
0.76	1.4236	1.6634	0.9569	2.6887	1.0866	0.8806	1.3144	1.3118	2.8121
0.78	1.4262	1.6329	0.9783	2.7438	1.0488	0.9328	1.3454	1.3427	2.9588
0.80	1.4265	1.5999	0.9989	2.8009	1.0080	0.9910	1.3739	1.3712	3.1138
0.82	1.4243	1.5642	1.0186	2.8606	0.9637	1.0569	1.3996	1.3968	3.2792
0.84	1.4195	1.5256	1.0374	2.9231	0.9156	1.1330	1.4221	1.4194	3.4578
0.86	1.4120	1.4839	1.0552	2.9893	0.8630	1.2227	1.4411	1.4386	3.6537
0.88	1.4012	1.4387	1.0719	3.0597	0.8050	1.3315	1.4562	1.4540	3.8733
0.90	1.3870	1.3895	1.0873	3.1358	0.7403	1.4688	1.4668	1.4649	4.1268
0.92	1.3685	1.3354	1.1014	3.2194	0.6669	1.6515	1.4720	1.4707	4.4325
0.94	1.3447	1.2749	1.1139	3.3135	0.5817	1.9149	1.4707	1.4701	4.8271
0.96	1.3136	1.2051	1.1246	3.4244	0.4783	2.3510	1.4607	1.4612	5.3998
0.98	1.2700	1.1183	1.1328	3.5679	0.3406	3.3261	1.4373	1.4392	6.4699
1.00	1.1631	0.9341	1.1374	3.9116	0.0000	-	1.3569	1.3627	-

1.50 Pipe arch (corrugated sheet metal)

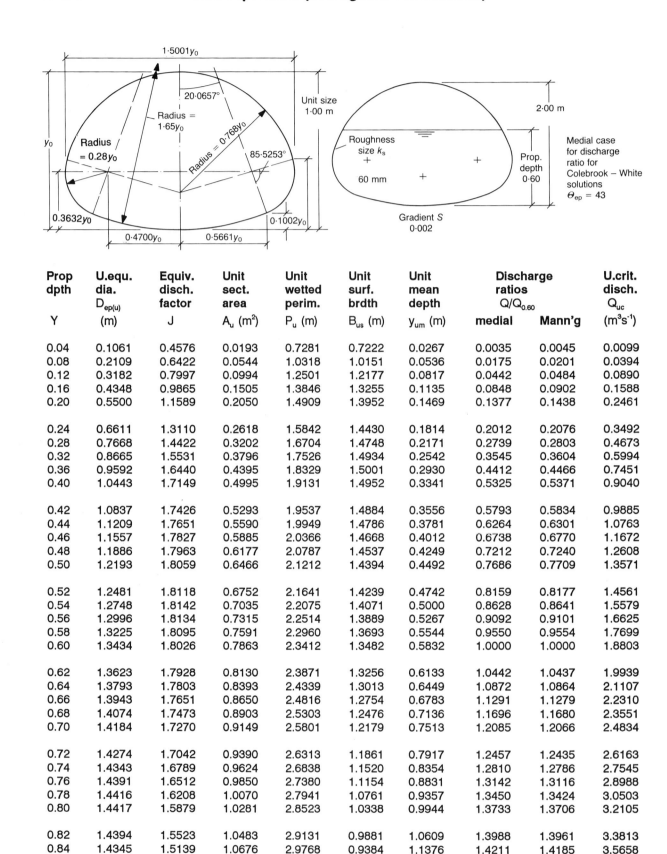

Prop dpth	U.equ. dia. $D_{ep(u)}$	Equiv. disch. factor	Unit sect. area	Unit wetted perim.	Unit surf. brdth	Unit mean depth	Discharge ratios $Q/Q_{0.60}$		U.crit. disch. Q_{uc}
Y	(m)	J	A_u (m²)	P_u (m)	B_{us} (m)	y_{um} (m)	medial	Mann'g	(m³s⁻¹)
0.04	0.1061	0.4576	0.0193	0.7281	0.7222	0.0267	0.0035	0.0045	0.0099
0.08	0.2109	0.6422	0.0544	1.0318	1.0151	0.0536	0.0175	0.0201	0.0394
0.12	0.3182	0.7997	0.0994	1.2501	1.2177	0.0817	0.0442	0.0484	0.0890
0.16	0.4348	0.9865	0.1505	1.3846	1.3255	0.1135	0.0848	0.0902	0.1588
0.20	0.5500	1.1589	0.2050	1.4909	1.3952	0.1469	0.1377	0.1438	0.2461
0.24	0.6611	1.3110	0.2618	1.5842	1.4430	0.1814	0.2012	0.2076	0.3492
0.28	0.7668	1.4422	0.3202	1.6704	1.4748	0.2171	0.2739	0.2803	0.4673
0.32	0.8665	1.5531	0.3796	1.7526	1.4934	0.2542	0.3545	0.3604	0.5994
0.36	0.9592	1.6440	0.4395	1.8329	1.5001	0.2930	0.4412	0.4466	0.7451
0.40	1.0443	1.7149	0.4995	1.9131	1.4952	0.3341	0.5325	0.5371	0.9040
0.42	1.0837	1.7426	0.5293	1.9537	1.4884	0.3556	0.5793	0.5834	0.9885
0.44	1.1209	1.7651	0.5590	1.9949	1.4786	0.3781	0.6264	0.6301	1.0763
0.46	1.1557	1.7827	0.5885	2.0366	1.4668	0.4012	0.6738	0.6770	1.1672
0.48	1.1886	1.7963	0.6177	2.0787	1.4537	0.4249	0.7212	0.7240	1.2608
0.50	1.2193	1.8059	0.6466	2.1212	1.4394	0.4492	0.7686	0.7709	1.3571
0.52	1.2481	1.8118	0.6752	2.1641	1.4239	0.4742	0.8159	0.8177	1.4561
0.54	1.2748	1.8142	0.7035	2.2075	1.4071	0.5000	0.8628	0.8641	1.5579
0.56	1.2996	1.8134	0.7315	2.2514	1.3889	0.5267	0.9092	0.9101	1.6625
0.58	1.3225	1.8095	0.7591	2.2960	1.3693	0.5544	0.9550	0.9554	1.7699
0.60	1.3434	1.8026	0.7863	2.3412	1.3482	0.5832	1.0000	1.0000	1.8803
0.62	1.3623	1.7928	0.8130	2.3871	1.3256	0.6133	1.0442	1.0437	1.9939
0.64	1.3793	1.7803	0.8393	2.4339	1.3013	0.6449	1.0872	1.0864	2.1107
0.66	1.3943	1.7651	0.8650	2.4816	1.2754	0.6783	1.1291	1.1279	2.2310
0.68	1.4074	1.7473	0.8903	2.5303	1.2476	0.7136	1.1696	1.1680	2.3551
0.70	1.4184	1.7270	0.9149	2.5801	1.2179	0.7513	1.2085	1.2066	2.4834
0.72	1.4274	1.7042	0.9390	2.6313	1.1861	0.7917	1.2457	1.2435	2.6163
0.74	1.4343	1.6789	0.9624	2.6838	1.1520	0.8354	1.2810	1.2786	2.7545
0.76	1.4391	1.6512	0.9850	2.7380	1.1154	0.8831	1.3142	1.3116	2.8988
0.78	1.4416	1.6208	1.0070	2.7941	1.0761	0.9357	1.3450	1.3424	3.0503
0.80	1.4417	1.5879	1.0281	2.8523	1.0338	0.9944	1.3733	1.3706	3.2105
0.82	1.4394	1.5523	1.0483	2.9131	0.9881	1.0609	1.3988	1.3961	3.3813
0.84	1.4345	1.5139	1.0676	2.9768	0.9384	1.1376	1.4211	1.4185	3.5658
0.86	1.4267	1.4723	1.0858	3.0442	0.8842	1.2280	1.4400	1.4375	3.7680
0.88	1.4157	1.4272	1.1029	3.1161	0.8244	1.3378	1.4548	1.4526	3.9947
0.90	1.4011	1.3782	1.1187	3.1938	0.7579	1.4761	1.4651	1.4634	4.2564
0.92	1.3823	1.3243	1.1332	3.2791	0.6826	1.6601	1.4701	1.4689	4.5721
0.94	1.3581	1.2641	1.1460	3.3752	0.5952	1.9254	1.4686	1.4681	4.9795
0.96	1.3265	1.1945	1.1568	3.4885	0.4892	2.3645	1.4584	1.4590	5.5706
0.98	1.2823	1.1082	1.1653	3.6351	0.3483	3.3461	1.4348	1.4368	6.6753
1.00	1.1739	0.9251	1.1700	3.9864	0.0000	-	1.3543	1.3601	-

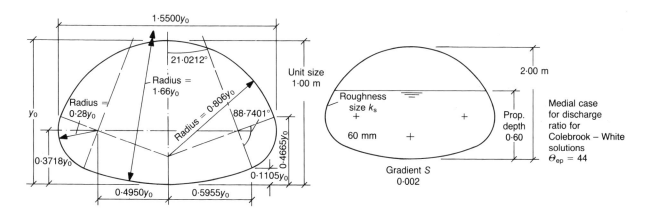

Prop dpth	U.equ. dia. $D_{ep(u)}$	Equiv. disch. factor	Unit sect. area	Unit wetted perim.	Unit surf. brdth	Unit mean depth	Discharge ratios $Q/Q_{0.60}$		U.crit. disch. Q_{uc}
Y	(m)	J	A_u (m²)	P_u (m)	B_{us} (m)	y_{um} (m)	medial	Mann'g	(m³s⁻¹)
0.04	0.1061	0.4563	0.0194	0.7303	0.7244	0.0267	0.0034	0.0044	0.0099
0.08	0.2109	0.6403	0.0546	1.0349	1.0182	0.0536	0.0169	0.0195	0.0396
0.12	0.3156	0.7831	0.0999	1.2659	1.2348	0.0809	0.0426	0.0467	0.0889
0.16	0.4305	0.9579	0.1519	1.4117	1.3562	0.1120	0.0821	0.0874	0.1592
0.20	0.5460	1.1269	0.2078	1.5222	1.4322	0.1451	0.1340	0.1400	0.2479
0.24	0.6582	1.2781	0.2662	1.6177	1.4841	0.1794	0.1968	0.2032	0.3530
0.28	0.7655	1.4103	0.3263	1.7051	1.5191	0.2148	0.2691	0.2754	0.4736
0.32	0.8670	1.5233	0.3875	1.7880	1.5404	0.2516	0.3495	0.3554	0.6087
0.36	0.9619	1.6173	0.4494	1.8686	1.5495	0.2900	0.4363	0.4417	0.7578
0.40	1.0496	1.6921	0.5113	1.9487	1.5472	0.3305	0.5281	0.5327	0.9205
0.42	1.0904	1.7222	0.5422	1.9891	1.5417	0.3517	0.5753	0.5795	1.0070
0.44	1.1291	1.7473	0.5730	2.0300	1.5332	0.3737	0.6230	0.6267	1.0969
0.46	1.1653	1.7672	0.6035	2.0716	1.5216	0.3967	0.6709	0.6742	1.1903
0.48	1.1992	1.7819	0.6338	2.1142	1.5071	0.4206	0.7189	0.7217	1.2872
0.50	1.2309	1.7926	0.6638	2.1572	1.4913	0.4451	0.7668	0.7691	1.3869
0.52	1.2605	1.7994	0.6935	2.2007	1.4743	0.4704	0.8144	0.8163	1.4894
0.54	1.2880	1.8026	0.7228	2.2447	1.4559	0.4964	0.8618	0.8631	1.5948
0.56	1.3134	1.8024	0.7517	2.2892	1.4362	0.5234	0.9085	0.9094	1.7030
0.58	1.3369	1.7990	0.7802	2.3345	1.4151	0.5513	0.9547	0.9551	1.8142
0.60	1.3582	1.7925	0.8083	2.3804	1.3926	0.5804	1.0000	1.0000	1.9284
0.62	1.3776	1.7831	0.8359	2.4271	1.3684	0.6108	1.0444	1.0440	2.0459
0.64	1.3950	1.7709	0.8630	2.4747	1.3427	0.6427	1.0877	1.0869	2.1667
0.66	1.4103	1.7559	0.8896	2.5232	1.3153	0.6764	1.1297	1.1285	2.2911
0.68	1.4236	1.7383	0.9156	2.5728	1.2860	0.7120	1.1704	1.1688	2.4194
0.70	1.4348	1.7181	0.9410	2.6235	1.2548	0.7500	1.2094	1.2075	2.5520
0.72	1.4439	1.6953	0.9658	2.6756	1.2214	0.7907	1.2467	1.2446	2.6894
0.74	1.4508	1.6700	0.9899	2.7292	1.1858	0.8348	1.2820	1.2797	2.8322
0.76	1.4555	1.6422	1.0132	2.7844	1.1477	0.8828	1.3152	1.3127	2.9813
0.78	1.4580	1.6119	1.0358	2.8416	1.1068	0.9358	1.3460	1.3434	3.1377
0.80	1.4580	1.5789	1.0575	2.9010	1.0628	0.9950	1.3743	1.3716	3.3031
0.82	1.4556	1.5432	1.0782	2.9631	1.0154	1.0619	1.3997	1.3970	3.4795
0.84	1.4504	1.5047	1.0980	3.0282	0.9640	1.1391	1.4219	1.4193	3.6699
0.86	1.4423	1.4630	1.1168	3.0971	0.9079	1.2300	1.4405	1.4381	3.8787
0.88	1.4311	1.4179	1.1343	3.1706	0.8463	1.3404	1.4552	1.4531	4.1126
0.90	1.4161	1.3688	1.1506	3.2500	0.7777	1.4795	1.4653	1.4636	4.3826
0.92	1.3968	1.3149	1.1654	3.3373	0.7002	1.6644	1.4701	1.4689	4.7082
0.94	1.3721	1.2546	1.1785	3.4356	0.6103	1.9310	1.4684	1.4679	5.1284
0.96	1.3399	1.1852	1.1897	3.5516	0.5015	2.3721	1.4580	1.4585	5.7379
0.98	1.2949	1.0990	1.1983	3.7017	0.3569	3.3579	1.4341	1.4361	6.8765
1.00	1.1849	0.9165	1.2031	4.0615	0.0000	-	1.3532	1.3589	-

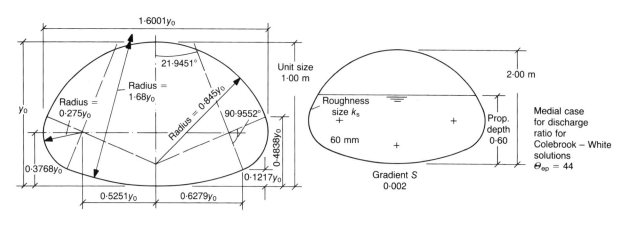

Prop dpth	U.equ. dia. $D_{ep(u)}$	Equiv. disch. factor	Unit sect. area	Unit wetted perim.	Unit surf. brdth	Unit mean depth	Discharge ratios $Q/Q_{0.60}$		U.crit. disch. Q_{uc}
Y	(m)	J	A_u (m²)	P_u (m)	B_{us} (m)	y_{um} (m)	medial	Mann'g	(m³s⁻¹)
0.04	0.1061	0.4536	0.0195	0.7347	0.7288	0.0267	0.0033	0.0043	0.0100
0.08	0.2110	0.6366	0.0549	1.0411	1.0245	0.0536	0.0165	0.0190	0.0398
0.12	0.3146	0.7737	0.1005	1.2776	1.2471	0.0806	0.0413	0.0453	0.0893
0.16	0.4263	0.9295	0.1535	1.4407	1.3885	0.1106	0.0796	0.0848	0.1599
0.20	0.5420	1.0942	0.2109	1.5561	1.4714	0.1433	0.1307	0.1367	0.2500
0.24	0.6552	1.2446	0.2709	1.6538	1.5273	0.1774	0.1929	0.1993	0.3573
0.28	0.7640	1.3775	0.3328	1.7424	1.5649	0.2127	0.2649	0.2712	0.4806
0.32	0.8674	1.4924	0.3959	1.8258	1.5883	0.2493	0.3451	0.3511	0.6190
0.36	0.9644	1.5891	0.4597	1.9066	1.5991	0.2875	0.4322	0.4376	0.7718
0.40	1.0544	1.6673	0.5237	1.9867	1.5982	0.3277	0.5244	0.5290	0.9387
0.42	1.0964	1.6993	0.5556	2.0270	1.5933	0.3487	0.5719	0.5761	1.0274
0.44	1.1363	1.7264	0.5874	2.0677	1.5854	0.3705	0.6200	0.6237	1.1196
0.46	1.1739	1.7483	0.6190	2.1093	1.5744	0.3932	0.6684	0.6717	1.2154
0.48	1.2089	1.7650	0.6503	2.1518	1.5600	0.4169	0.7169	0.7197	1.3150
0.50	1.2415	1.7766	0.6814	2.1953	1.5427	0.4417	0.7652	0.7675	1.4180
0.52	1.2718	1.7842	0.7120	2.2394	1.5242	0.4672	0.8132	0.8150	1.5240
0.54	1.3000	1.7881	0.7423	2.2841	1.5044	0.4934	0.8609	0.8622	1.6330
0.56	1.3261	1.7885	0.7722	2.3293	1.4832	0.5206	0.9080	0.9089	1.7448
0.58	1.3500	1.7856	0.8016	2.3752	1.4607	0.5488	0.9544	0.9548	1.8598
0.60	1.3719	1.7795	0.8306	2.4219	1.4367	0.5782	1.0000	1.0000	1.9778
0.62	1.3916	1.7704	0.8591	2.4694	1.4111	0.6088	1.0446	1.0442	2.0992
0.64	1.4093	1.7585	0.8871	2.5177	1.3839	0.6410	1.0881	1.0873	2.2240
0.66	1.4249	1.7437	0.9145	2.5671	1.3550	0.6749	1.1303	1.1291	2.3525
0.68	1.4384	1.7263	0.9412	2.6175	1.3242	0.7108	1.1711	1.1695	2.4850
0.70	1.4497	1.7062	0.9674	2.6692	1.2915	0.7491	1.2102	1.2083	2.6220
0.72	1.4589	1.6836	0.9929	2.7223	1.2567	0.7901	1.2475	1.2454	2.7638
0.74	1.4659	1.6584	1.0177	2.7769	1.2195	0.8345	1.2829	1.2805	2.9112
0.76	1.4706	1.6307	1.0417	2.8332	1.1798	0.8829	1.3160	1.3136	3.0650
0.78	1.4730	1.6003	1.0648	2.8916	1.1374	0.9362	1.3468	1.3442	3.2265
0.80	1.4730	1.5674	1.0871	2.9522	1.0918	0.9957	1.3750	1.3724	3.3971
0.82	1.4704	1.5318	1.1085	3.0155	1.0427	1.0631	1.4003	1.3976	3.5791
0.84	1.4650	1.4933	1.1288	3.0821	0.9895	1.1407	1.4224	1.4198	3.7755
0.86	1.4567	1.4517	1.1480	3.1524	0.9317	1.2322	1.4409	1.4385	3.9908
0.88	1.4451	1.4066	1.1660	3.2275	0.8681	1.3432	1.4554	1.4533	4.2320
0.90	1.4298	1.3576	1.1827	3.3087	0.7975	1.4830	1.4654	1.4637	4.5104
0.92	1.4101	1.3037	1.1979	3.3979	0.7178	1.6689	1.4700	1.4688	4.8460
0.94	1.3850	1.2436	1.2113	3.4985	0.6255	1.9367	1.4680	1.4676	5.2791
0.96	1.3522	1.1744	1.2228	3.6171	0.5138	2.3798	1.4574	1.4580	5.9071
0.98	1.3065	1.0885	1.2317	3.7708	0.3655	3.3697	1.4333	1.4353	7.0801
1.00	1.1949	0.9069	1.2365	4.1392	0.0000	-	1.3520	1.3578	-

1.65 Pipe arch (corrugated sheet metal)

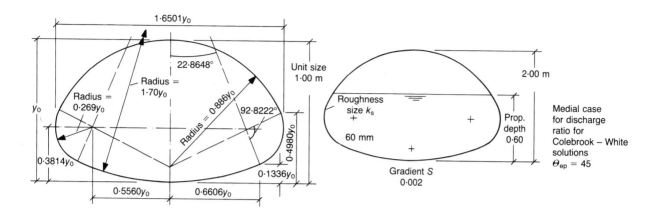

Prop dpth	U.equ. dia. $D_{ep(u)}$	Equiv. disch. factor	Unit sect. area	Unit wetted perim.	Unit surf. brdth	Unit mean depth	Discharge ratios $Q/Q_{0.60}$		U.crit. disch. Q_{uc}
Y	(m)	J	A_u (m²)	P_u (m)	B_{us} (m)	y_{um} (m)	medial	Mann'g	(m³s⁻¹)
0.04	0.1061	0.4509	0.0196	0.7390	0.7332	0.0267	0.0032	0.0041	0.0100
0.08	0.2110	0.6329	0.0552	1.0472	1.0307	0.0536	0.0160	0.0185	0.0400
0.12	0.3147	0.7693	0.1011	1.2851	1.2548	0.0806	0.0402	0.0442	0.0899
0.16	0.4223	0.9044	0.1549	1.4670	1.4175	0.1093	0.0771	0.0823	0.1603
0.20	0.5376	1.0629	0.2136	1.5889	1.5092	0.1415	0.1273	0.1333	0.2516
0.24	0.6517	1.2118	0.2752	1.6893	1.5697	0.1753	0.1890	0.1953	0.3609
0.28	0.7619	1.3452	0.3389	1.7792	1.6103	0.2104	0.2606	0.2669	0.4868
0.32	0.8670	1.4618	0.4038	1.8632	1.6358	0.2469	0.3407	0.3467	0.6284
0.36	0.9660	1.5609	0.4696	1.9443	1.6483	0.2849	0.4278	0.4333	0.7848
0.40	1.0582	1.6422	0.5355	2.0244	1.6488	0.3248	0.5204	0.5251	0.9558
0.42	1.1014	1.6759	0.5685	2.0646	1.6445	0.3457	0.5682	0.5724	1.0467
0.44	1.1425	1.7048	0.6013	2.1053	1.6371	0.3673	0.6167	0.6205	1.1412
0.46	1.1813	1.7287	0.6340	2.1467	1.6266	0.3897	0.6656	0.6689	1.2394
0.48	1.2176	1.7474	0.6663	2.1890	1.6126	0.4132	0.7146	0.7174	1.3414
0.50	1.2512	1.7605	0.6984	2.2328	1.5950	0.4379	0.7634	0.7657	1.4473
0.52	1.2824	1.7689	0.7301	2.2775	1.5750	0.4636	0.8119	0.8137	1.5568
0.54	1.3112	1.7734	0.7614	2.3228	1.5537	0.4901	0.8599	0.8612	1.6692
0.56	1.3379	1.7744	0.7923	2.3687	1.5311	0.5174	0.9074	0.9082	1.7847
0.58	1.3624	1.7719	0.8227	2.4154	1.5071	0.5459	0.9541	0.9545	1.9033
0.60	1.3847	1.7663	0.8525	2.4628	1.4816	0.5754	1.0000	1.0000	2.0252
0.62	1.4048	1.7575	0.8819	2.5111	1.4546	0.6063	1.0449	1.0445	2.1504
0.64	1.4228	1.7458	0.9107	2.5603	1.4259	0.6387	1.0886	1.0878	2.2792
0.66	1.4387	1.7313	0.9389	2.6105	1.3955	0.6728	1.1310	1.1298	2.4118
0.68	1.4524	1.7140	0.9665	2.6619	1.3633	0.7090	1.1719	1.1704	2.5485
0.70	1.4639	1.6941	0.9935	2.7146	1.3291	0.7475	1.2111	1.2093	2.6898
0.72	1.4732	1.6716	1.0197	2.7686	1.2927	0.7888	1.2486	1.2465	2.8360
0.74	1.4802	1.6465	1.0451	2.8243	1.2540	0.8335	1.2840	1.2817	2.9880
0.76	1.4850	1.6188	1.0698	2.8818	1.2127	0.8822	1.3172	1.3147	3.1466
0.78	1.4873	1.5885	1.0936	2.9413	1.1687	0.9358	1.3480	1.3454	3.3131
0.80	1.4871	1.5557	1.1165	3.0032	1.1214	0.9956	1.3761	1.3735	3.4889
0.82	1.4844	1.5200	1.1385	3.0678	1.0706	1.0634	1.4014	1.3987	3.6764
0.84	1.4789	1.4816	1.1593	3.1358	1.0157	1.1414	1.4234	1.4208	3.8788
0.86	1.4703	1.4400	1.1791	3.2077	0.9560	1.2333	1.4418	1.4394	4.1006
0.88	1.4584	1.3950	1.1975	3.2844	0.8905	1.3448	1.4562	1.4541	4.3490
0.90	1.4428	1.3460	1.2146	3.3674	0.8178	1.4853	1.4660	1.4643	4.6356
0.92	1.4228	1.2923	1.2302	3.4586	0.7358	1.6719	1.4704	1.4693	4.9812
0.94	1.3971	1.2324	1.2440	3.5616	0.6410	1.9407	1.4683	1.4679	5.4270
0.96	1.3638	1.1633	1.2557	3.6829	0.5264	2.3854	1.4575	1.4581	6.0733
0.98	1.3174	1.0777	1.2648	3.8402	0.3744	3.3784	1.4331	1.4351	7.2801
1.00	1.2043	0.8971	1.2698	4.2174	0.0000	-	1.3514	1.3571	-

C41

1.70 Pipe arch (corrugated sheet metal)

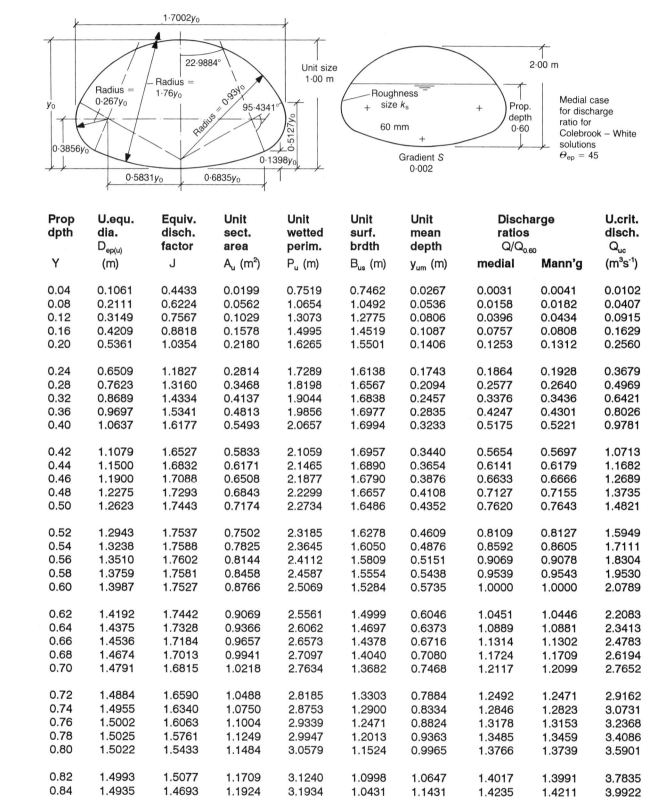

Prop dpth Y	U.equ. dia. $D_{ep(u)}$ (m)	Equiv. disch. factor J	Unit sect. area A_u (m²)	Unit wetted perim. P_u (m)	Unit surf. brdth B_{us} (m)	Unit mean depth y_{um} (m)	Discharge ratios $Q/Q_{0.60}$ medial	Mann'g	U.crit. disch. Q_{uc} (m³s⁻¹)
0.04	0.1061	0.4433	0.0199	0.7519	0.7462	0.0267	0.0031	0.0041	0.0102
0.08	0.2111	0.6224	0.0562	1.0654	1.0492	0.0536	0.0158	0.0182	0.0407
0.12	0.3149	0.7567	0.1029	1.3073	1.2775	0.0806	0.0396	0.0434	0.0915
0.16	0.4209	0.8818	0.1578	1.4995	1.4519	0.1087	0.0757	0.0808	0.1629
0.20	0.5361	1.0354	0.2180	1.6265	1.5501	0.1406	0.1253	0.1312	0.2560
0.24	0.6509	1.1827	0.2814	1.7289	1.6138	0.1743	0.1864	0.1928	0.3679
0.28	0.7623	1.3160	0.3468	1.8198	1.6567	0.2094	0.2577	0.2640	0.4969
0.32	0.8689	1.4334	0.4137	1.9044	1.6838	0.2457	0.3376	0.3436	0.6421
0.36	0.9697	1.5341	0.4813	1.9856	1.6977	0.2835	0.4247	0.4301	0.8026
0.40	1.0637	1.6177	0.5493	2.0657	1.6994	0.3233	0.5175	0.5221	0.9781
0.42	1.1079	1.6527	0.5833	2.1059	1.6957	0.3440	0.5654	0.5697	1.0713
0.44	1.1500	1.6832	0.6171	2.1465	1.6890	0.3654	0.6141	0.6179	1.1682
0.46	1.1900	1.7088	0.6508	2.1877	1.6790	0.3876	0.6633	0.6666	1.2689
0.48	1.2275	1.7293	0.6843	2.2299	1.6657	0.4108	0.7127	0.7155	1.3735
0.50	1.2623	1.7443	0.7174	2.2734	1.6486	0.4352	0.7620	0.7643	1.4821
0.52	1.2943	1.7537	0.7502	2.3185	1.6278	0.4609	0.8109	0.8127	1.5949
0.54	1.3238	1.7588	0.7825	2.3645	1.6050	0.4876	0.8592	0.8605	1.7111
0.56	1.3510	1.7602	0.8144	2.4112	1.5809	0.5151	0.9069	0.9078	1.8304
0.58	1.3759	1.7581	0.8458	2.4587	1.5554	0.5438	0.9539	0.9543	1.9530
0.60	1.3987	1.7527	0.8766	2.5069	1.5284	0.5735	1.0000	1.0000	2.0789
0.62	1.4192	1.7442	0.9069	2.5561	1.4999	0.6046	1.0451	1.0446	2.2083
0.64	1.4375	1.7328	0.9366	2.6062	1.4697	0.6373	1.0889	1.0881	2.3413
0.66	1.4536	1.7184	0.9657	2.6573	1.4378	0.6716	1.1314	1.1302	2.4783
0.68	1.4674	1.7013	0.9941	2.7097	1.4040	0.7080	1.1724	1.1709	2.6194
0.70	1.4791	1.6815	1.0218	2.7634	1.3682	0.7468	1.2117	1.2099	2.7652
0.72	1.4884	1.6590	1.0488	2.8185	1.3303	0.7884	1.2492	1.2471	2.9162
0.74	1.4955	1.6340	1.0750	2.8753	1.2900	0.8334	1.2846	1.2823	3.0731
0.76	1.5002	1.6063	1.1004	2.9339	1.2471	0.8824	1.3178	1.3153	3.2368
0.78	1.5025	1.5761	1.1249	2.9947	1.2013	0.9363	1.3485	1.3459	3.4086
0.80	1.5022	1.5433	1.1484	3.0579	1.1524	0.9965	1.3766	1.3739	3.5901
0.82	1.4993	1.5077	1.1709	3.1240	1.0998	1.0647	1.4017	1.3991	3.7835
0.84	1.4935	1.4693	1.1924	3.1934	1.0431	1.1431	1.4235	1.4211	3.9922
0.86	1.4847	1.4277	1.2126	3.2669	0.9814	1.2356	1.4418	1.4395	4.2210
0.88	1.4726	1.3828	1.2316	3.3454	0.9139	1.3476	1.4561	1.4540	4.4772
0.90	1.4566	1.3340	1.2491	3.4303	0.8390	1.4887	1.4657	1.4641	4.7728
0.92	1.4361	1.2804	1.2651	3.5236	0.7547	1.6762	1.4699	1.4688	5.1292
0.94	1.4100	1.2206	1.2792	3.6290	0.6573	1.9463	1.4676	1.4672	5.5887
0.96	1.3761	1.1518	1.2912	3.7533	0.5396	2.3928	1.4565	1.4572	6.2549
0.98	1.3290	1.0666	1.3006	3.9143	0.3837	3.3898	1.4319	1.4340	7.4986
1.00	1.2144	0.8871	1.3057	4.3008	0.0000	-	1.3499	1.3556	-

1.75 Pipe arch (corrugated sheet metal)

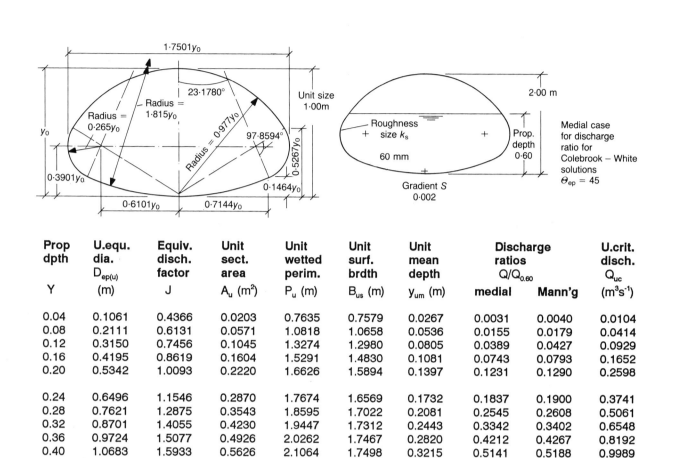

Prop dpth Y	U.equ. dia. $D_{ep(u)}$ (m)	Equiv. disch. factor J	Unit sect. area A_u (m²)	Unit wetted perim. P_u (m)	Unit surf. brdth B_{us} (m)	Unit mean depth y_{um} (m)	Discharge ratios $Q/Q_{0.60}$ medial	Mann'g	U.crit. disch. Q_{uc} (m³s⁻¹)
0.04	0.1061	0.4366	0.0203	0.7635	0.7579	0.0267	0.0031	0.0040	0.0104
0.08	0.2111	0.6131	0.0571	1.0818	1.0658	0.0536	0.0155	0.0179	0.0414
0.12	0.3150	0.7456	0.1045	1.3274	1.2980	0.0805	0.0389	0.0427	0.0929
0.16	0.4195	0.8619	0.1604	1.5291	1.4830	0.1081	0.0743	0.0793	0.1652
0.20	0.5342	1.0093	0.2220	1.6626	1.5894	0.1397	0.1231	0.1290	0.2598
0.24	0.6496	1.1546	0.2870	1.7674	1.6569	0.1732	0.1837	0.1900	0.3741
0.28	0.7621	1.2875	0.3543	1.8595	1.7022	0.2081	0.2545	0.2608	0.5061
0.32	0.8701	1.4055	0.4230	1.9447	1.7312	0.2443	0.3342	0.3402	0.6548
0.36	0.9724	1.5077	0.4926	2.0262	1.7467	0.2820	0.4212	0.4267	0.8192
0.40	1.0683	1.5933	0.5626	2.1064	1.7498	0.3215	0.5141	0.5188	0.9989
0.42	1.1135	1.6297	0.5975	2.1465	1.7467	0.3421	0.5622	0.5665	1.0944
0.44	1.1567	1.6616	0.6324	2.1870	1.7406	0.3633	0.6112	0.6150	1.1937
0.46	1.1977	1.6888	0.6671	2.2280	1.7314	0.3853	0.6607	0.6640	1.2968
0.48	1.2364	1.7111	0.7016	2.2700	1.7187	0.4082	0.7105	0.7133	1.4039
0.50	1.2725	1.7281	0.7359	2.3132	1.7024	0.4322	0.7602	0.7625	1.5150
0.52	1.3057	1.7394	0.7697	2.3581	1.6821	0.4576	0.8096	0.8114	1.6305
0.54	1.3359	1.7451	0.8031	2.4048	1.6580	0.4844	0.8583	0.8596	1.7504
0.56	1.3637	1.7469	0.8360	2.4523	1.6324	0.5122	0.9064	0.9072	1.8736
0.58	1.3891	1.7452	0.8684	2.5006	1.6053	0.5409	0.9537	0.9541	2.0001
0.60	1.4123	1.7401	0.9002	2.5497	1.5768	0.5709	1.0000	1.0000	2.1301
0.62	1.4332	1.7319	0.9315	2.5997	1.5468	0.6022	1.0453	1.0449	2.2636
0.64	1.4518	1.7205	0.9621	2.6508	1.5150	0.6350	1.0894	1.0885	2.4009
0.66	1.4681	1.7063	0.9921	2.7029	1.4816	0.6696	1.1320	1.1309	2.5422
0.68	1.4822	1.6893	1.0213	2.7563	1.4462	0.7062	1.1732	1.1717	2.6878
0.70	1.4939	1.6695	1.0499	2.8111	1.4088	0.7452	1.2126	1.2108	2.8382
0.72	1.5034	1.6471	1.0777	2.8673	1.3693	0.7870	1.2501	1.2480	2.9940
0.74	1.5105	1.6221	1.1046	2.9253	1.3273	0.8322	1.2856	1.2833	3.1558
0.76	1.5151	1.5945	1.1307	2.9852	1.2827	0.8815	1.3188	1.3163	3.3246
0.78	1.5173	1.5643	1.1559	3.0473	1.2353	0.9358	1.3495	1.3470	3.5017
0.80	1.5170	1.5314	1.1801	3.1119	1.1846	0.9963	1.3775	1.3749	3.6888
0.82	1.5139	1.4958	1.2033	3.1794	1.1302	1.0647	1.4026	1.4000	3.8882
0.84	1.5079	1.4574	1.2253	3.2504	1.0715	1.1435	1.4243	1.4219	4.1033
0.86	1.4988	1.4159	1.2461	3.3255	1.0079	1.2364	1.4425	1.4402	4.3390
0.88	1.4864	1.3710	1.2656	3.4059	0.9383	1.3489	1.4566	1.4546	4.6030
0.90	1.4700	1.3222	1.2836	3.4927	0.8612	1.4905	1.4661	1.4645	4.9075
0.92	1.4491	1.2687	1.3000	3.5883	0.7744	1.6787	1.4701	1.4691	5.2745
0.94	1.4225	1.2091	1.3145	3.6962	0.6742	1.9497	1.4676	1.4672	5.7477
0.96	1.3881	1.1405	1.3268	3.8235	0.5534	2.3976	1.4563	1.4570	6.4336
0.98	1.3402	1.0556	1.3364	3.9885	0.3933	3.3975	1.4315	1.4335	7.7137
1.00	1.2240	0.8770	1.3416	4.3845	0.0000	-	1.3489	1.3547	-

C43

1.35 Ellipse (corrugated sheet metal)

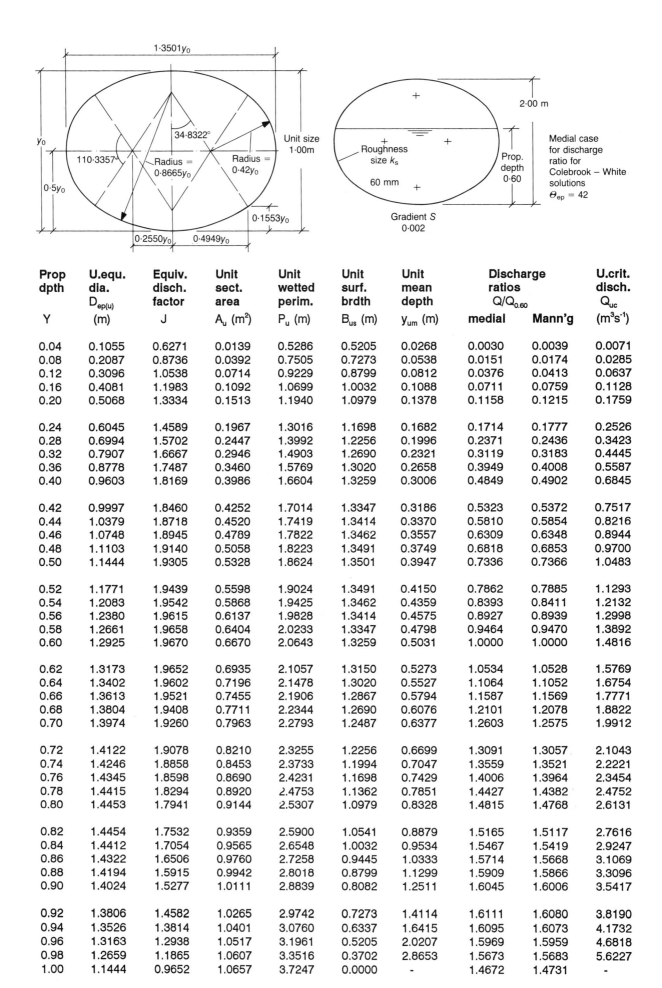

Prop dpth Y	U.equ. dia. $D_{ep(u)}$ (m)	Equiv. disch. factor J	Unit sect. area A_u (m²)	Unit wetted perim. P_u (m)	Unit surf. brdth B_{us} (m)	Unit mean depth y_{um} (m)	Discharge ratios $Q/Q_{0.60}$ medial	Mann'g	U.crit. disch. Q_{uc} (m³s⁻¹)
0.04	0.1055	0.6271	0.0139	0.5286	0.5205	0.0268	0.0030	0.0039	0.0071
0.08	0.2087	0.8736	0.0392	0.7505	0.7273	0.0538	0.0151	0.0174	0.0285
0.12	0.3096	1.0538	0.0714	0.9229	0.8799	0.0812	0.0376	0.0413	0.0637
0.16	0.4081	1.1983	0.1092	1.0699	1.0032	0.1088	0.0711	0.0759	0.1128
0.20	0.5068	1.3334	0.1513	1.1940	1.0979	0.1378	0.1158	0.1215	0.1759
0.24	0.6045	1.4589	0.1967	1.3016	1.1698	0.1682	0.1714	0.1777	0.2526
0.28	0.6994	1.5702	0.2447	1.3992	1.2256	0.1996	0.2371	0.2436	0.3423
0.32	0.7907	1.6667	0.2946	1.4903	1.2690	0.2321	0.3119	0.3183	0.4445
0.36	0.8778	1.7487	0.3460	1.5769	1.3020	0.2658	0.3949	0.4008	0.5587
0.40	0.9603	1.8169	0.3986	1.6604	1.3259	0.3006	0.4849	0.4902	0.6845
0.42	0.9997	1.8460	0.4252	1.7014	1.3347	0.3186	0.5323	0.5372	0.7517
0.44	1.0379	1.8718	0.4520	1.7419	1.3414	0.3370	0.5810	0.5854	0.8216
0.46	1.0748	1.8945	0.4789	1.7822	1.3462	0.3557	0.6309	0.6348	0.8944
0.48	1.1103	1.9140	0.5058	1.8223	1.3491	0.3749	0.6818	0.6853	0.9700
0.50	1.1444	1.9305	0.5328	1.8624	1.3501	0.3947	0.7336	0.7366	1.0483
0.52	1.1771	1.9439	0.5598	1.9024	1.3491	0.4150	0.7862	0.7885	1.1293
0.54	1.2083	1.9542	0.5868	1.9425	1.3462	0.4359	0.8393	0.8411	1.2132
0.56	1.2380	1.9615	0.6137	1.9828	1.3414	0.4575	0.8927	0.8939	1.2998
0.58	1.2661	1.9658	0.6404	2.0233	1.3347	0.4798	0.9464	0.9470	1.3892
0.60	1.2925	1.9670	0.6670	2.0643	1.3259	0.5031	1.0000	1.0000	1.4816
0.62	1.3173	1.9652	0.6935	2.1057	1.3150	0.5273	1.0534	1.0528	1.5769
0.64	1.3402	1.9602	0.7196	2.1478	1.3020	0.5527	1.1064	1.1052	1.6754
0.66	1.3613	1.9521	0.7455	2.1906	1.2867	0.5794	1.1587	1.1569	1.7771
0.68	1.3804	1.9408	0.7711	2.2344	1.2690	0.6076	1.2101	1.2078	1.8822
0.70	1.3974	1.9260	0.7963	2.2793	1.2487	0.6377	1.2603	1.2575	1.9912
0.72	1.4122	1.9078	0.8210	2.3255	1.2256	0.6699	1.3091	1.3057	2.1043
0.74	1.4246	1.8858	0.8453	2.3733	1.1994	0.7047	1.3559	1.3521	2.2221
0.76	1.4345	1.8598	0.8690	2.4231	1.1698	0.7429	1.4006	1.3964	2.3454
0.78	1.4415	1.8294	0.8920	2.4753	1.1362	0.7851	1.4427	1.4382	2.4752
0.80	1.4453	1.7941	0.9144	2.5307	1.0979	0.8328	1.4815	1.4768	2.6131
0.82	1.4454	1.7532	0.9359	2.5900	1.0541	0.8879	1.5165	1.5117	2.7616
0.84	1.4412	1.7054	0.9565	2.6548	1.0032	0.9534	1.5467	1.5419	2.9247
0.86	1.4322	1.6506	0.9760	2.7258	0.9445	1.0333	1.5714	1.5668	3.1069
0.88	1.4194	1.5915	0.9942	2.8018	0.8799	1.1299	1.5909	1.5866	3.3096
0.90	1.4024	1.5277	1.0111	2.8839	0.8082	1.2511	1.6045	1.6006	3.5417
0.92	1.3806	1.4582	1.0265	2.9742	0.7273	1.4114	1.6111	1.6080	3.8190
0.94	1.3526	1.3814	1.0401	3.0760	0.6337	1.6415	1.6095	1.6073	4.1732
0.96	1.3163	1.2938	1.0517	3.1961	0.5205	2.0207	1.5969	1.5959	4.6818
0.98	1.2659	1.1865	1.0607	3.3516	0.3702	2.8653	1.5673	1.5683	5.6227
1.00	1.1444	0.9652	1.0657	3.7247	0.0000	-	1.4672	1.4731	-

1.45 Ellipse (corrugated sheet metal)

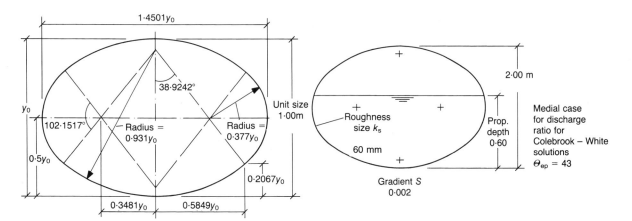

Prop dpth	U.equ. dia. $D_{ep(u)}$	Equiv. disch. factor	Unit sect. area	Unit wetted perim.	Unit surf. brdth	Unit mean depth	Discharge ratios $Q/Q_{0.60}$		U.crit. disch. Q_{uc}
Y	(m)	J	A_u (m²)	P_u (m)	B_{us} (m)	y_{um} (m)	medial	Mann'g	(m³s⁻¹)
0.04	0.1056	0.6056	0.0145	0.5478	0.5399	0.0268	0.0029	0.0038	0.0074
0.08	0.2090	0.8446	0.0406	0.7775	0.7551	0.0538	0.0145	0.0167	0.0295
0.12	0.3103	1.0199	0.0742	0.9558	0.9144	0.0811	0.0361	0.0397	0.0661
0.16	0.4094	1.1608	0.1134	1.1079	1.0437	0.1086	0.0683	0.0730	0.1170
0.20	0.5063	1.2791	0.1574	1.2435	1.1531	0.1365	0.1111	0.1167	0.1821
0.24	0.6026	1.3886	0.2054	1.3633	1.2421	0.1653	0.1648	0.1710	0.2615
0.28	0.6991	1.4968	0.2565	1.4673	1.3085	0.1960	0.2294	0.2358	0.3555
0.32	0.7935	1.5962	0.3098	1.5618	1.3587	0.2280	0.3036	0.3100	0.4634
0.36	0.8847	1.6841	0.3650	1.6502	1.3962	0.2614	0.3866	0.3925	0.5844
0.40	0.9717	1.7598	0.4214	1.7347	1.4231	0.2961	0.4771	0.4825	0.7181
0.42	1.0135	1.7929	0.4500	1.7759	1.4330	0.3140	0.5249	0.5299	0.7896
0.44	1.0541	1.8228	0.4787	1.8166	1.4405	0.3323	0.5741	0.5787	0.8642
0.46	1.0933	1.8497	0.5076	1.8570	1.4459	0.3510	0.6247	0.6287	0.9417
0.48	1.1313	1.8733	0.5365	1.8971	1.4491	0.3702	0.6763	0.6798	1.0223
0.50	1.1678	1.8938	0.5655	1.9371	1.4501	0.3900	0.7289	0.7319	1.1059
0.52	1.2028	1.9111	0.5945	1.9771	1.4491	0.4103	0.7823	0.7847	1.1925
0.54	1.2363	1.9253	0.6235	2.0173	1.4459	0.4312	0.8364	0.8382	1.2821
0.56	1.2681	1.9362	0.6523	2.0576	1.4405	0.4528	0.8908	0.8920	1.3747
0.58	1.2983	1.9438	0.6811	2.0983	1.4330	0.4753	0.9454	0.9460	1.4704
0.60	1.3267	1.9481	0.7096	2.1395	1.4231	0.4986	1.0000	1.0000	1.5693
0.62	1.3533	1.9489	0.7380	2.1814	1.4109	0.5230	1.0544	1.0538	1.6714
0.64	1.3778	1.9463	0.7661	2.2240	1.3962	0.5487	1.1083	1.1070	1.7770
0.66	1.4003	1.9400	0.7938	2.2676	1.3789	0.5757	1.1614	1.1596	1.8862
0.68	1.4205	1.9298	0.8212	2.3124	1.3587	0.6044	1.2135	1.2111	1.9993
0.70	1.4383	1.9156	0.8481	2.3587	1.3353	0.6352	1.2642	1.2613	2.1168
0.72	1.4534	1.8970	0.8746	2.4069	1.3085	0.6684	1.3131	1.3097	2.2392
0.74	1.4657	1.8737	0.9005	2.4574	1.2776	0.7048	1.3598	1.3560	2.3673
0.76	1.4746	1.8450	0.9257	2.5109	1.2421	0.7452	1.4037	1.3996	2.5024
0.78	1.4797	1.8100	0.9501	2.5683	1.2010	0.7911	1.4443	1.4399	2.6463
0.80	1.4804	1.7679	0.9737	2.6307	1.1531	0.8444	1.4806	1.4761	2.8018
0.82	1.4776	1.7212	0.9962	2.6968	1.1005	0.9052	1.5128	1.5083	2.9682
0.84	1.4715	1.6711	1.0176	2.7663	1.0437	0.9750	1.5410	1.5365	3.1468
0.86	1.4619	1.6172	1.0379	2.8398	0.9820	1.0569	1.5646	1.5603	3.3415
0.88	1.4486	1.5594	1.0569	2.9184	0.9144	1.1558	1.5831	1.5792	3.5582
0.90	1.4310	1.4969	1.0744	3.0033	0.8395	1.2798	1.5958	1.5924	3.8064
0.92	1.4085	1.4289	1.0904	3.0967	0.7551	1.4440	1.6018	1.5991	4.1032
0.94	1.3798	1.3537	1.1046	3.2021	0.6576	1.6796	1.5996	1.5977	4.4828
0.96	1.3427	1.2680	1.1166	3.3264	0.5399	2.0680	1.5866	1.5860	5.0283
0.98	1.2913	1.1632	1.1259	3.4876	0.3839	2.9330	1.5570	1.5583	6.0384
1.00	1.1678	0.9469	1.1310	3.8742	0.0000	-	1.4579	1.4638	-

C45

1.55 Ellipse (corrugated sheet metal)

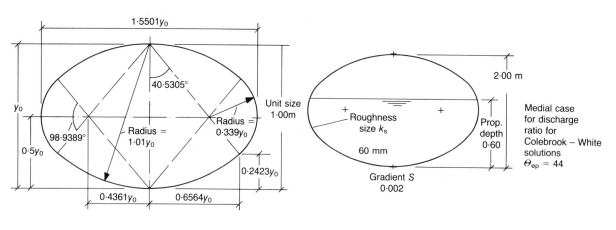

Prop dpth	U.equ. dia. $D_{ep(u)}$	Equiv. disch. factor	Unit sect. area	Unit wetted perim.	Unit surf. brdth	Unit mean depth	Discharge ratios $Q/Q_{0.60}$		U.crit. disch. Q_{uc}
Y	(m)	J	A_u (m²)	P_u (m)	B_{us} (m)	y_{um} (m)	medial	Mann'g	(m³s⁻¹)
0.04	0.1057	0.5820	0.0151	0.5704	0.5628	0.0268	0.0028	0.0037	0.0077
0.08	0.2094	0.8126	0.0424	0.8094	0.7879	0.0538	0.0140	0.0162	0.0308
0.12	0.3111	0.9824	0.0774	0.9947	0.9550	0.0810	0.0350	0.0385	0.0689
0.16	0.4108	1.1196	0.1184	1.1526	1.0911	0.1085	0.0663	0.0709	0.1221
0.20	0.5084	1.2351	0.1644	1.2932	1.2066	0.1362	0.1080	0.1135	0.1900
0.24	0.6040	1.3347	0.2147	1.4217	1.3072	0.1642	0.1602	0.1663	0.2725
0.28	0.6999	1.4320	0.2687	1.5355	1.3879	0.1936	0.2232	0.2296	0.3702
0.32	0.7963	1.5301	0.3254	1.6348	1.4466	0.2250	0.2967	0.3031	0.4834
0.36	0.8906	1.6212	0.3842	1.7257	1.4896	0.2579	0.3795	0.3855	0.6110
0.40	0.9815	1.7023	0.4444	1.8113	1.5199	0.2924	0.4704	0.4758	0.7526
0.42	1.0254	1.7386	0.4749	1.8528	1.5310	0.3102	0.5185	0.5235	0.8284
0.44	1.0681	1.7719	0.5057	1.8937	1.5394	0.3285	0.5682	0.5728	0.9075
0.46	1.1096	1.8022	0.5365	1.9341	1.5454	0.3472	0.6192	0.6233	0.9899
0.48	1.1497	1.8294	0.5675	1.9743	1.5489	0.3664	0.6715	0.6751	1.0756
0.50	1.1884	1.8534	0.5984	2.0143	1.5501	0.3861	0.7248	0.7278	1.1644
0.52	1.2256	1.8742	0.6294	2.0543	1.5489	0.4064	0.7790	0.7814	1.2565
0.54	1.2612	1.8916	0.6604	2.0945	1.5454	0.4273	0.8338	0.8357	1.3519
0.56	1.2951	1.9057	0.6912	2.1350	1.5394	0.4490	0.8891	0.8903	1.4505
0.58	1.3272	1.9162	0.7219	2.1758	1.5310	0.4716	0.9446	0.9452	1.5525
0.60	1.3574	1.9232	0.7525	2.2173	1.5199	0.4951	1.0000	1.0000	1.6580
0.62	1.3856	1.9263	0.7827	2.2596	1.5062	0.5197	1.0552	1.0546	1.7670
0.64	1.4116	1.9255	0.8127	2.3030	1.4896	0.5456	1.1098	1.1086	1.8798
0.66	1.4352	1.9205	0.8423	2.3476	1.4698	0.5731	1.1635	1.1617	1.9967
0.68	1.4562	1.9110	0.8715	2.3938	1.4466	0.6024	1.2160	1.2137	2.1181
0.70	1.4743	1.8965	0.9001	2.4422	1.4195	0.6341	1.2668	1.2640	2.2446
0.72	1.4892	1.8765	0.9282	2.4931	1.3879	0.6688	1.3155	1.3122	2.3771
0.74	1.5004	1.8501	0.9556	2.5476	1.3509	0.7074	1.3613	1.3577	2.5169
0.76	1.5071	1.8161	0.9822	2.6069	1.3072	0.7514	1.4035	1.3996	2.6662
0.78	1.5100	1.7767	1.0079	2.6699	1.2586	0.8008	1.4421	1.4380	2.8244
0.80	1.5098	1.7340	1.0325	2.7354	1.2066	0.8557	1.4773	1.4731	2.9910
0.82	1.5066	1.6879	1.0561	2.8040	1.1510	0.9176	1.5088	1.5046	3.1680
0.84	1.5000	1.6385	1.0785	2.8761	1.0911	0.9885	1.5362	1.5321	3.3581
0.86	1.4899	1.5854	1.0997	2.9524	1.0261	1.0718	1.5591	1.5551	3.5653
0.88	1.4760	1.5284	1.1195	3.0339	0.9550	1.1723	1.5769	1.5733	3.7959
0.90	1.4578	1.4668	1.1379	3.1222	0.8764	1.2984	1.5890	1.5859	4.0603
0.92	1.4345	1.3999	1.1545	3.2192	0.7879	1.4653	1.5944	1.5919	4.3765
0.94	1.4050	1.3260	1.1693	3.3289	0.6859	1.7049	1.5917	1.5901	4.7811
0.96	1.3670	1.2418	1.1818	3.4582	0.5629	2.0997	1.5783	1.5780	5.3628
0.98	1.3145	1.1388	1.1916	3.6260	0.4000	2.9789	1.5485	1.5500	6.4402
1.00	1.1884	0.9267	1.1969	4.0286	0.0000	-	1.4497	1.4557	-

1.65 Ellipse (corrugated sheet metal)

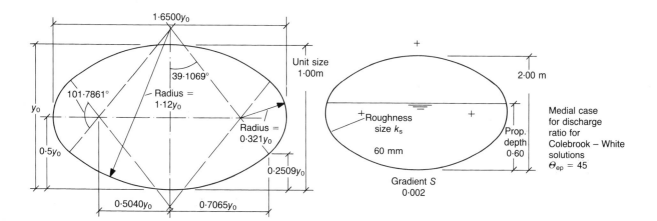

Prop dpth	U.equ. dia. $D_{ep(u)}$	Equiv. disch. factor	Unit sect. area	Unit wetted perim.	Unit surf. brdth	Unit mean depth	Discharge ratios $Q/Q_{0.60}$		U.crit. disch. Q_{uc}
Y	(m)	J	A_u (m²)	P_u (m)	B_{us} (m)	y_{um} (m)	medial	Mann'g	(m³s⁻¹)
0.04	0.1058	0.5534	0.0159	0.6005	0.5933	0.0268	0.0027	0.0036	0.0081
0.08	0.2098	0.7737	0.0447	0.8518	0.8314	0.0537	0.0137	0.0159	0.0324
0.12	0.3120	0.9365	0.0816	1.0464	1.0088	0.0809	0.0343	0.0377	0.0727
0.16	0.4123	1.0687	0.1249	1.2121	1.1538	0.1083	0.0651	0.0696	0.1288
0.20	0.5109	1.1806	0.1736	1.3594	1.2775	0.1359	0.1062	0.1116	0.2005
0.24	0.6076	1.2777	0.2269	1.4940	1.3856	0.1638	0.1577	0.1637	0.2876
0.28	0.7043	1.3705	0.2843	1.6144	1.4755	0.1927	0.2199	0.2262	0.3907
0.32	0.8029	1.4690	0.3446	1.7170	1.5395	0.2238	0.2930	0.2993	0.5106
0.36	0.9001	1.5628	0.4072	1.8094	1.5857	0.2568	0.3757	0.3817	0.6461
0.40	0.9944	1.6479	0.4713	1.8958	1.6180	0.2913	0.4667	0.4721	0.7965
0.42	1.0401	1.6865	0.5038	1.9375	1.6297	0.3091	0.5150	0.5200	0.8771
0.44	1.0846	1.7222	0.5365	1.9784	1.6387	0.3274	0.5648	0.5694	0.9612
0.46	1.1279	1.7551	0.5693	2.0189	1.6450	0.3461	0.6162	0.6202	1.0488
0.48	1.1699	1.7849	0.6022	2.0591	1.6487	0.3653	0.6687	0.6723	1.1399
0.50	1.2105	1.8115	0.6352	2.0992	1.6500	0.3850	0.7224	0.7254	1.2343
0.52	1.2495	1.8350	0.6682	2.1392	1.6487	0.4053	0.7770	0.7794	1.3322
0.54	1.2869	1.8551	0.7012	2.1794	1.6450	0.4263	0.8323	0.8341	1.4336
0.56	1.3226	1.8718	0.7340	2.2199	1.6387	0.4479	0.8880	0.8892	1.5384
0.58	1.3565	1.8849	0.7667	2.2609	1.6297	0.4705	0.9440	0.9446	1.6468
0.60	1.3883	1.8942	0.7992	2.3025	1.6180	0.4939	1.0000	1.0000	1.7589
0.62	1.4181	1.8997	0.8314	2.3451	1.6034	0.5185	1.0558	1.0551	1.8748
0.64	1.4455	1.9009	0.8633	2.3889	1.5857	0.5444	1.1109	1.1097	1.9948
0.66	1.4704	1.8977	0.8948	2.4342	1.5645	0.5719	1.1652	1.1634	2.1192
0.68	1.4925	1.8896	0.9259	2.4813	1.5395	0.6014	1.2181	1.2157	2.2484
0.70	1.5114	1.8760	0.9564	2.5310	1.5101	0.6333	1.2693	1.2664	2.3833
0.72	1.5267	1.8562	0.9862	2.5839	1.4755	0.6684	1.3180	1.3147	2.5250
0.74	1.5376	1.8288	1.0153	2.6414	1.4343	0.7079	1.3636	1.3600	2.6752
0.76	1.5435	1.7930	1.0435	2.7044	1.3856	0.7531	1.4052	1.4013	2.8360
0.78	1.5460	1.7532	1.0707	2.7703	1.3333	0.8031	1.4435	1.4394	3.0049
0.80	1.5455	1.7102	1.0969	2.8389	1.2775	0.8586	1.4783	1.4742	3.1828
0.82	1.5416	1.6639	1.1218	2.9107	1.2179	0.9211	1.5093	1.5052	3.3716
0.84	1.5344	1.6142	1.1455	2.9863	1.1538	0.9929	1.5363	1.5322	3.5745
0.86	1.5236	1.5609	1.1679	3.0663	1.0844	1.0770	1.5586	1.5548	3.7956
0.88	1.5088	1.5038	1.1889	3.1519	1.0088	1.1785	1.5759	1.5724	4.0417
0.90	1.4895	1.4422	1.2082	3.2446	0.9252	1.3059	1.5874	1.5844	4.3238
0.92	1.4652	1.3754	1.2258	3.3466	0.8314	1.4744	1.5922	1.5899	4.6612
0.94	1.4344	1.3017	1.2414	3.4618	0.7233	1.7162	1.5889	1.5875	5.0928
0.96	1.3948	1.2179	1.2546	3.5979	0.5933	2.1146	1.5750	1.5747	5.7132
0.98	1.3405	1.1157	1.2649	3.7744	0.4214	3.0014	1.5446	1.5461	6.8621
1.00	1.2105	0.9058	1.2705	4.1983	0.0000	-	1.4448	1.4509	-

C47   Trapezoidal - 0.125 to 1 side-slope

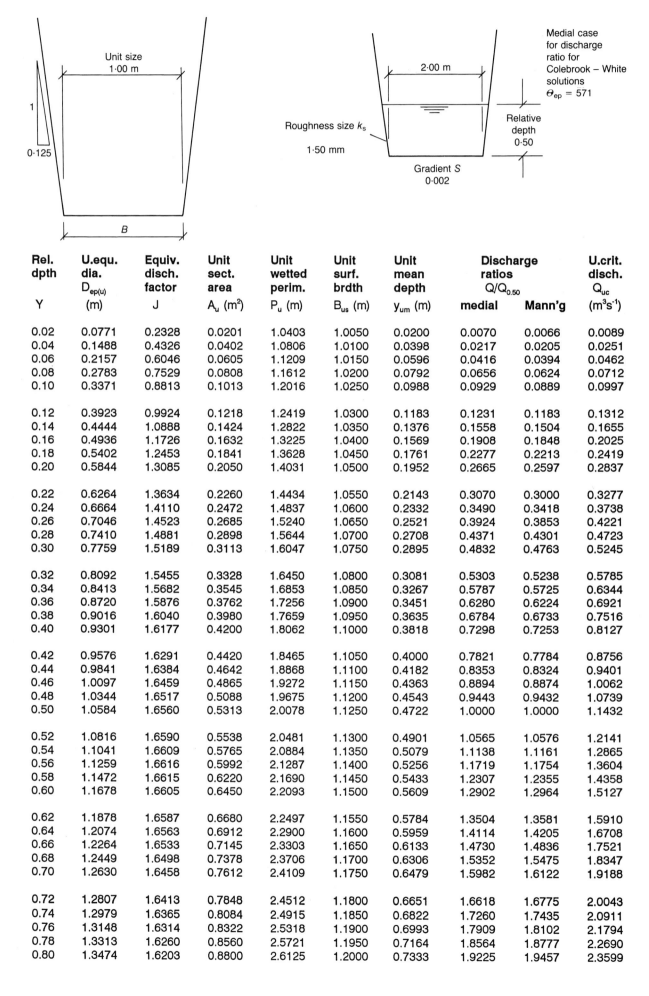

Rel. dpth Y	U.equ. dia. $D_{ep(u)}$ (m)	Equiv. disch. factor J	Unit sect. area A_u (m²)	Unit wetted perim. P_u (m)	Unit surf. brdth B_{us} (m)	Unit mean depth y_{um} (m)	Discharge ratios $Q/Q_{0.50}$ medial	Mann'g	U.crit. disch. Q_{uc} (m³s⁻¹)
0.02	0.0771	0.2328	0.0201	1.0403	1.0050	0.0200	0.0070	0.0066	0.0089
0.04	0.1488	0.4326	0.0402	1.0806	1.0100	0.0398	0.0217	0.0205	0.0251
0.06	0.2157	0.6046	0.0605	1.1209	1.0150	0.0596	0.0416	0.0394	0.0462
0.08	0.2783	0.7529	0.0808	1.1612	1.0200	0.0792	0.0656	0.0624	0.0712
0.10	0.3371	0.8813	0.1013	1.2016	1.0250	0.0988	0.0929	0.0889	0.0997
0.12	0.3923	0.9924	0.1218	1.2419	1.0300	0.1183	0.1231	0.1183	0.1312
0.14	0.4444	1.0888	0.1424	1.2822	1.0350	0.1376	0.1558	0.1504	0.1655
0.16	0.4936	1.1726	0.1632	1.3225	1.0400	0.1569	0.1908	0.1848	0.2025
0.18	0.5402	1.2453	0.1841	1.3628	1.0450	0.1761	0.2277	0.2213	0.2419
0.20	0.5844	1.3085	0.2050	1.4031	1.0500	0.1952	0.2665	0.2597	0.2837
0.22	0.6264	1.3634	0.2260	1.4434	1.0550	0.2143	0.3070	0.3000	0.3277
0.24	0.6664	1.4110	0.2472	1.4837	1.0600	0.2332	0.3490	0.3418	0.3738
0.26	0.7046	1.4523	0.2685	1.5240	1.0650	0.2521	0.3924	0.3853	0.4221
0.28	0.7410	1.4881	0.2898	1.5644	1.0700	0.2708	0.4371	0.4301	0.4723
0.30	0.7759	1.5189	0.3113	1.6047	1.0750	0.2895	0.4832	0.4763	0.5245
0.32	0.8092	1.5455	0.3328	1.6450	1.0800	0.3081	0.5303	0.5238	0.5785
0.34	0.8413	1.5682	0.3545	1.6853	1.0850	0.3267	0.5787	0.5725	0.6344
0.36	0.8720	1.5876	0.3762	1.7256	1.0900	0.3451	0.6280	0.6224	0.6921
0.38	0.9016	1.6040	0.3980	1.7659	1.0950	0.3635	0.6784	0.6733	0.7516
0.40	0.9301	1.6177	0.4200	1.8062	1.1000	0.3818	0.7298	0.7253	0.8127
0.42	0.9576	1.6291	0.4420	1.8465	1.1050	0.4000	0.7821	0.7784	0.8756
0.44	0.9841	1.6384	0.4642	1.8868	1.1100	0.4182	0.8353	0.8324	0.9401
0.46	1.0097	1.6459	0.4865	1.9272	1.1150	0.4363	0.8894	0.8874	1.0062
0.48	1.0344	1.6517	0.5088	1.9675	1.1200	0.4543	0.9443	0.9432	1.0739
0.50	1.0584	1.6560	0.5313	2.0078	1.1250	0.4722	1.0000	1.0000	1.1432
0.52	1.0816	1.6590	0.5538	2.0481	1.1300	0.4901	1.0565	1.0576	1.2141
0.54	1.1041	1.6609	0.5765	2.0884	1.1350	0.5079	1.1138	1.1161	1.2865
0.56	1.1259	1.6616	0.5992	2.1287	1.1400	0.5256	1.1719	1.1754	1.3604
0.58	1.1472	1.6615	0.6220	2.1690	1.1450	0.5433	1.2307	1.2355	1.4358
0.60	1.1678	1.6605	0.6450	2.2093	1.1500	0.5609	1.2902	1.2964	1.5127
0.62	1.1878	1.6587	0.6680	2.2497	1.1550	0.5784	1.3504	1.3581	1.5910
0.64	1.2074	1.6563	0.6912	2.2900	1.1600	0.5959	1.4114	1.4205	1.6708
0.66	1.2264	1.6533	0.7145	2.3303	1.1650	0.6133	1.4730	1.4836	1.7521
0.68	1.2449	1.6498	0.7378	2.3706	1.1700	0.6306	1.5352	1.5475	1.8347
0.70	1.2630	1.6458	0.7612	2.4109	1.1750	0.6479	1.5982	1.6122	1.9188
0.72	1.2807	1.6413	0.7848	2.4512	1.1800	0.6651	1.6618	1.6775	2.0043
0.74	1.2979	1.6365	0.8084	2.4915	1.1850	0.6822	1.7260	1.7435	2.0911
0.76	1.3148	1.6314	0.8322	2.5318	1.1900	0.6993	1.7909	1.8102	2.1794
0.78	1.3313	1.6260	0.8560	2.5721	1.1950	0.7164	1.8564	1.8777	2.2690
0.80	1.3474	1.6203	0.8800	2.6125	1.2000	0.7333	1.9225	1.9457	2.3599

Rel. dpth	U.equ. dia. $D_{ep(u)}$	Equiv. disch. factor	Unit sect. area	Unit wetted perim.	Unit surf. brdth	Unit mean depth	Discharge ratios $Q/Q_{0.50}$		U.crit. disch. Q_{uc}
Y	(m)	J	A_u (m^2)	P_u (m)	B_{us} (m)	y_{um} (m)	medial	Mann'g	(m^{3}s^{-1})
0.82	1.3632	1.6143	0.9041	2.6528	1.2050	0.7502	1.9892	2.0145	2.4522
0.84	1.3786	1.6082	0.9282	2.6931	1.2100	0.7671	2.0566	2.0839	2.5458
0.86	1.3938	1.6019	0.9525	2.7334	1.2150	0.7839	2.1246	2.1540	2.6408
0.88	1.4087	1.5955	0.9768	2.7737	1.2200	0.8007	2.1931	2.2248	2.7371
0.90	1.4232	1.5889	1.0012	2.8140	1.2250	0.8173	2.2623	2.2961	2.8347
0.92	1.4375	1.5822	1.0258	2.8543	1.2300	0.8340	2.3320	2.3682	2.9336
0.94	1.4516	1.5754	1.0504	2.8946	1.2350	0.8506	2.4024	2.4409	3.0338
0.96	1.4654	1.5685	1.0752	2.9349	1.2400	0.8671	2.4733	2.5142	3.1353
0.98	1.4789	1.5616	1.1000	2.9753	1.2450	0.8836	2.5448	2.5881	3.2381
1.00	1.4923	1.5546	1.1250	3.0156	1.2500	0.9000	2.6169	2.6627	3.3422
1.02	1.5054	1.5476	1.1500	3.0559	1.2550	0.9164	2.6896	2.7379	3.4476
1.04	1.5183	1.5405	1.1752	3.0962	1.2600	0.9327	2.7628	2.8137	3.5542
1.06	1.5309	1.5334	1.2004	3.1365	1.2650	0.9490	2.8366	2.8902	3.6621
1.08	1.5434	1.5263	1.2258	3.1768	1.2700	0.9652	2.9110	2.9672	3.7713
1.10	1.5557	1.5192	1.2512	3.2171	1.2750	0.9814	2.9859	3.0449	3.8817
1.12	1.5679	1.5121	1.2768	3.2574	1.2800	0.9975	3.0614	3.1232	3.9934
1.14	1.5798	1.5050	1.3024	3.2977	1.2850	1.0136	3.1375	3.2021	4.1063
1.16	1.5916	1.4979	1.3282	3.3381	1.2900	1.0296	3.2141	3.2816	4.2205
1.18	1.6032	1.4908	1.3540	3.3784	1.2950	1.0456	3.2913	3.3618	4.3359
1.20	1.6147	1.4838	1.3800	3.4187	1.3000	1.0615	3.3690	3.4425	4.4525
1.24	1.6371	1.4697	1.4322	3.4993	1.3100	1.0933	3.5262	3.6058	4.6895
1.28	1.6590	1.4559	1.4848	3.5799	1.3200	1.1248	3.6856	3.7715	4.9315
1.32	1.6804	1.4421	1.5378	3.6605	1.3300	1.1562	3.8472	3.9396	5.1783
1.36	1.7013	1.4286	1.5912	3.7412	1.3400	1.1875	4.0110	4.1101	5.4299
1.40	1.7217	1.4152	1.6450	3.8218	1.3500	1.2185	4.1770	4.2830	5.6865
1.44	1.7417	1.4021	1.6992	3.9024	1.3600	1.2494	4.3452	4.4583	5.9478
1.48	1.7613	1.3892	1.7538	3.9830	1.3700	1.2801	4.5155	4.6359	6.2140
1.52	1.7805	1.3764	1.8088	4.0637	1.3800	1.3107	4.6881	4.8160	6.4849
1.56	1.7993	1.3639	1.8642	4.1443	1.3900	1.3412	4.8629	4.9984	6.7607
1.60	1.8178	1.3517	1.9200	4.2249	1.4000	1.3714	5.0398	5.1833	7.0412
1.64	1.8360	1.3396	1.9762	4.3055	1.4100	1.4016	5.2190	5.3705	7.3265
1.68	1.8538	1.3278	2.0328	4.3861	1.4200	1.4315	5.4003	5.5601	7.6165
1.72	1.8714	1.3162	2.0898	4.4668	1.4300	1.4614	5.5839	5.7521	7.9113
1.76	1.8887	1.3048	2.1472	4.5474	1.4400	1.4911	5.7696	5.9465	8.2108
1.80	1.9058	1.2937	2.2050	4.6280	1.4500	1.5207	5.9575	6.1432	8.5151
1.84	1.9226	1.2827	2.2632	4.7086	1.4600	1.5501	6.1476	6.3424	8.8241
1.88	1.9392	1.2720	2.3218	4.7893	1.4700	1.5795	6.3399	6.5440	9.1377
1.92	1.9555	1.2615	2.3808	4.8699	1.4800	1.6086	6.5344	6.7480	9.4561
1.96	1.9717	1.2512	2.4402	4.9505	1.4900	1.6377	6.7312	6.9543	9.7792
2.00	1.9876	1.2411	2.5000	5.0311	1.5000	1.6667	6.9301	7.1631	10.107
2.05	2.0073	1.2288	2.5753	5.1319	1.5125	1.7027	7.1819	7.4275	10.523
2.10	2.0267	1.2167	2.6512	5.2327	1.5250	1.7385	7.4371	7.6957	10.947
2.15	2.0458	1.2050	2.7278	5.3335	1.5375	1.7742	7.6958	7.9677	11.378
2.20	2.0647	1.1936	2.8050	5.4342	1.5500	1.8097	7.9580	8.2434	11.817
2.25	2.0833	1.1824	2.8828	5.5350	1.5625	1.8450	8.2237	8.5230	12.262
2.30	2.1017	1.1716	2.9613	5.6358	1.5750	1.8802	8.4929	8.8065	12.715
2.35	2.1199	1.1609	3.0403	5.7366	1.5875	1.9152	8.7655	9.0937	13.176
2.40	2.1380	1.1506	3.1200	5.8374	1.6000	1.9500	9.0418	9.3848	13.644
2.45	2.1558	1.1405	3.2003	5.9381	1.6125	1.9847	9.3215	9.6798	14.119
2.50	2.1734	1.1306	3.2813	6.0389	1.6250	2.0192	9.6048	9.9787	14.601
2.55	2.1909	1.1210	3.3628	6.1397	1.6375	2.0536	9.8916	10.281	15.091
2.60	2.2082	1.1116	3.4450	6.2405	1.6500	2.0879	10.182	10.588	15.588
2.65	2.2253	1.1024	3.5278	6.3412	1.6625	2.1220	10.476	10.899	16.093
2.70	2.2423	1.0935	3.6113	6.4420	1.6750	2.1560	10.773	11.213	16.605
2.75	2.2592	1.0847	3.6953	6.5428	1.6875	2.1898	11.075	11.532	17.124

C48

Trapezoidal - 0.25 to 1 side-slope

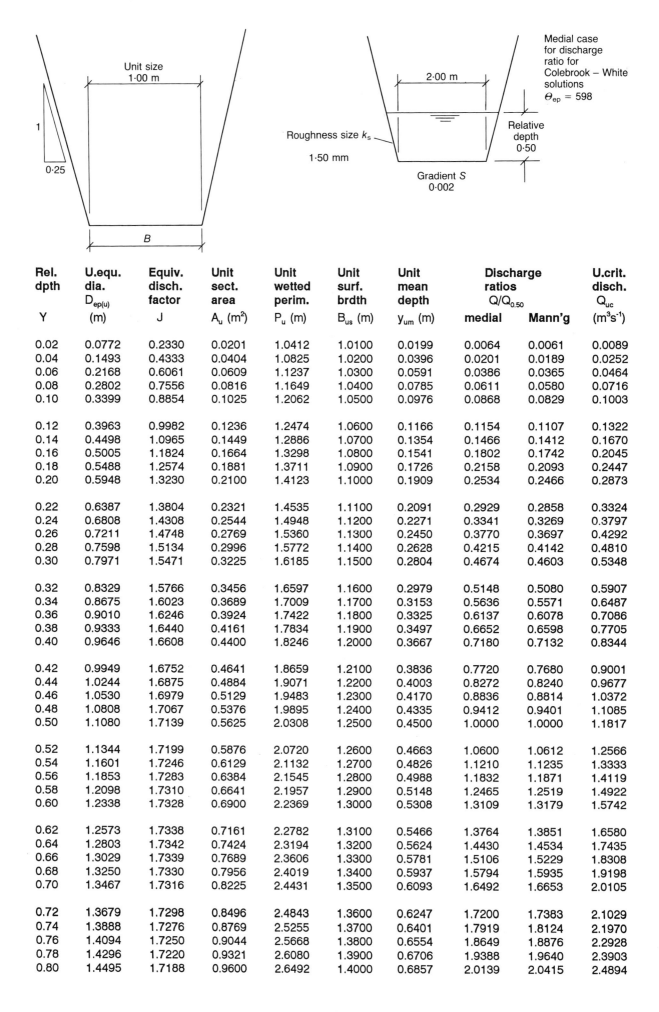

Rel. dpth Y	U.equ. dia. $D_{ep(u)}$ (m)	Equiv. disch. factor J	Unit sect. area A_u (m²)	Unit wetted perim. P_u (m)	Unit surf. brdth B_{us} (m)	Unit mean depth y_{um} (m)	Discharge ratios $Q/Q_{0.50}$ medial	Mann'g	U.crit. disch. Q_{uc} (m³s⁻¹)
0.02	0.0772	0.2330	0.0201	1.0412	1.0100	0.0199	0.0064	0.0061	0.0089
0.04	0.1493	0.4333	0.0404	1.0825	1.0200	0.0396	0.0201	0.0189	0.0252
0.06	0.2168	0.6061	0.0609	1.1237	1.0300	0.0591	0.0386	0.0365	0.0464
0.08	0.2802	0.7556	0.0816	1.1649	1.0400	0.0785	0.0611	0.0580	0.0716
0.10	0.3399	0.8854	0.1025	1.2062	1.0500	0.0976	0.0868	0.0829	0.1003
0.12	0.3963	0.9982	0.1236	1.2474	1.0600	0.1166	0.1154	0.1107	0.1322
0.14	0.4498	1.0965	0.1449	1.2886	1.0700	0.1354	0.1466	0.1412	0.1670
0.16	0.5005	1.1824	0.1664	1.3298	1.0800	0.1541	0.1802	0.1742	0.2045
0.18	0.5488	1.2574	0.1881	1.3711	1.0900	0.1726	0.2158	0.2093	0.2447
0.20	0.5948	1.3230	0.2100	1.4123	1.1000	0.1909	0.2534	0.2466	0.2873
0.22	0.6387	1.3804	0.2321	1.4535	1.1100	0.2091	0.2929	0.2858	0.3324
0.24	0.6808	1.4308	0.2544	1.4948	1.1200	0.2271	0.3341	0.3269	0.3797
0.26	0.7211	1.4748	0.2769	1.5360	1.1300	0.2450	0.3770	0.3697	0.4292
0.28	0.7598	1.5134	0.2996	1.5772	1.1400	0.2628	0.4215	0.4142	0.4810
0.30	0.7971	1.5471	0.3225	1.6185	1.1500	0.2804	0.4674	0.4603	0.5348
0.32	0.8329	1.5766	0.3456	1.6597	1.1600	0.2979	0.5148	0.5080	0.5907
0.34	0.8675	1.6023	0.3689	1.7009	1.1700	0.3153	0.5636	0.5571	0.6487
0.36	0.9010	1.6246	0.3924	1.7422	1.1800	0.3325	0.6137	0.6078	0.7086
0.38	0.9333	1.6440	0.4161	1.7834	1.1900	0.3497	0.6652	0.6598	0.7705
0.40	0.9646	1.6608	0.4400	1.8246	1.2000	0.3667	0.7180	0.7132	0.8344
0.42	0.9949	1.6752	0.4641	1.8659	1.2100	0.3836	0.7720	0.7680	0.9001
0.44	1.0244	1.6875	0.4884	1.9071	1.2200	0.4003	0.8272	0.8240	0.9677
0.46	1.0530	1.6979	0.5129	1.9483	1.2300	0.4170	0.8836	0.8814	1.0372
0.48	1.0808	1.7067	0.5376	1.9895	1.2400	0.4335	0.9412	0.9401	1.1085
0.50	1.1080	1.7139	0.5625	2.0308	1.2500	0.4500	1.0000	1.0000	1.1817
0.52	1.1344	1.7199	0.5876	2.0720	1.2600	0.4663	1.0600	1.0612	1.2566
0.54	1.1601	1.7246	0.6129	2.1132	1.2700	0.4826	1.1210	1.1235	1.3333
0.56	1.1853	1.7283	0.6384	2.1545	1.2800	0.4988	1.1832	1.1871	1.4119
0.58	1.2098	1.7310	0.6641	2.1957	1.2900	0.5148	1.2465	1.2519	1.4922
0.60	1.2338	1.7328	0.6900	2.2369	1.3000	0.5308	1.3109	1.3179	1.5742
0.62	1.2573	1.7338	0.7161	2.2782	1.3100	0.5466	1.3764	1.3851	1.6580
0.64	1.2803	1.7342	0.7424	2.3194	1.3200	0.5624	1.4430	1.4534	1.7435
0.66	1.3029	1.7339	0.7689	2.3606	1.3300	0.5781	1.5106	1.5229	1.8308
0.68	1.3250	1.7330	0.7956	2.4019	1.3400	0.5937	1.5794	1.5935	1.9198
0.70	1.3467	1.7316	0.8225	2.4431	1.3500	0.6093	1.6492	1.6653	2.0105
0.72	1.3679	1.7298	0.8496	2.4843	1.3600	0.6247	1.7200	1.7383	2.1029
0.74	1.3888	1.7276	0.8769	2.5255	1.3700	0.6401	1.7919	1.8124	2.1970
0.76	1.4094	1.7250	0.9044	2.5668	1.3800	0.6554	1.8649	1.8876	2.2928
0.78	1.4296	1.7220	0.9321	2.6080	1.3900	0.6706	1.9388	1.9640	2.3903
0.80	1.4495	1.7188	0.9600	2.6492	1.4000	0.6857	2.0139	2.0415	2.4894

Rel. dpth Y	U.equ. dia. $D_{ep(u)}$ (m)	Equiv. disch. factor J	Unit sect. area A_u (m²)	Unit wetted perim. P_u (m)	Unit surf. brdth B_{us} (m)	Unit mean depth y_{um} (m)	Discharge ratios $Q/Q_{0.50}$ medial	Mann'g	U.crit. disch. Q_{uc} (m³s⁻¹)
0.82	1.4690	1.7153	0.9881	2.6905	1.4100	0.7008	2.0900	2.1201	2.5903
0.84	1.4883	1.7116	1.0164	2.7317	1.4200	0.7158	2.1671	2.1998	2.6929
0.86	1.5073	1.7076	1.0449	2.7729	1.4300	0.7307	2.2453	2.2807	2.7971
0.88	1.5260	1.7035	1.0736	2.8142	1.4400	0.7456	2.3244	2.3627	2.9030
0.90	1.5444	1.6992	1.1025	2.8554	1.4500	0.7603	2.4047	2.4458	3.0105
0.92	1.5626	1.6948	1.1316	2.8966	1.4600	0.7751	2.4859	2.5301	3.1198
0.94	1.5806	1.6902	1.1609	2.9379	1.4700	0.7897	2.5682	2.6154	3.2307
0.96	1.5983	1.6855	1.1904	2.9791	1.4800	0.8043	2.6516	2.7019	3.3433
0.98	1.6159	1.6807	1.2201	3.0203	1.4900	0.8189	2.7359	2.7895	3.4575
1.00	1.6332	1.6758	1.2500	3.0616	1.5000	0.8333	2.8213	2.8782	3.5734
1.02	1.6503	1.6709	1.2801	3.1028	1.5100	0.8477	2.9078	2.9681	3.6909
1.04	1.6672	1.6658	1.3104	3.1440	1.5200	0.8621	2.9952	3.0591	3.8102
1.06	1.6839	1.6608	1.3409	3.1852	1.5300	0.8764	3.0837	3.1512	3.9311
1.08	1.7004	1.6557	1.3716	3.2265	1.5400	0.8906	3.1733	3.2444	4.0536
1.10	1.7168	1.6505	1.4025	3.2677	1.5500	0.9048	3.2638	3.3387	4.1778
1.12	1.7330	1.6453	1.4336	3.3089	1.5600	0.9190	3.3554	3.4342	4.3037
1.14	1.7490	1.6401	1.4649	3.3502	1.5700	0.9331	3.4481	3.5308	4.4312
1.16	1.7649	1.6349	1.4964	3.3914	1.5800	0.9471	3.5417	3.6285	4.5604
1.18	1.7807	1.6297	1.5281	3.4326	1.5900	0.9611	3.6365	3.7274	4.6913
1.20	1.7963	1.6244	1.5600	3.4739	1.6000	0.9750	3.7322	3.8274	4.8238
1.24	1.8271	1.6139	1.6244	3.5563	1.6200	1.0027	3.9269	4.0308	5.0938
1.28	1.8573	1.6035	1.6896	3.6388	1.6400	1.0302	4.1257	4.2388	5.3705
1.32	1.8871	1.5931	1.7556	3.7212	1.6600	1.0576	4.3287	4.4513	5.6539
1.36	1.9164	1.5828	1.8224	3.8037	1.6800	1.0848	4.5359	4.6684	5.9439
1.40	1.9454	1.5726	1.8900	3.8862	1.7000	1.1118	4.7474	4.8902	6.2406
1.44	1.9739	1.5625	1.9584	3.9686	1.7200	1.1386	4.9630	5.1166	6.5441
1.48	2.0020	1.5525	2.0276	4.0511	1.7400	1.1653	5.1830	5.3476	6.8542
1.52	2.0298	1.5427	2.0976	4.1336	1.7600	1.1918	5.4072	5.5833	7.1711
1.56	2.0573	1.5330	2.1684	4.2160	1.7800	1.2182	5.6357	5.8237	7.4948
1.60	2.0845	1.5234	2.2400	4.2985	1.8000	1.2444	5.8685	6.0689	7.8252
1.64	2.1113	1.5140	2.3124	4.3809	1.8200	1.2705	6.1057	6.3187	8.1624
1.68	2.1379	1.5047	2.3856	4.4634	1.8400	1.2965	6.3472	6.5734	8.5064
1.72	2.1642	1.4957	2.4596	4.5459	1.8600	1.3224	6.5930	6.8328	8.8573
1.76	2.1903	1.4867	2.5344	4.6283	1.8800	1.3481	6.8432	7.0971	9.2150
1.80	2.2162	1.4779	2.6100	4.7108	1.9000	1.3737	7.0979	7.3662	9.5795
1.84	2.2418	1.4693	2.6864	4.7933	1.9200	1.3992	7.3570	7.6401	9.9510
1.88	2.2672	1.4608	2.7636	4.8757	1.9400	1.4245	7.6205	7.9190	10.329
1.92	2.2925	1.4525	2.8416	4.9582	1.9600	1.4498	7.8884	8.2028	10.715
1.96	2.3175	1.4443	2.9204	5.0406	1.9800	1.4749	8.1609	8.4915	11.107
2.00	2.3423	1.4363	3.0000	5.1231	2.0000	1.5000	8.4379	8.7852	11.506
2.05	2.3731	1.4265	3.1006	5.2262	2.0250	1.5312	8.7905	9.1593	12.015
2.10	2.4037	1.4170	3.2025	5.3293	2.0500	1.5622	9.1502	9.5413	12.535
2.15	2.4340	1.4076	3.3056	5.4323	2.0750	1.5931	9.5171	9.9312	13.066
2.20	2.4641	1.3985	3.4100	5.5354	2.1000	1.6238	9.8911	10.329	13.608
2.25	2.4940	1.3896	3.5156	5.6385	2.1250	1.6544	10.272	10.735	14.161
2.30	2.5237	1.3809	3.6225	5.7416	2.1500	1.6849	10.661	11.149	14.725
2.35	2.5532	1.3723	3.7306	5.8446	2.1750	1.7152	11.057	11.571	15.300
2.40	2.5825	1.3640	3.8400	5.9477	2.2000	1.7455	11.460	12.001	15.887
2.45	2.6116	1.3559	3.9506	6.0508	2.2250	1.7756	11.871	12.440	16.485
2.50	2.6406	1.3480	4.0625	6.1539	2.2500	1.8056	12.289	12.886	17.095
2.55	2.6694	1.3403	4.1756	6.2570	2.2750	1.8354	12.715	13.341	17.715
2.60	2.6981	1.3327	4.2900	6.3600	2.3000	1.8652	13.148	13.805	18.348
2.65	2.7266	1.3253	4.4056	6.4631	2.3250	1.8949	13.589	14.277	18.992
2.70	2.7550	1.3181	4.5225	6.5662	2.3500	1.9245	14.037	14.757	19.647
2.75	2.7833	1.3111	4.6406	6.6693	2.3750	1.9539	14.493	15.246	20.314

C49

Trapezoidal - 0.375 to 1 side-slope

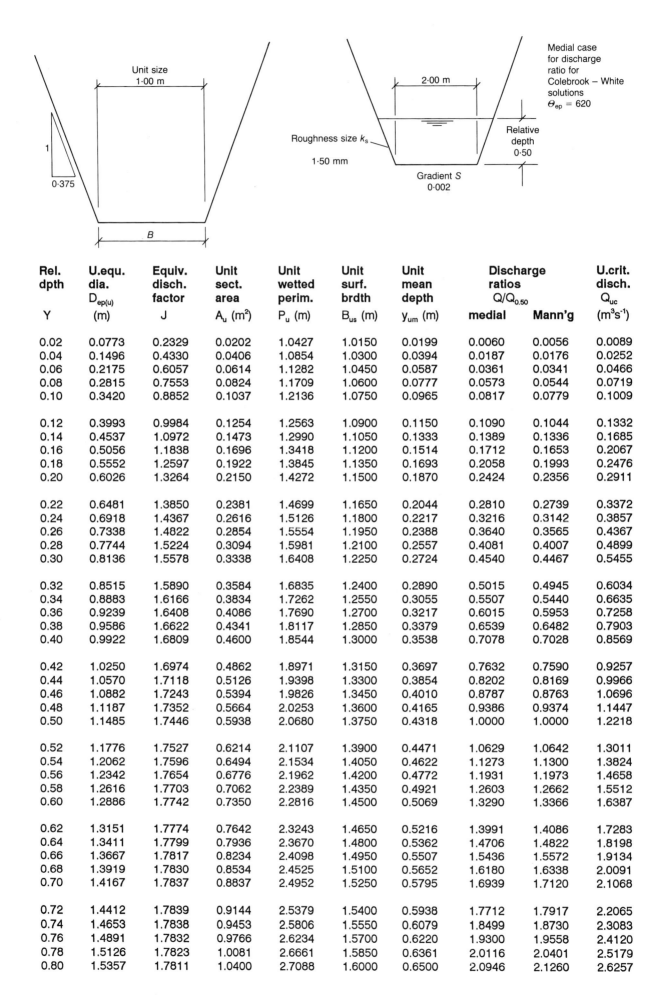

Rel. dpth	U.equ. dia. $D_{ep(u)}$	Equiv. disch. factor	Unit sect. area	Unit wetted perim.	Unit surf. brdth	Unit mean depth	Discharge ratios $Q/Q_{0.50}$		U.crit. disch. Q_{uc}
Y	(m)	J	A_u (m²)	P_u (m)	B_{us} (m)	y_{um} (m)	medial	Mann'g	(m³s⁻¹)
0.02	0.0773	0.2329	0.0202	1.0427	1.0150	0.0199	0.0060	0.0056	0.0089
0.04	0.1496	0.4330	0.0406	1.0854	1.0300	0.0394	0.0187	0.0176	0.0252
0.06	0.2175	0.6057	0.0614	1.1282	1.0450	0.0587	0.0361	0.0341	0.0466
0.08	0.2815	0.7553	0.0824	1.1709	1.0600	0.0777	0.0573	0.0544	0.0719
0.10	0.3420	0.8852	0.1037	1.2136	1.0750	0.0965	0.0817	0.0779	0.1009
0.12	0.3993	0.9984	0.1254	1.2563	1.0900	0.1150	0.1090	0.1044	0.1332
0.14	0.4537	1.0972	0.1473	1.2990	1.1050	0.1333	0.1389	0.1336	0.1685
0.16	0.5056	1.1838	0.1696	1.3418	1.1200	0.1514	0.1712	0.1653	0.2067
0.18	0.5552	1.2597	0.1922	1.3845	1.1350	0.1693	0.2058	0.1993	0.2476
0.20	0.6026	1.3264	0.2150	1.4272	1.1500	0.1870	0.2424	0.2356	0.2911
0.22	0.6481	1.3850	0.2381	1.4699	1.1650	0.2044	0.2810	0.2739	0.3372
0.24	0.6918	1.4367	0.2616	1.5126	1.1800	0.2217	0.3216	0.3142	0.3857
0.26	0.7338	1.4822	0.2854	1.5554	1.1950	0.2388	0.3640	0.3565	0.4367
0.28	0.7744	1.5224	0.3094	1.5981	1.2100	0.2557	0.4081	0.4007	0.4899
0.30	0.8136	1.5578	0.3338	1.6408	1.2250	0.2724	0.4540	0.4467	0.5455
0.32	0.8515	1.5890	0.3584	1.6835	1.2400	0.2890	0.5015	0.4945	0.6034
0.34	0.8883	1.6166	0.3834	1.7262	1.2550	0.3055	0.5507	0.5440	0.6635
0.36	0.9239	1.6408	0.4086	1.7690	1.2700	0.3217	0.6015	0.5953	0.7258
0.38	0.9586	1.6622	0.4341	1.8117	1.2850	0.3379	0.6539	0.6482	0.7903
0.40	0.9922	1.6809	0.4600	1.8544	1.3000	0.3538	0.7078	0.7028	0.8569
0.42	1.0250	1.6974	0.4862	1.8971	1.3150	0.3697	0.7632	0.7590	0.9257
0.44	1.0570	1.7118	0.5126	1.9398	1.3300	0.3854	0.8202	0.8169	0.9966
0.46	1.0882	1.7243	0.5394	1.9826	1.3450	0.4010	0.8787	0.8763	1.0696
0.48	1.1187	1.7352	0.5664	2.0253	1.3600	0.4165	0.9386	0.9374	1.1447
0.50	1.1485	1.7446	0.5938	2.0680	1.3750	0.4318	1.0000	1.0000	1.2218
0.52	1.1776	1.7527	0.6214	2.1107	1.3900	0.4471	1.0629	1.0642	1.3011
0.54	1.2062	1.7596	0.6494	2.1534	1.4050	0.4622	1.1273	1.1300	1.3824
0.56	1.2342	1.7654	0.6776	2.1962	1.4200	0.4772	1.1931	1.1973	1.4658
0.58	1.2616	1.7703	0.7062	2.2389	1.4350	0.4921	1.2603	1.2662	1.5512
0.60	1.2886	1.7742	0.7350	2.2816	1.4500	0.5069	1.3290	1.3366	1.6387
0.62	1.3151	1.7774	0.7642	2.3243	1.4650	0.5216	1.3991	1.4086	1.7283
0.64	1.3411	1.7799	0.7936	2.3670	1.4800	0.5362	1.4706	1.4822	1.8198
0.66	1.3667	1.7817	0.8234	2.4098	1.4950	0.5507	1.5436	1.5572	1.9134
0.68	1.3919	1.7830	0.8534	2.4525	1.5100	0.5652	1.6180	1.6338	2.0091
0.70	1.4167	1.7837	0.8837	2.4952	1.5250	0.5795	1.6939	1.7120	2.1068
0.72	1.4412	1.7839	0.9144	2.5379	1.5400	0.5938	1.7712	1.7917	2.2065
0.74	1.4653	1.7838	0.9453	2.5806	1.5550	0.6079	1.8499	1.8730	2.3083
0.76	1.4891	1.7832	0.9766	2.6234	1.5700	0.6220	1.9300	1.9558	2.4120
0.78	1.5126	1.7823	1.0081	2.6661	1.5850	0.6361	2.0116	2.0401	2.5179
0.80	1.5357	1.7811	1.0400	2.7088	1.6000	0.6500	2.0946	2.1260	2.6257

Rel. dpth	U.equ. dia. $D_{ep(u)}$	Equiv. disch. factor	Unit sect. area	Unit wetted perim.	Unit surf. brdth	Unit mean depth	Discharge ratios $Q/Q_{0.50}$		U.crit. disch. Q_{uc}
Y	(m)	J	A_u (m^2)	P_u (m)	B_{us} (m)	y_{um} (m)	medial	Mann'g	(m^{3}s^{-1})
0.82	1.5586	1.7795	1.0721	2.7515	1.6150	0.6639	2.1790	2.2135	2.7356
0.84	1.5813	1.7778	1.1046	2.7942	1.6300	0.6777	2.2649	2.3025	2.8476
0.86	1.6036	1.7758	1.1374	2.8370	1.6450	0.6914	2.3522	2.3930	2.9615
0.88	1.6257	1.7736	1.1704	2.8797	1.6600	0.7051	2.4409	2.4852	3.0776
0.90	1.6476	1.7712	1.2037	2.9224	1.6750	0.7187	2.5311	2.5789	3.1956
0.92	1.6693	1.7686	1.2374	2.9651	1.6900	0.7322	2.6228	2.6741	3.3158
0.94	1.6907	1.7659	1.2714	3.0078	1.7050	0.7457	2.7159	2.7710	3.4379
0.96	1.7119	1.7630	1.3056	3.0506	1.7200	0.7591	2.8104	2.8694	3.5621
0.98	1.7330	1.7600	1.3402	3.0933	1.7350	0.7724	2.9064	2.9694	3.6884
1.00	1.7538	1.7569	1.3750	3.1360	1.7500	0.7857	3.0039	3.0710	3.8168
1.02	1.7745	1.7537	1.4101	3.1787	1.7650	0.7990	3.1028	3.1742	3.9472
1.04	1.7950	1.7504	1.4456	3.2214	1.7800	0.8121	3.2032	3.2790	4.0796
1.06	1.8153	1.7471	1.4813	3.2642	1.7950	0.8253	3.3051	3.3854	4.2142
1.08	1.8354	1.7437	1.5174	3.3069	1.8100	0.8383	3.4084	3.4934	4.3508
1.10	1.8554	1.7402	1.5537	3.3496	1.8250	0.8514	3.5133	3.6030	4.4895
1.12	1.8753	1.7366	1.5904	3.3923	1.8400	0.8643	3.6196	3.7143	4.6303
1.14	1.8950	1.7331	1.6273	3.4350	1.8550	0.8773	3.7274	3.8271	4.7732
1.16	1.9146	1.7295	1.6646	3.4778	1.8700	0.8902	3.8367	3.9416	4.9182
1.18	1.9340	1.7258	1.7021	3.5205	1.8850	0.9030	3.9475	4.0578	5.0653
1.20	1.9533	1.7221	1.7400	3.5632	1.9000	0.9158	4.0598	4.1756	5.2144
1.24	1.9915	1.7147	1.8166	3.6486	1.9300	0.9412	4.2890	4.4161	5.5191
1.28	2.0293	1.7073	1.8944	3.7341	1.9600	0.9665	4.5242	4.6633	5.8323
1.32	2.0666	1.6998	1.9734	3.8195	1.9900	0.9917	4.7656	4.9172	6.1540
1.36	2.1036	1.6923	2.0536	3.9050	2.0200	1.0166	5.0131	5.1778	6.4842
1.40	2.1401	1.6849	2.1350	3.9904	2.0500	1.0415	5.2667	5.4452	6.8231
1.44	2.1763	1.6774	2.2176	4.0758	2.0800	1.0662	5.5267	5.7195	7.1706
1.48	2.2122	1.6701	2.3014	4.1613	2.1100	1.0907	5.7929	6.0006	7.5267
1.52	2.2478	1.6628	2.3864	4.2467	2.1400	1.1151	6.0654	6.2887	7.8917
1.56	2.2830	1.6556	2.4726	4.3322	2.1700	1.1394	6.3443	6.5839	8.2653
1.60	2.3180	1.6484	2.5600	4.4176	2.2000	1.1636	6.6296	6.8860	8.6479
1.64	2.3527	1.6414	2.6486	4.5030	2.2300	1.1877	6.9213	7.1953	9.0393
1.68	2.3872	1.6344	2.7384	4.5885	2.2600	1.2117	7.2195	7.5118	9.4396
1.72	2.4214	1.6275	2.8294	4.6739	2.2900	1.2355	7.5242	7.8355	9.8488
1.76	2.4555	1.6208	2.9216	4.7594	2.3200	1.2593	7.8356	8.1664	10.267
1.80	2.4893	1.6141	3.0150	4.8448	2.3500	1.2830	8.1535	8.5046	10.694
1.84	2.5229	1.6076	3.1096	4.9302	2.3800	1.3066	8.4781	8.8503	11.131
1.88	2.5563	1.6011	3.2054	5.0157	2.4100	1.3300	8.8094	9.2033	11.576
1.92	2.5895	1.5948	3.3024	5.1011	2.4400	1.3534	9.1474	9.5639	12.031
1.96	2.6226	1.5885	3.4006	5.1866	2.4700	1.3768	9.4922	9.9319	12.495
2.00	2.6555	1.5824	3.5000	5.2720	2.5000	1.4000	9.8438	10.308	12.969
2.05	2.6965	1.5749	3.6259	5.3788	2.5375	1.4289	10.293	10.788	13.573
2.10	2.7372	1.5675	3.7537	5.4856	2.5750	1.4578	10.753	11.280	14.193
2.15	2.7777	1.5603	3.8834	5.5924	2.6125	1.4865	11.224	11.785	14.827
2.20	2.8179	1.5533	4.0150	5.6992	2.6500	1.5151	11.706	12.302	15.476
2.25	2.8580	1.5464	4.1484	5.8060	2.6875	1.5436	12.199	12.831	16.140
2.30	2.8979	1.5397	4.2838	5.9128	2.7250	1.5720	12.703	13.372	16.820
2.35	2.9377	1.5331	4.4209	6.0196	2.7625	1.6003	13.218	13.926	17.514
2.40	2.9773	1.5267	4.5600	6.1264	2.8000	1.6286	13.745	14.493	18.223
2.45	3.0167	1.5204	4.7009	6.2332	2.8375	1.6567	14.283	15.073	18.948
2.50	3.0560	1.5143	4.8438	6.3400	2.8750	1.6848	14.833	15.665	19.689
2.55	3.0951	1.5083	4.9884	6.4468	2.9125	1.7128	15.394	16.271	20.444
2.60	3.1342	1.5024	5.1350	6.5536	2.9500	1.7407	15.967	16.889	21.216
2.65	3.1730	1.4966	5.2834	6.6604	2.9875	1.7685	16.551	17.521	22.003
2.70	3.2118	1.4910	5.4338	6.7672	3.0250	1.7963	17.147	18.166	22.806
2.75	3.2505	1.4855	5.5859	6.8740	3.0625	1.8240	17.756	18.824	23.625

C50

Trapezoidal - 0.50 to 1 side-slope

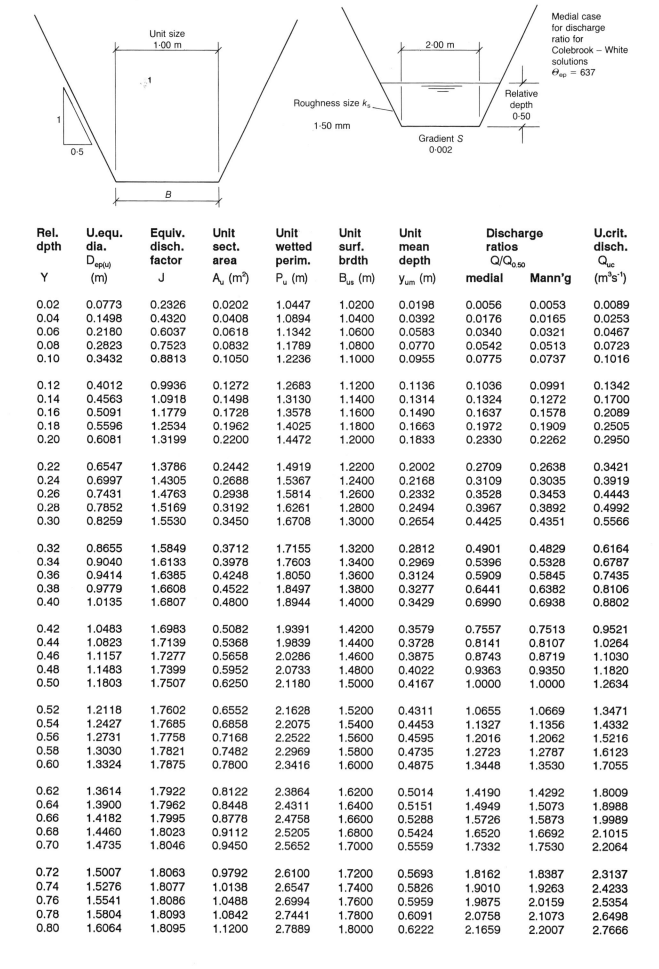

Rel. dpth	U.equ. dia. $D_{ep(u)}$	Equiv. disch. factor	Unit sect. area	Unit wetted perim.	Unit surf. brdth	Unit mean depth	Discharge ratios $Q/Q_{0.50}$		U.crit. disch. Q_{uc}
Y	(m)	J	A_u (m²)	P_u (m)	B_{us} (m)	y_{um} (m)	medial	Mann'g	(m³s⁻¹)
0.02	0.0773	0.2326	0.0202	1.0447	1.0200	0.0198	0.0056	0.0053	0.0089
0.04	0.1498	0.4320	0.0408	1.0894	1.0400	0.0392	0.0176	0.0165	0.0253
0.06	0.2180	0.6037	0.0618	1.1342	1.0600	0.0583	0.0340	0.0321	0.0467
0.08	0.2823	0.7523	0.0832	1.1789	1.0800	0.0770	0.0542	0.0513	0.0723
0.10	0.3432	0.8813	0.1050	1.2236	1.1000	0.0955	0.0775	0.0737	0.1016
0.12	0.4012	0.9936	0.1272	1.2683	1.1200	0.1136	0.1036	0.0991	0.1342
0.14	0.4563	1.0918	0.1498	1.3130	1.1400	0.1314	0.1324	0.1272	0.1700
0.16	0.5091	1.1779	0.1728	1.3578	1.1600	0.1490	0.1637	0.1578	0.2089
0.18	0.5596	1.2534	0.1962	1.4025	1.1800	0.1663	0.1972	0.1909	0.2505
0.20	0.6081	1.3199	0.2200	1.4472	1.2000	0.1833	0.2330	0.2262	0.2950
0.22	0.6547	1.3786	0.2442	1.4919	1.2200	0.2002	0.2709	0.2638	0.3421
0.24	0.6997	1.4305	0.2688	1.5367	1.2400	0.2168	0.3109	0.3035	0.3919
0.26	0.7431	1.4763	0.2938	1.5814	1.2600	0.2332	0.3528	0.3453	0.4443
0.28	0.7852	1.5169	0.3192	1.6261	1.2800	0.2494	0.3967	0.3892	0.4992
0.30	0.8259	1.5530	0.3450	1.6708	1.3000	0.2654	0.4425	0.4351	0.5566
0.32	0.8655	1.5849	0.3712	1.7155	1.3200	0.2812	0.4901	0.4829	0.6164
0.34	0.9040	1.6133	0.3978	1.7603	1.3400	0.2969	0.5396	0.5328	0.6787
0.36	0.9414	1.6385	0.4248	1.8050	1.3600	0.3124	0.5909	0.5845	0.7435
0.38	0.9779	1.6608	0.4522	1.8497	1.3800	0.3277	0.6441	0.6382	0.8106
0.40	1.0135	1.6807	0.4800	1.8944	1.4000	0.3429	0.6990	0.6938	0.8802
0.42	1.0483	1.6983	0.5082	1.9391	1.4200	0.3579	0.7557	0.7513	0.9521
0.44	1.0823	1.7139	0.5368	1.9839	1.4400	0.3728	0.8141	0.8107	1.0264
0.46	1.1157	1.7277	0.5658	2.0286	1.4600	0.3875	0.8743	0.8719	1.1030
0.48	1.1483	1.7399	0.5952	2.0733	1.4800	0.4022	0.9363	0.9350	1.1820
0.50	1.1803	1.7507	0.6250	2.1180	1.5000	0.4167	1.0000	1.0000	1.2634
0.52	1.2118	1.7602	0.6552	2.1628	1.5200	0.4311	1.0655	1.0669	1.3471
0.54	1.2427	1.7685	0.6858	2.2075	1.5400	0.4453	1.1327	1.1356	1.4332
0.56	1.2731	1.7758	0.7168	2.2522	1.5600	0.4595	1.2016	1.2062	1.5216
0.58	1.3030	1.7821	0.7482	2.2969	1.5800	0.4735	1.2723	1.2787	1.6123
0.60	1.3324	1.7875	0.7800	2.3416	1.6000	0.4875	1.3448	1.3530	1.7055
0.62	1.3614	1.7922	0.8122	2.3864	1.6200	0.5014	1.4190	1.4292	1.8009
0.64	1.3900	1.7962	0.8448	2.4311	1.6400	0.5151	1.4949	1.5073	1.8988
0.66	1.4182	1.7995	0.8778	2.4758	1.6600	0.5288	1.5726	1.5873	1.9989
0.68	1.4460	1.8023	0.9112	2.5205	1.6800	0.5424	1.6520	1.6692	2.1015
0.70	1.4735	1.8046	0.9450	2.5652	1.7000	0.5559	1.7332	1.7530	2.2064
0.72	1.5007	1.8063	0.9792	2.6100	1.7200	0.5693	1.8162	1.8387	2.3137
0.74	1.5276	1.8077	1.0138	2.6547	1.7400	0.5826	1.9010	1.9263	2.4233
0.76	1.5541	1.8086	1.0488	2.6994	1.7600	0.5959	1.9875	2.0159	2.5354
0.78	1.5804	1.8093	1.0842	2.7441	1.7800	0.6091	2.0758	2.1073	2.6498
0.80	1.6064	1.8095	1.1200	2.7889	1.8000	0.6222	2.1659	2.2007	2.7666

Trapezoidal - 0.50 to 1 side-slope

Rel. dpth Y	U.equ. dia. $D_{ep(u)}$ (m)	Equiv. disch. factor J	Unit sect. area A_u (m²)	Unit wetted perim. P_u (m)	Unit surf. brdth B_{us} (m)	Unit mean depth y_{um} (m)	Discharge ratios $Q/Q_{0.50}$ medial	Mann'g	U.crit. disch. Q_{uc} (m³s⁻¹)
0.82	1.6321	1.8095	1.1562	2.8336	1.8200	0.6353	2.2578	2.2961	2.8859
0.84	1.6576	1.8092	1.1928	2.8783	1.8400	0.6483	2.3515	2.3934	3.0075
0.86	1.6829	1.8087	1.2298	2.9230	1.8600	0.6612	2.4470	2.4926	3.1315
0.88	1.7080	1.8080	1.2672	2.9677	1.8800	0.6740	2.5443	2.5939	3.2580
0.90	1.7328	1.8070	1.3050	3.0125	1.9000	0.6868	2.6434	2.6971	3.3869
0.92	1.7574	1.8059	1.3432	3.0572	1.9200	0.6996	2.7444	2.8023	3.5182
0.94	1.7819	1.8046	1.3818	3.1019	1.9400	0.7123	2.8472	2.9095	3.6520
0.96	1.8061	1.8032	1.4208	3.1466	1.9600	0.7249	2.9519	3.0187	3.7882
0.98	1.8302	1.8016	1.4602	3.1913	1.9800	0.7375	3.0584	3.1299	3.9269
1.00	1.8541	1.7999	1.5000	3.2361	2.0000	0.7500	3.1668	3.2431	4.0680
1.02	1.8778	1.7981	1.5402	3.2808	2.0200	0.7625	3.2770	3.3584	4.2116
1.04	1.9014	1.7962	1.5808	3.3255	2.0400	0.7749	3.3891	3.4757	4.3577
1.06	1.9249	1.7942	1.6218	3.3702	2.0600	0.7873	3.5031	3.5951	4.5063
1.08	1.9481	1.7922	1.6632	3.4150	2.0800	0.7996	3.6190	3.7165	4.6574
1.10	1.9713	1.7900	1.7050	3.4597	2.1000	0.8119	3.7368	3.8401	4.8110
1.12	1.9943	1.7878	1.7472	3.5044	2.1200	0.8242	3.8565	3.9657	4.9671
1.14	2.0172	1.7855	1.7898	3.5491	2.1400	0.8364	3.9781	4.0934	5.1258
1.16	2.0399	1.7832	1.8328	3.5938	2.1600	0.8485	4.1016	4.2232	5.2870
1.18	2.0626	1.7808	1.8762	3.6386	2.1800	0.8606	4.2271	4.3551	5.4507
1.20	2.0851	1.7784	1.9200	3.6833	2.2000	0.8727	4.3545	4.4892	5.6170
1.24	2.1298	1.7735	2.0088	3.7727	2.2400	0.8968	4.6151	4.7637	5.9572
1.28	2.1741	1.7684	2.0992	3.8622	2.2800	0.9207	4.8837	5.0469	6.3077
1.32	2.2180	1.7633	2.1912	3.9516	2.3200	0.9445	5.1602	5.3388	6.6687
1.36	2.2616	1.7582	2.2848	4.0411	2.3600	0.9681	5.4446	5.6395	7.0401
1.40	2.3048	1.7530	2.3800	4.1305	2.4000	0.9917	5.7372	5.9491	7.4220
1.44	2.3477	1.7477	2.4768	4.2199	2.4400	1.0151	6.0378	6.2676	7.8145
1.48	2.3903	1.7425	2.5752	4.3094	2.4800	1.0384	6.3467	6.5952	8.2177
1.52	2.4327	1.7373	2.6752	4.3988	2.5200	1.0616	6.6639	6.9320	8.6317
1.56	2.4747	1.7322	2.7768	4.4883	2.5600	1.0847	6.9894	7.2780	9.0564
1.60	2.5165	1.7270	2.8800	4.5777	2.6000	1.1077	7.3233	7.6333	9.4921
1.64	2.5581	1.7219	2.9848	4.6672	2.6400	1.1306	7.6656	7.9980	9.9387
1.68	2.5995	1.7169	3.0912	4.7566	2.6800	1.1534	8.0166	8.3721	10.396
1.72	2.6407	1.7119	3.1992	4.8460	2.7200	1.1762	8.3761	8.7559	10.865
1.76	2.6816	1.7069	3.3088	4.9355	2.7600	1.1988	8.7443	9.1493	11.345
1.80	2.7224	1.7020	3.4200	5.0249	2.8000	1.2214	9.1213	9.5524	11.836
1.84	2.7630	1.6972	3.5328	5.1144	2.8400	1.2439	9.5071	9.9653	12.339
1.88	2.8035	1.6925	3.6472	5.2038	2.8800	1.2664	9.9017	10.388	12.853
1.92	2.8438	1.6878	3.7632	5.2933	2.9200	1.2888	10.305	10.821	13.378
1.96	2.8839	1.6831	3.8808	5.3827	2.9600	1.3111	10.718	11.264	13.915
2.00	2.9239	1.6786	4.0000	5.4721	3.0000	1.3333	11.140	11.717	14.464
2.05	2.9737	1.6730	4.1512	5.5839	3.0500	1.3611	11.680	12.298	15.166
2.10	3.0233	1.6675	4.3050	5.6957	3.1000	1.3887	12.234	12.895	15.887
2.15	3.0727	1.6622	4.4613	5.8075	3.1500	1.4163	12.803	13.508	16.626
2.20	3.1220	1.6569	4.6200	5.9193	3.2000	1.4438	13.387	14.138	17.384
2.25	3.1710	1.6517	4.7813	6.0312	3.2500	1.4712	13.985	14.784	18.161
2.30	3.2199	1.6467	4.9450	6.1430	3.3000	1.4985	14.599	15.447	18.956
2.35	3.2687	1.6417	5.1112	6.2548	3.3500	1.5257	15.227	16.127	19.771
2.40	3.3173	1.6369	5.2800	6.3666	3.4000	1.5529	15.871	16.824	20.605
2.45	3.3658	1.6322	5.4513	6.4784	3.4500	1.5801	16.530	17.539	21.458
2.50	3.4142	1.6275	5.6250	6.5902	3.5000	1.6071	17.204	18.271	22.331
2.55	3.4624	1.6230	5.8012	6.7020	3.5500	1.6342	17.894	19.021	23.224
2.60	3.5105	1.6185	5.9800	6.8138	3.6000	1.6611	18.600	19.788	24.136
2.65	3.5585	1.6142	6.1612	6.9256	3.6500	1.6880	19.321	20.573	25.068
2.70	3.6065	1.6099	6.3450	7.0374	3.7000	1.7149	20.058	21.376	26.020
2.75	3.6543	1.6058	6.5312	7.1492	3.7500	1.7417	20.812	22.198	26.992

C51

Trapezoidal - 0.625 to 1 side-slope

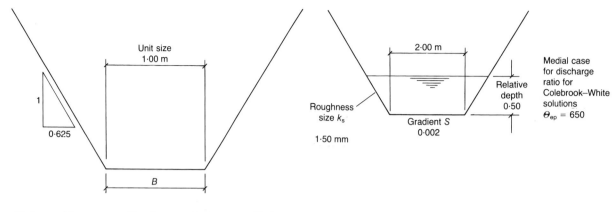

Unit size 1.00 m

Roughness size k_s 1.50 mm

2.00 m

Gradient S 0.002

Relative depth 0.50

Medial case for discharge ratio for Colebrook–White solutions $\Theta_{ep} = 650$

Rel. dpth	U.equ. dia. $D_{ep(u)}$	Equiv. disch. factor	Unit sect. area	Unit wetted perim.	Unit surf. brdth	Unit mean depth	Discharge ratios $Q/Q_{0.50}$		U.crit. disch. Q_{uc}
Y	(m)	J	A_u (m²)	P_u (m)	B_{us} (m)	y_{um} (m)	medial	Mann'g	(m³s⁻¹)
0.02	0.0774	0.2321	0.0203	1.0472	1.0250	0.0198	0.0053	0.0049	0.0089
0.04	0.1499	0.4302	0.0410	1.0943	1.0500	0.0390	0.0166	0.0156	0.0254
0.06	0.2181	0.6003	0.0623	1.1415	1.0750	0.0579	0.0323	0.0304	0.0469
0.08	0.2827	0.7470	0.0840	1.1887	1.1000	0.0764	0.0515	0.0487	0.0727
0.10	0.3439	0.8742	0.1063	1.2358	1.1250	0.0944	0.0738	0.0702	0.1023
0.12	0.4022	0.9847	0.1290	1.2830	1.1500	0.1122	0.0990	0.0946	0.1353
0.14	0.4578	1.0813	0.1523	1.3302	1.1750	0.1296	0.1268	0.1217	0.1716
0.16	0.5111	1.1658	0.1760	1.3774	1.2000	0.1467	0.1572	0.1514	0.2111
0.18	0.5623	1.2400	0.2003	1.4245	1.2250	0.1635	0.1899	0.1836	0.2535
0.20	0.6115	1.3054	0.2250	1.4717	1.2500	0.1800	0.2249	0.2182	0.2989
0.22	0.6590	1.3631	0.2503	1.5189	1.2750	0.1963	0.2622	0.2551	0.3472
0.24	0.7050	1.4142	0.2760	1.5660	1.3000	0.2123	0.3016	0.2943	0.3982
0.26	0.7494	1.4594	0.3023	1.6132	1.3250	0.2281	0.3432	0.3357	0.4521
0.28	0.7926	1.4996	0.3290	1.6604	1.3500	0.2437	0.3868	0.3793	0.5086
0.30	0.8345	1.5354	0.3563	1.7075	1.3750	0.2591	0.4325	0.4250	0.5679
0.32	0.8754	1.5672	0.3840	1.7547	1.4000	0.2743	0.4802	0.4730	0.6298
0.34	0.9152	1.5955	0.4123	1.8019	1.4250	0.2893	0.5300	0.5230	0.6944
0.36	0.9540	1.6208	0.4410	1.8491	1.4500	0.3041	0.5818	0.5752	0.7616
0.38	0.9920	1.6434	0.4702	1.8962	1.4750	0.3188	0.6356	0.6296	0.8315
0.40	1.0291	1.6636	0.5000	1.9434	1.5000	0.3333	0.6913	0.6860	0.9040
0.42	1.0655	1.6816	0.5302	1.9906	1.5250	0.3477	0.7491	0.7446	0.9791
0.44	1.1012	1.6977	0.5610	2.0377	1.5500	0.3619	0.8088	0.8052	1.0569
0.46	1.1363	1.7121	0.5922	2.0849	1.5750	0.3760	0.8706	0.8680	1.1373
0.48	1.1707	1.7250	0.6240	2.1321	1.6000	0.3900	0.9343	0.9330	1.2203
0.50	1.2045	1.7364	0.6563	2.1792	1.6250	0.4038	1.0000	1.0000	1.3060
0.52	1.2379	1.7466	0.6890	2.2264	1.6500	0.4176	1.0677	1.0692	1.3943
0.54	1.2707	1.7558	0.7223	2.2736	1.6750	0.4312	1.1375	1.1405	1.4852
0.56	1.3030	1.7639	0.7560	2.3208	1.7000	0.4447	1.2092	1.2140	1.5788
0.58	1.3349	1.7710	0.7902	2.3679	1.7250	0.4581	1.2829	1.2896	1.6750
0.60	1.3664	1.7774	0.8250	2.4151	1.7500	0.4714	1.3587	1.3674	1.7739
0.62	1.3975	1.7830	0.8603	2.4623	1.7750	0.4846	1.4364	1.4473	1.8754
0.64	1.4282	1.7879	0.8960	2.5094	1.8000	0.4978	1.5163	1.5295	1.9796
0.66	1.4586	1.7923	0.9323	2.5566	1.8250	0.5108	1.5981	1.6139	2.0865
0.68	1.4886	1.7960	0.9690	2.6038	1.8500	0.5238	1.6820	1.7004	2.1961
0.70	1.5183	1.7993	1.0063	2.6509	1.8750	0.5367	1.7680	1.7892	2.3084
0.72	1.5477	1.8021	1.0440	2.6981	1.9000	0.5495	1.8561	1.8803	2.4235
0.74	1.5769	1.8045	1.0822	2.7453	1.9250	0.5622	1.9462	1.9735	2.5412
0.76	1.6058	1.8065	1.1210	2.7925	1.9500	0.5749	2.0384	2.0691	2.6617
0.78	1.6344	1.8081	1.1602	2.8396	1.9750	0.5875	2.1327	2.1669	2.7849
0.80	1.6627	1.8095	1.2000	2.8868	2.0000	0.6000	2.2292	2.2670	2.9108

Rel. dpth Y	U.equ. dia. $D_{ep(u)}$ (m)	Equiv. disch. factor J	Unit sect. area A_u (m²)	Unit wetted perim. P_u (m)	Unit surf. brdth B_{us} (m)	Unit mean depth y_{um} (m)	Discharge ratios $Q/Q_{0.50}$ medial	Mann'g	U.crit. disch. Q_{uc} (m³s⁻¹)
0.82	1.6909	1.8105	1.2402	2.9340	2.0250	0.6125	2.3277	2.3694	3.0396
0.84	1.7188	1.8113	1.2810	2.9811	2.0500	0.6249	2.4284	2.4741	3.1711
0.86	1.7465	1.8118	1.3222	3.0283	2.0750	0.6372	2.5313	2.5811	3.3054
0.88	1.7740	1.8121	1.3640	3.0755	2.1000	0.6495	2.6363	2.6905	3.4425
0.90	1.8014	1.8122	1.4062	3.1226	2.1250	0.6618	2.7435	2.8023	3.5824
0.92	1.8285	1.8122	1.4490	3.1698	2.1500	0.6740	2.8528	2.9164	3.7251
0.94	1.8555	1.8119	1.4922	3.2170	2.1750	0.6861	2.9644	3.0329	3.8707
0.96	1.8823	1.8115	1.5360	3.2642	2.2000	0.6982	3.0781	3.1518	4.0192
0.98	1.9089	1.8110	1.5803	3.3113	2.2250	0.7102	3.1941	3.2731	4.1705
1.00	1.9354	1.8104	1.6250	3.3585	2.2500	0.7222	3.3123	3.3969	4.3246
1.02	1.9617	1.8096	1.6702	3.4057	2.2750	0.7342	3.4327	3.5231	4.4817
1.04	1.9879	1.8087	1.7160	3.4528	2.3000	0.7461	3.5554	3.6517	4.6416
1.06	2.0140	1.8077	1.7622	3.5000	2.3250	0.7580	3.6804	3.7829	4.8045
1.08	2.0399	1.8066	1.8090	3.5472	2.3500	0.7698	3.8076	3.9165	4.9703
1.10	2.0657	1.8055	1.8562	3.5943	2.3750	0.7816	3.9371	4.0526	5.1391
1.12	2.0914	1.8043	1.9040	3.6415	2.4000	0.7933	4.0689	4.1913	5.3107
1.14	2.1170	1.8030	1.9522	3.6887	2.4250	0.8051	4.2030	4.3325	5.4854
1.16	2.1425	1.8016	2.0010	3.7359	2.4500	0.8167	4.3395	4.4762	5.6630
1.18	2.1678	1.8002	2.0502	3.7830	2.4750	0.8284	4.4783	4.6225	5.8436
1.20	2.1931	1.7988	2.1000	3.8302	2.5000	0.8400	4.6194	4.7713	6.0273
1.24	2.2433	1.7957	2.2010	3.9245	2.5500	0.8631	4.9087	5.0769	6.4035
1.28	2.2932	1.7926	2.3040	4.0189	2.6000	0.8862	5.2076	5.3929	6.7920
1.32	2.3427	1.7893	2.4090	4.1132	2.6500	0.9091	5.5161	5.7196	7.1927
1.36	2.3919	1.7859	2.5160	4.2076	2.7000	0.9319	5.8344	6.0569	7.6058
1.40	2.4408	1.7824	2.6250	4.3019	2.7500	0.9545	6.1625	6.4052	8.0313
1.44	2.4894	1.7789	2.7360	4.3962	2.8000	0.9771	6.5004	6.7644	8.4695
1.48	2.5378	1.7754	2.8490	4.4906	2.8500	0.9996	6.8484	7.1347	8.9202
1.52	2.5859	1.7718	2.9640	4.5849	2.9000	1.0221	7.2065	7.5162	9.3838
1.56	2.6338	1.7682	3.0810	4.6793	2.9500	1.0444	7.5749	7.9090	9.8602
1.60	2.6814	1.7647	3.2000	4.7736	3.0000	1.0667	7.9535	8.3133	10.350
1.64	2.7289	1.7611	3.3210	4.8679	3.0500	1.0889	8.3425	8.7292	10.852
1.68	2.7761	1.7575	3.4440	4.9623	3.1000	1.1110	8.7420	9.1567	11.368
1.72	2.8232	1.7540	3.5690	5.0566	3.1500	1.1330	9.1520	9.5961	11.897
1.76	2.8701	1.7505	3.6960	5.1510	3.2000	1.1550	9.5728	10.047	12.439
1.80	2.9169	1.7470	3.8250	5.2453	3.2500	1.1769	10.004	10.511	12.995
1.84	2.9635	1.7435	3.9560	5.3396	3.3000	1.1988	10.447	10.986	13.564
1.88	3.0100	1.7401	4.0890	5.4340	3.3500	1.2206	10.900	11.474	14.147
1.92	3.0563	1.7368	4.2240	5.5283	3.4000	1.2424	11.364	11.974	14.744
1.96	3.1025	1.7334	4.3610	5.6227	3.4500	1.2641	11.840	12.486	15.354
2.00	3.1485	1.7301	4.5000	5.7170	3.5000	1.2857	12.327	13.012	15.979
2.05	3.2059	1.7261	4.6766	5.8349	3.5625	1.3127	12.951	13.686	16.779
2.10	3.2631	1.7221	4.8562	5.9528	3.6250	1.3397	13.593	14.381	17.602
2.15	3.3202	1.7182	5.0391	6.0708	3.6875	1.3665	14.253	15.095	18.447
2.20	3.3771	1.7143	5.2250	6.1887	3.7500	1.3933	14.932	15.831	19.314
2.25	3.4339	1.7105	5.4141	6.3066	3.8125	1.4201	15.628	16.587	20.204
2.30	3.4905	1.7068	5.6062	6.4245	3.8750	1.4468	16.344	17.364	21.117
2.35	3.5470	1.7032	5.8016	6.5425	3.9375	1.4734	17.078	18.162	22.053
2.40	3.6034	1.6996	6.0000	6.6604	4.0000	1.5000	17.831	18.982	23.012
2.45	3.6596	1.6961	6.2016	6.7783	4.0625	1.5265	18.602	19.823	23.995
2.50	3.7158	1.6927	6.4063	6.8962	4.1250	1.5530	19.394	20.687	25.001
2.55	3.7718	1.6893	6.6141	7.0142	4.1875	1.5795	20.204	21.572	26.031
2.60	3.8278	1.6860	6.8250	7.1321	4.2500	1.6059	21.034	22.479	27.084
2.65	3.8836	1.6828	7.0391	7.2500	4.3125	1.6322	21.884	23.409	28.162
2.70	3.9394	1.6797	7.2563	7.3679	4.3750	1.6586	22.754	24.362	29.264
2.75	3.9950	1.6766	7.4766	7.4859	4.4375	1.6849	23.643	25.338	30.391

C52

Trapezoidal - 0.75 to 1 side-slope

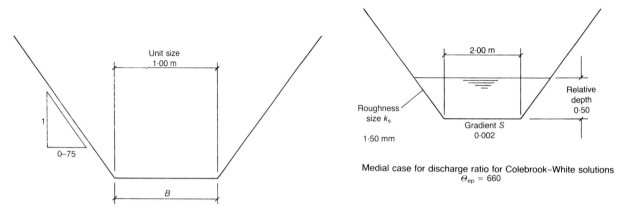

Medial case for discharge ratio for Colebrook–White solutions
$\Theta_{ep} = 660$

Rel. dpth Y	U.equ. dia. $D_{ep(u)}$ (m)	Equiv. disch. factor J	Unit sect. area A_u (m²)	Unit wetted perim. P_u (m)	Unit surf. brdth B_{us} (m)	Unit mean depth y_{um} (m)	Discharge ratios $Q/Q_{0.50}$ medial	Mann'g	U.crit. disch. Q_{uc} (m³s⁻¹)
0.02	0.0773	0.2314	0.0203	1.0500	1.0300	0.0197	0.0050	0.0047	0.0089
0.04	0.1498	0.4279	0.0412	1.1000	1.0600	0.0389	0.0158	0.0148	0.0254
0.06	0.2181	0.5958	0.0627	1.1500	1.0900	0.0575	0.0307	0.0289	0.0471
0.08	0.2827	0.7400	0.0848	1.2000	1.1200	0.0757	0.0492	0.0465	0.0731
0.10	0.3440	0.8645	0.1075	1.2500	1.1500	0.0935	0.0707	0.0672	0.1029
0.12	0.4025	0.9726	0.1308	1.3000	1.1800	0.1108	0.0950	0.0907	0.1364
0.14	0.4584	1.0667	0.1547	1.3500	1.2100	0.1279	0.1220	0.1170	0.1732
0.16	0.5120	1.1489	0.1792	1.4000	1.2400	0.1445	0.1516	0.1459	0.2133
0.18	0.5636	1.2210	0.2043	1.4500	1.2700	0.1609	0.1836	0.1774	0.2566
0.20	0.6133	1.2845	0.2300	1.5000	1.3000	0.1769	0.2179	0.2113	0.3030
0.22	0.6614	1.3406	0.2563	1.5500	1.3300	0.1927	0.2546	0.2476	0.3523
0.24	0.7080	1.3901	0.2832	1.6000	1.3600	0.2082	0.2936	0.2862	0.4047
0.26	0.7532	1.4341	0.3107	1.6500	1.3900	0.2235	0.3348	0.3273	0.4600
0.28	0.7972	1.4731	0.3388	1.7000	1.4200	0.2386	0.3782	0.3706	0.5182
0.30	0.8400	1.5079	0.3675	1.7500	1.4500	0.2534	0.4238	0.4163	0.5794
0.32	0.8818	1.5390	0.3968	1.8000	1.4800	0.2681	0.4716	0.4643	0.6434
0.34	0.9226	1.5667	0.4267	1.8500	1.5100	0.2826	0.5216	0.5145	0.7103
0.36	0.9625	1.5915	0.4572	1.9000	1.5400	0.2969	0.5737	0.5671	0.7801
0.38	1.0016	1.6137	0.4883	1.9500	1.5700	0.3110	0.6281	0.6220	0.8528
0.40	1.0400	1.6336	0.5200	2.0000	1.6000	0.3250	0.6846	0.6792	0.9283
0.42	1.0777	1.6515	0.5523	2.0500	1.6300	0.3388	0.7433	0.7387	1.0068
0.44	1.1147	1.6675	0.5852	2.1000	1.6600	0.3525	0.8042	0.8005	1.0881
0.46	1.1511	1.6819	0.6187	2.1500	1.6900	0.3661	0.8673	0.8647	1.1723
0.48	1.1869	1.6949	0.6528	2.2000	1.7200	0.3795	0.9325	0.9311	1.2594
0.50	1.2222	1.7065	0.6875	2.2500	1.7500	0.3929	1.0000	1.0000	1.3494
0.52	1.2570	1.7170	0.7228	2.3000	1.7800	0.4061	1.0697	1.0712	1.4424
0.54	1.2914	1.7264	0.7587	2.3500	1.8100	0.4192	1.1417	1.1448	1.5382
0.56	1.3253	1.7348	0.7952	2.4000	1.8400	0.4322	1.2158	1.2208	1.6371
0.58	1.3589	1.7424	0.8323	2.4500	1.8700	0.4451	1.2923	1.2992	1.7388
0.60	1.3920	1.7492	0.8700	2.5000	1.9000	0.4579	1.3710	1.3801	1.8436
0.62	1.4248	1.7553	0.9083	2.5500	1.9300	0.4706	1.4519	1.4634	1.9513
0.64	1.4572	1.7607	0.9472	2.6000	1.9600	0.4833	1.5352	1.5491	2.0620
0.66	1.4894	1.7656	0.9867	2.6500	1.9900	0.4958	1.6208	1.6374	2.1758
0.68	1.5212	1.7699	1.0268	2.7000	2.0200	0.5083	1.7087	1.7281	2.2925
0.70	1.5527	1.7738	1.0675	2.7500	2.0500	0.5207	1.7989	1.8213	2.4123
0.72	1.5840	1.7772	1.1088	2.8000	2.0800	0.5331	1.8915	1.9171	2.5352
0.74	1.6150	1.7802	1.1507	2.8500	2.1100	0.5454	1.9864	2.0155	2.6611
0.76	1.6458	1.7829	1.1932	2.9000	2.1400	0.5576	2.0838	2.1164	2.7901
0.78	1.6763	1.7852	1.2363	2.9500	2.1700	0.5697	2.1835	2.2199	2.9222
0.80	1.7067	1.7872	1.2800	3.0000	2.2000	0.5818	2.2856	2.3260	3.0575

Rel. dpth Y	U.equ. dia. $D_{ep(u)}$ (m)	Equiv. disch. factor J	Unit sect. area A_u (m²)	Unit wetted perim. P_u (m)	Unit surf. brdth B_{us} (m)	Unit mean depth y_{um} (m)	Discharge ratios $Q/Q_{0.50}$ medial	Mann'g	U.crit. disch. Q_{uc} (m³s⁻¹)
0.82	1.7368	1.7889	1.3243	3.0500	2.2300	0.5939	2.3901	2.4347	3.1959
0.84	1.7667	1.7904	1.3692	3.1000	2.2600	0.6058	2.4971	2.5461	3.3374
0.86	1.7964	1.7916	1.4147	3.1500	2.2900	0.6178	2.6066	2.6601	3.4821
0.88	1.8260	1.7926	1.4608	3.2000	2.3200	0.6297	2.7185	2.7768	3.6300
0.90	1.8554	1.7935	1.5075	3.2500	2.3500	0.6415	2.8329	2.8963	3.7810
0.92	1.8846	1.7941	1.5548	3.3000	2.3800	0.6533	2.9498	3.0184	3.9354
0.94	1.9137	1.7946	1.6027	3.3500	2.4100	0.6650	3.0692	3.1433	4.0929
0.96	1.9426	1.7949	1.6512	3.4000	2.4400	0.6767	3.1912	3.2710	4.2537
0.98	1.9714	1.7951	1.7003	3.4500	2.4700	0.6884	3.3157	3.4014	4.4177
1.00	2.0000	1.7952	1.7500	3.5000	2.5000	0.7000	3.4428	3.5347	4.5851
1.02	2.0285	1.7951	1.8003	3.5500	2.5300	0.7116	3.5724	3.6708	4.7557
1.04	2.0569	1.7949	1.8512	3.6000	2.5600	0.7231	3.7047	3.8097	4.9297
1.06	2.0852	1.7947	1.9027	3.6500	2.5900	0.7346	3.8395	3.9514	5.1070
1.08	2.1133	1.7943	1.9548	3.7000	2.6200	0.7461	3.9770	4.0961	5.2877
1.10	2.1413	1.7939	2.0075	3.7500	2.6500	0.7575	4.1172	4.2436	5.4717
1.12	2.1693	1.7934	2.0608	3.8000	2.6800	0.7690	4.2600	4.3941	5.6591
1.14	2.1971	1.7928	2.1147	3.8500	2.7100	0.7803	4.4054	4.5475	5.8499
1.16	2.2248	1.7921	2.1692	3.9000	2.7400	0.7917	4.5536	4.7039	6.0441
1.18	2.2525	1.7914	2.2243	3.9500	2.7700	0.8030	4.7045	4.8632	6.2418
1.20	2.2800	1.7907	2.2800	4.0000	2.8000	0.8143	4.8581	5.0256	6.4429
1.24	2.3348	1.7890	2.3932	4.1000	2.8600	0.8368	5.1735	5.3593	6.8556
1.28	2.3893	1.7872	2.5088	4.2000	2.9200	0.8592	5.5000	5.7053	7.2823
1.32	2.4435	1.7852	2.6268	4.3000	2.9800	0.8815	5.8377	6.0636	7.7231
1.36	2.4975	1.7831	2.7472	4.4000	3.0400	0.9037	6.1868	6.4345	8.1782
1.40	2.5511	1.7810	2.8700	4.5000	3.1000	0.9258	6.5473	6.8181	8.6477
1.44	2.6045	1.7787	2.9952	4.6000	3.1600	0.9478	6.9193	7.2145	9.1318
1.48	2.6577	1.7764	3.1228	4.7000	3.2200	0.9698	7.3031	7.6239	9.6305
1.52	2.7107	1.7741	3.2528	4.8000	3.2800	0.9917	7.6987	8.0464	10.144
1.56	2.7634	1.7717	3.3852	4.9000	3.3400	1.0135	8.1062	8.4822	10.672
1.60	2.8160	1.7693	3.5200	5.0000	3.4000	1.0353	8.5258	8.9315	11.216
1.64	2.8684	1.7669	3.6572	5.1000	3.4600	1.0570	8.9576	9.3944	11.775
1.68	2.9206	1.7645	3.7968	5.2000	3.5200	1.0786	9.4016	9.8710	12.349
1.72	2.9727	1.7620	3.9388	5.3000	3.5800	1.1002	9.8581	10.362	12.938
1.76	3.0246	1.7596	4.0832	5.4000	3.6400	1.1218	10.327	10.866	13.543
1.80	3.0764	1.7572	4.2300	5.5000	3.7000	1.1432	10.809	11.385	14.163
1.84	3.1280	1.7548	4.3792	5.6000	3.7600	1.1647	11.303	11.918	14.800
1.88	3.1795	1.7524	4.5308	5.7000	3.8200	1.1861	11.811	12.465	15.452
1.92	3.2309	1.7500	4.6848	5.8000	3.8800	1.2074	12.331	13.028	16.121
1.96	3.2822	1.7476	4.8412	5.9000	3.9400	1.2287	12.864	13.605	16.805
2.00	3.3333	1.7453	5.0000	6.0000	4.0000	1.2500	13.411	14.197	17.506
2.05	3.3971	1.7424	5.2019	6.1250	4.0750	1.2765	14.113	14.958	18.405
2.10	3.4608	1.7395	5.4075	6.2500	4.1500	1.3030	14.836	15.743	19.330
2.15	3.5243	1.7367	5.6169	6.3750	4.2250	1.3294	15.581	16.552	20.281
2.20	3.5877	1.7340	5.8300	6.5000	4.3000	1.3558	16.347	17.385	21.258
2.25	3.6509	1.7312	6.0469	6.6250	4.3750	1.3821	17.134	18.243	22.262
2.30	3.7141	1.7286	6.2675	6.7500	4.4500	1.4084	17.944	19.126	23.293
2.35	3.7771	1.7259	6.4919	6.8750	4.5250	1.4347	18.776	20.034	24.350
2.40	3.8400	1.7233	6.7200	7.0000	4.6000	1.4609	19.630	20.968	25.435
2.45	3.9028	1.7208	6.9519	7.1250	4.6750	1.4870	20.507	21.927	26.547
2.50	3.9655	1.7183	7.1875	7.2500	4.7500	1.5132	21.406	22.912	27.687
2.55	4.0281	1.7159	7.4269	7.3750	4.8250	1.5392	22.329	23.924	28.855
2.60	4.0907	1.7135	7.6700	7.5000	4.9000	1.5653	23.274	24.962	30.051
2.65	4.1531	1.7111	7.9169	7.6250	4.9750	1.5913	24.244	26.027	31.275
2.70	4.2155	1.7088	8.1675	7.7500	5.0500	1.6173	25.236	27.119	32.527
2.75	4.2778	1.7065	8.4219	7.8750	5.1250	1.6433	26.253	28.239	33.809

C53

Trapezoidal - 0.875 to 1 side-slope

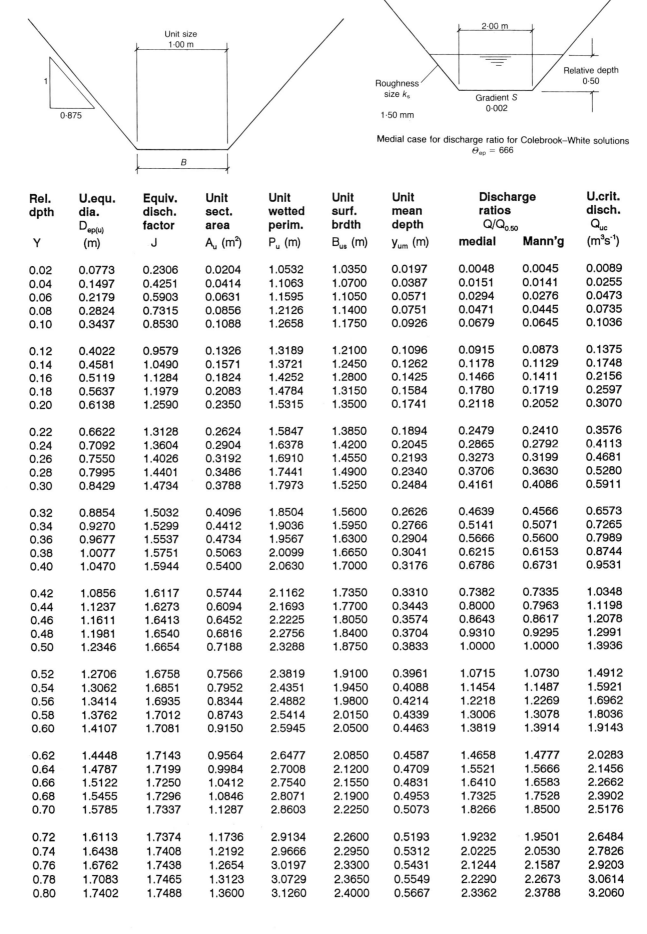

Unit size 1·00 m

0.875

1

B

2·00 m

Roughness size k_s
1·50 mm

Gradient S
0·002

Relative depth 0·50

Medial case for discharge ratio for Colebrook–White solutions
$\Theta_{ep} = 666$

Rel. dpth Y	U.equ. dia. $D_{ep(u)}$ (m)	Equiv. disch. factor J	Unit sect. area A_u (m²)	Unit wetted perim. P_u (m)	Unit surf. brdth B_{us} (m)	Unit mean depth y_{um} (m)	Discharge ratios $Q/Q_{0.50}$ medial	Mann'g	U.crit. disch. Q_{uc} (m³s⁻¹)
0.02	0.0773	0.2306	0.0204	1.0532	1.0350	0.0197	0.0048	0.0045	0.0089
0.04	0.1497	0.4251	0.0414	1.1063	1.0700	0.0387	0.0151	0.0141	0.0255
0.06	0.2179	0.5903	0.0631	1.1595	1.1050	0.0571	0.0294	0.0276	0.0473
0.08	0.2824	0.7315	0.0856	1.2126	1.1400	0.0751	0.0471	0.0445	0.0735
0.10	0.3437	0.8530	0.1088	1.2658	1.1750	0.0926	0.0679	0.0645	0.1036
0.12	0.4022	0.9579	0.1326	1.3189	1.2100	0.1096	0.0915	0.0873	0.1375
0.14	0.4581	1.0490	0.1571	1.3721	1.2450	0.1262	0.1178	0.1129	0.1748
0.16	0.5119	1.1284	0.1824	1.4252	1.2800	0.1425	0.1466	0.1411	0.2156
0.18	0.5637	1.1979	0.2083	1.4784	1.3150	0.1584	0.1780	0.1719	0.2597
0.20	0.6138	1.2590	0.2350	1.5315	1.3500	0.1741	0.2118	0.2052	0.3070
0.22	0.6622	1.3128	0.2624	1.5847	1.3850	0.1894	0.2479	0.2410	0.3576
0.24	0.7092	1.3604	0.2904	1.6378	1.4200	0.2045	0.2865	0.2792	0.4113
0.26	0.7550	1.4026	0.3192	1.6910	1.4550	0.2193	0.3273	0.3199	0.4681
0.28	0.7995	1.4401	0.3486	1.7441	1.4900	0.2340	0.3706	0.3630	0.5280
0.30	0.8429	1.4734	0.3788	1.7973	1.5250	0.2484	0.4161	0.4086	0.5911
0.32	0.8854	1.5032	0.4096	1.8504	1.5600	0.2626	0.4639	0.4566	0.6573
0.34	0.9270	1.5299	0.4412	1.9036	1.5950	0.2766	0.5141	0.5071	0.7265
0.36	0.9677	1.5537	0.4734	1.9567	1.6300	0.2904	0.5666	0.5600	0.7989
0.38	1.0077	1.5751	0.5063	2.0099	1.6650	0.3041	0.6215	0.6153	0.8744
0.40	1.0470	1.5944	0.5400	2.0630	1.7000	0.3176	0.6786	0.6731	0.9531
0.42	1.0856	1.6117	0.5744	2.1162	1.7350	0.3310	0.7382	0.7335	1.0348
0.44	1.1237	1.6273	0.6094	2.1693	1.7700	0.3443	0.8000	0.7963	1.1198
0.46	1.1611	1.6413	0.6452	2.2225	1.8050	0.3574	0.8643	0.8617	1.2078
0.48	1.1981	1.6540	0.6816	2.2756	1.8400	0.3704	0.9310	0.9295	1.2991
0.50	1.2346	1.6654	0.7188	2.3288	1.8750	0.3833	1.0000	1.0000	1.3936
0.52	1.2706	1.6758	0.7566	2.3819	1.9100	0.3961	1.0715	1.0730	1.4912
0.54	1.3062	1.6851	0.7952	2.4351	1.9450	0.4088	1.1454	1.1487	1.5921
0.56	1.3414	1.6935	0.8344	2.4882	1.9800	0.4214	1.2218	1.2269	1.6962
0.58	1.3762	1.7012	0.8743	2.5414	2.0150	0.4339	1.3006	1.3078	1.8036
0.60	1.4107	1.7081	0.9150	2.5945	2.0500	0.4463	1.3819	1.3914	1.9143
0.62	1.4448	1.7143	0.9564	2.6477	2.0850	0.4587	1.4658	1.4777	2.0283
0.64	1.4787	1.7199	0.9984	2.7008	2.1200	0.4709	1.5521	1.5666	2.1456
0.66	1.5122	1.7250	1.0412	2.7540	2.1550	0.4831	1.6410	1.6583	2.2662
0.68	1.5455	1.7296	1.0846	2.8071	2.1900	0.4953	1.7325	1.7528	2.3902
0.70	1.5785	1.7337	1.1287	2.8603	2.2250	0.5073	1.8266	1.8500	2.5176
0.72	1.6113	1.7374	1.1736	2.9134	2.2600	0.5193	1.9232	1.9501	2.6484
0.74	1.6438	1.7408	1.2192	2.9666	2.2950	0.5312	2.0225	2.0530	2.7826
0.76	1.6762	1.7438	1.2654	3.0197	2.3300	0.5431	2.1244	2.1587	2.9203
0.78	1.7083	1.7465	1.3123	3.0729	2.3650	0.5549	2.2290	2.2673	3.0614
0.80	1.7402	1.7488	1.3600	3.1260	2.4000	0.5667	2.3362	2.3788	3.2060

Rel. dpth Y	U.equ. dia. $D_{ep(u)}$ (m)	Equiv. disch. factor J	Unit sect. area A_u (m²)	Unit wetted perim. P_u (m)	Unit surf. brdth B_{us} (m)	Unit mean depth y_{um} (m)	Discharge ratios $Q/Q_{0.50}$ medial	Mann'g	U.crit. disch. Q_{uc} (m³s⁻¹)
0.82	1.7720	1.7510	1.4083	3.1792	2.4350	0.5784	2.4461	2.4932	3.3541
0.84	1.8035	1.7529	1.4574	3.2323	2.4700	0.5900	2.5588	2.6106	3.5057
0.86	1.8349	1.7545	1.5072	3.2855	2.5050	0.6017	2.6742	2.7310	3.6609
0.88	1.8662	1.7560	1.5576	3.3386	2.5400	0.6132	2.7924	2.8543	3.8197
0.90	1.8972	1.7572	1.6087	3.3918	2.5750	0.6248	2.9133	2.9807	3.9820
0.92	1.9282	1.7583	1.6606	3.4449	2.6100	0.6362	3.0370	3.1101	4.1480
0.94	1.9590	1.7593	1.7131	3.4981	2.6450	0.6477	3.1636	3.2426	4.3176
0.96	1.9896	1.7601	1.7664	3.5512	2.6800	0.6591	3.2930	3.3782	4.4908
0.98	2.0202	1.7607	1.8204	3.6044	2.7150	0.6705	3.4252	3.5169	4.6678
1.00	2.0506	1.7613	1.8750	3.6575	2.7500	0.6818	3.5603	3.6587	4.8484
1.02	2.0809	1.7617	1.9303	3.7107	2.7850	0.6931	3.6983	3.8037	5.0327
1.04	2.1110	1.7620	1.9864	3.7638	2.8200	0.7044	3.8393	3.9520	5.2208
1.06	2.1411	1.7622	2.0431	3.8170	2.8550	0.7156	3.9831	4.1034	5.4126
1.08	2.1711	1.7623	2.1006	3.8701	2.8900	0.7269	4.1299	4.2580	5.6082
1.10	2.2010	1.7624	2.1588	3.9233	2.9250	0.7380	4.2797	4.4160	5.8077
1.12	2.2307	1.7624	2.2176	3.9764	2.9600	0.7492	4.4325	4.5772	6.0109
1.14	2.2604	1.7623	2.2771	4.0296	2.9950	0.7603	4.5883	4.7417	6.2180
1.16	2.2900	1.7621	2.3374	4.0827	3.0300	0.7714	4.7471	4.9095	6.4289
1.18	2.3195	1.7619	2.3983	4.1359	3.0650	0.7825	4.9090	5.0808	6.6438
1.20	2.3490	1.7616	2.4600	4.1890	3.1000	0.7935	5.0739	5.2554	6.8625
1.24	2.4076	1.7609	2.5854	4.2953	3.1700	0.8156	5.4131	5.6148	7.3118
1.28	2.4660	1.7600	2.7136	4.4016	3.2400	0.8375	5.7648	5.9881	7.7769
1.32	2.5241	1.7590	2.8446	4.5079	3.3100	0.8594	6.1291	6.3753	8.2581
1.36	2.5819	1.7578	2.9784	4.6142	3.3800	0.8812	6.5062	6.7768	8.7554
1.40	2.6395	1.7566	3.1150	4.7206	3.4500	0.9029	6.8963	7.1927	9.2691
1.44	2.6969	1.7553	3.2544	4.8269	3.5200	0.9245	7.2996	7.6231	9.7993
1.48	2.7541	1.7539	3.3966	4.9332	3.5900	0.9461	7.7160	8.0682	10.346
1.52	2.8111	1.7524	3.5416	5.0395	3.6600	0.9677	8.1459	8.5283	10.910
1.56	2.8679	1.7509	3.6894	5.1458	3.7300	0.9891	8.5893	9.0036	11.491
1.60	2.9246	1.7493	3.8400	5.2521	3.8000	1.0105	9.0465	9.4941	12.088
1.64	2.9811	1.7477	3.9934	5.3584	3.8700	1.0319	9.5174	10.000	12.703
1.68	3.0374	1.7461	4.1496	5.4647	3.9400	1.0532	10.002	10.522	13.336
1.72	3.0936	1.7445	4.3086	5.5710	4.0100	1.0745	10.501	11.059	13.986
1.76	3.1497	1.7429	4.4704	5.6773	4.0800	1.0957	11.015	11.613	14.654
1.80	3.2056	1.7412	4.6350	5.7836	4.1500	1.1169	11.543	12.183	15.339
1.84	3.2615	1.7396	4.8024	5.8899	4.2200	1.1380	12.085	12.769	16.043
1.88	3.3172	1.7379	4.9726	5.9962	4.2900	1.1591	12.642	13.371	16.765
1.92	3.3728	1.7363	5.1456	6.1025	4.3600	1.1802	13.214	13.991	17.505
1.96	3.4283	1.7347	5.3214	6.2088	4.4300	1.2012	13.800	14.627	18.264
2.00	3.4837	1.7330	5.5000	6.3151	4.5000	1.2222	14.402	15.281	19.041
2.05	3.5529	1.7310	5.7272	6.4479	4.5875	1.2484	15.176	16.122	20.039
2.10	3.6219	1.7290	5.9587	6.5808	4.6750	1.2746	15.974	16.990	21.067
2.15	3.6908	1.7270	6.1947	6.7137	4.7625	1.3007	16.796	17.886	22.124
2.20	3.7595	1.7250	6.4350	6.8466	4.8500	1.3268	17.643	18.810	23.212
2.25	3.8282	1.7231	6.6797	6.9795	4.9375	1.3528	18.514	19.762	24.330
2.30	3.8968	1.7212	6.9287	7.1123	5.0250	1.3789	19.411	20.743	25.479
2.35	3.9652	1.7193	7.1822	7.2452	5.1125	1.4048	20.333	21.753	26.658
2.40	4.0336	1.7175	7.4400	7.3781	5.2000	1.4308	21.281	22.792	27.869
2.45	4.1018	1.7156	7.7022	7.5110	5.2875	1.4567	22.254	23.861	29.111
2.50	4.1700	1.7138	7.9688	7.6438	5.3750	1.4826	23.254	24.959	30.385
2.55	4.2381	1.7121	8.2397	7.7767	5.4625	1.5084	24.280	26.088	31.691
2.60	4.3062	1.7103	8.5150	7.9096	5.5500	1.5342	25.333	27.247	33.029
2.65	4.3741	1.7086	8.7947	8.0425	5.6375	1.5600	26.412	28.438	34.399
2.70	4.4420	1.7069	9.0788	8.1753	5.7250	1.5858	27.519	29.659	35.802
2.75	4.5098	1.7053	9.3672	8.3082	5.8125	1.6116	28.653	30.912	37.239

C54

Trapezoidal - 1.0 to 1 side-slope

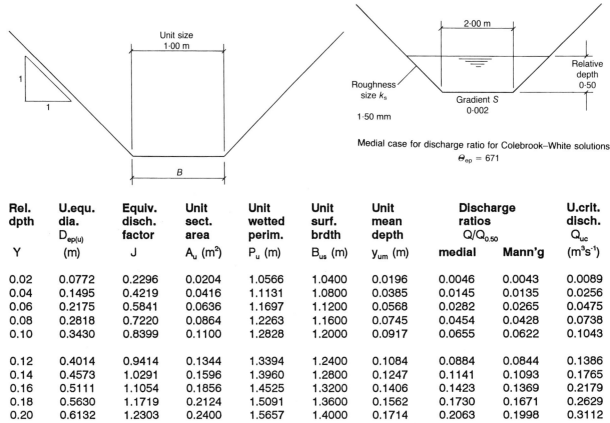

Unit size 1·00 m

B

2·00 m

Relative depth 0·50

Roughness size k_s
1·50 mm

Gradient S
0·002

Medial case for discharge ratio for Colebrook–White solutions
$\Theta_{ep} = 671$

Rel. dpth	U.equ. dia. $D_{ep(u)}$	Equiv. disch. factor	Unit sect. area	Unit wetted perim.	Unit surf. brdth	Unit mean depth	Discharge ratios $Q/Q_{0.50}$		U.crit. disch. Q_{uc}
Y	(m)	J	A_u (m²)	P_u (m)	B_{us} (m)	y_{um} (m)	medial	Mann'g	(m³s⁻¹)
0.02	0.0772	0.2296	0.0204	1.0566	1.0400	0.0196	0.0046	0.0043	0.0089
0.04	0.1495	0.4219	0.0416	1.1131	1.0800	0.0385	0.0145	0.0135	0.0256
0.06	0.2175	0.5841	0.0636	1.1697	1.1200	0.0568	0.0282	0.0265	0.0475
0.08	0.2818	0.7220	0.0864	1.2263	1.1600	0.0745	0.0454	0.0428	0.0738
0.10	0.3430	0.8399	0.1100	1.2828	1.2000	0.0917	0.0655	0.0622	0.1043
0.12	0.4014	0.9414	0.1344	1.3394	1.2400	0.1084	0.0884	0.0844	0.1386
0.14	0.4573	1.0291	0.1596	1.3960	1.2800	0.1247	0.1141	0.1093	0.1765
0.16	0.5111	1.1054	0.1856	1.4525	1.3200	0.1406	0.1423	0.1369	0.2179
0.18	0.5630	1.1719	0.2124	1.5091	1.3600	0.1562	0.1730	0.1671	0.2629
0.20	0.6132	1.2303	0.2400	1.5657	1.4000	0.1714	0.2063	0.1998	0.3112
0.22	0.6618	1.2816	0.2684	1.6223	1.4400	0.1864	0.2420	0.2351	0.3629
0.24	0.7091	1.3269	0.2976	1.6788	1.4800	0.2011	0.2802	0.2730	0.4179
0.26	0.7551	1.3669	0.3276	1.7354	1.5200	0.2155	0.3208	0.3134	0.4763
0.28	0.8000	1.4025	0.3584	1.7920	1.5600	0.2297	0.3638	0.3563	0.5380
0.30	0.8439	1.4342	0.3900	1.8485	1.6000	0.2437	0.4092	0.4018	0.6030
0.32	0.8869	1.4625	0.4224	1.9051	1.6400	0.2576	0.4571	0.4498	0.6713
0.34	0.9290	1.4878	0.4556	1.9617	1.6800	0.2712	0.5075	0.5004	0.7430
0.36	0.9704	1.5104	0.4896	2.0182	1.7200	0.2847	0.5603	0.5536	0.8180
0.38	1.0110	1.5308	0.5244	2.0748	1.7600	0.2980	0.6156	0.6094	0.8964
0.40	1.0510	1.5491	0.5600	2.1314	1.8000	0.3111	0.6733	0.6678	0.9782
0.42	1.0903	1.5655	0.5964	2.1879	1.8400	0.3241	0.7336	0.7288	1.0633
0.44	1.1292	1.5804	0.6336	2.2445	1.8800	0.3370	0.7964	0.7925	1.1519
0.46	1.1675	1.5939	0.6716	2.3011	1.9200	0.3498	0.8617	0.8590	1.2439
0.48	1.2053	1.6060	0.7104	2.3576	1.9600	0.3624	0.9296	0.9281	1.3393
0.50	1.2426	1.6170	0.7500	2.4142	2.0000	0.3750	1.0000	1.0000	1.4383
0.52	1.2796	1.6270	0.7904	2.4708	2.0400	0.3875	1.0731	1.0747	1.5407
0.54	1.3162	1.6360	0.8316	2.5274	2.0800	0.3998	1.1488	1.1521	1.6466
0.56	1.3524	1.6442	0.8736	2.5839	2.1200	0.4121	1.2271	1.2324	1.7561
0.58	1.3882	1.6516	0.9164	2.6405	2.1600	0.4243	1.3081	1.3155	1.8692
0.60	1.4238	1.6584	0.9600	2.6971	2.2000	0.4364	1.3918	1.4015	1.9859
0.62	1.4590	1.6646	1.0044	2.7536	2.2400	0.4484	1.4782	1.4905	2.1062
0.64	1.4940	1.6701	1.0496	2.8102	2.2800	0.4604	1.5674	1.5823	2.2301
0.66	1.5287	1.6752	1.0956	2.8668	2.3200	0.4722	1.6593	1.6772	2.3577
0.68	1.5631	1.6798	1.1424	2.9233	2.3600	0.4841	1.7540	1.7750	2.4890
0.70	1.5974	1.6840	1.1900	2.9799	2.4000	0.4958	1.8515	1.8758	2.6241
0.72	1.6314	1.6878	1.2384	3.0365	2.4400	0.5075	1.9518	1.9797	2.7628
0.74	1.6652	1.6913	1.2876	3.0930	2.4800	0.5192	2.0550	2.0867	2.9054
0.76	1.6988	1.6944	1.3376	3.1496	2.5200	0.5308	2.1610	2.1968	3.0518
0.78	1.7322	1.6972	1.3884	3.2062	2.5600	0.5423	2.2700	2.3100	3.2019
0.80	1.7654	1.6998	1.4400	3.2627	2.6000	0.5538	2.3819	2.4264	3.3560

Rel. dpth Y	U.equ. dia. $D_{ep(u)}$ (m)	Equiv. disch. factor J	Unit sect. area A_u (m²)	Unit wetted perim. P_u (m)	Unit surf. brdth B_{us} (m)	Unit mean depth y_{um} (m)	Discharge ratios $Q/Q_{0.50}$ medial	Mann'g	U.crit. disch. Q_{uc} (m³s⁻¹)
0.82	1.7984	1.7021	1.4924	3.3193	2.6400	0.5653	2.4967	2.5460	3.5139
0.84	1.8313	1.7042	1.5456	3.3759	2.6800	0.5767	2.6145	2.6688	3.6757
0.86	1.8641	1.7061	1.5996	3.4324	2.7200	0.5881	2.7353	2.7949	3.8414
0.88	1.8967	1.7078	1.6544	3.4890	2.7600	0.5994	2.8591	2.9242	4.0111
0.90	1.9292	1.7093	1.7100	3.5456	2.8000	0.6107	2.9860	3.0569	4.1848
0.92	1.9615	1.7107	1.7664	3.6022	2.8400	0.6220	3.1159	3.1929	4.3625
0.94	1.9937	1.7119	1.8236	3.6587	2.8800	0.6332	3.2489	3.3323	4.5442
0.96	2.0258	1.7129	1.8816	3.7153	2.9200	0.6444	3.3851	3.4751	4.7300
0.98	2.0578	1.7139	1.9404	3.7719	2.9600	0.6555	3.5243	3.6213	4.9198
1.00	2.0896	1.7147	2.0000	3.8284	3.0000	0.6667	3.6668	3.7710	5.1138
1.02	2.1214	1.7154	2.0604	3.8850	3.0400	0.6778	3.8124	3.9241	5.3119
1.04	2.1531	1.7160	2.1216	3.9416	3.0800	0.6888	3.9612	4.0808	5.5142
1.06	2.1846	1.7166	2.1836	3.9981	3.1200	0.6999	4.1133	4.2410	5.7206
1.08	2.2161	1.7170	2.2464	4.0547	3.1600	0.7109	4.2686	4.4047	5.9313
1.10	2.2475	1.7174	2.3100	4.1113	3.2000	0.7219	4.4271	4.5721	6.1462
1.12	2.2788	1.7176	2.3744	4.1678	3.2400	0.7328	4.5890	4.7431	6.3653
1.14	2.3100	1.7179	2.4396	4.2244	3.2800	0.7438	4.7542	4.9178	6.5887
1.16	2.3411	1.7180	2.5056	4.2810	3.3200	0.7547	4.9228	5.0961	6.8165
1.18	2.3722	1.7181	2.5724	4.3375	3.3600	0.7656	5.0947	5.2782	7.0485
1.20	2.4032	1.7182	2.6400	4.3941	3.4000	0.7765	5.2700	5.4640	7.2850
1.24	2.4650	1.7181	2.7776	4.5072	3.4800	0.7982	5.6308	5.8469	7.7710
1.28	2.5265	1.7179	2.9184	4.6204	3.5600	0.8198	6.0055	6.2451	8.2747
1.32	2.5878	1.7175	3.0624	4.7335	3.6400	0.8413	6.3942	6.6588	8.7963
1.36	2.6489	1.7170	3.2096	4.8467	3.7200	0.8628	6.7970	7.0882	9.3361
1.40	2.7098	1.7164	3.3600	4.9598	3.8000	0.8842	7.2142	7.5336	9.8941
1.44	2.7705	1.7157	3.5136	5.0729	3.8800	0.9056	7.6459	7.9952	10.471
1.48	2.8310	1.7149	3.6704	5.1861	3.9600	0.9269	8.0924	8.4732	11.066
1.52	2.8913	1.7140	3.8304	5.2992	4.0400	0.9481	8.5537	8.9677	11.680
1.56	2.9515	1.7131	3.9936	5.4123	4.1200	0.9693	9.0300	9.4791	12.313
1.60	3.0115	1.7122	4.1600	5.5255	4.2000	0.9905	9.5216	10.007	12.965
1.64	3.0714	1.7112	4.3296	5.6386	4.2800	1.0116	10.029	10.553	13.637
1.68	3.1311	1.7102	4.5024	5.7518	4.3600	1.0327	10.551	11.116	14.328
1.72	3.1908	1.7091	4.6784	5.8649	4.4400	1.0537	11.089	11.697	15.039
1.76	3.2503	1.7081	4.8576	5.9780	4.5200	1.0747	11.643	12.295	15.770
1.80	3.3097	1.7070	5.0400	6.0912	4.6000	1.0957	12.213	12.912	16.521
1.84	3.3690	1.7059	5.2256	6.2043	4.6800	1.1166	12.800	13.547	17.292
1.88	3.4282	1.7048	5.4144	6.3174	4.7600	1.1375	13.402	14.201	18.083
1.92	3.4873	1.7037	5.6064	6.4306	4.8400	1.1583	14.021	14.873	18.896
1.96	3.5464	1.7025	5.8016	6.5437	4.9200	1.1792	14.657	15.564	19.729
2.00	3.6053	1.7014	6.0000	6.6569	5.0000	1.2000	15.310	16.274	20.583
2.05	3.6789	1.7000	6.2525	6.7983	5.1000	1.2260	16.150	17.189	21.680
2.10	3.7523	1.6986	6.5100	6.9397	5.2000	1.2519	17.016	18.134	22.810
2.15	3.8257	1.6972	6.7725	7.0811	5.3000	1.2778	17.910	19.110	23.974
2.20	3.8989	1.6959	7.0400	7.2225	5.4000	1.3037	18.831	20.118	25.172
2.25	3.9720	1.6945	7.3125	7.3640	5.5000	1.3295	19.780	21.157	26.404
2.30	4.0451	1.6932	7.5900	7.5054	5.6000	1.3554	20.758	22.228	27.671
2.35	4.1181	1.6918	7.8725	7.6468	5.7000	1.3811	21.763	23.332	28.973
2.40	4.1909	1.6905	8.1600	7.7882	5.8000	1.4069	22.797	24.468	30.310
2.45	4.2637	1.6892	8.4525	7.9296	5.9000	1.4326	23.860	25.638	31.682
2.50	4.3365	1.6879	8.7500	8.0711	6.0000	1.4583	24.952	26.842	33.090
2.55	4.4091	1.6866	9.0525	8.2125	6.1000	1.4840	26.074	28.079	34.534
2.60	4.4817	1.6854	9.3600	8.3539	6.2000	1.5097	27.225	29.350	36.015
2.65	4.5543	1.6841	9.6725	8.4953	6.3000	1.5353	28.407	30.657	37.532
2.70	4.6267	1.6829	9.9900	8.6368	6.4000	1.5609	29.619	31.998	39.086
2.75	4.6992	1.6817	10.312	8.7782	6.5000	1.5865	30.861	33.375	40.677

C55

Trapezoidal - 1.25 to 1 side-slope

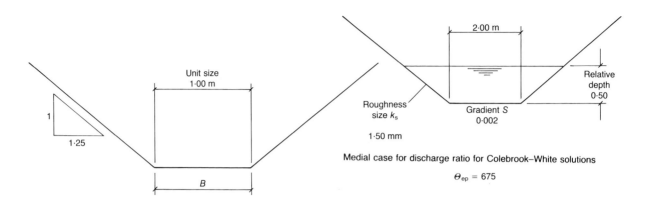

Medial case for discharge ratio for Colebrook–White solutions

$\Theta_{ep} = 675$

Rel. dpth	U.equ. dia. $D_{ep(u)}$	Equiv. disch. factor	Unit sect. area	Unit wetted perim.	Unit surf. brdth	Unit mean depth	Discharge ratios $Q/Q_{0.50}$		U.crit. disch. Q_{uc}
Y	(m)	J	A_u (m²)	P_u (m)	B_{us} (m)	y_{um} (m)	medial	Mann'g	(m³s⁻¹)
0.02	0.0771	0.2275	0.0205	1.0640	1.0500	0.0195	0.0042	0.0039	0.0090
0.04	0.1489	0.4147	0.0420	1.1281	1.1000	0.0382	0.0134	0.0125	0.0257
0.06	0.2164	0.5703	0.0645	1.1921	1.1500	0.0561	0.0263	0.0247	0.0478
0.08	0.2802	0.7008	0.0880	1.2561	1.2000	0.0733	0.0423	0.0400	0.0746
0.10	0.3409	0.8112	0.1125	1.3202	1.2500	0.0900	0.0614	0.0582	0.1057
0.12	0.3988	0.9051	0.1380	1.3842	1.3000	0.1062	0.0832	0.0793	0.1408
0.14	0.4544	0.9856	0.1645	1.4482	1.3500	0.1219	0.1077	0.1031	0.1798
0.16	0.5079	1.0550	0.1920	1.5122	1.4000	0.1371	0.1348	0.1297	0.2227
0.18	0.5595	1.1152	0.2205	1.5763	1.4500	0.1521	0.1646	0.1588	0.2693
0.20	0.6096	1.1676	0.2500	1.6403	1.5000	0.1667	0.1970	0.1907	0.3196
0.22	0.6583	1.2134	0.2805	1.7043	1.5500	0.1810	0.2319	0.2252	0.3737
0.24	0.7057	1.2537	0.3120	1.7684	1.6000	0.1950	0.2694	0.2624	0.4315
0.26	0.7520	1.2893	0.3445	1.8324	1.6500	0.2088	0.3095	0.3022	0.4929
0.28	0.7973	1.3207	0.3780	1.8964	1.7000	0.2224	0.3522	0.3448	0.5582
0.30	0.8416	1.3487	0.4125	1.9605	1.7500	0.2357	0.3975	0.3901	0.6272
0.32	0.8852	1.3735	0.4480	2.0245	1.8000	0.2489	0.4455	0.4381	0.6999
0.34	0.9279	1.3958	0.4845	2.0885	1.8500	0.2619	0.4961	0.4890	0.7765
0.36	0.9700	1.4157	0.5220	2.1526	1.9000	0.2747	0.5494	0.5426	0.8568
0.38	1.0115	1.4335	0.5605	2.2166	1.9500	0.2874	0.6054	0.5991	0.9410
0.40	1.0523	1.4496	0.6000	2.2806	2.0000	0.3000	0.6642	0.6585	1.0291
0.42	1.0927	1.4641	0.6405	2.3447	2.0500	0.3124	0.7257	0.7208	1.1211
0.44	1.1326	1.4771	0.6820	2.4087	2.1000	0.3248	0.7900	0.7861	1.2171
0.46	1.1720	1.4890	0.7245	2.4727	2.1500	0.3370	0.8571	0.8544	1.3170
0.48	1.2110	1.4997	0.7680	2.5367	2.2000	0.3491	0.9271	0.9257	1.4210
0.50	1.2496	1.5094	0.8125	2.6008	2.2500	0.3611	1.0000	1.0000	1.5290
0.52	1.2879	1.5183	0.8580	2.6648	2.3000	0.3730	1.0758	1.0775	1.6411
0.54	1.3258	1.5263	0.9045	2.7288	2.3500	0.3849	1.1546	1.1580	1.7573
0.56	1.3635	1.5337	0.9520	2.7929	2.4000	0.3967	1.2363	1.2418	1.8776
0.58	1.4008	1.5404	1.0005	2.8569	2.4500	0.4084	1.3211	1.3288	2.0022
0.60	1.4379	1.5465	1.0500	2.9209	2.5000	0.4200	1.4089	1.4190	2.1310
0.62	1.4747	1.5521	1.1005	2.9850	2.5500	0.4316	1.4997	1.5126	2.2640
0.64	1.5113	1.5572	1.1520	3.0490	2.6000	0.4431	1.5937	1.6095	2.4013
0.66	1.5477	1.5619	1.2045	3.1130	2.6500	0.4545	1.6908	1.7097	2.5430
0.68	1.5839	1.5661	1.2580	3.1771	2.7000	0.4659	1.7911	1.8133	2.6891
0.70	1.6198	1.5701	1.3125	3.2411	2.7500	0.4773	1.8946	1.9204	2.8395
0.72	1.6556	1.5737	1.3680	3.3051	2.8000	0.4886	2.0013	2.0310	2.9944
0.74	1.6912	1.5770	1.4245	3.3692	2.8500	0.4998	2.1113	2.1451	3.1538
0.76	1.7267	1.5800	1.4820	3.4332	2.9000	0.5110	2.2246	2.2628	3.3177
0.78	1.7620	1.5828	1.5405	3.4972	2.9500	0.5222	2.3412	2.3841	3.4861
0.80	1.7971	1.5853	1.6000	3.5612	3.0000	0.5333	2.4612	2.5090	3.6591

Rel. dpth	U.equ. dia. $D_{ep(u)}$	Equiv. disch. factor	Unit sect. area	Unit wetted perim.	Unit surf. brdth	Unit mean depth	Discharge ratios $Q/Q_{0.50}$		U.crit. disch. Q_{uc}
Y	(m)	J	A_u (m²)	P_u (m)	B_{us} (m)	y_{um} (m)	medial	Mann'g	(m³s⁻¹)
0.82	1.8321	1.5877	1.6605	3.6253	3.0500	0.5444	2.5845	2.6375	3.8368
0.84	1.8670	1.5898	1.7220	3.6893	3.1000	0.5555	2.7113	2.7698	4.0191
0.86	1.9018	1.5918	1.7845	3.7533	3.1500	0.5665	2.8415	2.9059	4.2061
0.88	1.9364	1.5936	1.8480	3.8174	3.2000	0.5775	2.9753	3.0457	4.3978
0.90	1.9709	1.5952	1.9125	3.8814	3.2500	0.5885	3.1125	3.1894	4.5943
0.92	2.0054	1.5967	1.9780	3.9454	3.3000	0.5994	3.2533	3.3369	4.7956
0.94	2.0397	1.5981	2.0445	4.0095	3.3500	0.6103	3.3976	3.4883	5.0017
0.96	2.0739	1.5994	2.1120	4.0735	3.4000	0.6212	3.5456	3.6437	5.2127
0.98	2.1080	1.6006	2.1805	4.1375	3.4500	0.6320	3.6971	3.8030	5.4286
1.00	2.1421	1.6016	2.2500	4.2016	3.5000	0.6429	3.8524	3.9664	5.6494
1.02	2.1760	1.6026	2.3205	4.2656	3.5500	0.6537	4.0113	4.1338	5.8751
1.04	2.2099	1.6035	2.3920	4.3296	3.6000	0.6644	4.1740	4.3053	6.1059
1.06	2.2437	1.6043	2.4645	4.3937	3.6500	0.6752	4.3404	4.4809	6.3417
1.08	2.2774	1.6050	2.5380	4.4577	3.7000	0.6859	4.5106	4.6606	6.5826
1.10	2.3111	1.6056	2.6125	4.5217	3.7500	0.6967	4.6846	4.8446	6.8286
1.12	2.3447	1.6062	2.6880	4.5857	3.8000	0.7074	4.8625	5.0328	7.0797
1.14	2.3782	1.6068	2.7645	4.6498	3.8500	0.7181	5.0442	5.2252	7.3359
1.16	2.4116	1.6072	2.8420	4.7138	3.9000	0.7287	5.2298	5.4219	7.5974
1.18	2.4450	1.6077	2.9205	4.7778	3.9500	0.7394	5.4194	5.6230	7.8641
1.20	2.4784	1.6080	3.0000	4.8419	4.0000	0.7500	5.6129	5.8285	8.1360
1.24	2.5449	1.6086	3.1620	4.9699	4.1000	0.7712	6.0119	6.2527	8.6958
1.28	2.6112	1.6091	3.3280	5.0980	4.2000	0.7924	6.4271	6.6948	9.2771
1.32	2.6774	1.6094	3.4980	5.2261	4.3000	0.8135	6.8586	7.1551	9.8800
1.36	2.7433	1.6096	3.6720	5.3541	4.4000	0.8345	7.3068	7.6338	10.505
1.40	2.8091	1.6097	3.8500	5.4822	4.5000	0.8556	7.7718	8.1313	11.152
1.44	2.8747	1.6097	4.0320	5.6102	4.6000	0.8765	8.2539	8.6479	11.821
1.48	2.9402	1.6097	4.2180	5.7383	4.7000	0.8974	8.7532	9.1837	12.513
1.52	3.0056	1.6095	4.4080	5.8664	4.8000	0.9183	9.2700	9.7391	13.228
1.56	3.0708	1.6093	4.6020	5.9944	4.9000	0.9392	9.8045	10.314	13.966
1.60	3.1360	1.6091	4.8000	6.1225	5.0000	0.9600	10.357	10.910	14.728
1.64	3.2010	1.6088	5.0020	6.2506	5.1000	0.9808	10.928	11.525	15.513
1.68	3.2659	1.6085	5.2080	6.3786	5.2000	1.0015	11.516	12.162	16.322
1.72	3.3307	1.6081	5.4180	6.5067	5.3000	1.0223	12.124	12.819	17.155
1.76	3.3955	1.6077	5.6320	6.6347	5.4000	1.0430	12.750	13.497	18.012
1.80	3.4601	1.6073	5.8500	6.7628	5.5000	1.0636	13.395	14.197	18.894
1.84	3.5247	1.6069	6.0720	6.8909	5.6000	1.0843	14.059	14.919	19.800
1.88	3.5891	1.6064	6.2980	7.0189	5.7000	1.1049	14.743	15.662	20.731
1.92	3.6536	1.6060	6.5280	7.1470	5.8000	1.1255	15.446	16.428	21.688
1.96	3.7179	1.6055	6.7620	7.2751	5.9000	1.1461	16.169	17.216	22.670
2.00	3.7822	1.6050	7.0000	7.4031	6.0000	1.1667	16.912	18.027	23.677
2.05	3.8625	1.6043	7.3031	7.5632	6.1250	1.1923	17.869	19.073	24.973
2.10	3.9426	1.6037	7.6125	7.7233	6.2500	1.2180	18.858	20.155	26.309
2.15	4.0227	1.6031	7.9281	7.8834	6.3750	1.2436	19.879	21.274	27.687
2.20	4.1027	1.6024	8.2500	8.0434	6.5000	1.2692	20.933	22.430	29.106
2.25	4.1827	1.6017	8.5781	8.2035	6.6250	1.2948	22.019	23.624	30.567
2.30	4.2625	1.6011	8.9125	8.3636	6.7500	1.3204	23.139	24.856	32.071
2.35	4.3423	1.6004	9.2531	8.5237	6.8750	1.3459	24.293	26.127	33.617
2.40	4.4221	1.5998	9.6000	8.6837	7.0000	1.3714	25.481	27.437	35.206
2.45	4.5017	1.5991	9.9531	8.8438	7.1250	1.3969	26.703	28.787	36.839
2.50	4.5813	1.5985	10.312	9.0039	7.2500	1.4224	27.960	30.177	38.516
2.55	4.6609	1.5978	10.678	9.1640	7.3750	1.4479	29.252	31.608	40.237
2.60	4.7404	1.5972	11.050	9.3241	7.5000	1.4733	30.579	33.080	42.002
2.65	4.8199	1.5965	11.428	9.4841	7.6250	1.4988	31.943	34.593	43.813
2.70	4.8993	1.5959	11.812	9.6442	7.7500	1.5242	33.343	36.148	45.669
2.75	4.9787	1.5953	12.203	9.8043	7.8750	1.5496	34.779	37.746	47.571

C56

Trapezoidal - 1.50 to 1 side-slope

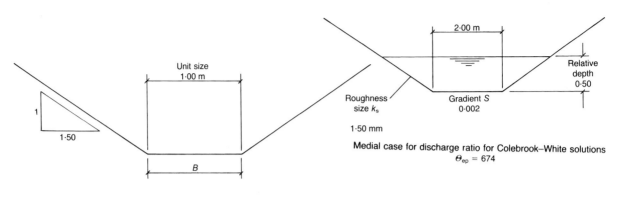

Unit size 1·00 m

Unit size 1·00 m

B

Roughness size k_s

1·50 mm

2·00 m

Relative depth 0·50

Gradient S 0·002

Medial case for discharge ratio for Colebrook–White solutions
$\Theta_{ep} = 674$

Rel. dpth Y	U.equ. dia. $D_{ep(u)}$ (m)	Equiv. disch. factor J	Unit sect. area A_u (m²)	Unit wetted perim. P_u (m)	Unit surf. brdth B_{us} (m)	Unit mean depth y_{um} (m)	Discharge ratios $Q/Q_{0.50}$ medial	Mann'g	U.crit. disch. Q_{uc} (m³s⁻¹)
0.02	0.0769	0.2252	0.0206	1.0721	1.0600	0.0194	0.0039	0.0037	0.0090
0.04	0.1482	0.4070	0.0424	1.1442	1.1200	0.0379	0.0125	0.0117	0.0258
0.06	0.2151	0.5555	0.0654	1.2163	1.1800	0.0554	0.0246	0.0231	0.0482
0.08	0.2782	0.6782	0.0896	1.2884	1.2400	0.0723	0.0399	0.0376	0.0754
0.10	0.3381	0.7807	0.1150	1.3606	1.3000	0.0885	0.0580	0.0550	0.1071
0.12	0.3953	0.8669	0.1416	1.4327	1.3600	0.1041	0.0788	0.0752	0.1431
0.14	0.4503	0.9401	0.1694	1.5048	1.4200	0.1193	0.1024	0.0981	0.1832
0.16	0.5033	1.0026	0.1984	1.5769	1.4800	0.1341	0.1287	0.1237	0.2275
0.18	0.5545	1.0564	0.2286	1.6490	1.5400	0.1484	0.1576	0.1521	0.2758
0.20	0.6043	1.1029	0.2600	1.7211	1.6000	0.1625	0.1892	0.1831	0.3282
0.22	0.6527	1.1434	0.2926	1.7932	1.6600	0.1763	0.2235	0.2170	0.3847
0.24	0.6999	1.1788	0.3264	1.8653	1.7200	0.1898	0.2605	0.2536	0.4453
0.26	0.7461	1.2098	0.3614	1.9374	1.7800	0.2030	0.3001	0.2930	0.5100
0.28	0.7914	1.2372	0.3976	2.0096	1.8400	0.2161	0.3426	0.3353	0.5788
0.30	0.8359	1.2614	0.4350	2.0817	1.9000	0.2289	0.3878	0.3804	0.6518
0.32	0.8796	1.2830	0.4736	2.1538	1.9600	0.2416	0.4358	0.4285	0.7290
0.34	0.9226	1.3021	0.5134	2.2259	2.0200	0.2542	0.4866	0.4795	0.8105
0.36	0.9650	1.3192	0.5544	2.2980	2.0800	0.2665	0.5403	0.5336	0.8963
0.38	1.0069	1.3346	0.5966	2.3701	2.1400	0.2788	0.5969	0.5907	0.9865
0.40	1.0482	1.3484	0.6400	2.4422	2.2000	0.2909	0.6565	0.6509	1.0810
0.42	1.0891	1.3608	0.6846	2.5143	2.2600	0.3029	0.7191	0.7142	1.1799
0.44	1.1296	1.3720	0.7304	2.5864	2.3200	0.3148	0.7847	0.7807	1.2834
0.46	1.1697	1.3821	0.7774	2.6586	2.3800	0.3266	0.8533	0.8505	1.3914
0.48	1.2094	1.3913	0.8256	2.7307	2.4400	0.3384	0.9251	0.9236	1.5039
0.50	1.2488	1.3997	0.8750	2.8028	2.5000	0.3500	1.0000	1.0000	1.6211
0.52	1.2878	1.4073	0.9256	2.8749	2.5600	0.3616	1.0781	1.0798	1.7429
0.54	1.3266	1.4142	0.9774	2.9470	2.6200	0.3731	1.1594	1.1630	1.8695
0.56	1.3652	1.4205	1.0304	3.0191	2.6800	0.3845	1.2440	1.2497	2.0008
0.58	1.4035	1.4263	1.0846	3.0912	2.7400	0.3958	1.3319	1.3399	2.1369
0.60	1.4415	1.4316	1.1400	3.1633	2.8000	0.4071	1.4232	1.4337	2.2779
0.62	1.4794	1.4364	1.1966	3.2354	2.8600	0.4184	1.5178	1.5311	2.4238
0.64	1.5170	1.4409	1.2544	3.3076	2.9200	0.4296	1.6158	1.6322	2.5747
0.66	1.5545	1.4449	1.3134	3.3797	2.9800	0.4407	1.7173	1.7370	2.7305
0.68	1.5918	1.4487	1.3736	3.4518	3.0400	0.4518	1.8223	1.8455	2.8914
0.70	1.6289	1.4521	1.4350	3.5239	3.1000	0.4629	1.9309	1.9579	3.0574
0.72	1.6659	1.4553	1.4976	3.5960	3.1600	0.4739	2.0430	2.0741	3.2286
0.74	1.7027	1.4582	1.5614	3.6681	3.2200	0.4849	2.1587	2.1942	3.4049
0.76	1.7394	1.4609	1.6264	3.7402	3.2800	0.4959	2.2781	2.3182	3.5864
0.78	1.7759	1.4634	1.6926	3.8123	3.3400	0.5068	2.4012	2.4463	3.7733
0.80	1.8124	1.4657	1.7600	3.8844	3.4000	0.5176	2.5280	2.5784	3.9654

Rel. dpth	U.equ. dia. $D_{ep(u)}$	Equiv. disch. factor	Unit sect. area	Unit wetted perim.	Unit surf. brdth	Unit mean depth	Discharge ratios $Q/Q_{0.50}$		U.crit. disch. Q_{uc}
Y	(m)	J	A_u (m²)	P_u (m)	B_{us} (m)	y_{um} (m)	medial	Mann'g	(m³s⁻¹)
0.82	1.8487	1.4679	1.8286	3.9566	3.4600	0.5285	2.6586	2.7146	4.1629
0.84	1.8849	1.4698	1.8984	4.0287	3.5200	0.5393	2.7929	2.8549	4.3659
0.86	1.9210	1.4716	1.9694	4.1008	3.5800	0.5501	2.9312	2.9993	4.5742
0.88	1.9570	1.4733	2.0416	4.1729	3.6400	0.5609	3.0733	3.1480	4.7881
0.90	1.9929	1.4749	2.1150	4.2450	3.7000	0.5716	3.2193	3.3010	5.0075
0.92	2.0288	1.4763	2.1896	4.3171	3.7600	0.5823	3.3692	3.4583	5.2326
0.94	2.0645	1.4776	2.2654	4.3892	3.8200	0.5930	3.5232	3.6199	5.4632
0.96	2.1002	1.4789	2.3424	4.4613	3.8800	0.6037	3.6812	3.7859	5.6995
0.98	2.1358	1.4800	2.4206	4.5334	3.9400	0.6144	3.8432	3.9564	5.9415
1.00	2.1713	1.4811	2.5000	4.6056	4.0000	0.6250	4.0093	4.1313	6.1893
1.02	2.2067	1.4820	2.5806	4.6777	4.0600	0.6356	4.1796	4.3108	6.4429
1.04	2.2421	1.4830	2.6624	4.7498	4.1200	0.6462	4.3540	4.4949	6.7023
1.06	2.2774	1.4838	2.7454	4.8219	4.1800	0.6568	4.5327	4.6836	6.9676
1.08	2.3127	1.4846	2.8296	4.8940	4.2400	0.6674	4.7155	4.8769	7.2388
1.10	2.3479	1.4853	2.9150	4.9661	4.3000	0.6779	4.9027	5.0750	7.5160
1.12	2.3831	1.4859	3.0016	5.0382	4.3600	0.6884	5.0942	5.2778	7.7991
1.14	2.4182	1.4865	3.0894	5.1103	4.4200	0.6990	5.2899	5.4853	8.0884
1.16	2.4532	1.4871	3.1784	5.1824	4.4800	0.7095	5.4901	5.6978	8.3837
1.18	2.4882	1.4876	3.2686	5.2546	4.5400	0.7200	5.6947	5.9150	8.6851
1.20	2.5232	1.4881	3.3600	5.3267	4.6000	0.7304	5.9037	6.1373	8.9927
1.24	2.5929	1.4889	3.5464	5.4709	4.7200	0.7514	6.3353	6.5966	9.6266
1.28	2.6625	1.4896	3.7376	5.6151	4.8400	0.7722	6.7850	7.0761	10.286
1.32	2.7320	1.4902	3.9336	5.7593	4.9600	0.7931	7.2531	7.5761	10.970
1.36	2.8013	1.4907	4.1344	5.9035	5.0800	0.8139	7.7400	8.0970	11.680
1.40	2.8705	1.4911	4.3400	6.0478	5.2000	0.8346	8.2459	8.6390	12.416
1.44	2.9395	1.4914	4.5504	6.1920	5.3200	0.8553	8.7710	9.2025	13.179
1.48	3.0085	1.4916	4.7656	6.3362	5.4400	0.8760	9.3155	9.7879	13.968
1.52	3.0773	1.4918	4.9856	6.4804	5.5600	0.8967	9.8799	10.395	14.784
1.56	3.1461	1.4919	5.2104	6.6247	5.6800	0.9173	10.464	11.025	15.628
1.60	3.2147	1.4920	5.4400	6.7689	5.8000	0.9379	11.069	11.678	16.499
1.64	3.2833	1.4920	5.6744	6.9131	5.9200	0.9585	11.694	12.354	17.397
1.68	3.3518	1.4920	5.9136	7.0573	6.0400	0.9791	12.340	13.053	18.324
1.72	3.4202	1.4920	6.1576	7.2015	6.1600	0.9996	13.006	13.776	19.279
1.76	3.4885	1.4919	6.4064	7.3458	6.2800	1.0201	13.695	14.523	20.263
1.80	3.5567	1.4918	6.6600	7.4900	6.4000	1.0406	14.404	15.294	21.276
1.84	3.6249	1.4917	6.9184	7.6342	6.5200	1.0611	15.135	16.090	22.317
1.88	3.6931	1.4915	7.1816	7.7784	6.6400	1.0816	15.888	16.910	23.389
1.92	3.7612	1.4914	7.4496	7.9227	6.7600	1.1020	16.664	17.756	24.490
1.96	3.8292	1.4912	7.7224	8.0669	6.8800	1.1224	17.462	18.628	25.621
2.00	3.8972	1.4910	8.0000	8.2111	7.0000	1.1429	18.282	19.525	26.782
2.05	3.9821	1.4908	8.3537	8.3914	7.1500	1.1684	19.340	20.684	28.277
2.10	4.0669	1.4905	8.7150	8.5717	7.3000	1.1938	20.434	21.883	29.819
2.15	4.1517	1.4902	9.0838	8.7519	7.4500	1.2193	21.564	23.125	31.411
2.20	4.2364	1.4900	9.4600	8.9322	7.6000	1.2447	22.732	24.409	33.051
2.25	4.3210	1.4897	9.8437	9.1125	7.7500	1.2702	23.937	25.737	34.742
2.30	4.4056	1.4894	10.235	9.2928	7.9000	1.2956	25.180	27.108	36.482
2.35	4.4901	1.4890	10.634	9.4730	8.0500	1.3210	26.461	28.523	38.273
2.40	4.5746	1.4887	11.040	9.6533	8.2000	1.3463	27.781	29.983	40.115
2.45	4.6590	1.4884	11.454	9.8336	8.3500	1.3717	29.141	31.488	42.009
2.50	4.7434	1.4881	11.875	10.014	8.5000	1.3971	30.540	33.039	43.954
2.55	4.8278	1.4878	12.304	10.194	8.6500	1.4224	31.979	34.637	45.952
2.60	4.9121	1.4874	12.740	10.374	8.8000	1.4477	33.458	36.281	48.004
2.65	4.9963	1.4871	13.184	10.555	8.9500	1.4730	34.979	37.973	50.108
2.70	5.0806	1.4868	13.635	10.735	9.1000	1.4984	36.540	39.713	52.266
2.75	5.1648	1.4865	14.094	10.915	9.2500	1.5236	38.144	41.502	54.479

C57

Trapezoidal - 1.75 to 1 side-slope

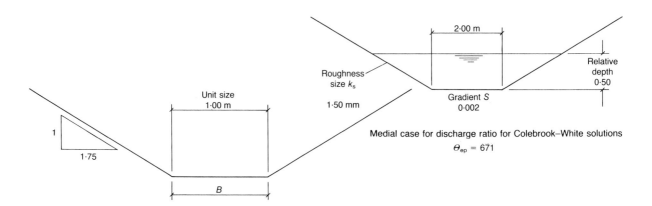

Rel. dpth	U.equ. dia. $D_{ep(u)}$	Equiv. disch. factor	Unit sect. area	Unit wetted perim.	Unit surf. brdth	Unit mean depth	Discharge ratios $Q/Q_{0.50}$		U.crit. disch. Q_{uc}
Y	(m)	J	A_u (m²)	P_u (m)	B_{us} (m)	y_{um} (m)	medial	Mann'g	(m³s⁻¹)
0.02	0.0766	0.2228	0.0207	1.0806	1.0700	0.0193	0.0037	0.0034	0.0090
0.04	0.1474	0.3988	0.0428	1.1612	1.1400	0.0375	0.0118	0.0110	0.0260
0.06	0.2135	0.5402	0.0663	1.2419	1.2100	0.0548	0.0233	0.0218	0.0486
0.08	0.2758	0.6553	0.0912	1.3225	1.2800	0.0712	0.0378	0.0356	0.0762
0.10	0.3350	0.7500	0.1175	1.4031	1.3500	0.0870	0.0551	0.0523	0.1086
0.12	0.3914	0.8288	0.1452	1.4837	1.4200	0.1023	0.0752	0.0717	0.1454
0.14	0.4457	0.8950	0.1743	1.5644	1.4900	0.1170	0.0980	0.0938	0.1867
0.16	0.4980	0.9511	0.2048	1.6450	1.5600	0.1313	0.1235	0.1187	0.2324
0.18	0.5487	0.9989	0.2367	1.7256	1.6300	0.1452	0.1517	0.1463	0.2825
0.20	0.5979	1.0400	0.2700	1.8062	1.7000	0.1588	0.1827	0.1768	0.3370
0.22	0.6459	1.0755	0.3047	1.8868	1.7700	0.1721	0.2164	0.2100	0.3959
0.24	0.6929	1.1063	0.3408	1.9675	1.8400	0.1852	0.2529	0.2461	0.4593
0.26	0.7388	1.1333	0.3783	2.0481	1.9100	0.1981	0.2922	0.2852	0.5272
0.28	0.7839	1.1569	0.4172	2.1287	1.9800	0.2107	0.3344	0.3272	0.5997
0.30	0.8283	1.1778	0.4575	2.2093	2.0500	0.2232	0.3795	0.3722	0.6768
0.32	0.8720	1.1962	0.4992	2.2900	2.1200	0.2355	0.4275	0.4203	0.7586
0.34	0.9150	1.2126	0.5423	2.3706	2.1900	0.2476	0.4785	0.4715	0.8451
0.36	0.9576	1.2272	0.5868	2.4512	2.2600	0.2596	0.5326	0.5258	0.9364
0.38	0.9996	1.2403	0.6327	2.5318	2.3300	0.2715	0.5897	0.5834	1.0325
0.40	1.0412	1.2520	0.6800	2.6125	2.4000	0.2833	0.6500	0.6443	1.1335
0.42	1.0823	1.2626	0.7287	2.6931	2.4700	0.2950	0.7135	0.7086	1.2395
0.44	1.1231	1.2721	0.7788	2.7737	2.5400	0.3066	0.7801	0.7762	1.3505
0.46	1.1636	1.2806	0.8303	2.8543	2.6100	0.3181	0.8501	0.8473	1.4665
0.48	1.2037	1.2884	0.8832	2.9349	2.6800	0.3296	0.9234	0.9218	1.5877
0.50	1.2435	1.2955	0.9375	3.0156	2.7500	0.3409	1.0000	1.0000	1.7142
0.52	1.2831	1.3019	0.9932	3.0962	2.8200	0.3522	1.0801	1.0818	1.8458
0.54	1.3225	1.3078	1.0503	3.1768	2.8900	0.3634	1.1636	1.1672	1.9828
0.56	1.3616	1.3131	1.1088	3.2574	2.9600	0.3746	1.2506	1.2564	2.1252
0.58	1.4005	1.3180	1.1687	3.3381	3.0300	0.3857	1.3412	1.3494	2.2730
0.60	1.4392	1.3225	1.2300	3.4187	3.1000	0.3968	1.4354	1.4462	2.4263
0.62	1.4777	1.3266	1.2927	3.4993	3.1700	0.4078	1.5332	1.5469	2.5851
0.64	1.5160	1.3304	1.3568	3.5799	3.2400	0.4188	1.6348	1.6516	2.7495
0.66	1.5542	1.3338	1.4223	3.6605	3.3100	0.4297	1.7400	1.7603	2.9197
0.68	1.5922	1.3370	1.4892	3.7412	3.3800	0.4406	1.8491	1.8730	3.0955
0.70	1.6301	1.3400	1.5575	3.8218	3.4500	0.4514	1.9620	1.9899	3.2771
0.72	1.6679	1.3427	1.6272	3.9024	3.5200	0.4623	2.0787	2.1109	3.4646
0.74	1.7055	1.3452	1.6983	3.9830	3.5900	0.4731	2.1994	2.2362	3.6579
0.76	1.7431	1.3475	1.7708	4.0637	3.6600	0.4838	2.3240	2.3657	3.8572
0.78	1.7805	1.3497	1.8447	4.1443	3.7300	0.4946	2.4526	2.4996	4.0625
0.80	1.8178	1.3517	1.9200	4.2249	3.8000	0.5053	2.5853	2.6379	4.2739

Trapezoidal - 1.75 to 1 side-slope

continued

Rel. dpth Y	U.equ. dia. $D_{ep(u)}$ (m)	Equiv. disch. factor J	Unit sect. area A_u (m²)	Unit wetted perim. P_u (m)	Unit surf. brdth B_{us} (m)	Unit mean depth y_{um} (m)	Discharge ratios $Q/Q_{0.50}$ medial	Mann'g	U.crit. disch. Q_{uc} (m³s⁻¹)
0.82	1.8550	1.3535	1.9967	4.3055	3.8700	0.5159	2.7221	2.7806	4.4913
0.84	1.8921	1.3552	2.0748	4.3861	3.9400	0.5266	2.8630	2.9277	4.7149
0.86	1.9292	1.3568	2.1543	4.4668	4.0100	0.5372	3.0081	3.0795	4.9448
0.88	1.9661	1.3583	2.2352	4.5474	4.0800	0.5478	3.1574	3.2358	5.1809
0.90	2.0030	1.3597	2.3175	4.6280	4.1500	0.5584	3.3110	3.3967	5.4233
0.92	2.0398	1.3609	2.4012	4.7086	4.2200	0.5690	3.4689	3.5624	5.6721
0.94	2.0766	1.3621	2.4863	4.7893	4.2900	0.5796	3.6311	3.7328	5.9274
0.96	2.1132	1.3632	2.5728	4.8699	4.3600	0.5901	3.7977	3.9080	6.1891
0.98	2.1498	1.3643	2.6607	4.9505	4.4300	0.6006	3.9688	4.0881	6.4573
1.00	2.1864	1.3652	2.7500	5.0311	4.5000	0.6111	4.1443	4.2730	6.7321
1.02	2.2229	1.3661	2.8407	5.1118	4.5700	0.6216	4.3244	4.4630	7.0136
1.04	2.2593	1.3669	2.9328	5.1924	4.6400	0.6321	4.5090	4.6579	7.3017
1.06	2.2957	1.3677	3.0263	5.2730	4.7100	0.6425	4.6981	4.8578	7.5966
1.08	2.3320	1.3684	3.1212	5.3536	4.7800	0.6530	4.8920	5.0629	7.8982
1.10	2.3683	1.3691	3.2175	5.4342	4.8500	0.6634	5.0904	5.2731	8.2067
1.12	2.4046	1.3697	3.3152	5.5149	4.9200	0.6738	5.2936	5.4885	8.5220
1.14	2.4408	1.3703	3.4143	5.5955	4.9900	0.6842	5.5016	5.7091	8.8443
1.16	2.4769	1.3709	3.5148	5.6761	5.0600	0.6946	5.7143	5.9351	9.1735
1.18	2.5130	1.3714	3.6167	5.7567	5.1300	0.7050	5.9319	6.1664	9.5098
1.20	2.5491	1.3719	3.7200	5.8374	5.2000	0.7154	6.1544	6.4031	9.8531
1.24	2.6211	1.3727	3.9308	5.9986	5.3400	0.7361	6.6140	6.8928	10.561
1.28	2.6931	1.3735	4.1472	6.1598	5.4800	0.7568	7.0936	7.4047	11.298
1.32	2.7648	1.3741	4.3692	6.3211	5.6200	0.7774	7.5935	7.9391	12.064
1.36	2.8365	1.3747	4.5968	6.4823	5.7600	0.7981	8.1139	8.4963	12.860
1.40	2.9081	1.3751	4.8300	6.6436	5.9000	0.8186	8.6551	9.0769	13.685
1.44	2.9795	1.3755	5.0688	6.8048	6.0400	0.8392	9.2175	9.6811	14.541
1.48	3.0509	1.3759	5.3132	6.9661	6.1800	0.8597	9.8012	10.309	15.428
1.52	3.1222	1.3762	5.5632	7.1273	6.3200	0.8803	10.407	10.962	16.345
1.56	3.1934	1.3764	5.8188	7.2886	6.4600	0.9007	11.034	11.639	17.294
1.60	3.2645	1.3766	6.0800	7.4498	6.6000	0.9212	11.684	12.342	18.274
1.64	3.3356	1.3768	6.3468	7.6111	6.7400	0.9417	12.356	13.069	19.287
1.68	3.4066	1.3769	6.6192	7.7723	6.8800	0.9621	13.052	13.823	20.332
1.72	3.4775	1.3770	6.8972	7.9335	7.0200	0.9825	13.770	14.603	21.409
1.76	3.5484	1.3771	7.1808	8.0948	7.1600	1.0029	14.512	15.409	22.520
1.80	3.6192	1.3771	7.4700	8.2560	7.3000	1.0233	15.277	16.242	23.664
1.84	3.6899	1.3772	7.7648	8.4173	7.4400	1.0437	16.067	17.103	24.841
1.88	3.7606	1.3772	8.0652	8.5785	7.5800	1.0640	16.880	17.991	26.052
1.92	3.8313	1.3772	8.3712	8.7398	7.7200	1.0844	17.718	18.906	27.298
1.96	3.9019	1.3771	8.6828	8.9010	7.8600	1.1047	18.581	19.850	28.578
2.00	3.9725	1.3771	9.0000	9.0623	8.0000	1.1250	19.469	20.823	29.894
2.05	4.0607	1.3771	9.4044	9.2638	8.1750	1.1504	20.614	22.079	31.587
2.10	4.1488	1.3770	9.8175	9.4654	8.3500	1.1757	21.799	23.381	33.336
2.15	4.2369	1.3769	10.239	9.6669	8.5250	1.2011	23.025	24.730	35.142
2.20	4.3249	1.3768	10.670	9.8685	8.7000	1.2264	24.292	26.126	37.004
2.25	4.4128	1.3767	11.109	10.070	8.8750	1.2518	25.600	27.569	38.923
2.30	4.5008	1.3765	11.557	10.272	9.0500	1.2771	26.950	29.061	40.901
2.35	4.5886	1.3764	12.014	10.473	9.2250	1.3024	28.343	30.602	42.937
2.40	4.6765	1.3763	12.480	10.675	9.4000	1.3277	29.778	32.192	45.032
2.45	4.7643	1.3761	12.954	10.876	9.5750	1.3529	31.257	33.833	47.186
2.50	4.8520	1.3760	13.437	11.078	9.7500	1.3782	32.780	35.524	49.401
2.55	4.9398	1.3758	13.929	11.279	9.9250	1.4035	34.347	37.267	51.676
2.60	5.0275	1.3757	14.430	11.481	10.100	1.4287	35.959	39.062	54.013
2.65	5.1151	1.3755	14.939	11.682	10.275	1.4540	37.616	40.909	56.412
2.70	5.2028	1.3753	15.457	11.884	10.450	1.4792	39.319	42.810	58.872
2.75	5.2904	1.3752	15.984	12.086	10.625	1.5044	41.068	44.765	61.396

C58 Trapezoidal - 2.0 to 1 side-slope

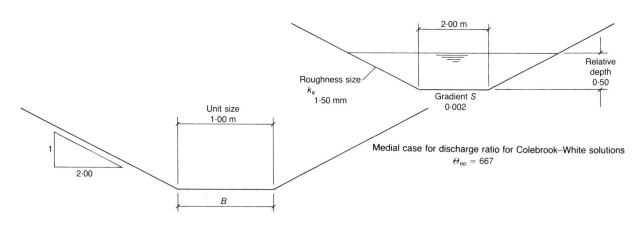

Unit size 1·00 m

2·00 m

Roughness size k_s 1·50 mm

Gradient S 0·002

Relative depth 0·50

Medial case for discharge ratio for Colebrook–White solutions $\Theta_{ep} = 667$

B

Rel. dpth	U.equ. dia. $D_{ep(u)}$	Equiv. disch. factor	Unit sect. area	Unit wetted perim.	Unit surf. brdth	Unit mean depth	Discharge ratios $Q/Q_{0.50}$		U.crit. disch. Q_{uc}
Y	(m)	J	A_u (m²)	P_u (m)	B_{us} (m)	y_{um} (m)	medial	Mann'g	(m³s⁻¹)
0.02	0.0764	0.2202	0.0208	1.0894	1.0800	0.0193	0.0035	0.0033	0.0090
0.04	0.1466	0.3906	0.0432	1.1789	1.1600	0.0372	0.0112	0.0104	0.0261
0.06	0.2119	0.5249	0.0672	1.2683	1.2400	0.0542	0.0221	0.0207	0.0490
0.08	0.2734	0.6325	0.0928	1.3578	1.3200	0.0703	0.0359	0.0339	0.0771
0.10	0.3317	0.7200	0.1200	1.4472	1.4000	0.0857	0.0526	0.0499	0.1100
0.12	0.3873	0.7919	0.1488	1.5367	1.4800	0.1005	0.0720	0.0686	0.1478
0.14	0.4408	0.8516	0.1792	1.6261	1.5600	0.1149	0.0941	0.0901	0.1902
0.16	0.4924	0.9018	0.2112	1.7155	1.6400	0.1288	0.1190	0.1143	0.2373
0.18	0.5425	0.9442	0.2448	1.8050	1.7200	0.1423	0.1466	0.1414	0.2892
0.20	0.5912	0.9804	0.2800	1.8944	1.8000	0.1556	0.1770	0.1712	0.3458
0.22	0.6388	1.0115	0.3168	1.9839	1.8800	0.1685	0.2102	0.2040	0.4072
0.24	0.6853	1.0383	0.3552	2.0733	1.9600	0.1812	0.2463	0.2397	0.4735
0.26	0.7309	1.0617	0.3952	2.1628	2.0400	0.1937	0.2853	0.2784	0.5447
0.28	0.7758	1.0821	0.4368	2.2522	2.1200	0.2060	0.3273	0.3202	0.6209
0.30	0.8199	1.1000	0.4800	2.3416	2.2000	0.2182	0.3723	0.3651	0.7021
0.32	0.8635	1.1158	0.5248	2.4311	2.2800	0.2302	0.4203	0.4132	0.7885
0.34	0.9065	1.1298	0.5712	2.5205	2.3600	0.2420	0.4715	0.4645	0.8800
0.36	0.9490	1.1422	0.6192	2.6100	2.4400	0.2538	0.5259	0.5192	0.9768
0.38	0.9910	1.1533	0.6688	2.6994	2.5200	0.2654	0.5835	0.5772	1.0790
0.40	1.0327	1.1633	0.7200	2.7889	2.6000	0.2769	0.6444	0.6387	1.1865
0.42	1.0740	1.1722	0.7728	2.8783	2.6800	0.2884	0.7086	0.7037	1.2996
0.44	1.1149	1.1802	0.8272	2.9677	2.7600	0.2997	0.7762	0.7722	1.4181
0.46	1.1556	1.1875	0.8832	3.0572	2.8400	0.3110	0.8473	0.8444	1.5424
0.48	1.1959	1.1940	0.9408	3.1466	2.9200	0.3222	0.9219	0.9203	1.6723
0.50	1.2361	1.2000	1.0000	3.2361	3.0000	0.3333	1.0000	1.0000	1.8080
0.52	1.2760	1.2054	1.0608	3.3255	3.0800	0.3444	1.0818	1.0835	1.9496
0.54	1.3156	1.2103	1.1232	3.4150	3.1600	0.3554	1.1672	1.1709	2.0970
0.56	1.3551	1.2148	1.1872	3.5044	3.2400	0.3664	1.2564	1.2622	2.2505
0.58	1.3944	1.2189	1.2528	3.5938	3.3200	0.3773	1.3493	1.3576	2.4100
0.60	1.4335	1.2227	1.3200	3.6833	3.4000	0.3882	1.4460	1.4571	2.5756
0.62	1.4725	1.2261	1.3888	3.7727	3.4800	0.3991	1.5467	1.5607	2.7475
0.64	1.5113	1.2293	1.4592	3.8622	3.5600	0.4099	1.6512	1.6685	2.9256
0.66	1.5500	1.2322	1.5312	3.9516	3.6400	0.4207	1.7598	1.7805	3.1100
0.68	1.5885	1.2349	1.6048	4.0411	3.7200	0.4314	1.8723	1.8969	3.3008
0.70	1.6269	1.2374	1.6800	4.1305	3.8000	0.4421	1.9890	2.0177	3.4981
0.72	1.6652	1.2397	1.7568	4.2199	3.8800	0.4528	2.1098	2.1429	3.7019
0.74	1.7034	1.2418	1.8352	4.3094	3.9600	0.4634	2.2348	2.2727	3.9124
0.76	1.7416	1.2438	1.9152	4.3988	4.0400	0.4741	2.3640	2.4070	4.1294
0.78	1.7796	1.2456	1.9968	4.4883	4.1200	0.4847	2.4974	2.5460	4.3532
0.80	1.8175	1.2473	2.0800	4.5777	4.2000	0.4952	2.6352	2.6896	4.5839

Rel. dpth	U.equ. dia. $D_{ep(u)}$	Equiv. disch. factor	Unit sect. area	Unit wetted perim.	Unit surf. brdth	Unit mean depth	Discharge ratios $Q/Q_{0.50}$		U.crit. disch. Q_{uc}
Y	(m)	J	A_u (m²)	P_u (m)	B_{us} (m)	y_{um} (m)	medial	Mann'g	(m³s⁻¹)
0.82	1.8553	1.2489	2.1648	4.6672	4.2800	0.5058	2.7774	2.8380	4.8213
0.84	1.8931	1.2503	2.2512	4.7566	4.3600	0.5163	2.9241	2.9911	5.0657
0.86	1.9308	1.2517	2.3392	4.8460	4.4400	0.5268	3.0752	3.1492	5.3170
0.88	1.9684	1.2529	2.4288	4.9355	4.5200	0.5373	3.2308	3.3122	5.5754
0.90	2.0060	1.2541	2.5200	5.0249	4.6000	0.5478	3.3910	3.4801	5.8409
0.92	2.0435	1.2552	2.6128	5.1144	4.6800	0.5583	3.5558	3.6531	6.1136
0.94	2.0809	1.2563	2.7072	5.2038	4.7600	0.5687	3.7252	3.8312	6.3935
0.96	2.1183	1.2572	2.8032	5.2933	4.8400	0.5792	3.8994	4.0144	6.6807
0.98	2.1556	1.2581	2.9008	5.3827	4.9200	0.5896	4.0783	4.2028	6.9752
1.00	2.1929	1.2589	3.0000	5.4721	5.0000	0.6000	4.2620	4.3965	7.2771
1.02	2.2302	1.2597	3.1008	5.5616	5.0800	0.6104	4.4506	4.5955	7.5865
1.04	2.2673	1.2605	3.2032	5.6510	5.1600	0.6208	4.6441	4.7999	7.9034
1.06	2.3045	1.2611	3.3072	5.7405	5.2400	0.6311	4.8425	5.0097	8.2278
1.08	2.3416	1.2618	3.4128	5.8299	5.3200	0.6415	5.0459	5.2250	8.5599
1.10	2.3786	1.2624	3.5200	5.9193	5.4000	0.6519	5.2543	5.4459	8.8997
1.12	2.4157	1.2630	3.6288	6.0088	5.4800	0.6622	5.4678	5.6723	9.2473
1.14	2.4526	1.2635	3.7392	6.0982	5.5600	0.6725	5.6864	5.9044	9.6026
1.16	2.4896	1.2640	3.8512	6.1877	5.6400	0.6828	5.9101	6.1421	9.9659
1.18	2.5265	1.2644	3.9648	6.2771	5.7200	0.6931	6.1391	6.3857	10.337
1.20	2.5634	1.2649	4.0800	6.3666	5.8000	0.7034	6.3732	6.6350	10.716
1.24	2.6371	1.2657	4.3152	6.5454	5.9600	0.7240	6.8575	7.1513	11.498
1.28	2.7106	1.2664	4.5568	6.7243	6.1200	0.7446	7.3633	7.6915	12.313
1.32	2.7841	1.2670	4.8048	6.9032	6.2800	0.7651	7.8909	8.2560	13.161
1.36	2.8575	1.2675	5.0592	7.0821	6.4400	0.7856	8.4406	8.8452	14.042
1.40	2.9307	1.2680	5.3200	7.2610	6.6000	0.8061	9.0129	9.4595	14.957
1.44	3.0039	1.2684	5.5872	7.4399	6.7600	0.8265	9.6079	10.099	15.907
1.48	3.0770	1.2688	5.8608	7.6188	6.9200	0.8469	10.226	10.765	16.891
1.52	3.1501	1.2691	6.1408	7.7976	7.0800	0.8673	10.868	11.457	17.909
1.56	3.2231	1.2694	6.4272	7.9765	7.2400	0.8877	11.533	12.176	18.964
1.60	3.2960	1.2696	6.7200	8.1554	7.4000	0.9081	12.223	12.922	20.054
1.64	3.3688	1.2698	7.0192	8.3343	7.5600	0.9285	12.936	13.696	21.180
1.68	3.4416	1.2700	7.3248	8.5132	7.7200	0.9488	13.675	14.497	22.343
1.72	3.5144	1.2702	7.6368	8.6921	7.8800	0.9691	14.439	15.327	23.543
1.76	3.5871	1.2703	7.9552	8.8710	8.0400	0.9895	15.227	16.185	24.780
1.80	3.6597	1.2704	8.2800	9.0498	8.2000	1.0098	16.042	17.073	26.055
1.84	3.7323	1.2705	8.6112	9.2287	8.3600	1.0300	16.882	17.990	27.369
1.88	3.8049	1.2706	8.9488	9.4076	8.5200	1.0503	17.749	18.937	28.720
1.92	3.8775	1.2707	9.2928	9.5865	8.6800	1.0706	18.642	19.914	30.111
1.96	3.9500	1.2707	9.6432	9.7654	8.8400	1.0909	19.562	20.921	31.540
2.00	4.0224	1.2707	10.000	9.9443	9.0000	1.1111	20.509	21.960	33.010
2.05	4.1130	1.2708	10.455	10.168	9.2000	1.1364	21.731	23.302	34.902
2.10	4.2034	1.2708	10.920	10.391	9.4000	1.1617	22.997	24.694	36.858
2.15	4.2939	1.2708	11.395	10.615	9.6000	1.1870	24.306	26.137	38.877
2.20	4.3843	1.2708	11.880	10.839	9.8000	1.2122	25.660	27.630	40.961
2.25	4.4747	1.2707	12.375	11.062	10.000	1.2375	27.059	29.176	43.110
2.30	4.5650	1.2707	12.880	11.286	10.200	1.2627	28.503	30.774	45.325
2.35	4.6553	1.2707	13.395	11.510	10.400	1.2880	29.994	32.425	47.606
2.40	4.7455	1.2706	13.920	11.733	10.600	1.3132	31.530	34.130	49.954
2.45	4.8358	1.2706	14.455	11.957	10.800	1.3384	33.114	35.889	52.369
2.50	4.9260	1.2705	15.000	12.180	11.000	1.3636	34.746	37.704	54.853
2.55	5.0161	1.2704	15.555	12.404	11.200	1.3888	36.426	39.575	57.406
2.60	5.1063	1.2704	16.120	12.628	11.400	1.4140	38.154	41.502	60.028
2.65	5.1964	1.2703	16.695	12.851	11.600	1.4392	39.931	43.487	62.721
2.70	5.2865	1.2702	17.280	13.075	11.800	1.4644	41.758	45.530	65.484
2.75	5.3766	1.2701	17.875	13.298	12.000	1.4896	43.636	47.631	68.319

C59

Trapezoidal - 2.5 to 1 side-slope

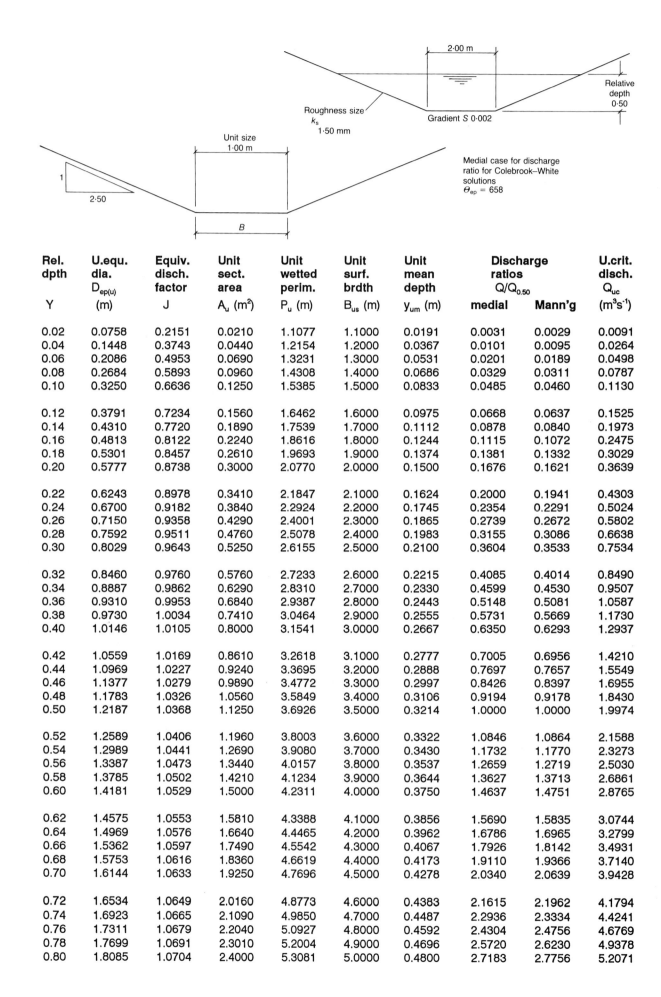

Rel. dpth Y	U.equ. dia. $D_{ep(u)}$ (m)	Equiv. disch. factor J	Unit sect. area A_u (m²)	Unit wetted perim. P_u (m)	Unit surf. brdth B_{us} (m)	Unit mean depth y_{um} (m)	Discharge ratios $Q/Q_{0.50}$ medial	Mann'g	U.crit. disch. Q_{uc} (m³s⁻¹)
0.02	0.0758	0.2151	0.0210	1.1077	1.1000	0.0191	0.0031	0.0029	0.0091
0.04	0.1448	0.3743	0.0440	1.2154	1.2000	0.0367	0.0101	0.0095	0.0264
0.06	0.2086	0.4953	0.0690	1.3231	1.3000	0.0531	0.0201	0.0189	0.0498
0.08	0.2684	0.5893	0.0960	1.4308	1.4000	0.0686	0.0329	0.0311	0.0787
0.10	0.3250	0.6636	0.1250	1.5385	1.5000	0.0833	0.0485	0.0460	0.1130
0.12	0.3791	0.7234	0.1560	1.6462	1.6000	0.0975	0.0668	0.0637	0.1525
0.14	0.4310	0.7720	0.1890	1.7539	1.7000	0.1112	0.0878	0.0840	0.1973
0.16	0.4813	0.8122	0.2240	1.8616	1.8000	0.1244	0.1115	0.1072	0.2475
0.18	0.5301	0.8457	0.2610	1.9693	1.9000	0.1374	0.1381	0.1332	0.3029
0.20	0.5777	0.8738	0.3000	2.0770	2.0000	0.1500	0.1676	0.1621	0.3639
0.22	0.6243	0.8978	0.3410	2.1847	2.1000	0.1624	0.2000	0.1941	0.4303
0.24	0.6700	0.9182	0.3840	2.2924	2.2000	0.1745	0.2354	0.2291	0.5024
0.26	0.7150	0.9358	0.4290	2.4001	2.3000	0.1865	0.2739	0.2672	0.5802
0.28	0.7592	0.9511	0.4760	2.5078	2.4000	0.1983	0.3155	0.3086	0.6638
0.30	0.8029	0.9643	0.5250	2.6155	2.5000	0.2100	0.3604	0.3533	0.7534
0.32	0.8460	0.9760	0.5760	2.7233	2.6000	0.2215	0.4085	0.4014	0.8490
0.34	0.8887	0.9862	0.6290	2.8310	2.7000	0.2330	0.4599	0.4530	0.9507
0.36	0.9310	0.9953	0.6840	2.9387	2.8000	0.2443	0.5148	0.5081	1.0587
0.38	0.9730	1.0034	0.7410	3.0464	2.9000	0.2555	0.5731	0.5669	1.1730
0.40	1.0146	1.0105	0.8000	3.1541	3.0000	0.2667	0.6350	0.6293	1.2937
0.42	1.0559	1.0169	0.8610	3.2618	3.1000	0.2777	0.7005	0.6956	1.4210
0.44	1.0969	1.0227	0.9240	3.3695	3.2000	0.2888	0.7697	0.7657	1.5549
0.46	1.1377	1.0279	0.9890	3.4772	3.3000	0.2997	0.8426	0.8397	1.6955
0.48	1.1783	1.0326	1.0560	3.5849	3.4000	0.3106	0.9194	0.9178	1.8430
0.50	1.2187	1.0368	1.1250	3.6926	3.5000	0.3214	1.0000	1.0000	1.9974
0.52	1.2589	1.0406	1.1960	3.8003	3.6000	0.3322	1.0846	1.0864	2.1588
0.54	1.2989	1.0441	1.2690	3.9080	3.7000	0.3430	1.1732	1.1770	2.3273
0.56	1.3387	1.0473	1.3440	4.0157	3.8000	0.3537	1.2659	1.2719	2.5030
0.58	1.3785	1.0502	1.4210	4.1234	3.9000	0.3644	1.3627	1.3713	2.6861
0.60	1.4181	1.0529	1.5000	4.2311	4.0000	0.3750	1.4637	1.4751	2.8765
0.62	1.4575	1.0553	1.5810	4.3388	4.1000	0.3856	1.5690	1.5835	3.0744
0.64	1.4969	1.0576	1.6640	4.4465	4.2000	0.3962	1.6786	1.6965	3.2799
0.66	1.5362	1.0597	1.7490	4.5542	4.3000	0.4067	1.7926	1.8142	3.4931
0.68	1.5753	1.0616	1.8360	4.6619	4.4000	0.4173	1.9110	1.9366	3.7140
0.70	1.6144	1.0633	1.9250	4.7696	4.5000	0.4278	2.0340	2.0639	3.9428
0.72	1.6534	1.0649	2.0160	4.8773	4.6000	0.4383	2.1615	2.1962	4.1794
0.74	1.6923	1.0665	2.1090	4.9850	4.7000	0.4487	2.2936	2.3334	4.4241
0.76	1.7311	1.0679	2.2040	5.0927	4.8000	0.4592	2.4304	2.4756	4.6769
0.78	1.7699	1.0691	2.3010	5.2004	4.9000	0.4696	2.5720	2.6230	4.9378
0.80	1.8085	1.0704	2.4000	5.3081	5.0000	0.4800	2.7183	2.7756	5.2071

Rel. dpth	U.equ. dia. $D_{ep(u)}$	Equiv. disch. factor	Unit sect. area	Unit wetted perim.	Unit surf. brdth	Unit mean depth	Discharge ratios $Q/Q_{0.50}$		U.crit. disch. Q_{uc}
Y	(m)	J	A_u (m²)	P_u (m)	B_{us} (m)	y_{um} (m)	medial	Mann'g	(m³s⁻¹)
0.82	1.8472	1.0715	2.5010	5.4158	5.1000	0.4904	2.8695	2.9334	5.4846
0.84	1.8857	1.0725	2.6040	5.5235	5.2000	0.5008	3.0257	3.0966	5.7706
0.86	1.9243	1.0735	2.7090	5.6312	5.3000	0.5111	3.1868	3.2652	6.0651
0.88	1.9627	1.0744	2.8160	5.7389	5.4000	0.5215	3.3529	3.4393	6.3681
0.90	2.0011	1.0753	2.9250	5.8466	5.5000	0.5318	3.5241	3.6189	6.6799
0.92	2.0395	1.0761	3.0360	5.9544	5.6000	0.5421	3.7005	3.8040	7.0003
0.94	2.0778	1.0768	3.1490	6.0621	5.7000	0.5525	3.8820	3.9949	7.3296
0.96	2.1161	1.0775	3.2640	6.1698	5.8000	0.5628	4.0688	4.1915	7.6678
0.98	2.1544	1.0781	3.3810	6.2775	5.9000	0.5731	4.2609	4.3939	8.0150
1.00	2.1926	1.0788	3.5000	6.3852	6.0000	0.5833	4.4583	4.6022	8.3712
1.02	2.2308	1.0793	3.6210	6.4929	6.1000	0.5936	4.6611	4.8164	8.7365
1.04	2.2689	1.0799	3.7440	6.6006	6.2000	0.6039	4.8694	5.0366	9.1110
1.06	2.3070	1.0804	3.8690	6.7083	6.3000	0.6141	5.0832	5.2629	9.4948
1.08	2.3451	1.0809	3.9960	6.8160	6.4000	0.6244	5.3026	5.4953	9.8880
1.10	2.3831	1.0813	4.1250	6.9237	6.5000	0.6346	5.5276	5.7339	10.291
1.12	2.4211	1.0817	4.2560	7.0314	6.6000	0.6448	5.7583	5.9787	10.703
1.14	2.4591	1.0821	4.3890	7.1391	6.7000	0.6551	5.9946	6.2299	11.124
1.16	2.4971	1.0825	4.5240	7.2468	6.8000	0.6653	6.2368	6.4874	11.556
1.18	2.5350	1.0829	4.6610	7.3545	6.9000	0.6755	6.4848	6.7514	11.996
1.20	2.5730	1.0832	4.8000	7.4622	7.0000	0.6857	6.7386	7.0219	12.447
1.24	2.6487	1.0838	5.0840	7.6776	7.2000	0.7061	7.2641	7.5827	13.378
1.28	2.7244	1.0844	5.3760	7.8930	7.4000	0.7265	7.8136	8.1703	14.349
1.32	2.8001	1.0849	5.6760	8.1084	7.6000	0.7468	8.3877	8.7851	15.361
1.36	2.8756	1.0853	5.9840	8.3238	7.8000	0.7672	8.9866	9.4276	16.413
1.40	2.9511	1.0857	6.3000	8.5392	8.0000	0.7875	9.6107	10.098	17.508
1.44	3.0265	1.0860	6.6240	8.7546	8.2000	0.8078	10.260	10.798	18.644
1.48	3.1019	1.0864	6.9560	8.9700	8.4000	0.8281	10.936	11.527	19.823
1.52	3.1772	1.0866	7.2960	9.1855	8.6000	0.8484	11.638	12.285	21.044
1.56	3.2525	1.0869	7.6440	9.4009	8.8000	0.8686	12.367	13.073	22.310
1.60	3.3277	1.0871	8.0000	9.6163	9.0000	0.8889	13.123	13.892	23.620
1.64	3.4029	1.0873	8.3640	9.8317	9.2000	0.9091	13.907	14.743	24.974
1.68	3.4780	1.0875	8.7360	10.047	9.4000	0.9294	14.718	15.624	26.373
1.72	3.5531	1.0877	9.1160	10.262	9.6000	0.9496	15.557	16.538	27.818
1.76	3.6282	1.0878	9.5040	10.478	9.8000	0.9698	16.425	17.483	29.309
1.80	3.7033	1.0880	9.9000	10.693	10.000	0.9900	17.322	18.462	30.847
1.84	3.7783	1.0881	10.304	10.909	10.200	1.0102	18.248	19.474	32.432
1.88	3.8533	1.0882	10.716	11.124	10.400	1.0304	19.204	20.520	34.064
1.92	3.9282	1.0883	11.136	11.340	10.600	1.0506	20.189	21.600	35.744
1.96	4.0031	1.0884	11.564	11.555	10.800	1.0707	21.205	22.714	37.472
2.00	4.0781	1.0884	12.000	11.770	11.000	1.0909	22.252	23.864	39.250
2.05	4.1717	1.0885	12.556	12.040	11.250	1.1161	23.603	25.351	41.541
2.10	4.2652	1.0886	13.125	12.309	11.500	1.1413	25.003	26.894	43.910
2.15	4.3588	1.0886	13.706	12.578	11.750	1.1665	26.453	28.494	46.357
2.20	4.4523	1.0887	14.300	12.847	12.000	1.1917	27.954	30.152	48.885
2.25	4.5458	1.0887	14.906	13.117	12.250	1.2168	29.505	31.869	51.493
2.30	4.6392	1.0888	15.525	13.386	12.500	1.2420	31.107	33.645	54.182
2.35	4.7327	1.0888	16.156	13.655	12.750	1.2672	32.762	35.482	56.953
2.40	4.8261	1.0888	16.800	13.924	13.000	1.2923	34.469	37.379	59.807
2.45	4.9195	1.0888	17.456	14.194	13.250	1.3175	36.230	39.339	62.745
2.50	5.0128	1.0888	18.125	14.463	13.500	1.3426	38.044	41.361	65.767
2.55	5.1062	1.0889	18.806	14.732	13.750	1.3677	39.913	43.447	68.875
2.60	5.1995	1.0889	19.500	15.001	14.000	1.3929	41.837	45.597	72.069
2.65	5.2928	1.0889	20.206	15.271	14.250	1.4180	43.817	47.812	75.350
2.70	5.3861	1.0888	20.925	15.540	14.500	1.4431	45.853	50.093	78.718
2.75	5.4794	1.0888	21.656	15.809	14.750	1.4682	47.945	52.440	82.175

C60

Trapezoidal - 3.0 to 1 side-slope

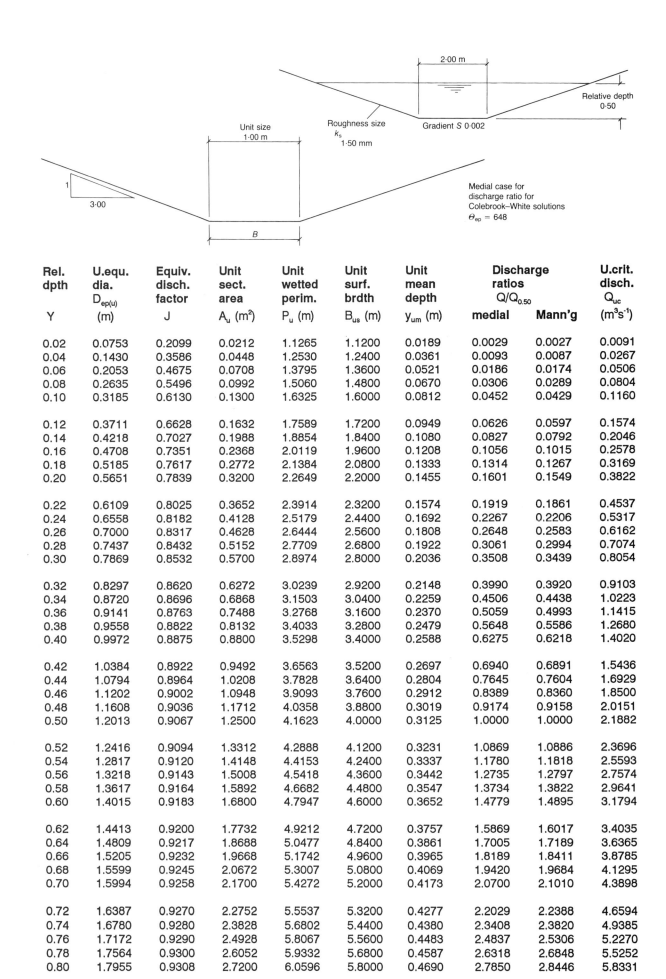

Rel. dpth Y	U.equ. dia. $D_{ep(u)}$ (m)	Equiv. disch. factor J	Unit sect. area A_u (m²)	Unit wetted perim. P_u (m)	Unit surf. brdth B_{us} (m)	Unit mean depth y_{um} (m)	Discharge ratios $Q/Q_{0.50}$		U.crit. disch. Q_{uc} (m³s⁻¹)
							medial	Mann'g	
0.02	0.0753	0.2099	0.0212	1.1265	1.1200	0.0189	0.0029	0.0027	0.0091
0.04	0.1430	0.3586	0.0448	1.2530	1.2400	0.0361	0.0093	0.0087	0.0267
0.06	0.2053	0.4675	0.0708	1.3795	1.3600	0.0521	0.0186	0.0174	0.0506
0.08	0.2635	0.5496	0.0992	1.5060	1.4800	0.0670	0.0306	0.0289	0.0804
0.10	0.3185	0.6130	0.1300	1.6325	1.6000	0.0812	0.0452	0.0429	0.1160
0.12	0.3711	0.6628	0.1632	1.7589	1.7200	0.0949	0.0626	0.0597	0.1574
0.14	0.4218	0.7027	0.1988	1.8854	1.8400	0.1080	0.0827	0.0792	0.2046
0.16	0.4708	0.7351	0.2368	2.0119	1.9600	0.1208	0.1056	0.1015	0.2578
0.18	0.5185	0.7617	0.2772	2.1384	2.0800	0.1333	0.1314	0.1267	0.3169
0.20	0.5651	0.7839	0.3200	2.2649	2.2000	0.1455	0.1601	0.1549	0.3822
0.22	0.6109	0.8025	0.3652	2.3914	2.3200	0.1574	0.1919	0.1861	0.4537
0.24	0.6558	0.8182	0.4128	2.5179	2.4400	0.1692	0.2267	0.2206	0.5317
0.26	0.7000	0.8317	0.4628	2.6444	2.5600	0.1808	0.2648	0.2583	0.6162
0.28	0.7437	0.8432	0.5152	2.7709	2.6800	0.1922	0.3061	0.2994	0.7074
0.30	0.7869	0.8532	0.5700	2.8974	2.8000	0.2036	0.3508	0.3439	0.8054
0.32	0.8297	0.8620	0.6272	3.0239	2.9200	0.2148	0.3990	0.3920	0.9103
0.34	0.8720	0.8696	0.6868	3.1503	3.0400	0.2259	0.4506	0.4438	1.0223
0.36	0.9141	0.8763	0.7488	3.2768	3.1600	0.2370	0.5059	0.4993	1.1415
0.38	0.9558	0.8822	0.8132	3.4033	3.2800	0.2479	0.5648	0.5586	1.2680
0.40	0.9972	0.8875	0.8800	3.5298	3.4000	0.2588	0.6275	0.6218	1.4020
0.42	1.0384	0.8922	0.9492	3.6563	3.5200	0.2697	0.6940	0.6891	1.5436
0.44	1.0794	0.8964	1.0208	3.7828	3.6400	0.2804	0.7645	0.7604	1.6929
0.46	1.1202	0.9002	1.0948	3.9093	3.7600	0.2912	0.8389	0.8360	1.8500
0.48	1.1608	0.9036	1.1712	4.0358	3.8800	0.3019	0.9174	0.9158	2.0151
0.50	1.2013	0.9067	1.2500	4.1623	4.0000	0.3125	1.0000	1.0000	2.1882
0.52	1.2416	0.9094	1.3312	4.2888	4.1200	0.3231	1.0869	1.0886	2.3696
0.54	1.2817	0.9120	1.4148	4.4153	4.2400	0.3337	1.1780	1.1818	2.5593
0.56	1.3218	0.9143	1.5008	4.5418	4.3600	0.3442	1.2735	1.2797	2.7574
0.58	1.3617	0.9164	1.5892	4.6682	4.4800	0.3547	1.3734	1.3822	2.9641
0.60	1.4015	0.9183	1.6800	4.7947	4.6000	0.3652	1.4779	1.4895	3.1794
0.62	1.4413	0.9200	1.7732	4.9212	4.7200	0.3757	1.5869	1.6017	3.4035
0.64	1.4809	0.9217	1.8688	5.0477	4.8400	0.3861	1.7005	1.7189	3.6365
0.66	1.5205	0.9232	1.9668	5.1742	4.9600	0.3965	1.8189	1.8411	3.8785
0.68	1.5599	0.9245	2.0672	5.3007	5.0800	0.4069	1.9420	1.9684	4.1295
0.70	1.5994	0.9258	2.1700	5.4272	5.2000	0.4173	2.0700	2.1010	4.3898
0.72	1.6387	0.9270	2.2752	5.5537	5.3200	0.4277	2.2029	2.2388	4.6594
0.74	1.6780	0.9280	2.3828	5.6802	5.4400	0.4380	2.3408	2.3820	4.9385
0.76	1.7172	0.9290	2.4928	5.8067	5.5600	0.4483	2.4837	2.5306	5.2270
0.78	1.7564	0.9300	2.6052	5.9332	5.6800	0.4587	2.6318	2.6848	5.5252
0.80	1.7955	0.9308	2.7200	6.0596	5.8000	0.4690	2.7850	2.8446	5.8331

Trapezoidal - 3.0 to 1 side-slope

Rel. dpth	U.equ. dia. $D_{ep(u)}$	Equiv. disch. factor	Unit sect. area	Unit wetted perim.	Unit surf. brdth	Unit mean depth	Discharge ratios $Q/Q_{0.50}$		U.crit. disch. Q_{uc}
Y	(m)	J	A_u (m²)	P_u (m)	B_{us} (m)	y_{um} (m)	medial	Mann'g	(m³s⁻¹)
0.82	1.8346	0.9316	2.8372	6.1861	5.9200	0.4793	2.9435	3.0101	6.1508
0.84	1.8736	0.9324	2.9568	6.3126	6.0400	0.4895	3.1073	3.1813	6.4785
0.86	1.9126	0.9331	3.0788	6.4391	6.1600	0.4998	3.2764	3.3583	6.8162
0.88	1.9515	0.9338	3.2032	6.5656	6.2800	0.5101	3.4510	3.5413	7.1640
0.90	1.9904	0.9344	3.3300	6.6921	6.4000	0.5203	3.6311	3.7302	7.5221
0.92	2.0293	0.9349	3.4592	6.8186	6.5200	0.5306	3.8167	3.9252	7.8904
0.94	2.0681	0.9355	3.5908	6.9451	6.6400	0.5408	4.0080	4.1264	8.2692
0.96	2.1069	0.9360	3.7248	7.0716	6.7600	0.5510	4.2049	4.3338	8.6585
0.98	2.1457	0.9365	3.8612	7.1981	6.8800	0.5612	4.4075	4.5474	9.0584
1.00	2.1844	0.9369	4.0000	7.3246	7.0000	0.5714	4.6160	4.7674	9.4689
1.02	2.2232	0.9373	4.1412	7.4510	7.1200	0.5816	4.8303	4.9939	9.8903
1.04	2.2618	0.9377	4.2848	7.5775	7.2400	0.5918	5.0506	5.2268	10.323
1.06	2.3005	0.9381	4.4308	7.7040	7.3600	0.6020	5.2768	5.4663	10.766
1.08	2.3392	0.9384	4.5792	7.8305	7.4800	0.6122	5.5091	5.7125	11.220
1.10	2.3778	0.9388	4.7300	7.9570	7.6000	0.6224	5.7474	5.9654	11.685
1.12	2.4164	0.9391	4.8832	8.0835	7.7200	0.6325	5.9919	6.2251	12.162
1.14	2.4550	0.9394	5.0388	8.2100	7.8400	0.6427	6.2426	6.4916	12.650
1.16	2.4935	0.9397	5.1968	8.3365	7.9600	0.6529	6.4996	6.7651	13.149
1.18	2.5321	0.9399	5.3572	8.4630	8.0800	0.6630	6.7629	7.0456	13.660
1.20	2.5706	0.9402	5.5200	8.5895	8.2000	0.6732	7.0326	7.3332	14.183
1.24	2.6476	0.9406	5.8528	8.8424	8.4400	0.6935	7.5913	7.9298	15.263
1.28	2.7245	0.9410	6.1952	9.0954	8.6800	0.7137	8.1762	8.5555	16.390
1.32	2.8014	0.9414	6.5472	9.3484	8.9200	0.7340	8.7877	9.2110	17.566
1.36	2.8782	0.9417	6.9088	9.6014	9.1600	0.7542	9.4263	9.8966	18.790
1.40	2.9550	0.9420	7.2800	9.8544	9.4000	0.7745	10.092	10.613	20.063
1.44	3.0318	0.9423	7.6608	10.107	9.6400	0.7947	10.786	11.361	21.386
1.48	3.1085	0.9426	8.0512	10.360	9.8800	0.8149	11.508	12.140	22.760
1.52	3.1851	0.9428	8.4512	10.613	10.120	0.8351	12.259	12.952	24.185
1.56	3.2618	0.9430	8.8608	10.866	10.360	0.8553	13.040	13.797	25.662
1.60	3.3383	0.9432	9.2800	11.119	10.600	0.8755	13.849	14.675	27.191
1.64	3.4149	0.9433	9.7088	11.372	10.840	0.8956	14.689	15.586	28.774
1.68	3.4914	0.9435	10.147	11.625	11.080	0.9158	15.559	16.533	30.410
1.72	3.5679	0.9436	10.595	11.878	11.320	0.9360	16.460	17.514	32.100
1.76	3.6444	0.9438	11.053	12.131	11.560	0.9561	17.392	18.531	33.845
1.80	3.7209	0.9439	11.520	12.384	11.800	0.9763	18.355	19.583	35.645
1.84	3.7973	0.9440	11.997	12.637	12.040	0.9964	19.350	20.672	37.501
1.88	3.8737	0.9441	12.483	12.890	12.280	1.0165	20.378	21.798	39.414
1.92	3.9501	0.9442	12.979	13.143	12.520	1.0367	21.438	22.961	41.384
1.96	4.0265	0.9442	13.485	13.396	12.760	1.0568	22.531	24.162	43.411
2.00	4.1028	0.9443	14.000	13.649	13.000	1.0769	23.658	25.401	45.497
2.05	4.1983	0.9444	14.657	13.965	13.300	1.1021	25.114	27.005	48.186
2.10	4.2936	0.9445	15.330	14.282	13.600	1.1272	26.624	28.670	50.969
2.15	4.3890	0.9445	16.018	14.598	13.900	1.1523	28.187	30.397	53.845
2.20	4.4844	0.9446	16.720	14.914	14.200	1.1775	29.806	32.189	56.816
2.25	4.5797	0.9446	17.438	15.230	14.500	1.2026	31.480	34.044	59.883
2.30	4.6750	0.9447	18.170	15.546	14.800	1.2277	33.211	35.965	63.047
2.35	4.7703	0.9447	18.917	15.863	15.100	1.2528	34.998	37.951	66.308
2.40	4.8656	0.9448	19.680	16.179	15.400	1.2779	36.844	40.005	69.669
2.45	4.9608	0.9448	20.458	16.495	15.700	1.3030	38.747	42.126	73.129
2.50	5.0561	0.9448	21.250	16.811	16.000	1.3281	40.710	44.317	76.690
2.55	5.1513	0.9448	22.057	17.128	16.300	1.3532	42.732	46.576	80.353
2.60	5.2465	0.9449	22.880	17.444	16.600	1.3783	44.814	48.907	84.118
2.65	5.3418	0.9449	23.718	17.760	16.900	1.4034	46.958	51.308	87.987
2.70	5.4370	0.9449	24.570	18.076	17.200	1.4285	49.163	53.782	91.961
2.75	5.5321	0.9449	25.437	18.393	17.500	1.4536	51.430	56.329	96.040

Regime trapezoidal - 1.5 to 1 side-slope

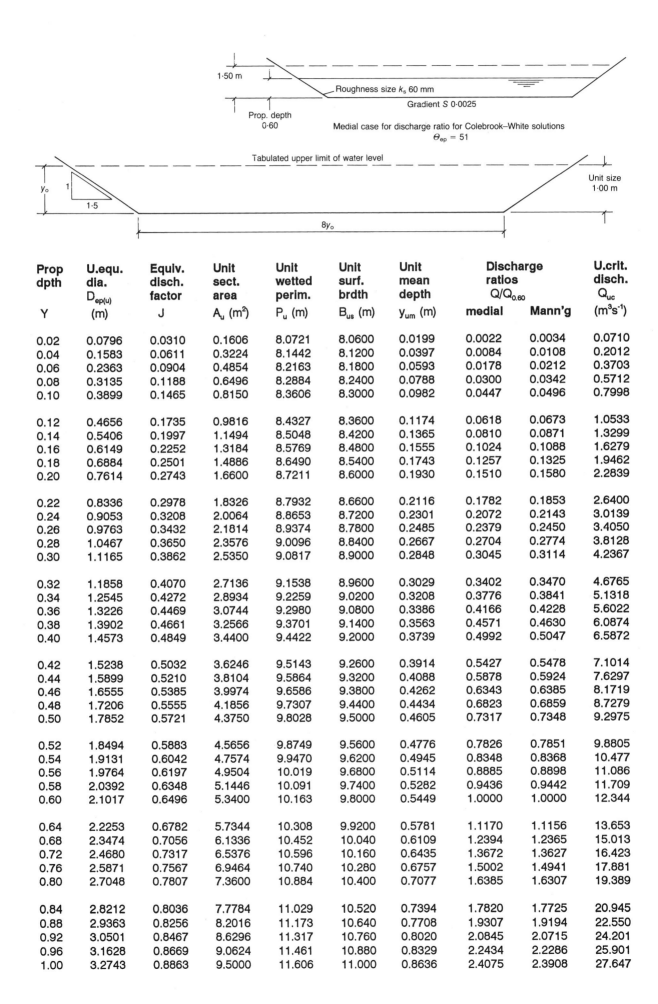

Prop dpth	U.equ. dia. $D_{ep(u)}$	Equiv. disch. factor	Unit sect. area	Unit wetted perim.	Unit surf. brdth	Unit mean depth	Discharge ratios $Q/Q_{0.60}$		U.crit. disch. Q_{uc}
Y	(m)	J	A_u (m^2)	P_u (m)	B_{us} (m)	y_{um} (m)	medial	Mann'g	(m^{3}s^{-1})
0.02	0.0796	0.0310	0.1606	8.0721	8.0600	0.0199	0.0022	0.0034	0.0710
0.04	0.1583	0.0611	0.3224	8.1442	8.1200	0.0397	0.0084	0.0108	0.2012
0.06	0.2363	0.0904	0.4854	8.2163	8.1800	0.0593	0.0178	0.0212	0.3703
0.08	0.3135	0.1188	0.6496	8.2884	8.2400	0.0788	0.0300	0.0342	0.5712
0.10	0.3899	0.1465	0.8150	8.3606	8.3000	0.0982	0.0447	0.0496	0.7998
0.12	0.4656	0.1735	0.9816	8.4327	8.3600	0.1174	0.0618	0.0673	1.0533
0.14	0.5406	0.1997	1.1494	8.5048	8.4200	0.1365	0.0810	0.0871	1.3299
0.16	0.6149	0.2252	1.3184	8.5769	8.4800	0.1555	0.1024	0.1088	1.6279
0.18	0.6884	0.2501	1.4886	8.6490	8.5400	0.1743	0.1257	0.1325	1.9462
0.20	0.7614	0.2743	1.6600	8.7211	8.6000	0.1930	0.1510	0.1580	2.2839
0.22	0.8336	0.2978	1.8326	8.7932	8.6600	0.2116	0.1782	0.1853	2.6400
0.24	0.9053	0.3208	2.0064	8.8653	8.7200	0.2301	0.2072	0.2143	3.0139
0.26	0.9763	0.3432	2.1814	8.9374	8.7800	0.2485	0.2379	0.2450	3.4050
0.28	1.0467	0.3650	2.3576	9.0096	8.8400	0.2667	0.2704	0.2774	3.8128
0.30	1.1165	0.3862	2.5350	9.0817	8.9000	0.2848	0.3045	0.3114	4.2367
0.32	1.1858	0.4070	2.7136	9.1538	8.9600	0.3029	0.3402	0.3470	4.6765
0.34	1.2545	0.4272	2.8934	9.2259	9.0200	0.3208	0.3776	0.3841	5.1318
0.36	1.3226	0.4469	3.0744	9.2980	9.0800	0.3386	0.4166	0.4228	5.6022
0.38	1.3902	0.4661	3.2566	9.3701	9.1400	0.3563	0.4571	0.4630	6.0874
0.40	1.4573	0.4849	3.4400	9.4422	9.2000	0.3739	0.4992	0.5047	6.5872
0.42	1.5238	0.5032	3.6246	9.5143	9.2600	0.3914	0.5427	0.5478	7.1014
0.44	1.5899	0.5210	3.8104	9.5864	9.3200	0.4088	0.5878	0.5924	7.6297
0.46	1.6555	0.5385	3.9974	9.6586	9.3800	0.4262	0.6343	0.6385	8.1719
0.48	1.7206	0.5555	4.1856	9.7307	9.4400	0.4434	0.6823	0.6859	8.7279
0.50	1.7852	0.5721	4.3750	9.8028	9.5000	0.4605	0.7317	0.7348	9.2975
0.52	1.8494	0.5883	4.5656	9.8749	9.5600	0.4776	0.7826	0.7851	9.8805
0.54	1.9131	0.6042	4.7574	9.9470	9.6200	0.4945	0.8348	0.8368	10.477
0.56	1.9764	0.6197	4.9504	10.019	9.6800	0.5114	0.8885	0.8898	11.086
0.58	2.0392	0.6348	5.1446	10.091	9.7400	0.5282	0.9436	0.9442	11.709
0.60	2.1017	0.6496	5.3400	10.163	9.8000	0.5449	1.0000	1.0000	12.344
0.64	2.2253	0.6782	5.7344	10.308	9.9200	0.5781	1.1170	1.1156	13.653
0.68	2.3474	0.7056	6.1336	10.452	10.040	0.6109	1.2394	1.2365	15.013
0.72	2.4680	0.7317	6.5376	10.596	10.160	0.6435	1.3672	1.3627	16.423
0.76	2.5871	0.7567	6.9464	10.740	10.280	0.6757	1.5002	1.4941	17.881
0.80	2.7048	0.7807	7.3600	10.884	10.400	0.7077	1.6385	1.6307	19.389
0.84	2.8212	0.8036	7.7784	11.029	10.520	0.7394	1.7820	1.7725	20.945
0.88	2.9363	0.8256	8.2016	11.173	10.640	0.7708	1.9307	1.9194	22.550
0.92	3.0501	0.8467	8.6296	11.317	10.760	0.8020	2.0845	2.0715	24.201
0.96	3.1628	0.8669	9.0624	11.461	10.880	0.8329	2.2434	2.2286	25.901
1.00	3.2743	0.8863	9.5000	11.606	11.000	0.8636	2.4075	2.3908	27.647

Regime trapezoidal - 2.0 to 1 side-slope

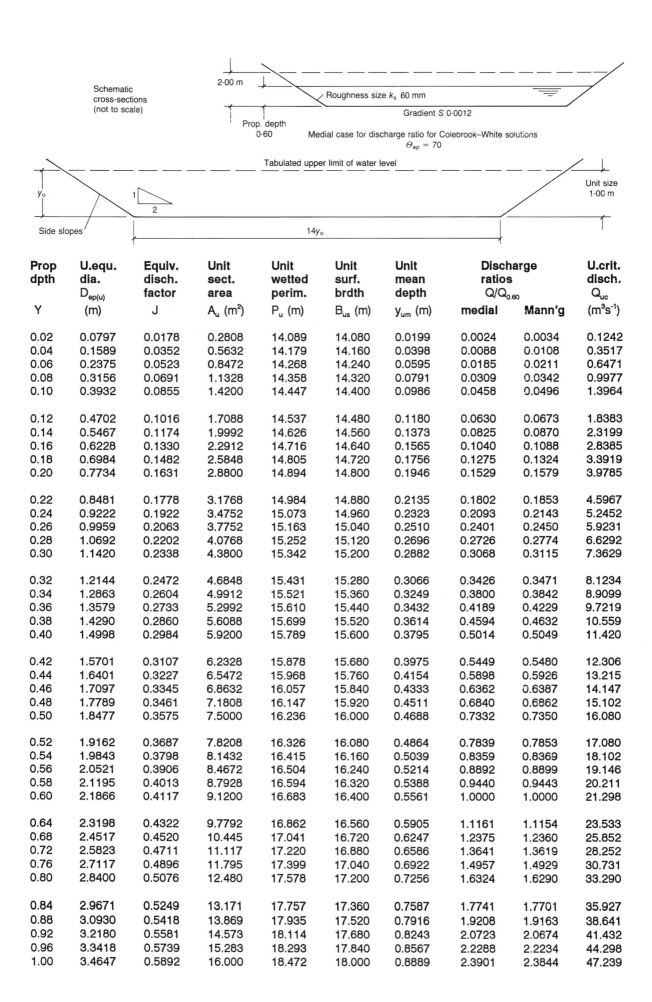

Schematic cross-sections (not to scale)

2·00 m

Roughness size k_s 60 mm

Gradient S 0·0012

Prop. depth 0·60

Medial case for discharge ratio for Colebrook–White solutions
$\Theta_{ep} = 70$

Tabulated upper limit of water level

y_o

Unit size 1·00 m

1 over 2

Side slopes

$14y_o$

Prop dpth Y	U.equ. dia. $D_{ep(u)}$ (m)	Equiv. disch. factor J	Unit sect. area A_u (m²)	Unit wetted perim. P_u (m)	Unit surf. brdth B_{us} (m)	Unit mean depth y_{um} (m)	Discharge ratios $Q/Q_{0.60}$ medial	Mann'g	U.crit. disch. Q_{uc} (m³s⁻¹)
0.02	0.0797	0.0178	0.2808	14.089	14.080	0.0199	0.0024	0.0034	0.1242
0.04	0.1589	0.0352	0.5632	14.179	14.160	0.0398	0.0088	0.0108	0.3517
0.06	0.2375	0.0523	0.8472	14.268	14.240	0.0595	0.0185	0.0211	0.6471
0.08	0.3156	0.0691	1.1328	14.358	14.320	0.0791	0.0309	0.0342	0.9977
0.10	0.3932	0.0855	1.4200	14.447	14.400	0.0986	0.0458	0.0496	1.3964
0.12	0.4702	0.1016	1.7088	14.537	14.480	0.1180	0.0630	0.0673	1.8383
0.14	0.5467	0.1174	1.9992	14.626	14.560	0.1373	0.0825	0.0870	2.3199
0.16	0.6228	0.1330	2.2912	14.716	14.640	0.1565	0.1040	0.1088	2.8385
0.18	0.6984	0.1482	2.5848	14.805	14.720	0.1756	0.1275	0.1324	3.3919
0.20	0.7734	0.1631	2.8800	14.894	14.800	0.1946	0.1529	0.1579	3.9785
0.22	0.8481	0.1778	3.1768	14.984	14.880	0.2135	0.1802	0.1853	4.5967
0.24	0.9222	0.1922	3.4752	15.073	14.960	0.2323	0.2093	0.2143	5.2452
0.26	0.9959	0.2063	3.7752	15.163	15.040	0.2510	0.2401	0.2450	5.9231
0.28	1.0692	0.2202	4.0768	15.252	15.120	0.2696	0.2726	0.2774	6.6292
0.30	1.1420	0.2338	4.3800	15.342	15.200	0.2882	0.3068	0.3115	7.3629
0.32	1.2144	0.2472	4.6848	15.431	15.280	0.3066	0.3426	0.3471	8.1234
0.34	1.2863	0.2604	4.9912	15.521	15.360	0.3249	0.3800	0.3842	8.9099
0.36	1.3579	0.2733	5.2992	15.610	15.440	0.3432	0.4189	0.4229	9.7219
0.38	1.4290	0.2860	5.6088	15.699	15.520	0.3614	0.4594	0.4632	10.559
0.40	1.4998	0.2984	5.9200	15.789	15.600	0.3795	0.5014	0.5049	11.420
0.42	1.5701	0.3107	6.2328	15.878	15.680	0.3975	0.5449	0.5480	12.306
0.44	1.6401	0.3227	6.5472	15.968	15.760	0.4154	0.5898	0.5926	13.215
0.46	1.7097	0.3345	6.8632	16.057	15.840	0.4333	0.6362	0.6387	14.147
0.48	1.7789	0.3461	7.1808	16.147	15.920	0.4511	0.6840	0.6862	15.102
0.50	1.8477	0.3575	7.5000	16.236	16.000	0.4688	0.7332	0.7350	16.080
0.52	1.9162	0.3687	7.8208	16.326	16.080	0.4864	0.7839	0.7853	17.080
0.54	1.9843	0.3798	8.1432	16.415	16.160	0.5039	0.8359	0.8369	18.102
0.56	2.0521	0.3906	8.4672	16.504	16.240	0.5214	0.8892	0.8899	19.146
0.58	2.1195	0.4013	8.7928	16.594	16.320	0.5388	0.9440	0.9443	20.211
0.60	2.1866	0.4117	9.1200	16.683	16.400	0.5561	1.0000	1.0000	21.298
0.64	2.3198	0.4322	9.7792	16.862	16.560	0.5905	1.1161	1.1154	23.533
0.68	2.4517	0.4520	10.445	17.041	16.720	0.6247	1.2375	1.2360	25.852
0.72	2.5823	0.4711	11.117	17.220	16.880	0.6586	1.3641	1.3619	28.252
0.76	2.7117	0.4896	11.795	17.399	17.040	0.6922	1.4957	1.4929	30.731
0.80	2.8400	0.5076	12.480	17.578	17.200	0.7256	1.6324	1.6290	33.290
0.84	2.9671	0.5249	13.171	17.757	17.360	0.7587	1.7741	1.7701	35.927
0.88	3.0930	0.5418	13.869	17.935	17.520	0.7916	1.9208	1.9163	38.641
0.92	3.2180	0.5581	14.573	18.114	17.680	0.8243	2.0723	2.0674	41.432
0.96	3.3418	0.5739	15.283	18.293	17.840	0.8567	2.2288	2.2234	44.298
1.00	3.4647	0.5892	16.000	18.472	18.000	0.8889	2.3901	2.3844	47.239

C63 Regime trapezoidal - 2.5 to 1 side-slope

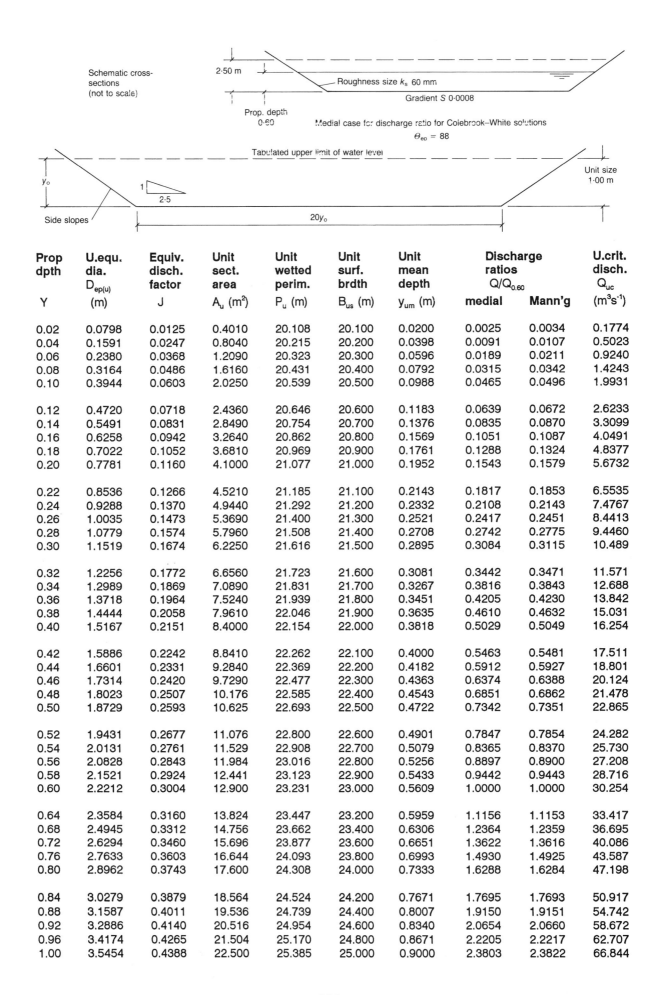

Schematic cross-sections (not to scale)

2·50 m

Roughness size k_s 60 mm

Gradient S 0·0008

Prop. depth 0·60

Medial case for discharge ratio for Colebrook–White solutions $\Theta_{ep} = 88$

Tabulated upper limit of water level

Unit size 1·00 m

y_o

Side slopes 1 / 2·5

$20y_o$

Prop dpth Y	U.equ. dia. $D_{ep(u)}$ (m)	Equiv. disch. factor J	Unit sect. area A_u (m^2)	Unit wetted perim. P_u (m)	Unit surf. brdth B_{us} (m)	Unit mean depth y_{um} (m)	Discharge ratios $Q/Q_{0.60}$ medial	Mann'g	U.crit. disch. Q_{uc} (m^{3}s^{-1})
0.02	0.0798	0.0125	0.4010	20.108	20.100	0.0200	0.0025	0.0034	0.1774
0.04	0.1591	0.0247	0.8040	20.215	20.200	0.0398	0.0091	0.0107	0.5023
0.06	0.2380	0.0368	1.2090	20.323	20.300	0.0596	0.0189	0.0211	0.9240
0.08	0.3164	0.0486	1.6160	20.431	20.400	0.0792	0.0315	0.0342	1.4243
0.10	0.3944	0.0603	2.0250	20.539	20.500	0.0988	0.0465	0.0496	1.9931
0.12	0.4720	0.0718	2.4360	20.646	20.600	0.1183	0.0639	0.0672	2.6233
0.14	0.5491	0.0831	2.8490	20.754	20.700	0.1376	0.0835	0.0870	3.3099
0.16	0.6258	0.0942	3.2640	20.862	20.800	0.1569	0.1051	0.1087	4.0491
0.18	0.7022	0.1052	3.6810	20.969	20.900	0.1761	0.1288	0.1324	4.8377
0.20	0.7781	0.1160	4.1000	21.077	21.000	0.1952	0.1543	0.1579	5.6732
0.22	0.8536	0.1266	4.5210	21.185	21.100	0.2143	0.1817	0.1853	6.5535
0.24	0.9288	0.1370	4.9440	21.292	21.200	0.2332	0.2108	0.2143	7.4767
0.26	1.0035	0.1473	5.3690	21.400	21.300	0.2521	0.2417	0.2451	8.4413
0.28	1.0779	0.1574	5.7960	21.508	21.400	0.2708	0.2742	0.2775	9.4460
0.30	1.1519	0.1674	6.2250	21.616	21.500	0.2895	0.3084	0.3115	10.489
0.32	1.2256	0.1772	6.6560	21.723	21.600	0.3081	0.3442	0.3471	11.571
0.34	1.2989	0.1869	7.0890	21.831	21.700	0.3267	0.3816	0.3843	12.688
0.36	1.3718	0.1964	7.5240	21.939	21.800	0.3451	0.4205	0.4230	13.842
0.38	1.4444	0.2058	7.9610	22.046	21.900	0.3635	0.4610	0.4632	15.031
0.40	1.5167	0.2151	8.4000	22.154	22.000	0.3818	0.5029	0.5049	16.254
0.42	1.5886	0.2242	8.8410	22.262	22.100	0.4000	0.5463	0.5481	17.511
0.44	1.6601	0.2331	9.2840	22.369	22.200	0.4182	0.5912	0.5927	18.801
0.46	1.7314	0.2420	9.7290	22.477	22.300	0.4363	0.6374	0.6388	20.124
0.48	1.8023	0.2507	10.176	22.585	22.400	0.4543	0.6851	0.6862	21.478
0.50	1.8729	0.2593	10.625	22.693	22.500	0.4722	0.7342	0.7351	22.865
0.52	1.9431	0.2677	11.076	22.800	22.600	0.4901	0.7847	0.7854	24.282
0.54	2.0131	0.2761	11.529	22.908	22.700	0.5079	0.8365	0.8370	25.730
0.56	2.0828	0.2843	11.984	23.016	22.800	0.5256	0.8897	0.8900	27.208
0.58	2.1521	0.2924	12.441	23.123	22.900	0.5433	0.9442	0.9443	28.716
0.60	2.2212	0.3004	12.900	23.231	23.000	0.5609	1.0000	1.0000	30.254
0.64	2.3584	0.3160	13.824	23.447	23.200	0.5959	1.1156	1.1153	33.417
0.68	2.4945	0.3312	14.756	23.662	23.400	0.6306	1.2364	1.2359	36.695
0.72	2.6294	0.3460	15.696	23.877	23.600	0.6651	1.3622	1.3616	40.086
0.76	2.7633	0.3603	16.644	24.093	23.800	0.6993	1.4930	1.4925	43.587
0.80	2.8962	0.3743	17.600	24.308	24.000	0.7333	1.6288	1.6284	47.198
0.84	3.0279	0.3879	18.564	24.524	24.200	0.7671	1.7695	1.7693	50.917
0.88	3.1587	0.4011	19.536	24.739	24.400	0.8007	1.9150	1.9151	54.742
0.92	3.2886	0.4140	20.516	24.954	24.600	0.8340	2.0654	2.0660	58.672
0.96	3.4174	0.4265	21.504	25.170	24.800	0.8671	2.2205	2.2217	62.707
1.00	3.5454	0.4388	22.500	25.385	25.000	0.9000	2.3803	2.3822	66.844

Regime trapezoidal - 3.0 to 1 side-slope

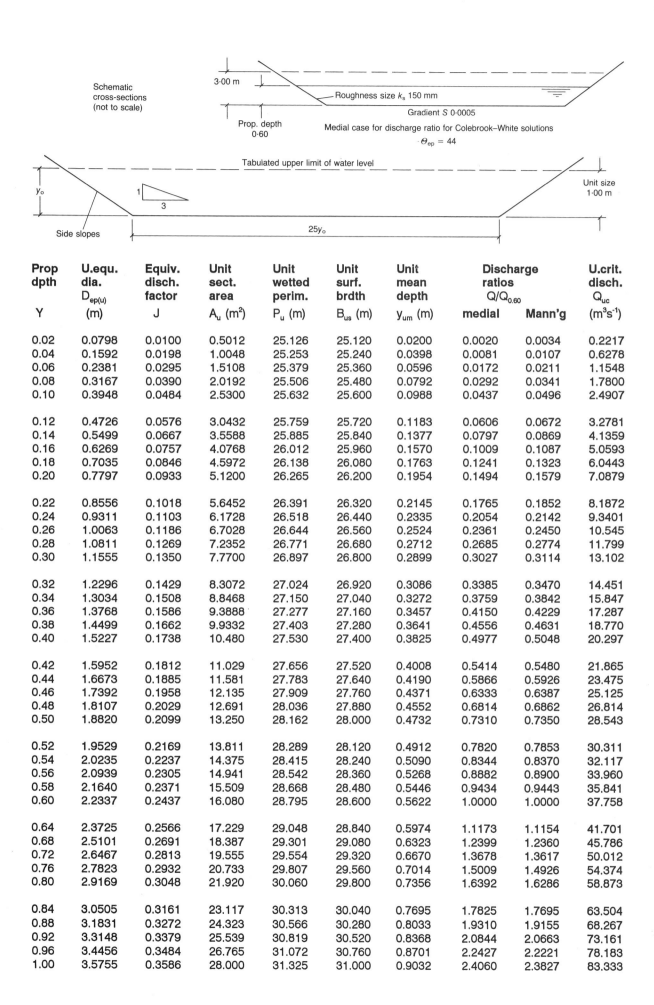

Schematic cross-sections (not to scale)

3·00 m

Roughness size k_s 150 mm

Gradient S 0·0005

Prop. depth 0·60

Medial case for discharge ratio for Colebrook–White solutions

$\Theta_{ep} = 44$

Tabulated upper limit of water level

y_o

Unit size 1·00 m

1 / 3

Side slopes

$25y_o$

Prop dpth	U.equ. dia. $D_{ep(u)}$	Equiv. disch. factor	Unit sect. area	Unit wetted perim.	Unit surf. brdth	Unit mean depth	Discharge ratios $Q/Q_{0.60}$		U.crit. disch. Q_{uc}
Y	(m)	J	A_u (m²)	P_u (m)	B_{us} (m)	y_{um} (m)	medial	Mann'g	(m³s⁻¹)
0.02	0.0798	0.0100	0.5012	25.126	25.120	0.0200	0.0020	0.0034	0.2217
0.04	0.1592	0.0198	1.0048	25.253	25.240	0.0398	0.0081	0.0107	0.6278
0.06	0.2381	0.0295	1.5108	25.379	25.360	0.0596	0.0172	0.0211	1.1548
0.08	0.3167	0.0390	2.0192	25.506	25.480	0.0792	0.0292	0.0341	1.7800
0.10	0.3948	0.0484	2.5300	25.632	25.600	0.0988	0.0437	0.0496	2.4907
0.12	0.4726	0.0576	3.0432	25.759	25.720	0.1183	0.0606	0.0672	3.2781
0.14	0.5499	0.0667	3.5588	25.885	25.840	0.1377	0.0797	0.0869	4.1359
0.16	0.6269	0.0757	4.0768	26.012	25.960	0.1570	0.1009	0.1087	5.0593
0.18	0.7035	0.0846	4.5972	26.138	26.080	0.1763	0.1241	0.1323	6.0443
0.20	0.7797	0.0933	5.1200	26.265	26.200	0.1954	0.1494	0.1579	7.0879
0.22	0.8556	0.1018	5.6452	26.391	26.320	0.2145	0.1765	0.1852	8.1872
0.24	0.9311	0.1103	6.1728	26.518	26.440	0.2335	0.2054	0.2142	9.3401
0.26	1.0063	0.1186	6.7028	26.644	26.560	0.2524	0.2361	0.2450	10.545
0.28	1.0811	0.1269	7.2352	26.771	26.680	0.2712	0.2685	0.2774	11.799
0.30	1.1555	0.1350	7.7700	26.897	26.800	0.2899	0.3027	0.3114	13.102
0.32	1.2296	0.1429	8.3072	27.024	26.920	0.3086	0.3385	0.3470	14.451
0.34	1.3034	0.1508	8.8468	27.150	27.040	0.3272	0.3759	0.3842	15.847
0.36	1.3768	0.1586	9.3888	27.277	27.160	0.3457	0.4150	0.4229	17.287
0.38	1.4499	0.1662	9.9332	27.403	27.280	0.3641	0.4556	0.4631	18.770
0.40	1.5227	0.1738	10.480	27.530	27.400	0.3825	0.4977	0.5048	20.297
0.42	1.5952	0.1812	11.029	27.656	27.520	0.4008	0.5414	0.5480	21.865
0.44	1.6673	0.1885	11.581	27.783	27.640	0.4190	0.5866	0.5926	23.475
0.46	1.7392	0.1958	12.135	27.909	27.760	0.4371	0.6333	0.6387	25.125
0.48	1.8107	0.2029	12.691	28.036	27.880	0.4552	0.6814	0.6862	26.814
0.50	1.8820	0.2099	13.250	28.162	28.000	0.4732	0.7310	0.7350	28.543
0.52	1.9529	0.2169	13.811	28.289	28.120	0.4912	0.7820	0.7853	30.311
0.54	2.0235	0.2237	14.375	28.415	28.240	0.5090	0.8344	0.8370	32.117
0.56	2.0939	0.2305	14.941	28.542	28.360	0.5268	0.8882	0.8900	33.960
0.58	2.1640	0.2371	15.509	28.668	28.480	0.5446	0.9434	0.9443	35.841
0.60	2.2337	0.2437	16.080	28.795	28.600	0.5622	1.0000	1.0000	37.758
0.64	2.3725	0.2566	17.229	29.048	28.840	0.5974	1.1173	1.1154	41.701
0.68	2.5101	0.2691	18.387	29.301	29.080	0.6323	1.2399	1.2360	45.786
0.72	2.6467	0.2813	19.555	29.554	29.320	0.6670	1.3678	1.3617	50.012
0.76	2.7823	0.2932	20.733	29.807	29.560	0.7014	1.5009	1.4926	54.374
0.80	2.9169	0.3048	21.920	30.060	29.800	0.7356	1.6392	1.6286	58.873
0.84	3.0505	0.3161	23.117	30.313	30.040	0.7695	1.7825	1.7695	63.504
0.88	3.1831	0.3272	24.323	30.566	30.280	0.8033	1.9310	1.9155	68.267
0.92	3.3148	0.3379	25.539	30.819	30.520	0.8368	2.0844	2.0663	73.161
0.96	3.4456	0.3484	26.765	31.072	30.760	0.8701	2.2427	2.2221	78.183
1.00	3.5755	0.3586	28.000	31.325	31.000	0.9032	2.4060	2.3827	83.333

C65

Regime trapezoidal – 4.0 to 1 side-slope

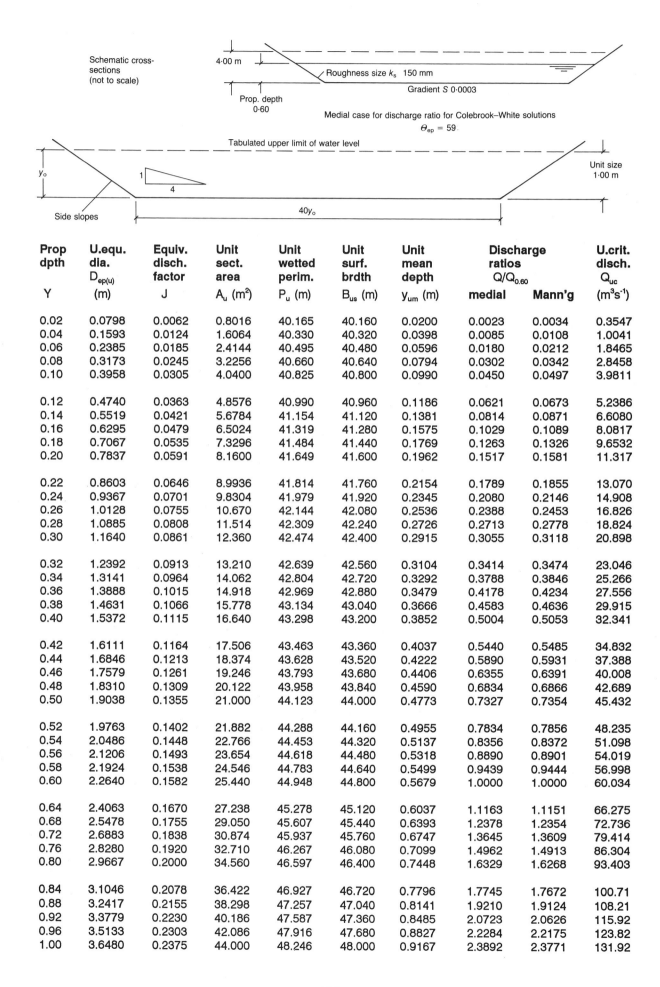

Schematic cross-sections (not to scale)

4·00 m

Roughness size k_s 150 mm

Gradient S 0·0003

Prop. depth 0·60

Medial case for discharge ratio for Colebrook–White solutions
$\Theta_{ep} = 59$

Tabulated upper limit of water level

y_o

Unit size 1·00 m

Side slopes

1 : 4

$40y_o$

Prop dpth	U.equ. dia. $D_{ep(u)}$	Equiv. disch. factor	Unit sect. area	Unit wetted perim.	Unit surf. brdth	Unit mean depth	Discharge ratios $Q/Q_{0.60}$		U.crit. disch. Q_{uc}
Y	(m)	J	A_u (m²)	P_u (m)	B_{us} (m)	y_{um} (m)	medial	Mann'g	(m³s⁻¹)
0.02	0.0798	0.0062	0.8016	40.165	40.160	0.0200	0.0023	0.0034	0.3547
0.04	0.1593	0.0124	1.6064	40.330	40.320	0.0398	0.0085	0.0108	1.0041
0.06	0.2385	0.0185	2.4144	40.495	40.480	0.0596	0.0180	0.0212	1.8465
0.08	0.3173	0.0245	3.2256	40.660	40.640	0.0794	0.0302	0.0342	2.8458
0.10	0.3958	0.0305	4.0400	40.825	40.800	0.0990	0.0450	0.0497	3.9811
0.12	0.4740	0.0363	4.8576	40.990	40.960	0.1186	0.0621	0.0673	5.2386
0.14	0.5519	0.0421	5.6784	41.154	41.120	0.1381	0.0814	0.0871	6.6080
0.16	0.6295	0.0479	6.5024	41.319	41.280	0.1575	0.1029	0.1089	8.0817
0.18	0.7067	0.0535	7.3296	41.484	41.440	0.1769	0.1263	0.1326	9.6532
0.20	0.7837	0.0591	8.1600	41.649	41.600	0.1962	0.1517	0.1581	11.317
0.22	0.8603	0.0646	8.9936	41.814	41.760	0.2154	0.1789	0.1855	13.070
0.24	0.9367	0.0701	9.8304	41.979	41.920	0.2345	0.2080	0.2146	14.908
0.26	1.0128	0.0755	10.670	42.144	42.080	0.2536	0.2388	0.2453	16.826
0.28	1.0885	0.0808	11.514	42.309	42.240	0.2726	0.2713	0.2778	18.824
0.30	1.1640	0.0861	12.360	42.474	42.400	0.2915	0.3055	0.3118	20.898
0.32	1.2392	0.0913	13.210	42.639	42.560	0.3104	0.3414	0.3474	23.046
0.34	1.3141	0.0964	14.062	42.804	42.720	0.3292	0.3788	0.3846	25.266
0.36	1.3888	0.1015	14.918	42.969	42.880	0.3479	0.4178	0.4234	27.556
0.38	1.4631	0.1066	15.778	43.134	43.040	0.3666	0.4583	0.4636	29.915
0.40	1.5372	0.1115	16.640	43.298	43.200	0.3852	0.5004	0.5053	32.341
0.42	1.6111	0.1164	17.506	43.463	43.360	0.4037	0.5440	0.5485	34.832
0.44	1.6846	0.1213	18.374	43.628	43.520	0.4222	0.5890	0.5931	37.388
0.46	1.7579	0.1261	19.246	43.793	43.680	0.4406	0.6355	0.6391	40.008
0.48	1.8310	0.1309	20.122	43.958	43.840	0.4590	0.6834	0.6866	42.689
0.50	1.9038	0.1355	21.000	44.123	44.000	0.4773	0.7327	0.7354	45.432
0.52	1.9763	0.1402	21.882	44.288	44.160	0.4955	0.7834	0.7856	48.235
0.54	2.0486	0.1448	22.766	44.453	44.320	0.5137	0.8356	0.8372	51.098
0.56	2.1206	0.1493	23.654	44.618	44.480	0.5318	0.8890	0.8901	54.019
0.58	2.1924	0.1538	24.546	44.783	44.640	0.5499	0.9439	0.9444	56.998
0.60	2.2640	0.1582	25.440	44.948	44.800	0.5679	1.0000	1.0000	60.034
0.64	2.4063	0.1670	27.238	45.278	45.120	0.6037	1.1163	1.1151	66.275
0.68	2.5478	0.1755	29.050	45.607	45.440	0.6393	1.2378	1.2354	72.736
0.72	2.6883	0.1838	30.874	45.937	45.760	0.6747	1.3645	1.3609	79.414
0.76	2.8280	0.1920	32.710	46.267	46.080	0.7099	1.4962	1.4913	86.304
0.80	2.9667	0.2000	34.560	46.597	46.400	0.7448	1.6329	1.6268	93.403
0.84	3.1046	0.2078	36.422	46.927	46.720	0.7796	1.7745	1.7672	100.71
0.88	3.2417	0.2155	38.298	47.257	47.040	0.8141	1.9210	1.9124	108.21
0.92	3.3779	0.2230	40.186	47.587	47.360	0.8485	2.0723	2.0626	115.92
0.96	3.5133	0.2303	42.086	47.916	47.680	0.8827	2.2284	2.2175	123.82
1.00	3.6480	0.2375	44.000	48.246	48.000	0.9167	2.3892	2.3771	131.92

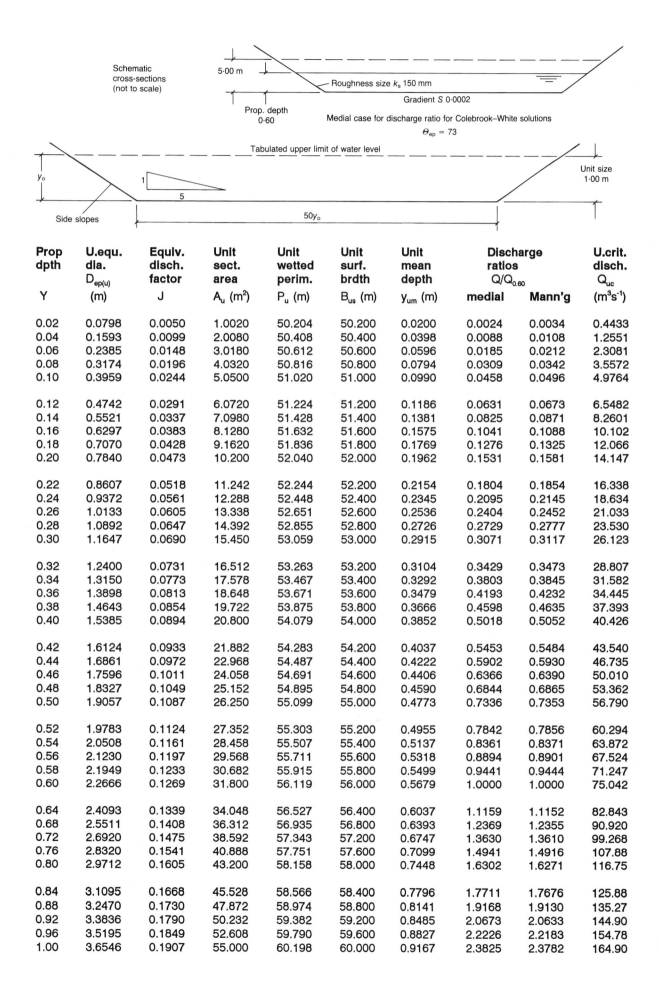

Schematic cross-sections (not to scale)

5·00 m

Roughness size k_s 150 mm

Gradient S 0·0002

Prop. depth 0·60

Medial case for discharge ratio for Colebrook–White solutions

$\Theta_{ep} = 73$

Tabulated upper limit of water level

y_o

Unit size 1·00 m

1

5

Side slopes

$50y_o$

Prop dpth	U.equ. dia. $D_{ep(u)}$	Equiv. disch. factor	Unit sect. area	Unit wetted perim.	Unit surf. brdth	Unit mean depth	Discharge ratios $Q/Q_{0.60}$		U.crit. disch. Q_{uc}
Y	(m)	J	A_u (m²)	P_u (m)	B_{us} (m)	y_{um} (m)	medial	Mann'g	(m³s⁻¹)
0.02	0.0798	0.0050	1.0020	50.204	50.200	0.0200	0.0024	0.0034	0.4433
0.04	0.1593	0.0099	2.0080	50.408	50.400	0.0398	0.0088	0.0108	1.2551
0.06	0.2385	0.0148	3.0180	50.612	50.600	0.0596	0.0185	0.0212	2.3081
0.08	0.3174	0.0196	4.0320	50.816	50.800	0.0794	0.0309	0.0342	3.5572
0.10	0.3959	0.0244	5.0500	51.020	51.000	0.0990	0.0458	0.0496	4.9764
0.12	0.4742	0.0291	6.0720	51.224	51.200	0.1186	0.0631	0.0673	6.5482
0.14	0.5521	0.0337	7.0980	51.428	51.400	0.1381	0.0825	0.0871	8.2601
0.16	0.6297	0.0383	8.1280	51.632	51.600	0.1575	0.1041	0.1088	10.102
0.18	0.7070	0.0428	9.1620	51.836	51.800	0.1769	0.1276	0.1325	12.066
0.20	0.7840	0.0473	10.200	52.040	52.000	0.1962	0.1531	0.1581	14.147
0.22	0.8607	0.0518	11.242	52.244	52.200	0.2154	0.1804	0.1854	16.338
0.24	0.9372	0.0561	12.288	52.448	52.400	0.2345	0.2095	0.2145	18.634
0.26	1.0133	0.0605	13.338	52.651	52.600	0.2536	0.2404	0.2452	21.033
0.28	1.0892	0.0647	14.392	52.855	52.800	0.2726	0.2729	0.2777	23.530
0.30	1.1647	0.0690	15.450	53.059	53.000	0.2915	0.3071	0.3117	26.123
0.32	1.2400	0.0731	16.512	53.263	53.200	0.3104	0.3429	0.3473	28.807
0.34	1.3150	0.0773	17.578	53.467	53.400	0.3292	0.3803	0.3845	31.582
0.36	1.3898	0.0813	18.648	53.671	53.600	0.3479	0.4193	0.4232	34.445
0.38	1.4643	0.0854	19.722	53.875	53.800	0.3666	0.4598	0.4635	37.393
0.40	1.5385	0.0894	20.800	54.079	54.000	0.3852	0.5018	0.5052	40.426
0.42	1.6124	0.0933	21.882	54.283	54.200	0.4037	0.5453	0.5484	43.540
0.44	1.6861	0.0972	22.968	54.487	54.400	0.4222	0.5902	0.5930	46.735
0.46	1.7596	0.1011	24.058	54.691	54.600	0.4406	0.6366	0.6390	50.010
0.48	1.8327	0.1049	25.152	54.895	54.800	0.4590	0.6844	0.6865	53.362
0.50	1.9057	0.1087	26.250	55.099	55.000	0.4773	0.7336	0.7353	56.790
0.52	1.9783	0.1124	27.352	55.303	55.200	0.4955	0.7842	0.7856	60.294
0.54	2.0508	0.1161	28.458	55.507	55.400	0.5137	0.8361	0.8371	63.872
0.56	2.1230	0.1197	29.568	55.711	55.600	0.5318	0.8894	0.8901	67.524
0.58	2.1949	0.1233	30.682	55.915	55.800	0.5499	0.9441	0.9444	71.247
0.60	2.2666	0.1269	31.800	56.119	56.000	0.5679	1.0000	1.0000	75.042
0.64	2.4093	0.1339	34.048	56.527	56.400	0.6037	1.1159	1.1152	82.843
0.68	2.5511	0.1408	36.312	56.935	56.800	0.6393	1.2369	1.2355	90.920
0.72	2.6920	0.1475	38.592	57.343	57.200	0.6747	1.3630	1.3610	99.268
0.76	2.8320	0.1541	40.888	57.751	57.600	0.7099	1.4941	1.4916	107.88
0.80	2.9712	0.1605	43.200	58.158	58.000	0.7448	1.6302	1.6271	116.75
0.84	3.1095	0.1668	45.528	58.566	58.400	0.7796	1.7711	1.7676	125.88
0.88	3.2470	0.1730	47.872	58.974	58.800	0.8141	1.9168	1.9130	135.27
0.92	3.3836	0.1790	50.232	59.382	59.200	0.8485	2.0673	2.0633	144.90
0.96	3.5195	0.1849	52.608	59.790	59.600	0.8827	2.2226	2.2183	154.78
1.00	3.6546	0.1907	55.000	60.198	60.000	0.9167	2.3825	2.3782	164.90

C67   Wide rectangular channel (free surface)

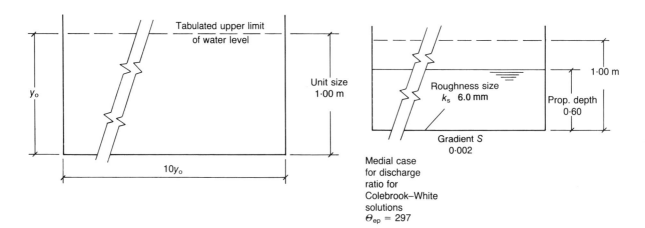

Medial case for discharge ratio for Colebrook–White solutions $\Theta_{ep} = 297$

Prop dpth Y	U.equ. dia. $D_{ep(u)}$ (m)	Equiv. disch. factor J	Unit sect. area A_u (m²)	Unit wetted perim. P_u (m)	Unit surf. brdth B_{us} (m)	Unit mean depth y_{um} (m)	Discharge ratios $Q/Q_{0.60}$ medial	Mann'g	U.crit. disch. Q_{uc} (m³s⁻¹)
0.02	0.0797	0.0249	0.2000	10.040	10.000	0.0200	0.0035	0.0037	0.0886
0.04	0.1587	0.0495	0.4000	10.080	10.000	0.0400	0.0115	0.0118	0.2505
0.06	0.2372	0.0736	0.6000	10.120	10.000	0.0600	0.0230	0.0231	0.4602
0.08	0.3150	0.0974	0.8000	10.160	10.000	0.0800	0.0374	0.0371	0.7086
0.10	0.3922	0.1208	1.0000	10.200	10.000	0.1000	0.0544	0.0537	0.9903
0.12	0.4687	0.1438	1.2000	10.240	10.000	0.1200	0.0737	0.0726	1.3018
0.14	0.5447	0.1665	1.4000	10.280	10.000	0.1400	0.0951	0.0936	1.6404
0.16	0.6202	0.1888	1.6000	10.320	10.000	0.1600	0.1186	0.1167	2.0042
0.18	0.6950	0.2107	1.8000	10.360	10.000	0.1800	0.1440	0.1416	2.3915
0.20	0.7692	0.2324	2.0000	10.400	10.000	0.2000	0.1711	0.1684	2.8009
0.22	0.8429	0.2536	2.2000	10.440	10.000	0.2200	0.2000	0.1968	3.2314
0.24	0.9160	0.2746	2.4000	10.480	10.000	0.2400	0.2305	0.2270	3.6819
0.26	0.9886	0.2952	2.6000	10.520	10.000	0.2600	0.2625	0.2587	4.1516
0.28	1.0606	0.3155	2.8000	10.560	10.000	0.2800	0.2960	0.2920	4.6398
0.30	1.1321	0.3355	3.0000	10.600	10.000	0.3000	0.3310	0.3268	5.1457
0.32	1.2030	0.3552	3.2000	10.640	10.000	0.3200	0.3674	0.3630	5.6687
0.34	1.2734	0.3746	3.4000	10.680	10.000	0.3400	0.4051	0.4005	6.2084
0.36	1.3433	0.3937	3.6000	10.720	10.000	0.3600	0.4441	0.4395	6.7642
0.38	1.4126	0.4124	3.8000	10.760	10.000	0.3800	0.4843	0.4797	7.3356
0.40	1.4815	0.4309	4.0000	10.800	10.000	0.4000	0.5257	0.5212	7.9223
0.42	1.5498	0.4491	4.2000	10.840	10.000	0.4200	0.5684	0.5640	8.5238
0.44	1.6176	0.4671	4.4000	10.880	10.000	0.4400	0.6121	0.6080	9.1399
0.46	1.6850	0.4847	4.6000	10.920	10.000	0.4600	0.6570	0.6531	9.7701
0.48	1.7518	0.5021	4.8000	10.960	10.000	0.4800	0.7030	0.6994	10.414
0.50	1.8182	0.5193	5.0000	11.000	10.000	0.5000	0.7500	0.7469	11.072
0.52	1.8841	0.5361	5.2000	11.040	10.000	0.5200	0.7981	0.7954	11.743
0.54	1.9495	0.5527	5.4000	11.080	10.000	0.5400	0.8471	0.8450	12.427
0.56	2.0144	0.5691	5.6000	11.120	10.000	0.5600	0.8971	0.8956	13.123
0.58	2.0789	0.5852	5.8000	11.160	10.000	0.5800	0.9481	0.9473	13.833
0.60	2.1429	0.6011	6.0000	11.200	10.000	0.6000	1.0000	1.0000	14.554
0.64	2.2695	0.6321	6.4000	11.280	10.000	0.6400	1.1066	1.1083	16.034
0.68	2.3944	0.6621	6.8000	11.360	10.000	0.6800	1.2166	1.2204	17.560
0.72	2.5175	0.6913	7.2000	11.440	10.000	0.7200	1.3300	1.3361	19.132
0.76	2.6389	0.7196	7.6000	11.520	10.000	0.7600	1.4466	1.4553	20.748
0.80	2.7586	0.7471	8.0000	11.600	10.000	0.8000	1.5662	1.5779	22.408
0.84	2.8767	0.7737	8.4000	11.680	10.000	0.8400	1.6889	1.7037	24.109
0.88	2.9932	0.7996	8.8000	11.760	10.000	0.8800	1.8144	1.8327	25.851
0.92	3.1081	0.8247	9.2000	11.840	10.000	0.9200	1.9426	1.9647	27.634
0.96	3.2215	0.8490	9.6000	11.920	10.000	0.9600	2.0735	2.0997	29.456
1.00	3.3333	0.8726	10.000	12.000	10.000	1.0000	2.2070	2.2375	31.316

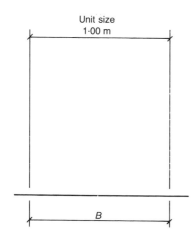

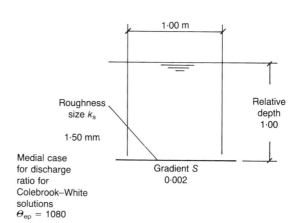

Prop dpth	U.equ. dia. $D_{ep(u)}$	Equiv. disch. factor	Unit sect. area	Unit wetted perim.	Unit surf. brdth	Unit mean depth	Discharge ratios $Q/Q_{1.00}$		U.crit. disch. Q_{uc}
Y	(m)	J	A_u (m²)	P_u (m)	B_{us} (m)	y_{um} (m)	medial	Mann'g	(m³s⁻¹)
0.04	0.1600	0.5026	0.0400	1.0000	1.0000	0.0400	0.0051	0.0047	0.0251
0.08	0.3200	1.0053	0.0800	1.0000	1.0000	0.0800	0.0163	0.0149	0.0709
0.12	0.4800	1.5079	0.1200	1.0000	1.0000	0.1200	0.0318	0.0292	0.1302
0.16	0.6400	2.0106	0.1600	1.0000	1.0000	0.1600	0.0510	0.0472	0.2004
0.20	0.8000	2.5132	0.2000	1.0000	1.0000	0.2000	0.0736	0.0684	0.2801
0.24	0.9600	3.0159	0.2400	1.0000	1.0000	0.2400	0.0991	0.0927	0.3682
0.28	1.1200	3.5185	0.2800	1.0000	1.0000	0.2800	0.1274	0.1198	0.4640
0.32	1.2800	4.0211	0.3200	1.0000	1.0000	0.3200	0.1583	0.1497	0.5669
0.36	1.4400	4.5238	0.3600	1.0000	1.0000	0.3600	0.1917	0.1822	0.6764
0.40	1.6000	5.0264	0.4000	1.0000	1.0000	0.4000	0.2275	0.2172	0.7922
0.44	1.7600	5.5291	0.4400	1.0000	1.0000	0.4400	0.2655	0.2545	0.9140
0.48	1.9200	6.0317	0.4800	1.0000	1.0000	0.4800	0.3057	0.2943	1.0414
0.52	2.0800	6.5344	0.5200	1.0000	1.0000	0.5200	0.3480	0.3363	1.1743
0.56	2.2400	7.0370	0.5600	1.0000	1.0000	0.5600	0.3923	0.3805	1.3123
0.60	2.4000	7.5396	0.6000	1.0000	1.0000	0.6000	0.4386	0.4268	1.4554
0.64	2.5600	8.0423	0.6400	1.0000	1.0000	0.6400	0.4869	0.4753	1.6034
0.68	2.7200	8.5449	0.6800	1.0000	1.0000	0.6800	0.5370	0.5258	1.7560
0.72	2.8800	9.0476	0.7200	1.0000	1.0000	0.7200	0.5889	0.5784	1.9132
0.76	3.0400	9.5502	0.7600	1.0000	1.0000	0.7600	0.6426	0.6329	2.0748
0.80	3.2000	10.053	0.8000	1.0000	1.0000	0.8000	0.6980	0.6894	2.2408
0.84	3.3600	10.556	0.8400	1.0000	1.0000	0.8400	0.7551	0.7478	2.4109
0.88	3.5200	11.058	0.8800	1.0000	1.0000	0.8800	0.8139	0.8081	2.5851
0.92	3.6800	11.561	0.9200	1.0000	1.0000	0.9200	0.8744	0.8703	2.7634
0.96	3.8400	12.063	0.9600	1.0000	1.0000	0.9600	0.9364	0.9342	2.9456
1.00	4.0000	12.566	1.0000	1.0000	1.0000	1.0000	1.0000	1.0000	3.1316
1.10	4.4000	13.823	1.1000	1.0000	1.0000	1.1000	1.1658	1.1722	3.6128
1.20	4.8000	15.079	1.2000	1.0000	1.0000	1.2000	1.3409	1.3551	4.1165
1.30	5.2000	16.336	1.3000	1.0000	1.0000	1.3000	1.5250	1.5485	4.6417
1.40	5.6000	17.593	1.4000	1.0000	1.0000	1.4000	1.7178	1.7521	5.1874
1.50	6.0000	18.849	1.5000	1.0000	1.0000	1.5000	1.9190	1.9656	5.7530
1.60	6.4000	20.106	1.6000	1.0000	1.0000	1.6000	2.1283	2.1888	6.3378
1.70	6.8000	21.362	1.7000	1.0000	1.0000	1.7000	2.3457	2.4215	6.9412
1.80	7.2000	22.619	1.8000	1.0000	1.0000	1.8000	2.5708	2.6635	7.5626
1.90	7.6000	23.876	1.9000	1.0000	1.0000	1.9000	2.8035	2.9147	8.2015
2.00	8.0000	25.132	2.0000	1.0000	1.0000	2.0000	3.0437	3.1748	8.8574
2.10	8.4000	26.389	2.1000	1.0000	1.0000	2.1000	3.2910	3.4438	9.5299
2.20	8.8000	27.645	2.2000	1.0000	1.0000	2.2000	3.5455	3.7214	10.219
2.30	9.2000	28.902	2.3000	1.0000	1.0000	2.3000	3.8070	4.0076	10.923
2.40	9.6000	30.159	2.4000	1.0000	1.0000	2.4000	4.0753	4.3021	11.643
2.50	10.000	31.415	2.5000	1.0000	1.0000	2.5000	4.3503	4.6050	12.379

C69

Arc invert (free surface)

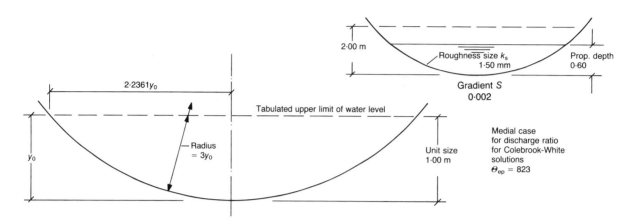

Prop dpth	U.equ. dia. $D_{ep(u)}$	Equiv. disch. factor	Unit sect. area	Unit wetted perim.	Unit surf. brdth	Unit mean depth	Discharge ratios $Q/Q_{0.60}$		U.crit. disch. Q_{uc}
Y	(m)	J	A_u (m^2)	P_u (m)	B_{us} (m)	y_{um} (m)	medial	Mann'g	(m^{3}s^{-1})
0.02	0.0533	0.2413	0.0092	0.6932	0.6917	0.0133	0.0007	0.0007	0.0033
0.04	0.1063	0.3406	0.0261	0.9809	0.9765	0.0267	0.0033	0.0030	0.0133
0.06	0.1593	0.4162	0.0479	1.2020	1.1940	0.0401	0.0078	0.0072	0.0300
0.08	0.2120	0.4796	0.0736	1.3887	1.3764	0.0535	0.0145	0.0134	0.0533
0.10	0.2646	0.5350	0.1028	1.5535	1.5362	0.0669	0.0233	0.0217	0.0832
0.12	0.3170	0.5849	0.1349	1.7028	1.6800	0.0803	0.0343	0.0322	0.1198
0.14	0.3693	0.6304	0.1699	1.8402	1.8115	0.0938	0.0476	0.0449	0.1629
0.16	0.4213	0.6724	0.2073	1.9684	1.9333	0.1072	0.0632	0.0598	0.2126
0.18	0.4733	0.7117	0.2472	2.0890	2.0470	0.1207	0.0811	0.0770	0.2689
0.20	0.5250	0.7486	0.2892	2.2032	2.1541	0.1342	0.1013	0.0966	0.3318
0.22	0.5766	0.7834	0.3333	2.3121	2.2553	0.1478	0.1239	0.1185	0.4012
0.24	0.6280	0.8165	0.3794	2.4163	2.3515	0.1613	0.1488	0.1427	0.4772
0.26	0.6792	0.8480	0.4273	2.5164	2.4433	0.1749	0.1760	0.1694	0.5596
0.28	0.7303	0.8781	0.4771	2.6129	2.5311	0.1885	0.2056	0.1985	0.6486
0.30	0.7812	0.9069	0.5285	2.7062	2.6153	0.2021	0.2376	0.2300	0.7441
0.32	0.8320	0.9346	0.5817	2.7965	2.6964	0.2157	0.2719	0.2640	0.8460
0.34	0.8825	0.9613	0.6364	2.8843	2.7745	0.2294	0.3086	0.3004	0.9544
0.36	0.9329	0.9869	0.6926	2.9696	2.8498	0.2430	0.3477	0.3393	1.0693
0.38	0.9832	1.0118	0.7503	3.0528	2.9227	0.2567	0.3891	0.3806	1.1906
0.40	1.0332	1.0357	0.8095	3.1339	2.9933	0.2704	0.4329	0.4245	1.3183
0.42	1.0831	1.0590	0.8701	3.2132	3.0618	0.2842	0.4791	0.4708	1.4524
0.44	1.1328	1.0815	0.9320	3.2907	3.1282	0.2979	0.5276	0.5196	1.5930
0.46	1.1824	1.1033	0.9952	3.3666	3.1927	0.3117	0.5784	0.5709	1.7399
0.48	1.2318	1.1246	1.0597	3.4411	3.2555	0.3255	0.6317	0.6247	1.8932
0.50	1.2810	1.1452	1.1254	3.5141	3.3166	0.3393	0.6872	0.6810	2.0529
0.52	1.3300	1.1652	1.1923	3.5858	3.3762	0.3532	0.7451	0.7398	2.2189
0.54	1.3789	1.1848	1.2604	3.6563	3.4342	0.3670	0.8054	0.8011	2.3912
0.56	1.4276	1.2038	1.3297	3.7256	3.4908	0.3809	0.8679	0.8649	2.5699
0.58	1.4761	1.2223	1.4000	3.7938	3.5460	0.3948	0.9328	0.9312	2.7549
0.60	1.5245	1.2404	1.4715	3.8610	3.6000	0.4088	1.0000	1.0000	2.9461
0.64	1.6207	1.2753	1.6176	3.9924	3.7043	0.4367	1.1413	1.1451	3.3475
0.68	1.7162	1.3085	1.7678	4.1203	3.8040	0.4647	1.2918	1.3001	3.7739
0.72	1.8110	1.3403	1.9219	4.2449	3.8995	0.4928	1.4513	1.4650	4.2251
0.76	1.9051	1.3706	2.0797	4.3666	3.9912	0.5211	1.6199	1.6397	4.7012
0.80	1.9985	1.3997	2.2411	4.4855	4.0792	0.5494	1.7974	1.8243	5.2020
0.84	2.0913	1.4276	2.4060	4.6020	4.1638	0.5778	1.9837	2.0186	5.7274
0.88	2.1833	1.4544	2.5742	4.7161	4.2453	0.6064	2.1788	2.2227	6.2772
0.92	2.2746	1.4800	2.7456	4.8282	4.3237	0.6350	2.3825	2.4363	6.8515
0.96	2.3653	1.5047	2.9200	4.9382	4.3993	0.6638	2.5948	2.6595	7.4500
1.00	2.4552	1.5284	3.0975	5.0464	4.4721	0.6926	2.8156	2.8922	8.0726

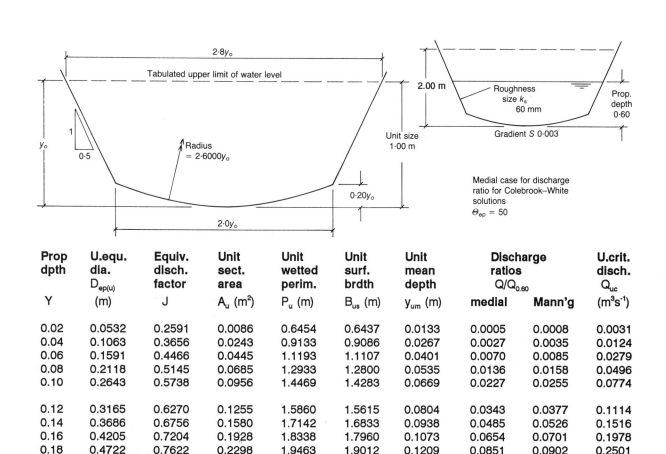

Prop dpth	U.equ. dia. $D_{ep(u)}$	Equiv. disch. factor	Unit sect. area	Unit wetted perim.	Unit surf. brdth	Unit mean depth	Discharge ratios $Q/Q_{0.60}$		U.crit. disch. Q_{uc}
Y	(m)	J	A_u (m²)	P_u (m)	B_{us} (m)	y_{um} (m)	medial	Mann'g	(m³s⁻¹)
0.02	0.0532	0.2591	0.0086	0.6454	0.6437	0.0133	0.0005	0.0008	0.0031
0.04	0.1063	0.3656	0.0243	0.9133	0.9086	0.0267	0.0027	0.0035	0.0124
0.06	0.1591	0.4466	0.0445	1.1193	1.1107	0.0401	0.0070	0.0085	0.0279
0.08	0.2118	0.5145	0.0685	1.2933	1.2800	0.0535	0.0136	0.0158	0.0496
0.10	0.2643	0.5738	0.0956	1.4469	1.4283	0.0669	0.0227	0.0255	0.0774
0.12	0.3165	0.6270	0.1255	1.5860	1.5615	0.0804	0.0343	0.0377	0.1114
0.14	0.3686	0.6756	0.1580	1.7142	1.6833	0.0938	0.0485	0.0526	0.1516
0.16	0.4205	0.7204	0.1928	1.8338	1.7960	0.1073	0.0654	0.0701	0.1978
0.18	0.4722	0.7622	0.2298	1.9463	1.9012	0.1209	0.0851	0.0902	0.2501
0.20	0.5237	0.8014	0.2688	2.0529	2.0000	0.1344	0.1075	0.1131	0.3086
0.22	0.5892	0.8824	0.3090	2.0976	2.0200	0.1530	0.1347	0.1406	0.3784
0.24	0.6527	0.9571	0.3496	2.1424	2.0400	0.1714	0.1643	0.1703	0.4532
0.26	0.7144	1.0261	0.3906	2.1871	2.0600	0.1896	0.1960	0.2021	0.5326
0.28	0.7742	1.0898	0.4320	2.2318	2.0800	0.2077	0.2297	0.2358	0.6165
0.30	0.8325	1.1488	0.4738	2.2765	2.1000	0.2256	0.2654	0.2714	0.7047
0.32	0.8892	1.2034	0.5160	2.3212	2.1200	0.2434	0.3030	0.3089	0.7972
0.34	0.9444	1.2539	0.5586	2.3660	2.1400	0.2610	0.3424	0.3481	0.8937
0.36	0.9982	1.3008	0.6016	2.4107	2.1600	0.2785	0.3835	0.3890	0.9942
0.38	1.0507	1.3443	0.6450	2.4554	2.1800	0.2959	0.4263	0.4316	1.0987
0.40	1.1020	1.3847	0.6888	2.5001	2.2000	0.3131	0.4708	0.4758	1.2069
0.42	1.1521	1.4222	0.7330	2.5448	2.2200	0.3302	0.5169	0.5215	1.3190
0.44	1.2011	1.4571	0.7776	2.5896	2.2400	0.3471	0.5646	0.5688	1.4347
0.46	1.2490	1.4895	0.8226	2.6343	2.2600	0.3640	0.6139	0.6177	1.5541
0.48	1.2960	1.5197	0.8680	2.6790	2.2800	0.3807	0.6647	0.6680	1.6771
0.50	1.3420	1.5478	0.9138	2.7237	2.3000	0.3973	0.7169	0.7198	1.8037
0.52	1.3870	1.5739	0.9600	2.7685	2.3200	0.4138	0.7707	0.7730	1.9338
0.54	1.4312	1.5983	1.0066	2.8132	2.3400	0.4302	0.8259	0.8276	2.0674
0.56	1.4746	1.6210	1.0536	2.8579	2.3600	0.4464	0.8825	0.8837	2.2045
0.58	1.5172	1.6421	1.1010	2.9026	2.3800	0.4626	0.9406	0.9412	2.3450
0.60	1.5591	1.6618	1.1488	2.9473	2.4000	0.4787	1.0000	1.0000	2.4889
0.64	1.6407	1.6973	1.2456	3.0368	2.4400	0.5105	1.1231	1.1218	2.7869
0.68	1.7196	1.7280	1.3440	3.1262	2.4800	0.5419	1.2516	1.2489	3.0983
0.72	1.7962	1.7548	1.4440	3.2157	2.5200	0.5730	1.3855	1.3814	3.4230
0.76	1.8705	1.7780	1.5456	3.3051	2.5600	0.6037	1.5248	1.5191	3.7608
0.80	1.9429	1.7980	1.6488	3.3946	2.6000	0.6341	1.6693	1.6620	4.1117
0.84	2.0133	1.8154	1.7536	3.4840	2.6400	0.6642	1.8190	1.8101	4.4756
0.88	2.0820	1.8304	1.8600	3.5734	2.6800	0.6940	1.9740	1.9634	4.8524
0.92	2.1491	1.8432	1.9680	3.6629	2.7200	0.7235	2.1341	2.1218	5.2422
0.96	2.2147	1.8542	2.0776	3.7523	2.7600	0.7527	2.2994	2.2853	5.6448
1.00	2.2789	1.8636	2.1888	3.8418	2.8000	0.7817	2.4699	2.4540	6.0602

C71

5 % concave bed river

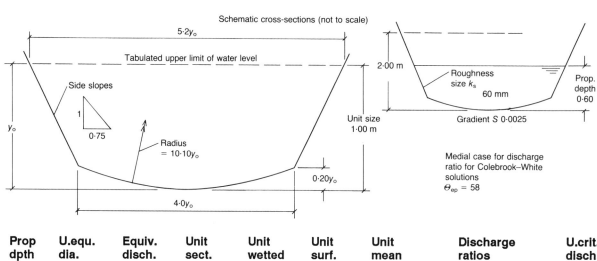

Schematic cross-sections (not to scale)

5·2y₀

Tabulated upper limit of water level

Side slopes

1

0·75

Radius = 10·10y₀

y₀

0·20y₀

4·0y₀

2·00 m

Roughness size k_s 60 mm

Unit size 1·00 m

Gradient S 0·0025

Prop. depth 0·60

Medial case for discharge ratio for Colebrook–White solutions $\Theta_{ep} = 58$

Prop dpth	U.equ. dia. $D_{ep(u)}$	Equiv. disch. factor	Unit sect. area	Unit wetted perim.	Unit surf. brdth	Unit mean depth	Discharge ratios $Q/Q_{0.60}$		U.crit. disch. Q_{uc}
Y	(m)	J	A_u (m²)	P_u (m)	B_{us} (m)	y_{um} (m)	medial	Mann'g	(m³s⁻¹)
0.02	0.0533	0.1317	0.0169	1.2714	1.2706	0.0133	0.0005	0.0007	0.0061
0.04	0.1066	0.1862	0.0479	1.7984	1.7960	0.0267	0.0025	0.0032	0.0245
0.06	0.1598	0.2279	0.0880	2.2029	2.1985	0.0400	0.0064	0.0078	0.0551
0.08	0.2129	0.2629	0.1354	2.5441	2.5374	0.0534	0.0125	0.0145	0.0980
0.10	0.2661	0.2938	0.1892	2.8449	2.8355	0.0667	0.0209	0.0235	0.1531
0.12	0.3191	0.3216	0.2487	3.1169	3.1046	0.0801	0.0316	0.0349	0.2204
0.14	0.3721	0.3472	0.3133	3.3672	3.3517	0.0935	0.0448	0.0487	0.2999
0.16	0.4251	0.3709	0.3826	3.6003	3.5813	0.1068	0.0606	0.0650	0.3916
0.18	0.4780	0.3932	0.4564	3.8193	3.7966	0.1202	0.0789	0.0838	0.4956
0.20	0.5309	0.4142	0.5344	4.0266	4.0000	0.1336	0.0999	0.1053	0.6117
0.22	0.6031	0.4648	0.6147	4.0766	4.0300	0.1525	0.1262	0.1318	0.7518
0.24	0.6743	0.5133	0.6956	4.1266	4.0600	0.1713	0.1549	0.1607	0.9016
0.26	0.7442	0.5598	0.7771	4.1766	4.0900	0.1900	0.1858	0.1917	1.0607
0.28	0.8131	0.6044	0.8592	4.2266	4.1200	0.2085	0.2190	0.2249	1.2287
0.30	0.8810	0.6471	0.9419	4.2766	4.1500	0.2270	0.2542	0.2601	1.4052
0.32	0.9478	0.6882	1.0252	4.3266	4.1800	0.2453	0.2915	0.2972	1.5900
0.34	1.0137	0.7276	1.1091	4.3766	4.2100	0.2634	0.3307	0.3362	1.7827
0.36	1.0786	0.7654	1.1936	4.4266	4.2400	0.2815	0.3718	0.3772	1.9832
0.38	1.1426	0.8018	1.2787	4.4766	4.2700	0.2995	0.4148	0.4199	2.1913
0.40	1.2057	0.8367	1.3644	4.5266	4.3000	0.3173	0.4596	0.4644	2.4068
0.42	1.2679	0.8703	1.4507	4.5766	4.3300	0.3350	0.5062	0.5106	2.6296
0.44	1.3294	0.9026	1.5376	4.6266	4.3600	0.3527	0.5545	0.5585	2.8594
0.46	1.3900	0.9337	1.6251	4.6766	4.3900	0.3702	0.6045	0.6081	3.0963
0.48	1.4498	0.9636	1.7132	4.7266	4.4200	0.3876	0.6562	0.6593	3.3401
0.50	1.5089	0.9924	1.8019	4.7766	4.4500	0.4049	0.7096	0.7122	3.5907
0.52	1.5673	1.0201	1.8912	4.8266	4.4800	0.4221	0.7645	0.7667	3.8479
0.54	1.6250	1.0468	1.9811	4.8766	4.5100	0.4393	0.8210	0.8227	4.1118
0.56	1.6820	1.0725	2.0716	4.9266	4.5400	0.4563	0.8792	0.8803	4.3822
0.58	1.7383	1.0973	2.1627	4.9766	4.5700	0.4732	0.9388	0.9394	4.6590
0.60	1.7940	1.1212	2.2544	5.0266	4.6000	0.4901	1.0000	1.0000	4.9423
0.64	1.9035	1.1664	2.4396	5.1266	4.6600	0.5235	1.1270	1.1258	5.5277
0.68	2.0106	1.2085	2.6272	5.2266	4.7200	0.5566	1.2598	1.2574	6.1380
0.72	2.1156	1.2477	2.8172	5.3266	4.7800	0.5894	1.3985	1.3948	6.7729
0.76	2.2184	1.2843	3.0096	5.4266	4.8400	0.6218	1.5430	1.5380	7.4319
0.80	2.3192	1.3183	3.2044	5.5266	4.9000	0.6540	1.6931	1.6868	8.1149
0.84	2.4182	1.3502	3.4016	5.6266	4.9600	0.6858	1.8488	1.8412	8.8215
0.88	2.5154	1.3799	3.6012	5.7266	5.0200	0.7174	2.0100	2.0011	9.5517
0.92	2.6109	1.4077	3.8032	5.8266	5.0800	0.7487	2.1767	2.1666	10.305
0.96	2.7048	1.4337	4.0076	5.9266	5.1400	0.7797	2.3489	2.3374	11.082
1.00	2.7972	1.4581	4.2144	6.0266	5.2000	0.8105	2.5265	2.5137	11.881

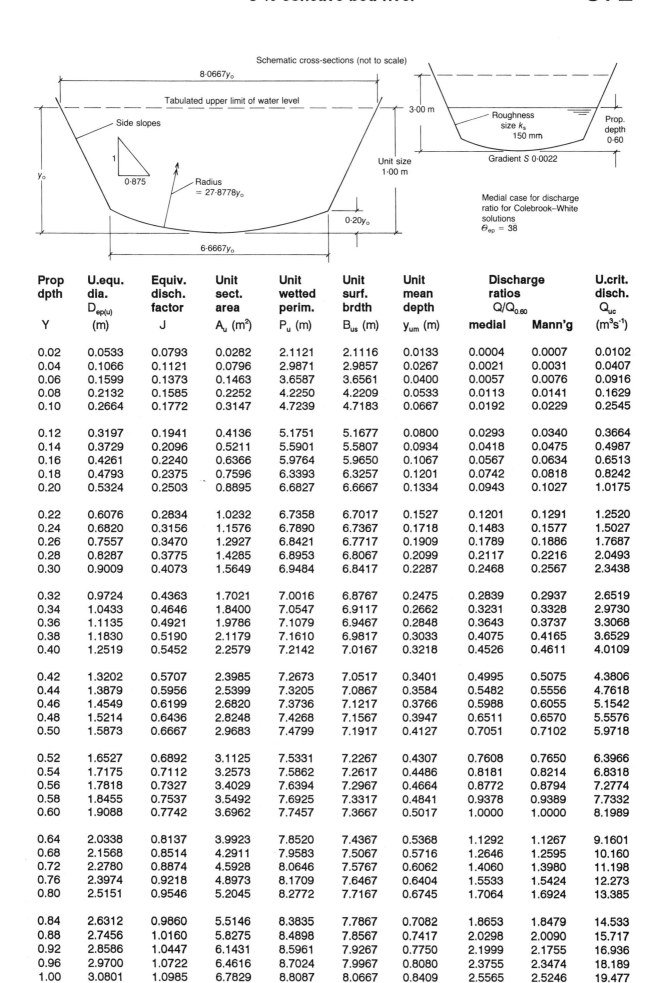

Schematic cross-sections (not to scale)

Prop dpth	U.equ. dia. $D_{ep(u)}$	Equiv. disch. factor	Unit sect. area	Unit wetted perim.	Unit surf. brdth	Unit mean depth	Discharge ratios $Q/Q_{0.60}$		U.crit. disch. Q_{uc}
Y	(m)	J	A_u (m²)	P_u (m)	B_{us} (m)	y_{um} (m)	medial	Mann'g	(m³s⁻¹)
0.02	0.0533	0.0793	0.0282	2.1121	2.1116	0.0133	0.0004	0.0007	0.0102
0.04	0.1066	0.1121	0.0796	2.9871	2.9857	0.0267	0.0021	0.0031	0.0407
0.06	0.1599	0.1373	0.1463	3.6587	3.6561	0.0400	0.0057	0.0076	0.0916
0.08	0.2132	0.1585	0.2252	4.2250	4.2209	0.0533	0.0113	0.0141	0.1629
0.10	0.2664	0.1772	0.3147	4.7239	4.7183	0.0667	0.0192	0.0229	0.2545
0.12	0.3197	0.1941	0.4136	5.1751	5.1677	0.0800	0.0293	0.0340	0.3664
0.14	0.3729	0.2096	0.5211	5.5901	5.5807	0.0934	0.0418	0.0475	0.4987
0.16	0.4261	0.2240	0.6366	5.9764	5.9650	0.1067	0.0567	0.0634	0.6513
0.18	0.4793	0.2375	0.7596	6.3393	6.3257	0.1201	0.0742	0.0818	0.8242
0.20	0.5324	0.2503	0.8895	6.6827	6.6667	0.1334	0.0943	0.1027	1.0175
0.22	0.6076	0.2834	1.0232	6.7358	6.7017	0.1527	0.1201	0.1291	1.2520
0.24	0.6820	0.3156	1.1576	6.7890	6.7367	0.1718	0.1483	0.1577	1.5027
0.26	0.7557	0.3470	1.2927	6.8421	6.7717	0.1909	0.1789	0.1886	1.7687
0.28	0.8287	0.3775	1.4285	6.8953	6.8067	0.2099	0.2117	0.2216	2.0493
0.30	0.9009	0.4073	1.5649	6.9484	6.8417	0.2287	0.2468	0.2567	2.3438
0.32	0.9724	0.4363	1.7021	7.0016	6.8767	0.2475	0.2839	0.2937	2.6519
0.34	1.0433	0.4646	1.8400	7.0547	6.9117	0.2662	0.3231	0.3328	2.9730
0.36	1.1135	0.4921	1.9786	7.1079	6.9467	0.2848	0.3643	0.3737	3.3068
0.38	1.1830	0.5190	2.1179	7.1610	6.9817	0.3033	0.4075	0.4165	3.6529
0.40	1.2519	0.5452	2.2579	7.2142	7.0167	0.3218	0.4526	0.4611	4.0109
0.42	1.3202	0.5707	2.3985	7.2673	7.0517	0.3401	0.4995	0.5075	4.3806
0.44	1.3879	0.5956	2.5399	7.3205	7.0867	0.3584	0.5482	0.5556	4.7618
0.46	1.4549	0.6199	2.6820	7.3736	7.1217	0.3766	0.5988	0.6055	5.1542
0.48	1.5214	0.6436	2.8248	7.4268	7.1567	0.3947	0.6511	0.6570	5.5576
0.50	1.5873	0.6667	2.9683	7.4799	7.1917	0.4127	0.7051	0.7102	5.9718
0.52	1.6527	0.6892	3.1125	7.5331	7.2267	0.4307	0.7608	0.7650	6.3966
0.54	1.7175	0.7112	3.2573	7.5862	7.2617	0.4486	0.8181	0.8214	6.8318
0.56	1.7818	0.7327	3.4029	7.6394	7.2967	0.4664	0.8772	0.8794	7.2774
0.58	1.8455	0.7537	3.5492	7.6925	7.3317	0.4841	0.9378	0.9389	7.7332
0.60	1.9088	0.7742	3.6962	7.7457	7.3667	0.5017	1.0000	1.0000	8.1989
0.64	2.0338	0.8137	3.9923	7.8520	7.4367	0.5368	1.1292	1.1267	9.1601
0.68	2.1568	0.8514	4.2911	7.9583	7.5067	0.5716	1.2646	1.2595	10.160
0.72	2.2780	0.8874	4.5928	8.0646	7.5767	0.6062	1.4060	1.3980	11.198
0.76	2.3974	0.9218	4.8973	8.1709	7.6467	0.6404	1.5533	1.5424	12.273
0.80	2.5151	0.9546	5.2045	8.2772	7.7167	0.6745	1.7064	1.6924	13.385
0.84	2.6312	0.9860	5.5146	8.3835	7.7867	0.7082	1.8653	1.8479	14.533
0.88	2.7456	1.0160	5.8275	8.4898	7.8567	0.7417	2.0298	2.0090	15.717
0.92	2.8586	1.0447	6.1431	8.5961	7.9267	0.7750	2.1999	2.1755	16.936
0.96	2.9700	1.0722	6.4616	8.7024	7.9967	0.8080	2.3755	2.3474	18.189
1.00	3.0801	1.0985	6.7829	8.8087	8.0667	0.8409	2.5565	2.5246	19.477

2 % concave bed river

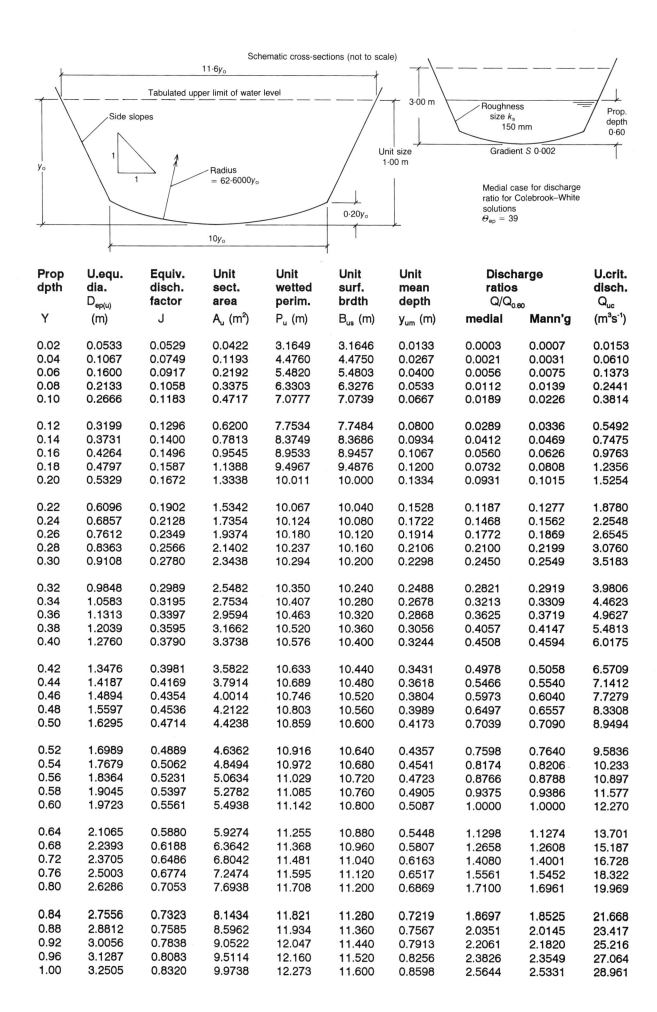

Schematic cross-sections (not to scale)

11·6y_o

Tabulated upper limit of water level

Side slopes

y_o

1 / 1

Radius = 62·6000y_o

0·20y_o

10y_o

3·00 m

Roughness size k_s 150 mm

Prop. depth 0·60

Unit size 1·00 m

Gradient S 0·002

Medial case for discharge ratio for Colebrook–White solutions $\Theta_{ep} = 39$

Prop dpth Y	U.equ. dia. $D_{ep(u)}$ (m)	Equiv. disch. factor J	Unit sect. area A_u (m²)	Unit wetted perim. P_u (m)	Unit surf. brdth B_{us} (m)	Unit mean depth y_{um} (m)	Discharge ratios $Q/Q_{0.60}$ medial	Mann'g	U.crit. disch. Q_{uc} (m³s⁻¹)
0.02	0.0533	0.0529	0.0422	3.1649	3.1646	0.0133	0.0003	0.0007	0.0153
0.04	0.1067	0.0749	0.1193	4.4760	4.4750	0.0267	0.0021	0.0031	0.0610
0.06	0.1600	0.0917	0.2192	5.4820	5.4803	0.0400	0.0056	0.0075	0.1373
0.08	0.2133	0.1058	0.3375	6.3303	6.3276	0.0533	0.0112	0.0139	0.2441
0.10	0.2666	0.1183	0.4717	7.0777	7.0739	0.0667	0.0189	0.0226	0.3814
0.12	0.3199	0.1296	0.6200	7.7534	7.7484	0.0800	0.0289	0.0336	0.5492
0.14	0.3731	0.1400	0.7813	8.3749	8.3686	0.0934	0.0412	0.0469	0.7475
0.16	0.4264	0.1496	0.9545	8.9533	8.9457	0.1067	0.0560	0.0626	0.9763
0.18	0.4797	0.1587	1.1388	9.4967	9.4876	0.1200	0.0732	0.0808	1.2356
0.20	0.5329	0.1672	1.3338	10.011	10.000	0.1334	0.0931	0.1015	1.5254
0.22	0.6096	0.1902	1.5342	10.067	10.040	0.1528	0.1187	0.1277	1.8780
0.24	0.6857	0.2128	1.7354	10.124	10.080	0.1722	0.1468	0.1562	2.2548
0.26	0.7612	0.2349	1.9374	10.180	10.120	0.1914	0.1772	0.1869	2.6545
0.28	0.8363	0.2566	2.1402	10.237	10.160	0.2106	0.2100	0.2199	3.0760
0.30	0.9108	0.2780	2.3438	10.294	10.200	0.2298	0.2450	0.2549	3.5183
0.32	0.9848	0.2989	2.5482	10.350	10.240	0.2488	0.2821	0.2919	3.9806
0.34	1.0583	0.3195	2.7534	10.407	10.280	0.2678	0.3213	0.3309	4.4623
0.36	1.1313	0.3397	2.9594	10.463	10.320	0.2868	0.3625	0.3719	4.9627
0.38	1.2039	0.3595	3.1662	10.520	10.360	0.3056	0.4057	0.4147	5.4813
0.40	1.2760	0.3790	3.3738	10.576	10.400	0.3244	0.4508	0.4594	6.0175
0.42	1.3476	0.3981	3.5822	10.633	10.440	0.3431	0.4978	0.5058	6.5709
0.44	1.4187	0.4169	3.7914	10.689	10.480	0.3618	0.5466	0.5540	7.1412
0.46	1.4894	0.4354	4.0014	10.746	10.520	0.3804	0.5973	0.6040	7.7279
0.48	1.5597	0.4536	4.2122	10.803	10.560	0.3989	0.6497	0.6557	8.3308
0.50	1.6295	0.4714	4.4238	10.859	10.600	0.4173	0.7039	0.7090	8.9494
0.52	1.6989	0.4889	4.6362	10.916	10.640	0.4357	0.7598	0.7640	9.5836
0.54	1.7679	0.5062	4.8494	10.972	10.680	0.4541	0.8174	0.8206	10.233
0.56	1.8364	0.5231	5.0634	11.029	10.720	0.4723	0.8766	0.8788	10.897
0.58	1.9045	0.5397	5.2782	11.085	10.760	0.4905	0.9375	0.9386	11.577
0.60	1.9723	0.5561	5.4938	11.142	10.800	0.5087	1.0000	1.0000	12.270
0.64	2.1065	0.5880	5.9274	11.255	10.880	0.5448	1.1298	1.1274	13.701
0.68	2.2393	0.6188	6.3642	11.368	10.960	0.5807	1.2658	1.2608	15.187
0.72	2.3705	0.6486	6.8042	11.481	11.040	0.6163	1.4080	1.4001	16.728
0.76	2.5003	0.6774	7.2474	11.595	11.120	0.6517	1.5561	1.5452	18.322
0.80	2.6286	0.7053	7.6938	11.708	11.200	0.6869	1.7100	1.6961	19.969
0.84	2.7556	0.7323	8.1434	11.821	11.280	0.7219	1.8697	1.8525	21.668
0.88	2.8812	0.7585	8.5962	11.934	11.360	0.7567	2.0351	2.0145	23.417
0.92	3.0056	0.7838	9.0522	12.047	11.440	0.7913	2.2061	2.1820	25.216
0.96	3.1287	0.8083	9.5114	12.160	11.520	0.8256	2.3826	2.3549	27.064
1.00	3.2505	0.8320	9.9738	12.273	11.600	0.8598	2.5644	2.5331	28.961

1.5 % concave bed river

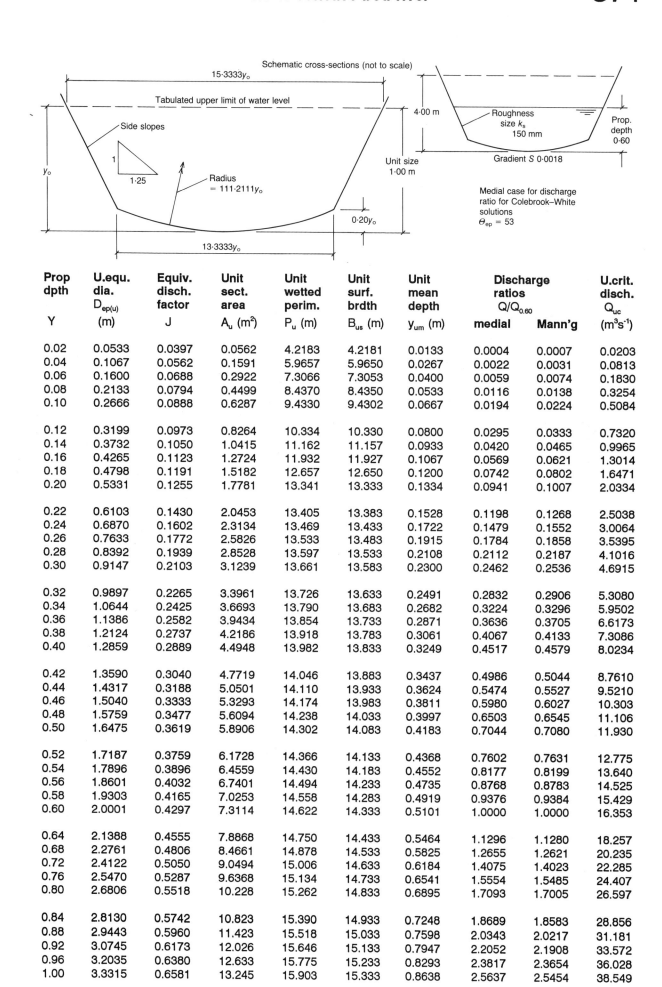

Schematic cross-sections (not to scale)

15·3333y_o

Tabulated upper limit of water level

Side slopes

1 / 1·25

Radius = 111·2111y_o

y_o

0·20y_o

13·3333y_o

4·00 m

Unit size 1·00 m

Roughness size k_s 150 mm

Prop. depth 0·60

Gradient S 0·0018

Medial case for discharge ratio for Colebrook–White solutions $\Theta_{ep} = 53$

Prop dpth Y	U.equ. dia. $D_{ep(u)}$ (m)	Equiv. disch. factor J	Unit sect. area A_u (m²)	Unit wetted perim. P_u (m)	Unit surf. brdth B_{us} (m)	Unit mean depth y_{um} (m)	Discharge ratios $Q/Q_{0.60}$ medial	Mann'g	U.crit. disch. Q_{uc} (m³s⁻¹)
0.02	0.0533	0.0397	0.0562	4.2183	4.2181	0.0133	0.0004	0.0007	0.0203
0.04	0.1067	0.0562	0.1591	5.9657	5.9650	0.0267	0.0022	0.0031	0.0813
0.06	0.1600	0.0688	0.2922	7.3066	7.3053	0.0400	0.0059	0.0074	0.1830
0.08	0.2133	0.0794	0.4499	8.4370	8.4350	0.0533	0.0116	0.0138	0.3254
0.10	0.2666	0.0888	0.6287	9.4330	9.4302	0.0667	0.0194	0.0224	0.5084
0.12	0.3199	0.0973	0.8264	10.334	10.330	0.0800	0.0295	0.0333	0.7320
0.14	0.3732	0.1050	1.0415	11.162	11.157	0.0933	0.0420	0.0465	0.9965
0.16	0.4265	0.1123	1.2724	11.932	11.927	0.1067	0.0569	0.0621	1.3014
0.18	0.4798	0.1191	1.5182	12.657	12.650	0.1200	0.0742	0.0802	1.6471
0.20	0.5331	0.1255	1.7781	13.341	13.333	0.1334	0.0941	0.1007	2.0334
0.22	0.6103	0.1430	2.0453	13.405	13.383	0.1528	0.1198	0.1268	2.5038
0.24	0.6870	0.1602	2.3134	13.469	13.433	0.1722	0.1479	0.1552	3.0064
0.26	0.7633	0.1772	2.5826	13.533	13.483	0.1915	0.1784	0.1858	3.5395
0.28	0.8392	0.1939	2.8528	13.597	13.533	0.2108	0.2112	0.2187	4.1016
0.30	0.9147	0.2103	3.1239	13.661	13.583	0.2300	0.2462	0.2536	4.6915
0.32	0.9897	0.2265	3.3961	13.726	13.633	0.2491	0.2832	0.2906	5.3080
0.34	1.0644	0.2425	3.6693	13.790	13.683	0.2682	0.3224	0.3296	5.9502
0.36	1.1386	0.2582	3.9434	13.854	13.733	0.2871	0.3636	0.3705	6.6173
0.38	1.2124	0.2737	4.2186	13.918	13.783	0.3061	0.4067	0.4133	7.3086
0.40	1.2859	0.2889	4.4948	13.982	13.833	0.3249	0.4517	0.4579	8.0234
0.42	1.3590	0.3040	4.7719	14.046	13.883	0.3437	0.4986	0.5044	8.7610
0.44	1.4317	0.3188	5.0501	14.110	13.933	0.3624	0.5474	0.5527	9.5210
0.46	1.5040	0.3333	5.3293	14.174	13.983	0.3811	0.5980	0.6027	10.303
0.48	1.5759	0.3477	5.6094	14.238	14.033	0.3997	0.6503	0.6545	11.106
0.50	1.6475	0.3619	5.8906	14.302	14.083	0.4183	0.7044	0.7080	11.930
0.52	1.7187	0.3759	6.1728	14.366	14.133	0.4368	0.7602	0.7631	12.775
0.54	1.7896	0.3896	6.4559	14.430	14.183	0.4552	0.8177	0.8199	13.640
0.56	1.8601	0.4032	6.7401	14.494	14.233	0.4735	0.8768	0.8783	14.525
0.58	1.9303	0.4165	7.0253	14.558	14.283	0.4919	0.9376	0.9384	15.429
0.60	2.0001	0.4297	7.3114	14.622	14.333	0.5101	1.0000	1.0000	16.353
0.64	2.1388	0.4555	7.8868	14.750	14.433	0.5464	1.1296	1.1280	18.257
0.68	2.2761	0.4806	8.4661	14.878	14.533	0.5825	1.2655	1.2621	20.235
0.72	2.4122	0.5050	9.0494	15.006	14.633	0.6184	1.4075	1.4023	22.285
0.76	2.5470	0.5287	9.6368	15.134	14.733	0.6541	1.5554	1.5485	24.407
0.80	2.6806	0.5518	10.228	15.262	14.833	0.6895	1.7093	1.7005	26.597
0.84	2.8130	0.5742	10.823	15.390	14.933	0.7248	1.8689	1.8583	28.856
0.88	2.9443	0.5960	11.423	15.518	15.033	0.7598	2.0343	2.0217	31.181
0.92	3.0745	0.6173	12.026	15.646	15.133	0.7947	2.2052	2.1908	33.572
0.96	3.2035	0.6380	12.633	15.775	15.233	0.8293	2.3817	2.3654	36.028
1.00	3.3315	0.6581	13.245	15.903	15.333	0.8638	2.5637	2.5454	38.549

C75

1.25 % concave bed river

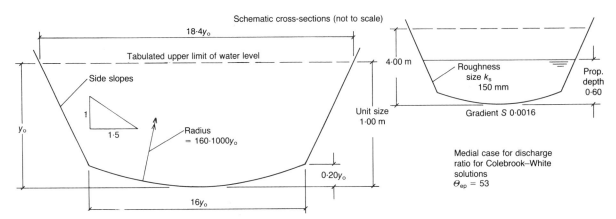

Schematic cross-sections (not to scale)

Prop dpth Y	U.equ. dia. $D_{ep(u)}$ (m)	Equiv. disch. factor J	Unit sect. area A_u (m²)	Unit wetted perim. P_u (m)	Unit surf. brdth B_{us} (m)	Unit mean depth y_{um} (m)	Discharge ratios $Q/Q_{0.60}$ medial	Mann'g	U.crit. disch. Q_{uc} (m³s⁻¹)
0.02	0.0533	0.0331	0.0675	5.0613	5.0611	0.0133	0.0004	0.0007	0.0244
0.04	0.1067	0.0468	0.1909	7.1578	7.1572	0.0267	0.0022	0.0031	0.0976
0.06	0.1600	0.0573	0.3506	8.7666	8.7655	0.0400	0.0059	0.0074	0.2196
0.08	0.2133	0.0662	0.5398	10.123	10.121	0.0533	0.0115	0.0138	0.3904
0.10	0.2666	0.0740	0.7544	11.318	11.315	0.0667	0.0193	0.0224	0.6100
0.12	0.3200	0.0811	0.9917	12.398	12.395	0.0800	0.0294	0.0332	0.8785
0.14	0.3733	0.0876	1.2497	13.392	13.388	0.0933	0.0418	0.0463	1.1957
0.16	0.4266	0.0936	1.5268	14.316	14.312	0.1067	0.0567	0.0619	1.5617
0.18	0.4799	0.0993	1.8217	15.185	15.179	0.1200	0.0739	0.0799	1.9762
0.20	0.5332	0.1046	2.1336	16.007	16.000	0.1334	0.0938	0.1004	2.4399
0.22	0.6105	0.1193	2.4542	16.079	16.060	0.1528	0.1194	0.1263	3.0044
0.24	0.6875	0.1337	2.7760	16.151	16.120	0.1722	0.1475	0.1547	3.6075
0.26	0.7641	0.1480	3.0990	16.223	16.180	0.1915	0.1779	0.1853	4.2472
0.28	0.8403	0.1620	3.4232	16.295	16.240	0.2108	0.2106	0.2181	4.9217
0.30	0.9161	0.1758	3.7486	16.367	16.300	0.2300	0.2455	0.2529	5.6295
0.32	0.9916	0.1895	4.0752	16.439	16.360	0.2491	0.2825	0.2899	6.3693
0.34	1.0667	0.2029	4.4030	16.511	16.420	0.2681	0.3217	0.3288	7.1400
0.36	1.1414	0.2162	4.7320	16.584	16.480	0.2871	0.3628	0.3697	7.9405
0.38	1.2157	0.2293	5.0622	16.656	16.540	0.3061	0.4059	0.4125	8.7700
0.40	1.2897	0.2422	5.3936	16.728	16.600	0.3249	0.4509	0.4571	9.6277
0.42	1.3634	0.2549	5.7262	16.800	16.660	0.3437	0.4979	0.5036	10.513
0.44	1.4367	0.2675	6.0600	16.872	16.720	0.3624	0.5467	0.5519	11.425
0.46	1.5097	0.2799	6.3950	16.944	16.780	0.3811	0.5973	0.6020	12.363
0.48	1.5823	0.2921	6.7312	17.016	16.840	0.3997	0.6497	0.6538	13.327
0.50	1.6546	0.3042	7.0686	17.088	16.900	0.4183	0.7038	0.7074	14.316
0.52	1.7266	0.3161	7.4072	17.160	16.960	0.4367	0.7597	0.7626	15.330
0.54	1.7982	0.3278	7.7470	17.233	17.020	0.4552	0.8173	0.8195	16.367
0.56	1.8696	0.3394	8.0880	17.305	17.080	0.4735	0.8765	0.8780	17.429
0.58	1.9406	0.3508	8.4302	17.377	17.140	0.4918	0.9375	0.9382	18.514
0.60	2.0113	0.3621	8.7736	17.449	17.200	0.5101	1.0000	1.0000	19.623
0.64	2.1518	0.3842	9.4640	17.593	17.320	0.5464	1.1300	1.1284	21.908
0.68	2.2910	0.4058	10.159	17.737	17.440	0.5825	1.2663	1.2630	24.282
0.72	2.4291	0.4268	10.859	17.882	17.560	0.6184	1.4088	1.4037	26.742
0.76	2.5661	0.4472	11.564	18.026	17.680	0.6541	1.5574	1.5505	29.287
0.80	2.7019	0.4672	12.274	18.170	17.800	0.6895	1.7120	1.7032	31.916
0.84	2.8367	0.4866	12.988	18.314	17.920	0.7248	1.8724	1.8618	34.626
0.88	2.9704	0.5055	13.707	18.458	18.040	0.7598	2.0386	2.0261	37.417
0.92	3.1030	0.5240	14.431	18.603	18.160	0.7947	2.2106	2.1962	40.286
0.96	3.2347	0.5421	15.160	18.747	18.280	0.8293	2.3882	2.3719	43.234
1.00	3.3653	0.5596	15.894	18.891	18.400	0.8638	2.5713	2.5532	46.258

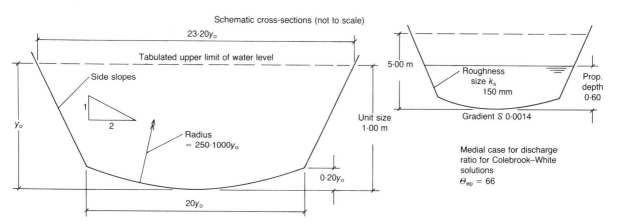

Schematic cross-sections (not to scale)

23·20y_o — Tabulated upper limit of water level — Side slopes — 1/2 — Radius = 250·1000y_o — y_o — 20y_o — 0·20y_o

5·00 m — Roughness size k_s 150 mm — Prop. depth 0·60 — Unit size 1·00 m — Gradient S 0·0014

Medial case for discharge ratio for Colebrook–White solutions $\Theta_{ep} = 66$

Prop dpth Y	U.equ. dia. $D_{ep(u)}$ (m)	Equiv. disch. factor J	Unit sect. area A_u (m²)	Unit wetted perim. P_u (m)	Unit surf. brdth B_{us} (m)	Unit mean depth y_{um} (m)	Discharge ratios $Q/Q_{0.60}$ medial	Mann'g	U.crit. disch. Q_{uc} (m³s⁻¹)
0.02	0.0533	0.0265	0.0844	6.3259	6.3257	0.0133	0.0004	0.0007	0.0305
0.04	0.1067	0.0375	0.2385	8.9462	8.9457	0.0267	0.0023	0.0031	0.1220
0.06	0.1600	0.0459	0.4383	10.957	10.956	0.0400	0.0061	0.0074	0.2745
0.08	0.2133	0.0530	0.6747	12.652	12.651	0.0533	0.0118	0.0137	0.4879
0.10	0.2666	0.0592	0.9428	14.145	14.144	0.0667	0.0198	0.0223	0.7623
0.12	0.3200	0.0649	1.2395	15.496	15.493	0.0800	0.0299	0.0331	1.0979
0.14	0.3733	0.0701	1.5621	16.737	16.734	0.0933	0.0425	0.0462	1.4946
0.16	0.4266	0.0749	1.9084	17.893	17.889	0.1067	0.0574	0.0617	1.9520
0.18	0.4799	0.0794	2.2770	18.979	18.974	0.1200	0.0748	0.0796	2.4702
0.20	0.5332	0.0837	2.6669	20.005	20.000	0.1333	0.0947	0.1000	3.0497
0.22	0.6106	0.0955	3.0677	20.095	20.080	0.1528	0.1204	0.1259	3.7549
0.24	0.6877	0.1070	3.4701	20.184	20.160	0.1721	0.1485	0.1542	4.5084
0.26	0.7644	0.1184	3.8741	20.274	20.240	0.1914	0.1789	0.1847	5.3077
0.28	0.8407	0.1297	4.2797	20.363	20.320	0.2106	0.2116	0.2174	6.1506
0.30	0.9166	0.1408	4.6869	20.453	20.400	0.2297	0.2465	0.2522	7.0351
0.32	0.9922	0.1517	5.0957	20.542	20.480	0.2488	0.2835	0.2891	7.9597
0.34	1.0675	0.1625	5.5061	20.631	20.560	0.2678	0.3226	0.3280	8.9230
0.36	1.1424	0.1732	5.9181	20.721	20.640	0.2867	0.3637	0.3688	9.9238
0.38	1.2170	0.1837	6.3317	20.810	20.720	0.3056	0.4067	0.4116	10.961
0.40	1.2913	0.1941	6.7469	20.900	20.800	0.3244	0.4517	0.4562	12.033
0.42	1.3652	0.2043	7.1637	20.989	20.880	0.3431	0.4986	0.5027	13.140
0.44	1.4388	0.2144	7.5821	21.079	20.960	0.3617	0.5473	0.5510	14.281
0.46	1.5121	0.2244	8.0021	21.168	21.040	0.3803	0.5978	0.6012	15.454
0.48	1.5851	0.2342	8.4237	21.258	21.120	0.3988	0.6501	0.6530	16.660
0.50	1.6577	0.2440	8.8469	21.347	21.200	0.4173	0.7042	0.7066	17.897
0.52	1.7301	0.2535	9.2717	21.436	21.280	0.4357	0.7600	0.7620	19.165
0.54	1.8021	0.2630	9.6981	21.526	21.360	0.4540	0.8175	0.8190	20.464
0.56	1.8739	0.2723	10.126	21.615	21.440	0.4723	0.8767	0.8777	21.793
0.58	1.9453	0.2816	10.556	21.705	21.520	0.4905	0.9375	0.9380	23.151
0.60	2.0165	0.2907	10.987	21.794	21.600	0.5087	1.0000	1.0000	24.538
0.64	2.1579	0.3085	11.854	21.973	21.760	0.5448	1.1299	1.1288	27.399
0.68	2.2982	0.3259	12.728	22.152	21.920	0.5806	1.2661	1.2640	30.371
0.72	2.4375	0.3429	13.608	22.331	22.080	0.6163	1.4085	1.4054	33.453
0.76	2.5756	0.3595	14.494	22.510	22.240	0.6517	1.5571	1.5530	36.642
0.80	2.7127	0.3756	15.387	22.689	22.400	0.6869	1.7117	1.7067	39.936
0.84	2.8488	0.3914	16.286	22.867	22.560	0.7219	1.8723	1.8663	43.333
0.88	2.9838	0.4067	17.192	23.046	22.720	0.7567	2.0387	2.0319	46.831
0.92	3.1179	0.4217	18.104	23.225	22.880	0.7912	2.2109	2.2033	50.429
0.96	3.2511	0.4364	19.022	23.404	23.040	0.8256	2.3888	2.3805	54.126
1.00	3.3833	0.4507	19.947	23.583	23.200	0.8598	2.5723	2.5635	57.920

C77

0.75 % concave bed river

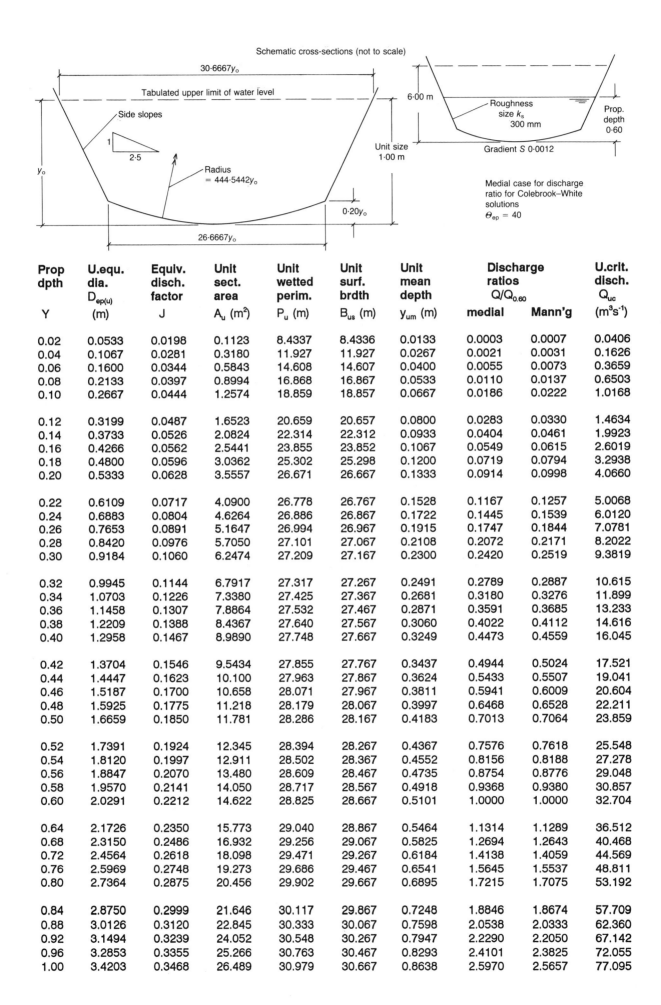

Schematic cross-sections (not to scale)

Prop dpth Y	U.equ. dia. $D_{ep(u)}$ (m)	Equiv. disch. factor J	Unit sect. area A_u (m²)	Unit wetted perim. P_u (m)	Unit surf. brdth B_{us} (m)	Unit mean depth y_{um} (m)	Discharge ratios $Q/Q_{0.60}$ medial	Mann'g	U.crit. disch. Q_{uc} (m³s⁻¹)
0.02	0.0533	0.0198	0.1123	8.4337	8.4336	0.0133	0.0003	0.0007	0.0406
0.04	0.1067	0.0281	0.3180	11.927	11.927	0.0267	0.0021	0.0031	0.1626
0.06	0.1600	0.0344	0.5843	14.608	14.607	0.0400	0.0055	0.0073	0.3659
0.08	0.2133	0.0397	0.8994	16.868	16.867	0.0533	0.0110	0.0137	0.6503
0.10	0.2667	0.0444	1.2574	18.859	18.857	0.0667	0.0186	0.0222	1.0168
0.12	0.3199	0.0487	1.6523	20.659	20.657	0.0800	0.0283	0.0330	1.4634
0.14	0.3733	0.0526	2.0824	22.314	22.312	0.0933	0.0404	0.0461	1.9923
0.16	0.4266	0.0562	2.5441	23.855	23.852	0.1067	0.0549	0.0615	2.6019
0.18	0.4800	0.0596	3.0362	25.302	25.298	0.1200	0.0719	0.0794	3.2938
0.20	0.5333	0.0628	3.5557	26.671	26.667	0.1333	0.0914	0.0998	4.0660
0.22	0.6109	0.0717	4.0900	26.778	26.767	0.1528	0.1167	0.1257	5.0068
0.24	0.6883	0.0804	4.6264	26.886	26.867	0.1722	0.1445	0.1539	6.0120
0.26	0.7653	0.0891	5.1647	26.994	26.967	0.1915	0.1747	0.1844	7.0781
0.28	0.8420	0.0976	5.7050	27.101	27.067	0.2108	0.2072	0.2171	8.2022
0.30	0.9184	0.1060	6.2474	27.209	27.167	0.2300	0.2420	0.2519	9.3819
0.32	0.9945	0.1144	6.7917	27.317	27.267	0.2491	0.2789	0.2887	10.615
0.34	1.0703	0.1226	7.3380	27.425	27.367	0.2681	0.3180	0.3276	11.899
0.36	1.1458	0.1307	7.8864	27.532	27.467	0.2871	0.3591	0.3685	13.233
0.38	1.2209	0.1388	8.4367	27.640	27.567	0.3060	0.4022	0.4112	14.616
0.40	1.2958	0.1467	8.9890	27.748	27.667	0.3249	0.4473	0.4559	16.045
0.42	1.3704	0.1546	9.5434	27.855	27.767	0.3437	0.4944	0.5024	17.521
0.44	1.4447	0.1623	10.100	27.963	27.867	0.3624	0.5433	0.5507	19.041
0.46	1.5187	0.1700	10.658	28.071	27.967	0.3811	0.5941	0.6009	20.604
0.48	1.5925	0.1775	11.218	28.179	28.067	0.3997	0.6468	0.6528	22.211
0.50	1.6659	0.1850	11.781	28.286	28.167	0.4183	0.7013	0.7064	23.859
0.52	1.7391	0.1924	12.345	28.394	28.267	0.4367	0.7576	0.7618	25.548
0.54	1.8120	0.1997	12.911	28.502	28.367	0.4552	0.8156	0.8188	27.278
0.56	1.8847	0.2070	13.480	28.609	28.467	0.4735	0.8754	0.8776	29.048
0.58	1.9570	0.2141	14.050	28.717	28.567	0.4918	0.9368	0.9380	30.857
0.60	2.0291	0.2212	14.622	28.825	28.667	0.5101	1.0000	1.0000	32.704
0.64	2.1726	0.2350	15.773	29.040	28.867	0.5464	1.1314	1.1289	36.512
0.68	2.3150	0.2486	16.932	29.256	29.067	0.5825	1.2694	1.2643	40.468
0.72	2.4564	0.2618	18.098	29.471	29.267	0.6184	1.4138	1.4059	44.569
0.76	2.5969	0.2748	19.273	29.686	29.467	0.6541	1.5645	1.5537	48.811
0.80	2.7364	0.2875	20.456	29.902	29.667	0.6895	1.7215	1.7075	53.192
0.84	2.8750	0.2999	21.646	30.117	29.867	0.7248	1.8846	1.8674	57.709
0.88	3.0126	0.3120	22.845	30.333	30.067	0.7598	2.0538	2.0333	62.360
0.92	3.1494	0.3239	24.052	30.548	30.267	0.7947	2.2290	2.2050	67.142
0.96	3.2853	0.3355	25.266	30.763	30.467	0.8293	2.4101	2.3825	72.055
1.00	3.4203	0.3468	26.489	30.979	30.667	0.8638	2.5970	2.5657	77.095

0.50 % concave bed river

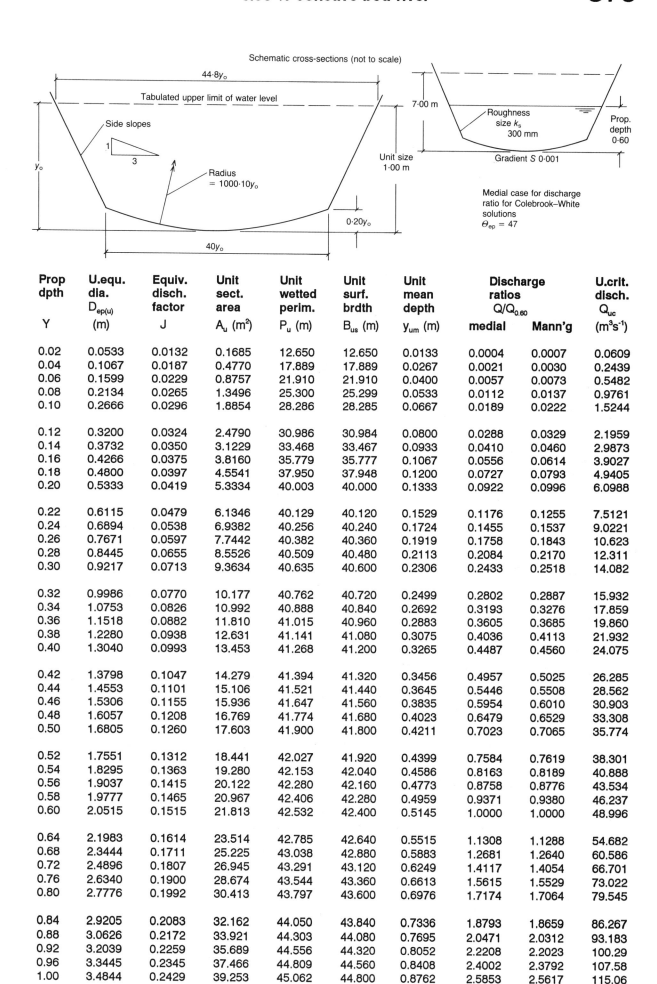

Schematic cross-sections (not to scale)

44·8y_o

Tabulated upper limit of water level

Side slopes

1 / 3

y_o

Radius = 1000·10y_o

0·20y_o

40y_o

7·00 m

Roughness size k_s 300 mm

Prop. depth 0·60

Unit size 1·00 m

Gradient S 0·001

Medial case for discharge ratio for Colebrook–White solutions $\Theta_{ep} = 47$

Prop dpth Y	U.equ. dia. $D_{ep(u)}$ (m)	Equiv. disch. factor J	Unit sect. area A_u (m²)	Unit wetted perim. P_u (m)	Unit surf. brdth B_{us} (m)	Unit mean depth y_{um} (m)	Discharge ratios $Q/Q_{0.60}$ medial	Mann'g	U.crit. disch. Q_{uc} (m³s⁻¹)
0.02	0.0533	0.0132	0.1685	12.650	12.650	0.0133	0.0004	0.0007	0.0609
0.04	0.1067	0.0187	0.4770	17.889	17.889	0.0267	0.0021	0.0030	0.2439
0.06	0.1599	0.0229	0.8757	21.910	21.910	0.0400	0.0057	0.0073	0.5482
0.08	0.2134	0.0265	1.3496	25.300	25.299	0.0533	0.0112	0.0137	0.9761
0.10	0.2666	0.0296	1.8854	28.286	28.285	0.0667	0.0189	0.0222	1.5244
0.12	0.3200	0.0324	2.4790	30.986	30.984	0.0800	0.0288	0.0329	2.1959
0.14	0.3732	0.0350	3.1229	33.468	33.467	0.0933	0.0410	0.0460	2.9873
0.16	0.4266	0.0375	3.8160	35.779	35.777	0.1067	0.0556	0.0614	3.9027
0.18	0.4800	0.0397	4.5541	37.950	37.948	0.1200	0.0727	0.0793	4.9405
0.20	0.5333	0.0419	5.3334	40.003	40.000	0.1333	0.0922	0.0996	6.0988
0.22	0.6115	0.0479	6.1346	40.129	40.120	0.1529	0.1176	0.1255	7.5121
0.24	0.6894	0.0538	6.9382	40.256	40.240	0.1724	0.1455	0.1537	9.0221
0.26	0.7671	0.0597	7.7442	40.382	40.360	0.1919	0.1758	0.1843	10.623
0.28	0.8445	0.0655	8.5526	40.509	40.480	0.2113	0.2084	0.2170	12.311
0.30	0.9217	0.0713	9.3634	40.635	40.600	0.2306	0.2433	0.2518	14.082
0.32	0.9986	0.0770	10.177	40.762	40.720	0.2499	0.2802	0.2887	15.932
0.34	1.0753	0.0826	10.992	40.888	40.840	0.2692	0.3193	0.3276	17.859
0.36	1.1518	0.0882	11.810	41.015	40.960	0.2883	0.3605	0.3685	19.860
0.38	1.2280	0.0938	12.631	41.141	41.080	0.3075	0.4036	0.4113	21.932
0.40	1.3040	0.0993	13.453	41.268	41.200	0.3265	0.4487	0.4560	24.075
0.42	1.3798	0.1047	14.279	41.394	41.320	0.3456	0.4957	0.5025	26.285
0.44	1.4553	0.1101	15.106	41.521	41.440	0.3645	0.5446	0.5508	28.562
0.46	1.5306	0.1155	15.936	41.647	41.560	0.3835	0.5954	0.6010	30.903
0.48	1.6057	0.1208	16.769	41.774	41.680	0.4023	0.6479	0.6529	33.308
0.50	1.6805	0.1260	17.603	41.900	41.800	0.4211	0.7023	0.7065	35.774
0.52	1.7551	0.1312	18.441	42.027	41.920	0.4399	0.7584	0.7619	38.301
0.54	1.8295	0.1363	19.280	42.153	42.040	0.4586	0.8163	0.8189	40.888
0.56	1.9037	0.1415	20.122	42.280	42.160	0.4773	0.8758	0.8776	43.534
0.58	1.9777	0.1465	20.967	42.406	42.280	0.4959	0.9371	0.9380	46.237
0.60	2.0515	0.1515	21.813	42.532	42.400	0.5145	1.0000	1.0000	48.996
0.64	2.1983	0.1614	23.514	42.785	42.640	0.5515	1.1308	1.1288	54.682
0.68	2.3444	0.1711	25.225	43.038	42.880	0.5883	1.2681	1.2640	60.586
0.72	2.4896	0.1807	26.945	43.291	43.120	0.6249	1.4117	1.4054	66.701
0.76	2.6340	0.1900	28.674	43.544	43.360	0.6613	1.5615	1.5529	73.022
0.80	2.7776	0.1992	30.413	43.797	43.600	0.6976	1.7174	1.7064	79.545
0.84	2.9205	0.2083	32.162	44.050	43.840	0.7336	1.8793	1.8659	86.267
0.88	3.0626	0.2172	33.921	44.303	44.080	0.7695	2.0471	2.0312	93.183
0.92	3.2039	0.2259	35.689	44.556	44.320	0.8052	2.2208	2.2023	100.29
0.96	3.3445	0.2345	37.466	44.809	44.560	0.8408	2.4002	2.3792	107.58
1.00	3.4844	0.2429	39.253	45.062	44.800	0.8762	2.5853	2.5617	115.06

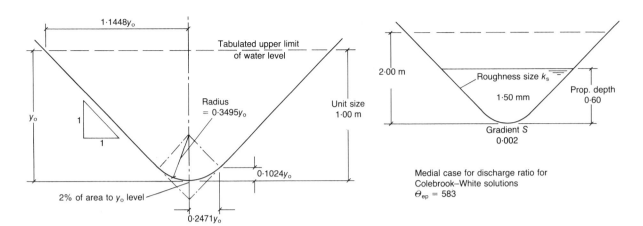

Prop dpth	U.equ. dia. $D_{ep(u)}$	Equiv. disch. factor	Unit sect. area	Unit wetted perim.	Unit surf. brdth	Unit mean depth	Discharge ratios $Q/Q_{0.60}$		U.crit. disch. Q_{uc}
Y	(m)	J	A_u (m²)	P_u (m)	B_{us} (m)	y_{um} (m)	medial	Mann'g	(m³s⁻¹)
0.02	0.0526	0.6957	0.0031	0.2376	0.2331	0.0134	0.0008	0.0008	0.0011
0.04	0.1038	0.9657	0.0088	0.3377	0.3247	0.0270	0.0037	0.0035	0.0045
0.06	0.1535	1.1603	0.0160	0.4157	0.3916	0.0407	0.0087	0.0082	0.0101
0.08	0.2018	1.3140	0.0243	0.4824	0.4450	0.0547	0.0159	0.0150	0.0178
0.10	0.2486	1.4401	0.0337	0.5422	0.4895	0.0688	0.0252	0.0239	0.0277
0.12	0.2931	1.5378	0.0439	0.5988	0.5295	0.0829	0.0366	0.0348	0.0396
0.14	0.3349	1.6053	0.0549	0.6554	0.5695	0.0964	0.0498	0.0476	0.0533
0.16	0.3745	1.6526	0.0667	0.7120	0.6095	0.1094	0.0649	0.0623	0.0690
0.18	0.4125	1.6861	0.0793	0.7686	0.6495	0.1220	0.0821	0.0789	0.0867
0.20	0.4491	1.7100	0.0926	0.8251	0.6895	0.1344	0.1013	0.0977	0.1064
0.22	0.4847	1.7270	0.1068	0.8817	0.7295	0.1465	0.1225	0.1185	0.1280
0.24	0.5194	1.7390	0.1218	0.9383	0.7695	0.1583	0.1460	0.1415	0.1518
0.26	0.5533	1.7473	0.1376	0.9948	0.8095	0.1700	0.1716	0.1667	0.1777
0.28	0.5867	1.7530	0.1542	1.0514	0.8495	0.1815	0.1995	0.1942	0.2057
0.30	0.6195	1.7566	0.1716	1.1080	0.8895	0.1929	0.2297	0.2241	0.2360
0.32	0.6519	1.7586	0.1898	1.1645	0.9295	0.2042	0.2623	0.2564	0.2686
0.34	0.6839	1.7595	0.2088	1.2211	0.9695	0.2153	0.2973	0.2913	0.3034
0.36	0.7156	1.7595	0.2286	1.2777	1.0095	0.2264	0.3348	0.3286	0.3406
0.38	0.7470	1.7588	0.2492	1.3342	1.0495	0.2374	0.3749	0.3687	0.3802
0.40	0.7781	1.7576	0.2705	1.3908	1.0895	0.2483	0.4176	0.4113	0.4222
0.42	0.8090	1.7560	0.2927	1.4474	1.1295	0.2592	0.4630	0.4568	0.4667
0.44	0.8397	1.7541	0.3157	1.5039	1.1695	0.2700	0.5110	0.5051	0.5137
0.46	0.8703	1.7520	0.3395	1.5605	1.2095	0.2807	0.5619	0.5562	0.5633
0.48	0.9007	1.7497	0.3641	1.6171	1.2495	0.2914	0.6156	0.6103	0.6155
0.50	0.9309	1.7473	0.3895	1.6736	1.2895	0.3021	0.6721	0.6674	0.6704
0.52	0.9610	1.7449	0.4157	1.7302	1.3295	0.3127	0.7316	0.7275	0.7279
0.54	0.9910	1.7424	0.4427	1.7868	1.3695	0.3232	0.7941	0.7908	0.7882
0.56	1.0209	1.7399	0.4705	1.8434	1.4095	0.3338	0.8596	0.8573	0.8512
0.58	1.0507	1.7373	0.4991	1.8999	1.4495	0.3443	0.9283	0.9270	0.9170
0.60	1.0804	1.7348	0.5285	1.9565	1.4895	0.3548	1.0000	1.0000	0.9857
0.64	1.1396	1.7298	0.5896	2.0696	1.5695	0.3757	1.1532	1.1562	1.1317
0.68	1.1985	1.7249	0.6540	2.1828	1.6495	0.3965	1.3195	1.3262	1.2896
0.72	1.2572	1.7202	0.7216	2.2959	1.7295	0.4172	1.4994	1.5106	1.4596
0.76	1.3157	1.7157	0.7924	2.4090	1.8095	0.4379	1.6932	1.7099	1.6420
0.80	1.3740	1.7114	0.8664	2.5222	1.8895	0.4585	1.9013	1.9244	1.8371
0.84	1.4321	1.7072	0.9435	2.6353	1.9695	0.4791	2.1242	2.1545	2.0451
0.88	1.4902	1.7033	1.0239	2.7485	2.0495	0.4996	2.3620	2.4008	2.2664
0.92	1.5481	1.6995	1.1075	2.8616	2.1295	0.5201	2.6153	2.6636	2.5011
0.96	1.6059	1.6959	1.1943	2.9747	2.2095	0.5405	2.8843	2.9434	2.7496
1.00	1.6636	1.6925	1.2843	3.0879	2.2895	0.5609	3.1695	3.2406	3.0121

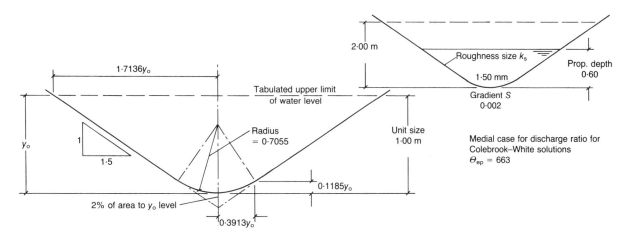

Prop dpth	U.equ. dia. $D_{ep(u)}$	Equiv. disch. factor	Unit sect. area	Unit wetted perim.	Unit surf. brdth	Unit mean depth	Discharge ratios $Q/Q_{0.60}$		U.crit. disch. Q_{uc}
Y	(m)	J	A_u (m²)	P_u (m)	B_{us} (m)	y_{um} (m)	medial	Mann'g	(m³s⁻¹)
0.02	0.0530	0.4942	0.0045	0.3368	0.3336	0.0134	0.0007	0.0007	0.0016
0.04	0.1053	0.6926	0.0126	0.4774	0.4684	0.0268	0.0033	0.0031	0.0064
0.06	0.1568	0.8405	0.0230	0.5861	0.5694	0.0404	0.0079	0.0074	0.0145
0.08	0.2077	0.9615	0.0352	0.6785	0.6526	0.0540	0.0146	0.0137	0.0256
0.10	0.2578	1.0649	0.0490	0.7604	0.7242	0.0677	0.0233	0.0220	0.0399
0.12	0.3072	1.1555	0.0641	0.8351	0.7872	0.0815	0.0341	0.0323	0.0573
0.14	0.3548	1.2287	0.0805	0.9072	0.8472	0.0950	0.0470	0.0447	0.0777
0.16	0.4003	1.2842	0.0980	0.9793	0.9072	0.1080	0.0618	0.0590	0.1009
0.18	0.4442	1.3272	0.1168	1.0514	0.9672	0.1207	0.0786	0.0753	0.1270
0.20	0.4867	1.3608	0.1367	1.1236	1.0272	0.1331	0.0975	0.0937	0.1562
0.22	0.5281	1.3875	0.1579	1.1957	1.0872	0.1452	0.1185	0.1142	0.1884
0.24	0.5685	1.4088	0.1802	1.2678	1.1472	0.1571	0.1417	0.1369	0.2236
0.26	0.6082	1.4261	0.2037	1.3399	1.2072	0.1688	0.1672	0.1620	0.2621
0.28	0.6473	1.4401	0.2285	1.4120	1.2672	0.1803	0.1949	0.1893	0.3038
0.30	0.6857	1.4515	0.2544	1.4841	1.3272	0.1917	0.2250	0.2191	0.3489
0.32	0.7237	1.4610	0.2816	1.5562	1.3872	0.2030	0.2576	0.2513	0.3973
0.34	0.7613	1.4688	0.3099	1.6283	1.4472	0.2141	0.2926	0.2861	0.4491
0.36	0.7985	1.4752	0.3395	1.7004	1.5072	0.2252	0.3302	0.3235	0.5045
0.38	0.8354	1.4806	0.3702	1.7726	1.5672	0.2362	0.3703	0.3636	0.5635
0.40	0.8720	1.4851	0.4022	1.8447	1.6272	0.2471	0.4132	0.4065	0.6261
0.42	0.9084	1.4888	0.4353	1.9168	1.6872	0.2580	0.4587	0.4521	0.6924
0.44	0.9445	1.4919	0.4696	1.9889	1.7472	0.2688	0.5070	0.5006	0.7625
0.46	0.9805	1.4945	0.5052	2.0610	1.8072	0.2795	0.5582	0.5521	0.8364
0.48	1.0162	1.4966	0.5419	2.1331	1.8672	0.2902	0.6122	0.6066	0.9143
0.50	1.0518	1.4984	0.5799	2.2052	1.9272	0.3009	0.6692	0.6641	0.9961
0.52	1.0873	1.4998	0.6190	2.2773	1.9872	0.3115	0.7291	0.7248	1.0819
0.54	1.1226	1.5010	0.6594	2.3494	2.0472	0.3221	0.7922	0.7886	1.1718
0.56	1.1578	1.5020	0.7009	2.4216	2.1072	0.3326	0.8583	0.8557	1.2659
0.58	1.1929	1.5028	0.7436	2.4937	2.1672	0.3431	0.9275	0.9262	1.3641
0.60	1.2278	1.5034	0.7876	2.5658	2.2272	0.3536	1.0000	1.0000	1.4667
0.64	1.2975	1.5041	0.8791	2.7100	2.3472	0.3745	1.1548	1.1580	1.6847
0.68	1.3669	1.5045	0.9754	2.8542	2.4672	0.3953	1.3230	1.3303	1.9205
0.72	1.4360	1.5045	1.0765	2.9984	2.5872	0.4161	1.5051	1.5172	2.1744
0.76	1.5049	1.5043	1.1823	3.1427	2.7072	0.4367	1.7014	1.7193	2.4469
0.80	1.5736	1.5040	1.2930	3.2869	2.8272	0.4574	1.9123	1.9370	2.7384
0.84	1.6421	1.5035	1.4085	3.4311	2.9472	0.4779	2.1381	2.1708	3.0493
0.88	1.7104	1.5029	1.5288	3.5753	3.0672	0.4984	2.3794	2.4212	3.3800
0.92	1.7786	1.5022	1.6539	3.7196	3.1872	0.5189	2.6364	2.6884	3.7309
0.96	1.8467	1.5015	1.7838	3.8638	3.3072	0.5394	2.9095	2.9731	4.1025
1.00	1.9146	1.5007	1.9185	4.0080	3.4272	0.5598	3.1990	3.2756	4.4950

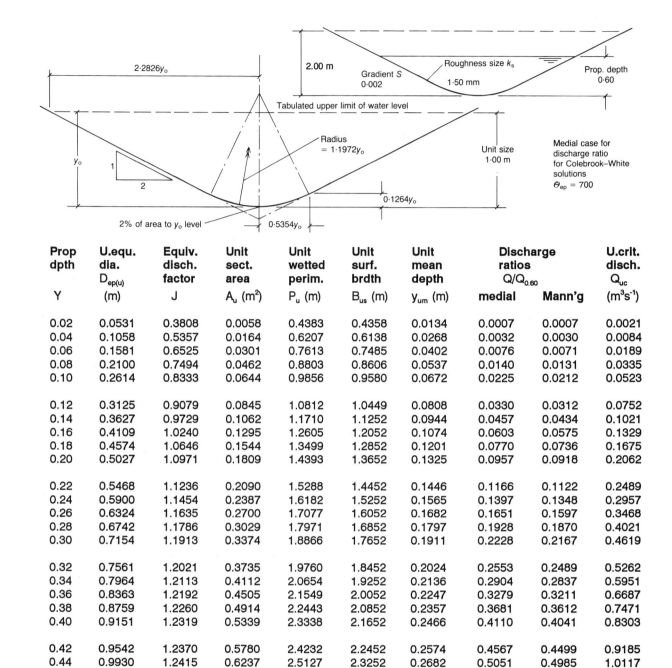

Prop dpth	U.equ. dia. $D_{ep(u)}$	Equiv. disch. factor	Unit sect. area	Unit wetted perim.	Unit surf. brdth	Unit mean depth	Discharge ratios $Q/Q_{0.60}$		U.crit. disch. Q_{uc}
Y	(m)	J	A_u (m²)	P_u (m)	B_{us} (m)	y_{um} (m)	medial	Mann'g	(m³s⁻¹)
0.02	0.0531	0.3808	0.0058	0.4383	0.4358	0.0134	0.0007	0.0007	0.0021
0.04	0.1058	0.5357	0.0164	0.6207	0.6138	0.0268	0.0032	0.0030	0.0084
0.06	0.1581	0.6525	0.0301	0.7613	0.7485	0.0402	0.0076	0.0071	0.0189
0.08	0.2100	0.7494	0.0462	0.8803	0.8606	0.0537	0.0140	0.0131	0.0335
0.10	0.2614	0.8333	0.0644	0.9856	0.9580	0.0672	0.0225	0.0212	0.0523
0.12	0.3125	0.9079	0.0845	1.0812	1.0449	0.0808	0.0330	0.0312	0.0752
0.14	0.3627	0.9729	0.1062	1.1710	1.1252	0.0944	0.0457	0.0434	0.1021
0.16	0.4109	1.0240	0.1295	1.2605	1.2052	0.1074	0.0603	0.0575	0.1329
0.18	0.4574	1.0646	0.1544	1.3499	1.2852	0.1201	0.0770	0.0736	0.1675
0.20	0.5027	1.0971	0.1809	1.4393	1.3652	0.1325	0.0957	0.0918	0.2062
0.22	0.5468	1.1236	0.2090	1.5288	1.4452	0.1446	0.1166	0.1122	0.2489
0.24	0.5900	1.1454	0.2387	1.6182	1.5252	0.1565	0.1397	0.1348	0.2957
0.26	0.6324	1.1635	0.2700	1.7077	1.6052	0.1682	0.1651	0.1597	0.3468
0.28	0.6742	1.1786	0.3029	1.7971	1.6852	0.1797	0.1928	0.1870	0.4021
0.30	0.7154	1.1913	0.3374	1.8866	1.7652	0.1911	0.2228	0.2167	0.4619
0.32	0.7561	1.2021	0.3735	1.9760	1.8452	0.2024	0.2553	0.2489	0.5262
0.34	0.7964	1.2113	0.4112	2.0654	1.9252	0.2136	0.2904	0.2837	0.5951
0.36	0.8363	1.2192	0.4505	2.1549	2.0052	0.2247	0.3279	0.3211	0.6687
0.38	0.8759	1.2260	0.4914	2.2443	2.0852	0.2357	0.3681	0.3612	0.7471
0.40	0.9151	1.2319	0.5339	2.3338	2.1652	0.2466	0.4110	0.4041	0.8303
0.42	0.9542	1.2370	0.5780	2.4232	2.2452	0.2574	0.4567	0.4499	0.9185
0.44	0.9930	1.2415	0.6237	2.5127	2.3252	0.2682	0.5051	0.4985	1.0117
0.46	1.0315	1.2454	0.6710	2.6021	2.4052	0.2790	0.5564	0.5501	1.1100
0.48	1.0699	1.2488	0.7199	2.6915	2.4852	0.2897	0.6106	0.6048	1.2135
0.50	1.1082	1.2518	0.7705	2.7810	2.5652	0.3003	0.6678	0.6625	1.3223
0.52	1.1463	1.2545	0.8226	2.8704	2.6452	0.3110	0.7279	0.7234	1.4364
0.54	1.1842	1.2569	0.8763	2.9599	2.7252	0.3215	0.7912	0.7876	1.5560
0.56	1.2220	1.2590	0.9316	3.0493	2.8052	0.3321	0.8576	0.8550	1.6811
0.58	1.2597	1.2608	0.9885	3.1388	2.8852	0.3426	0.9272	0.9258	1.8118
0.60	1.2973	1.2625	1.0470	3.2282	2.9652	0.3531	1.0000	1.0000	1.9482
0.64	1.3722	1.2652	1.1688	3.4071	3.1252	0.3740	1.1556	1.1589	2.2383
0.68	1.4467	1.2674	1.2970	3.5860	3.2852	0.3948	1.3247	1.3322	2.5520
0.72	1.5210	1.2692	1.4316	3.7649	3.4452	0.4155	1.5079	1.5204	2.8899
0.76	1.5951	1.2706	1.5726	3.9437	3.6052	0.4362	1.7053	1.7239	3.2526
0.80	1.6689	1.2717	1.7200	4.1226	3.7652	0.4568	1.9176	1.9432	3.6405
0.84	1.7425	1.2726	1.8738	4.3015	3.9252	0.4774	2.1450	2.1788	4.0544
0.88	1.8160	1.2733	2.0340	4.4804	4.0852	0.4979	2.3879	2.4311	4.4946
0.92	1.8893	1.2738	2.2007	4.6593	4.2452	0.5184	2.6468	2.7005	4.9618
0.96	1.9625	1.2743	2.3737	4.8382	4.4052	0.5388	2.9218	2.9876	5.4564
1.00	2.0355	1.2746	2.5531	5.0171	4.5652	0.5592	3.2135	3.2927	5.9789

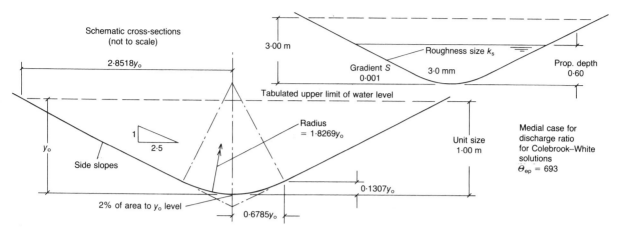

Prop dpth	U.equ. dia. $D_{ep(u)}$	Equiv. disch. factor	Unit sect. area	Unit wetted perim.	Unit surf. brdth	Unit mean depth	Discharge ratios $Q/Q_{0.60}$		U.crit. disch. Q_{uc}
Y	(m)	J	A_u (m²)	P_u (m)	B_{us} (m)	y_{um} (m)	medial	Mann'g	(m³s⁻¹)
0.02	0.0532	0.3088	0.0072	0.5412	0.5392	0.0133	0.0007	0.0006	0.0026
0.04	0.1061	0.4352	0.0203	0.7660	0.7604	0.0267	0.0031	0.0029	0.0104
0.06	0.1588	0.5312	0.0373	0.9390	0.9287	0.0401	0.0073	0.0069	0.0234
0.08	0.2111	0.6112	0.0573	1.0853	1.0694	0.0536	0.0136	0.0128	0.0415
0.10	0.2633	0.6809	0.0799	1.2145	1.1923	0.0670	0.0219	0.0207	0.0648
0.12	0.3151	0.7433	0.1049	1.3317	1.3024	0.0805	0.0323	0.0307	0.0932
0.14	0.3664	0.7991	0.1320	1.4406	1.4037	0.0940	0.0448	0.0427	0.1267
0.16	0.4160	0.8442	0.1610	1.5483	1.5037	0.1071	0.0593	0.0567	0.1650
0.18	0.4640	0.8803	0.1921	1.6560	1.6037	0.1198	0.0758	0.0727	0.2082
0.20	0.5107	0.9097	0.2252	1.7637	1.7037	0.1322	0.0945	0.0909	0.2564
0.22	0.5563	0.9339	0.2603	1.8714	1.8037	0.1443	0.1153	0.1112	0.3096
0.24	0.6009	0.9539	0.2973	1.9791	1.9037	0.1562	0.1383	0.1337	0.3680
0.26	0.6448	0.9707	0.3364	2.0868	2.0037	0.1679	0.1636	0.1586	0.4317
0.28	0.6880	0.9850	0.3775	2.1945	2.1037	0.1794	0.1912	0.1858	0.5007
0.30	0.7307	0.9971	0.4206	2.3022	2.2037	0.1908	0.2212	0.2155	0.5753
0.32	0.7729	1.0075	0.4656	2.4099	2.3037	0.2021	0.2537	0.2476	0.6556
0.34	0.8146	1.0165	0.5127	2.5176	2.4037	0.2133	0.2887	0.2824	0.7415
0.36	0.8559	1.0242	0.5618	2.6253	2.5037	0.2244	0.3263	0.3198	0.8333
0.38	0.8970	1.0310	0.6128	2.7330	2.6037	0.2354	0.3665	0.3600	0.9311
0.40	0.9377	1.0370	0.6659	2.8407	2.7037	0.2463	0.4094	0.4029	1.0349
0.42	0.9781	1.0422	0.7210	2.9484	2.8037	0.2572	0.4551	0.4487	1.1450
0.44	1.0184	1.0468	0.7781	3.0561	2.9037	0.2680	0.5036	0.4974	1.2613
0.46	1.0584	1.0509	0.8371	3.1638	3.0037	0.2787	0.5550	0.5491	1.3840
0.48	1.0982	1.0546	0.8982	3.2715	3.1037	0.2894	0.6093	0.6038	1.5132
0.50	1.1379	1.0578	0.9613	3.3792	3.2037	0.3001	0.6666	0.6617	1.6490
0.52	1.1774	1.0607	1.0264	3.4869	3.3037	0.3107	0.7270	0.7227	1.7915
0.54	1.2167	1.0634	1.0934	3.5946	3.4037	0.3213	0.7905	0.7870	1.9408
0.56	1.2560	1.0657	1.1625	3.7023	3.5037	0.3318	0.8571	0.8546	2.0970
0.58	1.2951	1.0678	1.2336	3.8100	3.6037	0.3423	0.9269	0.9256	2.2602
0.60	1.3341	1.0698	1.3067	3.9177	3.7037	0.3528	1.0000	1.0000	2.4304
0.64	1.4118	1.0731	1.4588	4.1332	3.9037	0.3737	1.1562	1.1594	2.7927
0.68	1.4892	1.0758	1.6189	4.3486	4.1037	0.3945	1.3261	1.3333	3.1844
0.72	1.5663	1.0781	1.7871	4.5640	4.3037	0.4152	1.5101	1.5221	3.6063
0.76	1.6431	1.0800	1.9632	4.7794	4.5037	0.4359	1.7086	1.7264	4.0592
0.80	1.7197	1.0816	2.1474	4.9948	4.7037	0.4565	1.9220	1.9465	4.5437
0.84	1.7961	1.0830	2.3395	5.2102	4.9037	0.4771	2.1507	2.1831	5.0605
0.88	1.8724	1.0841	2.5397	5.4256	5.1037	0.4976	2.3951	2.4364	5.6103
0.92	1.9485	1.0851	2.7478	5.6410	5.3037	0.5181	2.6555	2.7071	6.1938
0.96	2.0244	1.0860	2.9640	5.8564	5.5037	0.5385	2.9324	2.9954	6.8116
1.00	2.1003	1.0867	3.1881	6.0718	5.7037	0.5590	3.2260	3.3019	7.4642

C83

3.0 to 1 tangent river

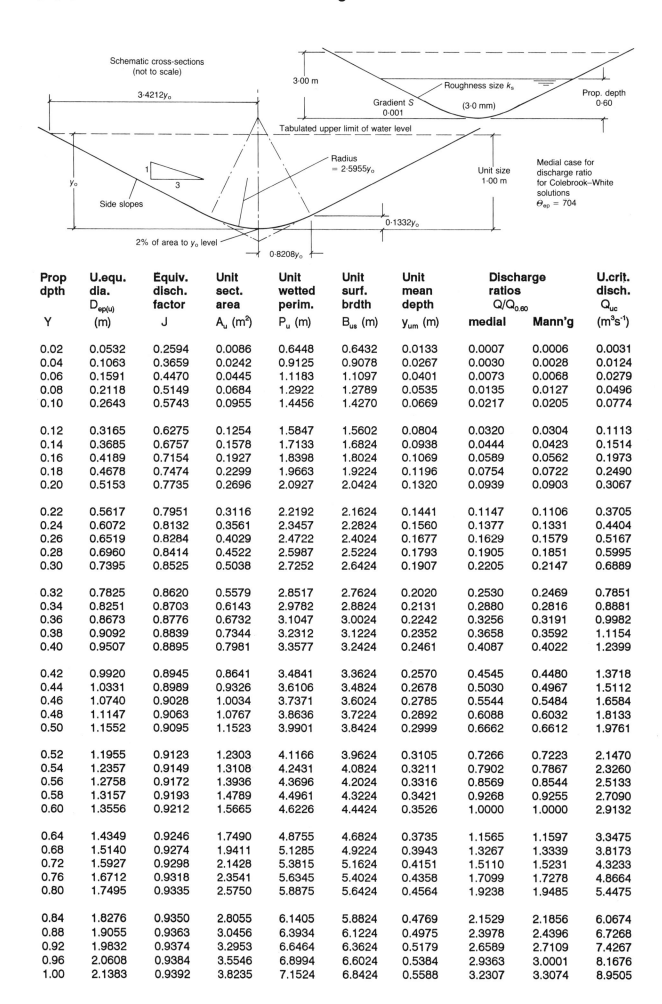

Schematic cross-sections (not to scale)

3·4212y₀

3·00 m

Roughness size k_s (3·0 mm)

Gradient S 0·001

Prop. depth 0·60

Tabulated upper limit of water level

Radius = 2·5955y₀

y_0

1/3

Side slopes

Unit size 1·00 m

Medial case for discharge ratio for Colebrook–White solutions $\Theta_{ep} = 704$

2% of area to y₀ level

0·1332y₀

0·8208y₀

Prop dpth	U.equ. dia. $D_{ep(u)}$	Equiv. disch. factor	Unit sect. area	Unit wetted perim.	Unit surf. brdth	Unit mean depth	Discharge ratios $Q/Q_{0.60}$		U.crit. disch. Q_{uc}
Y	(m)	J	A_u (m²)	P_u (m)	B_{us} (m)	y_{um} (m)	medial	Mann'g	(m³s⁻¹)
0.02	0.0532	0.2594	0.0086	0.6448	0.6432	0.0133	0.0007	0.0006	0.0031
0.04	0.1063	0.3659	0.0242	0.9125	0.9078	0.0267	0.0030	0.0028	0.0124
0.06	0.1591	0.4470	0.0445	1.1183	1.1097	0.0401	0.0073	0.0068	0.0279
0.08	0.2118	0.5149	0.0684	1.2922	1.2789	0.0535	0.0135	0.0127	0.0496
0.10	0.2643	0.5743	0.0955	1.4456	1.4270	0.0669	0.0217	0.0205	0.0774
0.12	0.3165	0.6275	0.1254	1.5847	1.5602	0.0804	0.0320	0.0304	0.1113
0.14	0.3685	0.6757	0.1578	1.7133	1.6824	0.0938	0.0444	0.0423	0.1514
0.16	0.4189	0.7154	0.1927	1.8398	1.8024	0.1069	0.0589	0.0562	0.1973
0.18	0.4678	0.7474	0.2299	1.9663	1.9224	0.1196	0.0754	0.0722	0.2490
0.20	0.5153	0.7735	0.2696	2.0927	2.0424	0.1320	0.0939	0.0903	0.3067
0.22	0.5617	0.7951	0.3116	2.2192	2.1624	0.1441	0.1147	0.1106	0.3705
0.24	0.6072	0.8132	0.3561	2.3457	2.2824	0.1560	0.1377	0.1331	0.4404
0.26	0.6519	0.8284	0.4029	2.4722	2.4024	0.1677	0.1629	0.1579	0.5167
0.28	0.6960	0.8414	0.4522	2.5987	2.5224	0.1793	0.1905	0.1851	0.5995
0.30	0.7395	0.8525	0.5038	2.7252	2.6424	0.1907	0.2205	0.2147	0.6889
0.32	0.7825	0.8620	0.5579	2.8517	2.7624	0.2020	0.2530	0.2469	0.7851
0.34	0.8251	0.8703	0.6143	2.9782	2.8824	0.2131	0.2880	0.2816	0.8881
0.36	0.8673	0.8776	0.6732	3.1047	3.0024	0.2242	0.3256	0.3191	0.9982
0.38	0.9092	0.8839	0.7344	3.2312	3.1224	0.2352	0.3658	0.3592	1.1154
0.40	0.9507	0.8895	0.7981	3.3577	3.2424	0.2461	0.4087	0.4022	1.2399
0.42	0.9920	0.8945	0.8641	3.4841	3.3624	0.2570	0.4545	0.4480	1.3718
0.44	1.0331	0.8989	0.9326	3.6106	3.4824	0.2678	0.5030	0.4967	1.5112
0.46	1.0740	0.9028	1.0034	3.7371	3.6024	0.2785	0.5544	0.5484	1.6584
0.48	1.1147	0.9063	1.0767	3.8636	3.7224	0.2892	0.6088	0.6032	1.8133
0.50	1.1552	0.9095	1.1523	3.9901	3.8424	0.2999	0.6662	0.6612	1.9761
0.52	1.1955	0.9123	1.2303	4.1166	3.9624	0.3105	0.7266	0.7223	2.1470
0.54	1.2357	0.9149	1.3108	4.2431	4.0824	0.3211	0.7902	0.7867	2.3260
0.56	1.2758	0.9172	1.3936	4.3696	4.2024	0.3316	0.8569	0.8544	2.5133
0.58	1.3157	0.9193	1.4789	4.4961	4.3224	0.3421	0.9268	0.9255	2.7090
0.60	1.3556	0.9212	1.5665	4.6226	4.4424	0.3526	1.0000	1.0000	2.9132
0.64	1.4349	0.9246	1.7490	4.8755	4.6824	0.3735	1.1565	1.1597	3.3475
0.68	1.5140	0.9274	1.9411	5.1285	4.9224	0.3943	1.3267	1.3339	3.8173
0.72	1.5927	0.9298	2.1428	5.3815	5.1624	0.4151	1.5110	1.5231	4.3233
0.76	1.6712	0.9318	2.3541	5.6345	5.4024	0.4358	1.7099	1.7278	4.8664
0.80	1.7495	0.9335	2.5750	5.8875	5.6424	0.4564	1.9238	1.9485	5.4475
0.84	1.8276	0.9350	2.8055	6.1405	5.8824	0.4769	2.1529	2.1856	6.0674
0.88	1.9055	0.9363	3.0456	6.3934	6.1224	0.4975	2.3978	2.4396	6.7268
0.92	1.9832	0.9374	3.2953	6.6464	6.3624	0.5179	2.6589	2.7109	7.4267
0.96	2.0608	0.9384	3.5546	6.8994	6.6024	0.5384	2.9363	3.0001	8.1676
1.00	2.1383	0.9392	3.8235	7.1524	6.8424	0.5588	3.2307	3.3074	8.9505

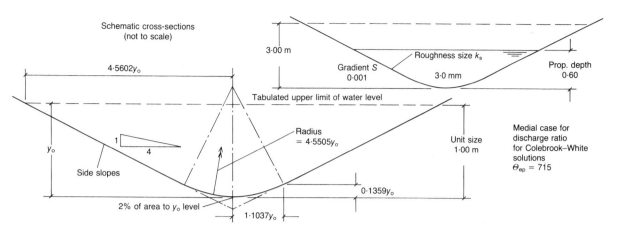

Prop dpth	U.equ. dia. $D_{ep(u)}$	Equiv. disch. factor	Unit sect. area	Unit wetted perim.	Unit surf. brdth	Unit mean depth	Discharge ratios $Q/Q_{0.60}$		U.crit. disch. Q_{uc}
Y	(m)	J	A_u (m²)	P_u (m)	B_{us} (m)	y_{um} (m)	medial	Mann'g	(m³s⁻¹)
0.02	0.0533	0.1961	0.0114	0.8536	0.8523	0.0133	0.0007	0.0006	0.0041
0.04	0.1064	0.2769	0.0321	1.2076	1.2041	0.0267	0.0030	0.0028	0.0164
0.06	0.1595	0.3387	0.0590	1.4795	1.4730	0.0401	0.0072	0.0067	0.0370
0.08	0.2125	0.3905	0.0908	1.7091	1.6990	0.0534	0.0133	0.0125	0.0657
0.10	0.2653	0.4360	0.1268	1.9115	1.8975	0.0668	0.0215	0.0203	0.1026
0.12	0.3180	0.4770	0.1665	2.0947	2.0763	0.0802	0.0317	0.0300	0.1477
0.14	0.3706	0.5143	0.2097	2.2636	2.2404	0.0936	0.0440	0.0419	0.2010
0.16	0.4219	0.5457	0.2561	2.4286	2.4004	0.1067	0.0584	0.0558	0.2620
0.18	0.4716	0.5712	0.3057	2.5935	2.5604	0.1194	0.0748	0.0717	0.3309
0.20	0.5199	0.5922	0.3586	2.7584	2.7204	0.1318	0.0934	0.0897	0.4076
0.22	0.5672	0.6096	0.4146	2.9233	2.8804	0.1439	0.1141	0.1099	0.4925
0.24	0.6136	0.6242	0.4738	3.0883	3.0404	0.1558	0.1370	0.1324	0.5857
0.26	0.6593	0.6366	0.5362	3.2532	3.2004	0.1675	0.1622	0.1572	0.6873
0.28	0.7042	0.6472	0.6018	3.4181	3.3604	0.1791	0.1898	0.1843	0.7975
0.30	0.7486	0.6564	0.6706	3.5830	3.5204	0.1905	0.2198	0.2140	0.9165
0.32	0.7925	0.6643	0.7426	3.7480	3.6804	0.2018	0.2522	0.2461	1.0446
0.34	0.8360	0.6712	0.8178	3.9129	3.8404	0.2129	0.2872	0.2809	1.1818
0.36	0.8791	0.6773	0.8962	4.0778	4.0004	0.2240	0.3248	0.3183	1.3284
0.38	0.9219	0.6826	0.9778	4.2427	4.1604	0.2350	0.3651	0.3584	1.4845
0.40	0.9644	0.6873	1.0626	4.4076	4.3204	0.2460	0.4080	0.4014	1.6503
0.42	1.0066	0.6915	1.1506	4.5726	4.4804	0.2568	0.4538	0.4472	1.8260
0.44	1.0485	0.6953	1.2418	4.7375	4.6404	0.2676	0.5024	0.4960	2.0118
0.46	1.0903	0.6987	1.3363	4.9024	4.8004	0.2784	0.5538	0.5478	2.2078
0.48	1.1318	0.7017	1.4339	5.0673	4.9604	0.2891	0.6083	0.6026	2.4141
0.50	1.1732	0.7044	1.5347	5.2323	5.1204	0.2997	0.6657	0.6606	2.6311
0.52	1.2145	0.7069	1.6387	5.3972	5.2804	0.3103	0.7262	0.7218	2.8587
0.54	1.2556	0.7091	1.7459	5.5621	5.4404	0.3209	0.7899	0.7863	3.0972
0.56	1.2965	0.7112	1.8563	5.7270	5.6004	0.3315	0.8567	0.8541	3.3467
0.58	1.3373	0.7131	1.9699	5.8920	5.7604	0.3420	0.9267	0.9253	3.6075
0.60	1.3781	0.7148	2.0867	6.0569	5.9204	0.3525	1.0000	1.0000	3.8795
0.64	1.4592	0.7178	2.3299	6.3867	6.2404	0.3734	1.1567	1.1600	4.4583
0.68	1.5400	0.7203	2.5859	6.7166	6.5604	0.3942	1.3273	1.3345	5.0842
0.72	1.6205	0.7225	2.8548	7.0464	6.8804	0.4149	1.5120	1.5242	5.7585
0.76	1.7008	0.7244	3.1364	7.3763	7.2004	0.4356	1.7113	1.7294	6.4822
0.80	1.7808	0.7260	3.4308	7.7061	7.5204	0.4562	1.9256	1.9506	7.2566
0.84	1.8606	0.7274	3.7380	8.0360	7.8404	0.4768	2.1553	2.1883	8.0826
0.88	1.9403	0.7286	4.0580	8.3658	8.1604	0.4973	2.4007	2.4429	8.9614
0.92	2.0198	0.7297	4.3908	8.6957	8.4804	0.5178	2.6624	2.7150	9.8940
0.96	2.0991	0.7306	4.7364	9.0255	8.8004	0.5382	2.9405	3.0049	10.881
1.00	2.1784	0.7315	5.0949	9.3554	9.1204	0.5586	3.2356	3.3132	11.925

5 to 1 tangent river

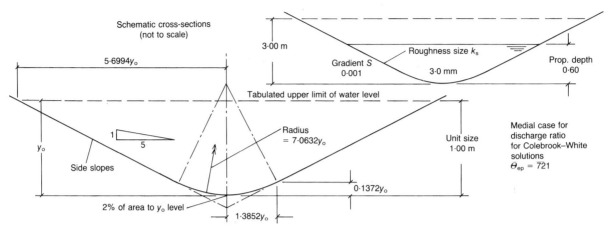

Prop dpth	U.equ. dia. $D_{ep(u)}$	Equiv. disch. factor	Unit sect. area	Unit wetted perim.	Unit surf. brdth	Unit mean depth	Discharge ratios $Q/Q_{0.60}$		U.crit. disch. Q_{uc}
Y	(m)	J	A_u (m²)	P_u (m)	B_{us} (m)	y_{um} (m)	medial	Mann'g	(m³s⁻¹)
0.02	0.0533	0.1575	0.0142	1.0633	1.0623	0.0133	0.0007	0.0006	0.0051
0.04	0.1065	0.2225	0.0401	1.5041	1.5013	0.0267	0.0030	0.0028	0.0205
0.06	0.1597	0.2723	0.0736	1.8426	1.8374	0.0400	0.0071	0.0067	0.0461
0.08	0.2128	0.3141	0.1132	2.1281	2.1201	0.0534	0.0132	0.0124	0.0819
0.10	0.2658	0.3508	0.1581	2.3799	2.3687	0.0668	0.0213	0.0201	0.1280
0.12	0.3187	0.3840	0.2078	2.6077	2.5929	0.0801	0.0315	0.0299	0.1842
0.14	0.3716	0.4143	0.2617	2.8174	2.7988	0.0935	0.0438	0.0417	0.2506
0.16	0.4233	0.4401	0.3197	3.0214	2.9988	0.1066	0.0582	0.0555	0.3269
0.18	0.4734	0.4610	0.3817	3.2253	3.1988	0.1193	0.0746	0.0714	0.4129
0.20	0.5222	0.4783	0.4477	3.4293	3.3988	0.1317	0.0931	0.0894	0.5088
0.22	0.5699	0.4927	0.5176	3.6333	3.5988	0.1438	0.1138	0.1096	0.6148
0.24	0.6167	0.5049	0.5916	3.8372	3.7988	0.1557	0.1367	0.1321	0.7311
0.26	0.6628	0.5152	0.6696	4.0412	3.9988	0.1674	0.1619	0.1568	0.8580
0.28	0.7082	0.5241	0.7516	4.2452	4.1988	0.1790	0.1895	0.1840	0.9957
0.30	0.7530	0.5317	0.8375	4.4491	4.3988	0.1904	0.2194	0.2136	1.1445
0.32	0.7973	0.5383	0.9275	4.6531	4.5988	0.2017	0.2519	0.2457	1.3044
0.34	0.8412	0.5441	1.0215	4.8570	4.7988	0.2129	0.2869	0.2805	1.4759
0.36	0.8848	0.5492	1.1195	5.0610	4.9988	0.2239	0.3245	0.3179	1.6590
0.38	0.9280	0.5537	1.2214	5.2650	5.1988	0.2349	0.3647	0.3580	1.8540
0.40	0.9709	0.5577	1.3274	5.4689	5.3988	0.2459	0.4077	0.4010	2.0612
0.42	1.0135	0.5613	1.4374	5.6729	5.5988	0.2567	0.4535	0.4469	2.2807
0.44	1.0559	0.5645	1.5514	5.8768	5.7988	0.2675	0.5021	0.4957	2.5128
0.46	1.0981	0.5673	1.6693	6.0808	5.9988	0.2783	0.5536	0.5475	2.7577
0.48	1.1401	0.5699	1.7913	6.2848	6.1988	0.2890	0.6080	0.6023	3.0155
0.50	1.1819	0.5722	1.9173	6.4887	6.3988	0.2996	0.6655	0.6604	3.2866
0.52	1.2236	0.5743	2.0473	6.6927	6.5988	0.3102	0.7260	0.7216	3.5710
0.54	1.2651	0.5763	2.1812	6.8966	6.7988	0.3208	0.7897	0.7861	3.8690
0.56	1.3065	0.5780	2.3192	7.1006	6.9988	0.3314	0.8566	0.8540	4.1808
0.58	1.3478	0.5796	2.4612	7.3046	7.1988	0.3419	0.9266	0.9253	4.5066
0.60	1.3889	0.5811	2.6072	7.5085	7.3988	0.3524	1.0000	1.0000	4.8466
0.64	1.4709	0.5837	2.9111	7.9164	7.7988	0.3733	1.1569	1.1601	5.5698
0.68	1.5526	0.5859	3.2311	8.3244	8.1988	0.3941	1.3275	1.3348	6.3519
0.72	1.6339	0.5878	3.5670	8.7323	8.5988	0.4148	1.5124	1.5247	7.1945
0.76	1.7150	0.5895	3.9190	9.1402	8.9988	0.4355	1.7119	1.7301	8.0989
0.80	1.7959	0.5909	4.2869	9.5481	9.3988	0.4561	1.9264	1.9516	9.0666
0.84	1.8766	0.5921	4.6709	9.9561	9.7988	0.4767	2.1564	2.1896	10.099
0.88	1.9571	0.5932	5.0708	10.364	10.199	0.4972	2.4021	2.4446	11.197
0.92	2.0374	0.5942	5.4868	10.772	10.599	0.5177	2.6641	2.7170	12.363
0.96	2.1177	0.5951	5.9187	11.180	10.999	0.5381	2.9426	3.0073	13.597
1.00	2.1977	0.5958	6.3667	11.588	11.399	0.5585	3.2380	3.3160	14.901

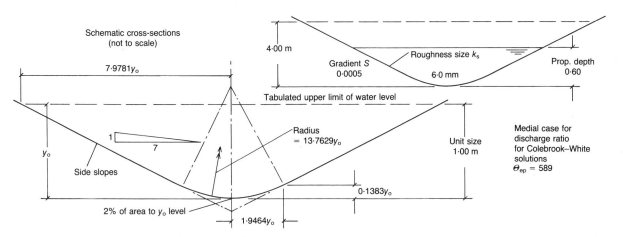

Schematic cross-sections (not to scale)

7·9781y₀

4·00 m

Gradient S 0·0005

Roughness size k_s

6·0 mm

Prop. depth 0·60

Tabulated upper limit of water level

Radius = 13·7629y₀

Unit size 1·00 m

Medial case for discharge ratio for Colebrook–White solutions Θ_{ep} = 589

1 / 7

y_o

Side slopes

2% of area to y₀ level

0·1383y₀

1·9464y₀

Prop dpth	U.equ. dia. $D_{ep(u)}$	Equiv. disch. factor	Unit sect. area	Unit wetted perim.	Unit surf. brdth	Unit mean depth	Discharge ratios $Q/Q_{0.60}$		U.crit. disch. Q_{uc}
Y	(m)	J	A_u (m²)	P_u (m)	B_{us} (m)	y_{um} (m)	medial	Mann'g	(m³s⁻¹)
0.02	0.0533	0.1129	0.0198	1.4841	1.4834	0.0133	0.0006	0.0006	0.0072
0.04	0.1066	0.1595	0.0559	2.0991	2.0971	0.0267	0.0029	0.0028	0.0286
0.06	0.1598	0.1953	0.1027	2.5712	2.5674	0.0400	0.0070	0.0066	0.0644
0.08	0.2130	0.2254	0.1581	2.9693	2.9636	0.0534	0.0130	0.0124	0.1144
0.10	0.2662	0.2519	0.2210	3.3202	3.3121	0.0667	0.0210	0.0200	0.1787
0.12	0.3193	0.2758	0.2904	3.6375	3.6270	0.0801	0.0311	0.0297	0.2573
0.14	0.3724	0.2978	0.3659	3.9295	3.9162	0.0934	0.0433	0.0415	0.3502
0.16	0.4245	0.3166	0.4470	4.2124	4.1962	0.1065	0.0576	0.0553	0.4569
0.18	0.4749	0.3319	0.5337	4.4952	4.4762	0.1192	0.0739	0.0712	0.5771
0.20	0.5241	0.3446	0.6261	4.7781	4.7562	0.1316	0.0923	0.0892	0.7113
0.22	0.5722	0.3552	0.7240	5.0609	5.0362	0.1438	0.1129	0.1094	0.8596
0.24	0.6194	0.3641	0.8275	5.3437	5.3162	0.1557	0.1358	0.1318	1.0224
0.26	0.6659	0.3718	0.9366	5.6266	5.5962	0.1674	0.1609	0.1565	1.1999
0.28	0.7116	0.3783	1.0513	5.9094	5.8762	0.1789	0.1884	0.1837	1.3926
0.30	0.7569	0.3840	1.1717	6.1923	6.1562	0.1903	0.2183	0.2133	1.6007
0.32	0.8016	0.3889	1.2976	6.4751	6.4362	0.2016	0.2508	0.2454	1.8245
0.34	0.8459	0.3932	1.4291	6.7580	6.7162	0.2128	0.2857	0.2801	2.0644
0.36	0.8898	0.3970	1.5662	7.0408	6.9962	0.2239	0.3233	0.3176	2.3207
0.38	0.9334	0.4004	1.7090	7.3236	7.2762	0.2349	0.3636	0.3577	2.5936
0.40	0.9767	0.4034	1.8573	7.6065	7.5562	0.2458	0.4066	0.4007	2.8836
0.42	1.0197	0.4060	2.0112	7.8893	7.8362	0.2567	0.4524	0.4465	3.1908
0.44	1.0625	0.4084	2.1707	8.1722	8.1162	0.2675	0.5010	0.4954	3.5156
0.46	1.1051	0.4106	2.3359	8.4550	8.3962	0.2782	0.5526	0.5472	3.8582
0.48	1.1475	0.4125	2.5066	8.7379	8.6762	0.2889	0.6071	0.6021	4.2191
0.50	1.1897	0.4143	2.6829	9.0207	8.9562	0.2996	0.6647	0.6602	4.5984
0.52	1.2317	0.4159	2.8648	9.3035	9.2362	0.3102	0.7253	0.7214	4.9965
0.54	1.2736	0.4174	3.0524	9.5864	9.5162	0.3208	0.7892	0.7860	5.4135
0.56	1.3154	0.4187	3.2455	9.8692	9.7962	0.3313	0.8562	0.8539	5.8499
0.58	1.3570	0.4199	3.4442	10.152	10.076	0.3418	0.9264	0.9252	6.3059
0.60	1.3986	0.4211	3.6485	10.435	10.356	0.3523	1.0000	1.0000	6.7817
0.64	1.4814	0.4230	4.0740	11.001	10.916	0.3732	1.1574	1.1602	7.7939
0.68	1.5638	0.4247	4.5218	11.566	11.476	0.3940	1.3286	1.3351	8.8886
0.72	1.6459	0.4262	4.9921	12.132	12.036	0.4148	1.5141	1.5251	10.068
0.76	1.7278	0.4275	5.4847	12.698	12.596	0.4354	1.7144	1.7308	11.334
0.80	1.8094	0.4286	5.9998	13.263	13.156	0.4560	1.9299	1.9525	12.688
0.84	1.8909	0.4295	6.5372	13.829	13.716	0.4766	2.1609	2.1907	14.133
0.88	1.9721	0.4304	7.0971	14.395	14.276	0.4971	2.4078	2.4460	15.670
0.92	2.0532	0.4312	7.6793	14.960	14.836	0.5176	2.6710	2.7188	17.301
0.96	2.1342	0.4318	8.2840	15.526	15.396	0.5381	2.9510	3.0094	19.029
1.00	2.2150	0.4324	8.9110	16.092	15.956	0.5585	3.2480	3.3185	20.854

10 to 1 tangent river

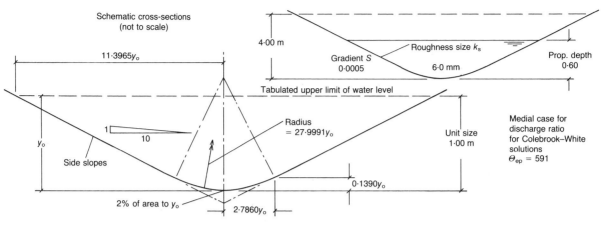

Prop dpth	U.equ. dia. $D_{ep(u)}$	Equiv. disch. factor	Unit sect. area	Unit wetted perim.	Unit surf. brdth	Unit mean depth	Discharge ratios $Q/Q_{0.60}$		U.crit. disch. Q_{uc}
Y	(m)	J	A_u (m²)	P_u (m)	B_{us} (m)	y_{um} (m)	medial	Mann'g	(m³s⁻¹)
0.02	0.0533	0.0791	0.0282	2.1167	2.1162	0.0133	0.0006	0.0006	0.0102
0.04	0.1066	0.1119	0.0798	2.9936	2.9922	0.0267	0.0029	0.0027	0.0408
0.06	0.1599	0.1370	0.1466	3.6667	3.6640	0.0400	0.0069	0.0066	0.0918
0.08	0.2132	0.1582	0.2257	4.2341	4.2301	0.0533	0.0129	0.0123	0.1632
0.10	0.2664	0.1768	0.3154	4.7342	4.7286	0.0667	0.0210	0.0200	0.2550
0.12	0.3197	0.1936	0.4145	5.1864	5.1789	0.0800	0.0310	0.0297	0.3672
0.14	0.3729	0.2091	0.5223	5.6023	5.5929	0.0934	0.0432	0.0414	0.4998
0.16	0.4251	0.2224	0.6381	6.0043	5.9929	0.1065	0.0575	0.0552	0.6521
0.18	0.4758	0.2333	0.7620	6.4063	6.3929	0.1192	0.0738	0.0711	0.8238
0.20	0.5252	0.2423	0.8938	6.8083	6.7929	0.1316	0.0922	0.0891	1.0154
0.22	0.5735	0.2499	1.0337	7.2103	7.1929	0.1437	0.1128	0.1092	1.2272
0.24	0.6209	0.2562	1.1816	7.6123	7.5929	0.1556	0.1356	0.1316	1.4596
0.26	0.6675	0.2617	1.3374	8.0143	7.9929	0.1673	0.1608	0.1564	1.7132
0.28	0.7135	0.2663	1.5013	8.4162	8.3929	0.1789	0.1883	0.1835	1.9884
0.30	0.7589	0.2704	1.6731	8.8182	8.7929	0.1903	0.2182	0.2131	2.2855
0.32	0.8039	0.2739	1.8530	9.2202	9.1929	0.2016	0.2506	0.2452	2.6052
0.34	0.8484	0.2770	2.0409	9.6222	9.5929	0.2127	0.2856	0.2799	2.9478
0.36	0.8925	0.2797	2.2367	10.024	9.9929	0.2238	0.3231	0.3174	3.3138
0.38	0.9363	0.2821	2.4406	10.426	10.393	0.2348	0.3634	0.3575	3.7036
0.40	0.9798	0.2843	2.6524	10.828	10.793	0.2458	0.4064	0.4005	4.1177
0.42	1.0231	0.2862	2.8723	11.230	11.193	0.2566	0.4522	0.4464	4.5565
0.44	1.0661	0.2879	3.1001	11.632	11.593	0.2674	0.5009	0.4952	5.0204
0.46	1.1088	0.2895	3.3360	12.034	11.993	0.2782	0.5524	0.5470	5.5098
0.48	1.1514	0.2909	3.5799	12.436	12.393	0.2889	0.6070	0.6020	6.0252
0.50	1.1939	0.2921	3.8317	12.838	12.793	0.2995	0.6646	0.6600	6.5670
0.52	1.2361	0.2933	4.0916	13.240	13.193	0.3101	0.7252	0.7213	7.1355
0.54	1.2782	0.2943	4.3594	13.642	13.593	0.3207	0.7891	0.7859	7.7313
0.56	1.3202	0.2953	4.6353	14.044	13.993	0.3313	0.8561	0.8538	8.3545
0.58	1.3621	0.2962	4.9192	14.446	14.393	0.3418	0.9264	0.9252	9.0058
0.60	1.4038	0.2970	5.2110	14.848	14.793	0.3523	1.0000	1.0000	9.6854
0.64	1.4870	0.2985	5.8187	15.652	15.593	0.3732	1.1574	1.1603	11.131
0.68	1.5699	0.2997	6.4585	16.456	16.393	0.3940	1.3287	1.3353	12.695
0.72	1.6524	0.3008	7.1302	17.260	17.193	0.4147	1.5144	1.5254	14.379
0.76	1.7347	0.3017	7.8339	18.064	17.993	0.4354	1.7148	1.7311	16.187
0.80	1.8167	0.3025	8.5696	18.868	18.793	0.4560	1.9303	1.9530	18.122
0.84	1.8986	0.3032	9.3373	19.672	19.593	0.4766	2.1614	2.1914	20.186
0.88	1.9803	0.3038	10.137	20.476	20.393	0.4971	2.4085	2.4468	22.381
0.92	2.0618	0.3044	10.969	21.280	21.193	0.5176	2.6718	2.7197	24.712
0.96	2.1432	0.3049	11.832	22.084	21.993	0.5380	2.9519	3.0106	27.179
1.00	2.2244	0.3053	12.728	22.888	22.793	0.5584	3.2491	3.3198	29.786

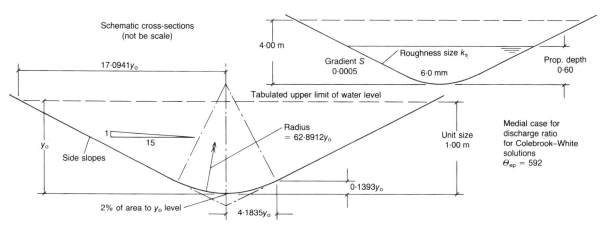

Prop dpth	U.equ. dia. $D_{ep(u)}$	Equiv. disch. factor	Unit sect. area	Unit wetted perim.	Unit surf. brdth	Unit mean depth	Discharge ratios $Q/Q_{0.60}$		U.crit. disch. Q_{uc}
Y	(m)	J	A_u (m²)	P_u (m)	B_{us} (m)	y_{um} (m)	medial	Mann'g	(m³s⁻¹)
0.02	0.0533	0.0528	0.0423	3.1722	3.1719	0.0133	0.0006	0.0006	0.0153
0.04	0.1067	0.0747	0.1196	4.4863	4.4854	0.0267	0.0029	0.0027	0.0612
0.06	0.1600	0.0915	0.2197	5.4948	5.4930	0.0400	0.0069	0.0066	0.1376
0.08	0.2133	0.1056	0.3383	6.3450	6.3423	0.0533	0.0129	0.0123	0.2447
0.10	0.2666	0.1180	0.4728	7.0941	7.0903	0.0667	0.0209	0.0200	0.3823
0.12	0.3199	0.1293	0.6214	7.7714	7.7665	0.0800	0.0310	0.0296	0.5505
0.14	0.3731	0.1396	0.7831	8.3943	8.3881	0.0934	0.0432	0.0414	0.7492
0.16	0.4255	0.1486	0.9568	8.9957	8.9881	0.1065	0.0574	0.0552	0.9776
0.18	0.4762	0.1559	1.1426	9.5970	9.5881	0.1192	0.0737	0.0710	1.2352
0.20	0.5257	0.1619	1.3403	10.198	10.188	0.1316	0.0921	0.0890	1.5224
0.22	0.5741	0.1670	1.5501	10.800	10.788	0.1437	0.1127	0.1091	1.8401
0.24	0.6217	0.1713	1.7719	11.401	11.388	0.1556	0.1355	0.1315	2.1887
0.26	0.6684	0.1750	2.0056	12.002	11.988	0.1673	0.1607	0.1563	2.5690
0.28	0.7145	0.1781	2.2514	12.604	12.588	0.1789	0.1882	0.1834	2.9817
0.30	0.7601	0.1808	2.5092	13.205	13.188	0.1903	0.2181	0.2130	3.4274
0.32	0.8051	0.1832	2.7789	13.806	13.788	0.2015	0.2505	0.2451	3.9068
0.34	0.8497	0.1853	3.0607	14.408	14.388	0.2127	0.2855	0.2799	4.4206
0.36	0.8940	0.1871	3.3544	15.009	14.988	0.2238	0.3230	0.3173	4.9696
0.38	0.9379	0.1887	3.6602	15.610	15.588	0.2348	0.3633	0.3574	5.5542
0.40	0.9815	0.1902	3.9780	16.212	16.188	0.2457	0.4063	0.4004	6.1752
0.42	1.0249	0.1915	4.3077	16.813	16.788	0.2566	0.4521	0.4463	6.8333
0.44	1.0680	0.1927	4.6495	17.414	17.388	0.2674	0.5008	0.4951	7.5291
0.46	1.1109	0.1937	5.0033	18.016	17.988	0.2781	0.5524	0.5469	8.2632
0.48	1.1536	0.1947	5.3690	18.617	18.588	0.2888	0.6069	0.6019	9.0362
0.50	1.1961	0.1955	5.7468	19.218	19.188	0.2995	0.6645	0.6600	9.8487
0.52	1.2385	0.1963	6.1365	19.820	19.788	0.3101	0.7252	0.7213	10.701
0.54	1.2807	0.1970	6.5383	20.421	20.388	0.3207	0.7890	0.7859	11.595
0.56	1.3228	0.1977	6.9521	21.022	20.988	0.3312	0.8561	0.8538	12.530
0.58	1.3648	0.1983	7.3778	21.624	21.588	0.3418	0.9264	0.9252	13.507
0.60	1.4066	0.1988	7.8156	22.225	22.188	0.3522	1.0000	1.0000	14.526
0.64	1.4901	0.1998	8.7271	23.428	23.388	0.3731	1.1575	1.1604	16.694
0.68	1.5731	0.2006	9.6866	24.630	24.588	0.3940	1.3288	1.3354	19.040
0.72	1.6559	0.2014	10.694	25.833	25.788	0.4147	1.5145	1.5255	21.566
0.76	1.7384	0.2020	11.750	27.036	26.988	0.4354	1.7149	1.7313	24.278
0.80	1.8207	0.2026	12.853	28.238	28.188	0.4560	1.9305	1.9532	27.180
0.84	1.9028	0.2030	14.005	29.441	29.388	0.4765	2.1617	2.1917	30.275
0.88	1.9847	0.2035	15.204	30.644	30.588	0.4971	2.4088	2.4472	33.568
0.92	2.0664	0.2038	16.452	31.846	31.788	0.5175	2.6723	2.7202	37.064
0.96	2.1480	0.2042	17.747	33.049	32.988	0.5380	2.9525	3.0112	40.764
1.00	2.2295	0.2045	19.091	34.252	34.188	0.5584	3.2498	3.3206	44.675

C89

Triangular open channel (modified table)

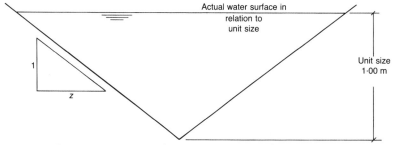

Actual water surface in relation to unit size

Unit size 1·00 m

For this shape, the values in column 1 are of *z* (as above). Thus all data relates to triangular channels of specified side slopes which are full to unit depth

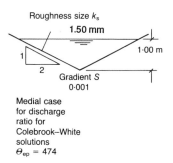

Roughness size k_s
1.50 mm
1·00 m

Gradient *S*
0·001

Medial case for discharge ratio for Colebrook–White solutions
$\Theta_{ep} = 474$

Side slope	U.equ. dia. $D_{ep(u)}$	Equiv. disch. factor	Unit sect. area	Unit wetted perim.	Unit surf. brdth	Unit mean depth	Discharge ratios $Q/Q_{2.00}$		U.crit. disch. Q_{uc}
z	(m)	J	A_u (m²)	P_u (m)	B_{us} (m)	y_{um} (m)	medial	Mann'g	(m³s⁻¹)
0.25	0.4851	0.7392	0.2500	2.0616	0.5000	0.5000	0.0550	0.0524	0.5536
0.375	0.7022	1.0328	0.3750	2.1360	0.7500	0.5000	0.1045	0.1005	0.8304
0.50	0.8944	1.2566	0.5000	2.2361	1.0000	0.5000	0.1623	0.1575	1.1072
0.625	1.0600	1.4119	0.6250	2.3585	1.2500	0.5000	0.2257	0.2205	1.3840
0.75	1.2000	1.5079	0.7500	2.5000	1.5000	0.5000	0.2927	0.2874	1.6608
0.875	1.3170	1.5569	0.8750	2.6575	1.7500	0.5000	0.3619	0.3567	1.9376
1.00	1.4142	1.5708	1.0000	2.8284	2.0000	0.5000	0.4323	0.4275	2.2143
1.125	1.4948	1.5599	1.1250	3.0104	2.2500	0.5000	0.5034	0.4990	2.4911
1.25	1.5617	1.5324	1.2500	3.2016	2.5000	0.5000	0.5747	0.5709	2.7679
1.375	1.6175	1.4943	1.3750	3.4004	2.7500	0.5000	0.6460	0.6429	3.0447
1.50	1.6641	1.4499	1.5000	3.6056	3.0000	0.5000	0.7172	0.7147	3.3215
1.75	1.7365	1.3533	1.7500	4.0311	3.5000	0.5000	0.8591	0.8578	3.8751
2.00	1.7889	1.2566	2.0000	4.4721	4.0000	0.5000	1.0000	1.0000	4.4287
2.50	1.8570	1.0833	2.5000	5.3852	5.0000	0.5000	1.2792	1.2815	5.5359
3.00	1.8974	0.9425	3.0000	6.3246	6.0000	0.5000	1.5555	1.5601	6.6430
3.50	1.9230	0.8298	3.5000	7.2801	7.0000	0.5000	1.8299	1.8365	7.7502
4.00	1.9403	0.7392	4.0000	8.2462	8.0000	0.5000	2.1029	2.1113	8.8574
4.50	1.9524	0.6653	4.5000	9.2195	9.0000	0.5000	2.3748	2.3851	9.9646
5.00	1.9612	0.6041	5.0000	10.198	10.000	0.5000	2.6460	2.6581	11.072
6.00	1.9728	0.5094	6.0000	12.166	12.000	0.5000	3.1867	3.2023	13.286
7.00	1.9799	0.4398	7.0000	14.142	14.000	0.5000	3.7261	3.7450	15.500
8.00	1.9846	0.3866	8.0000	16.125	16.000	0.5000	4.2646	4.2867	17.715
9.00	1.9878	0.3448	9.0000	18.111	18.000	0.5000	4.8024	4.8277	19.929
10.00	1.9901	0.3110	10.000	20.100	20.000	0.5000	5.3398	5.3683	22.143
12.00	1.9931	0.2600	12.000	24.083	24.000	0.5000	6.4138	6.4484	26.572
14.00	1.9949	0.2233	14.000	28.071	28.000	0.5000	7.4870	7.5277	31.001
16.00	1.9961	0.1956	16.000	32.062	32.000	0.5000	8.5597	8.6065	35.430
18.00	1.9969	0.1740	18.000	36.056	36.000	0.5000	9.6320	9.6850	39.858
20.00	1.9975	0.1567	20.000	40.050	40.000	0.5000	10.704	10.763	44.287
25.00	1.9984	0.1255	25.000	50.040	50.000	0.5000	13.384	13.458	55.359
30.00	1.9989	0.1046	30.000	60.033	60.000	0.5000	16.063	16.152	66.430
35.00	1.9992	0.0897	35.000	70.029	70.000	0.5000	18.742	18.846	77.502
40.00	1.9994	0.0785	40.000	80.025	80.000	0.5000	21.421	21.540	88.574
45.00	1.9995	0.0698	45.000	90.022	90.000	0.5000	24.099	24.233	99.646
50.00	1.9996	0.0628	50.000	100.02	100.00	0.5000	26.778	26.927	110.72
60.00	1.9997	0.0523	60.000	120.02	120.00	0.5000	32.135	32.314	132.86
70.00	1.9998	0.0449	70.000	140.01	140.00	0.5000	37.491	37.700	155.00
80.00	1.9998	0.0393	80.000	160.01	160.00	0.5000	42.848	43.086	177.15
90.00	1.9999	0.0349	90.000	180.01	180.00	0.5000	48.204	48.473	199.29
100.0	1.9999	0.0314	100.00	200.01	200.00	0.5000	53.560	53.859	221.43

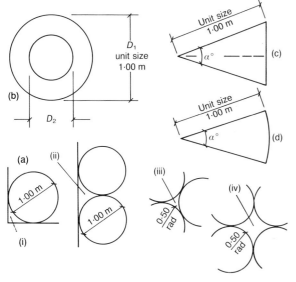

(b) Annulus

D₂/D₁	U.equ. dia. $D_{ep(u)}$ (m)	Equiv. disch. factor J	Unit sect. area A_u (m²)	Unit wetted perim. P_u (m)
0.10	0.9000	0.8182	0.7775	3.4558
0.20	0.8000	0.6667	0.7540	3.7699
0.30	0.7000	0.5385	0.7147	4.0841
0.40	0.6000	0.4286	0.6597	4.3982
0.50	0.5000	0.3333	0.5890	4.7124
0.55	0.4500	0.2903	0.5478	4.8695
0.60	0.4000	0.2500	0.5027	5.0265
0.65	0.3500	0.2121	0.4536	5.1836
0.70	0.3000	0.1765	0.4006	5.3407
0.75	0.2500	0.1429	0.3436	5.4978
0.80	0.2000	0.1111	0.2827	5.6549
0.85	0.1500	0.0811	0.2179	5.8119
0.90	0.1000	0.0526	0.1492	5.9690
0.95	0.0500	0.0256	0.0756	6.1261
0.98	0.0200	0.0101	0.0311	6.2204

(a) Tube interstices
(1.00 m dia. unit size throughout)

Shape	U.equ. dia. $D_{ep(u)}$ (m)	Equiv. disch. factor J	Unit sect. area A_u (m²)	Unit wetted perim. P_u (m)
(i)	0.1202	0.2115	0.0537	1.7854
(ii)	0.1670	0.2040	0.1073	2.5708
(iii)	0.1027	0.2053	0.0403	1.5708
(iv)	0.2732	0.2732	0.2146	3.1416

(c) Isosceles triangle

α°	U.equ. dia. $D_{ep(u)}$ (m)	Equiv. disch. factor J	Unit sect. area A_u (m²)	Unit wetted perim. P_u (m)
10	0.1597	0.2308	0.0868	2.1743
20	0.2914	0.3900	0.1710	2.3473
30	0.3972	0.4956	0.2500	2.5176
40	0.4790	0.5606	0.3239	2.6840
50	0.5385	0.5946	0.3830	2.8452
60	0.5774	0.6046	0.4330	3.0000
70	0.5972	0.5961	0.4698	3.1472
80	0.5995	0.5732	0.4924	3.2856
90	0.5858	0.5390	0.5000	3.4142
100	0.5576	0.4960	0.4924	3.5321
110	0.5166	0.4460	0.4698	3.6383
120	0.4641	0.3907	0.4330	3.7321
130	0.4018	0.3311	0.3830	3.8126
140	0.3314	0.2687	0.3214	3.8794
150	0.2543	0.2032	0.2500	3.9319

(d) Sector of circle

α°	U.equ. dia. $D_{ep(u)}$ (m)	Equiv. disch. factor J	Unit sect. area A_u (m²)	Unit wetted perim. P_u (m)
10	0.1605	0.2319	0.0873	2.1745
20	0.2972	0.3975	0.1745	2.3491
30	0.4150	0.5166	0.2618	2.5236
40	0.5175	0.6025	0.3491	2.6981
50	0.6076	0.6644	0.4363	2.8727
60	0.6873	0.7086	0.5236	3.0472
70	0.7584	0.7396	0.6109	3.2217
80	0.8222	0.7606	0.6981	3.3963
90	0.8798	0.7741	0.7854	3.5708
100	0.9320	0.7818	0.8727	3.7453
110	0.9796	0.7851	0.9599	3.9199
120	1.0231	0.7850	1.0472	4.0944
140	1.0998	0.7776	1.2217	4.4435
160	1.1654	0.7639	1.3963	4.7925
180	1.2220	0.7467	1.5708	5.1416

D1

n is Manning coefficient
S = 0.00015 to 0.00070

ie hydraulic gradient =
1 in 6667 to 1 in 1429

Water at normal temperature;
full bore conditions

nV in ms⁻¹
$nQ \times 1000$ in m³s⁻¹
ie nQ in litres/sec.

| Gradient | (Equivalent) Pipe diameters in m | | | | | | | | | | | |
	0.350	0.375	0.400	0.450	0.500	0.525	0.600	0.675	0.700	0.750	0.800	0.825
0.00015	0.0024	0.0025	0.0026	0.0029	0.0031	0.0032	0.0035	0.0037	0.0038	0.0040	0.0042	0.0043
1/ 6667	0.2322	0.2792	0.3316	0.4539	0.6012	0.6847	0.9776	1.3384	1.4747	1.7725	2.1054	2.2855
0.00016	0.0025	0.0026	0.0027	0.0029	0.0032	0.0033	0.0036	0.0039	0.0040	0.0041	0.0043	0.0044
1/ 6250	0.2399	0.2883	0.3425	0.4688	0.6209	0.7072	1.0097	1.3822	1.5230	1.8307	2.1744	2.3604
0.00017	0.0026	0.0027	0.0028	0.0030	0.0033	0.0034	0.0037	0.0040	0.0041	0.0043	0.0045	0.0046
1/ 5882	0.2472	0.2972	0.3530	0.4833	0.6400	0.7290	1.0407	1.4248	1.5699	1.8870	2.2414	2.4330
0.00018	0.0026	0.0028	0.0029	0.0031	0.0034	0.0035	0.0038	0.0041	0.0042	0.0044	0.0046	0.0047
1/ 5556	0.2544	0.3058	0.3632	0.4973	0.6586	0.7501	1.0709	1.4661	1.6154	1.9417	2.3064	2.5036
0.00019	0.0027	0.0028	0.0030	0.0032	0.0034	0.0036	0.0039	0.0042	0.0043	0.0045	0.0047	0.0048
1/ 5263	0.2614	0.3142	0.3732	0.5109	0.6766	0.7706	1.1003	1.5063	1.6597	1.9949	2.3696	2.5722
0.00020	0.0028	0.0029	0.0030	0.0033	0.0035	0.0037	0.0040	0.0043	0.0044	0.0046	0.0048	0.0049
1/ 5000	0.2682	0.3223	0.3829	0.5242	0.6942	0.7907	1.1288	1.5454	1.7028	2.0467	2.4311	2.6390
0.00022	0.0029	0.0031	0.0032	0.0035	0.0037	0.0038	0.0042	0.0045	0.0046	0.0049	0.0051	0.0052
1/ 4545	0.2813	0.3381	0.4016	0.5497	0.7281	0.8292	1.1839	1.6208	1.7859	2.1466	2.5498	2.7678
0.00024	0.0031	0.0032	0.0033	0.0036	0.0039	0.0040	0.0044	0.0047	0.0048	0.0051	0.0053	0.0054
1/ 4167	0.2938	0.3531	0.4194	0.5742	0.7605	0.8661	1.2366	1.6929	1.8653	2.2421	2.6631	2.8909
0.00026	0.0032	0.0033	0.0035	0.0038	0.0040	0.0042	0.0046	0.0049	0.0050	0.0053	0.0055	0.0056
1/ 3846	0.3058	0.3675	0.4365	0.5976	0.7915	0.9015	1.2871	1.7620	1.9415	2.3336	2.7719	3.0089
0.00028	0.0033	0.0035	0.0036	0.0039	0.0042	0.0043	0.0047	0.0051	0.0052	0.0055	0.0057	0.0058
1/ 3571	0.3173	0.3814	0.4530	0.6202	0.8214	0.9355	1.3357	1.8285	2.0148	2.4217	2.8765	3.1225
0.00030	0.0034	0.0036	0.0037	0.0040	0.0043	0.0045	0.0049	0.0053	0.0054	0.0057	0.0059	0.0060
1/ 3333	0.3284	0.3948	0.4689	0.6420	0.8502	0.9684	1.3825	1.8927	2.0855	2.5067	2.9775	3.2321
0.00032	0.0035	0.0037	0.0039	0.0042	0.0045	0.0046	0.0051	0.0055	0.0056	0.0059	0.0061	0.0062
1/ 3125	0.3392	0.4077	0.4843	0.6630	0.8781	1.0001	1.4279	1.9548	2.1539	2.5889	3.0751	3.3381
0.00034	0.0036	0.0038	0.0040	0.0043	0.0046	0.0048	0.0052	0.0056	0.0058	0.0060	0.0063	0.0064
1/ 2941	0.3497	0.4203	0.4992	0.6834	0.9051	1.0309	1.4718	2.0150	2.2202	2.6686	3.1698	3.4409
0.00036	0.0037	0.0039	0.0041	0.0044	0.0047	0.0049	0.0054	0.0058	0.0059	0.0062	0.0065	0.0066
1/ 2778	0.3598	0.4325	0.5137	0.7032	0.9314	1.0608	1.5145	2.0734	2.2845	2.7460	3.2617	3.5406
0.00038	0.0038	0.0040	0.0042	0.0045	0.0049	0.0050	0.0055	0.0060	0.0061	0.0064	0.0067	0.0068
1/ 2632	0.3696	0.4443	0.5278	0.7225	0.9569	1.0898	1.5560	2.1302	2.3471	2.8212	3.3511	3.6376
0.00040	0.0039	0.0041	0.0043	0.0047	0.0050	0.0052	0.0056	0.0061	0.0063	0.0066	0.0068	0.0070
1/ 2500	0.3793	0.4559	0.5415	0.7413	0.9817	1.1182	1.5964	2.1855	2.4081	2.8945	3.4381	3.7321
0.00042	0.0040	0.0042	0.0044	0.0048	0.0051	0.0053	0.0058	0.0063	0.0064	0.0067	0.0070	0.0072
1/ 2381	0.3886	0.4671	0.5548	0.7596	1.0060	1.1458	1.6359	2.2395	2.4676	2.9660	3.5230	3.8243
0.00044	0.0041	0.0043	0.0045	0.0049	0.0052	0.0054	0.0059	0.0064	0.0066	0.0069	0.0072	0.0073
1/ 2273	0.3978	0.4781	0.5679	0.7775	1.0297	1.1727	1.6743	2.2922	2.5256	3.0358	3.6059	3.9143
0.00046	0.0042	0.0044	0.0046	0.0050	0.0054	0.0055	0.0061	0.0065	0.0067	0.0070	0.0073	0.0075
1/ 2174	0.4067	0.4889	0.5807	0.7949	1.0528	1.1991	1.7120	2.3437	2.5824	3.1040	3.6870	4.0023
0.00048	0.0043	0.0045	0.0047	0.0051	0.0055	0.0057	0.0062	0.0067	0.0069	0.0072	0.0075	0.0076
1/ 2083	0.4154	0.4994	0.5931	0.8120	1.0754	1.2249	1.7488	2.3941	2.6379	3.1708	3.7663	4.0883
0.00050	0.0044	0.0046	0.0048	0.0052	0.0056	0.0058	0.0063	0.0068	0.0070	0.0073	0.0076	0.0078
1/ 2000	0.4240	0.5097	0.6054	0.8288	1.0976	1.2501	1.7849	2.4435	2.6923	3.2362	3.8439	4.1726
0.00055	0.0046	0.0048	0.0051	0.0055	0.0059	0.0061	0.0066	0.0072	0.0073	0.0077	0.0080	0.0082
1/ 1818	0.4447	0.5345	0.6349	0.8692	1.1512	1.3112	1.8720	2.5628	2.8237	3.3941	4.0315	4.3763
0.00060	0.0048	0.0051	0.0053	0.0057	0.0061	0.0063	0.0069	0.0075	0.0077	0.0080	0.0084	0.0086
1/ 1667	0.4645	0.5583	0.6632	0.9079	1.2024	1.3695	1.9552	2.6767	2.9493	3.5450	4.2108	4.5709
0.00065	0.0050	0.0053	0.0055	0.0059	0.0064	0.0066	0.0072	0.0078	0.0080	0.0084	0.0087	0.0089
1/ 1538	0.4835	0.5811	0.6902	0.9449	1.2515	1.4254	2.0351	2.7860	3.0697	3.6898	4.3827	4.7575
0.00070	0.0052	0.0055	0.0057	0.0062	0.0066	0.0068	0.0075	0.0081	0.0083	0.0087	0.0090	0.0092
1/ 1429	0.5017	0.6030	0.7163	0.9806	1.2987	1.4792	2.1119	2.8912	3.1856	3.8291	4.5482	4.9371

S = 0.00015 to 0.00070

n is Manning coefficient
S = 0.00075 to 0.00380

ie hydraulic gradient =
1 in 1333 to 1 in 263

Water at normal temperature;
full bore conditions.

nV in ms^{-1}
$nQ \times 1000$ in m^3s^{-1}
ie nQ in litres/sec.

Gradient	(Equivalent) Pipe diameters in m											
	0.350	0.375	0.400	0.450	0.500	0.525	0.600	0.675	0.700	0.750	0.800	0.825
0.00075 1/ 1333	0.0054 0.5193	0.0057 0.6242	0.0059 0.7414	0.0064 1.0150	0.0068 1.3443	0.0071 1.5311	0.0077 2.1860	0.0084 2.9927	0.0086 3.2974	0.0090 3.9635	0.0094 4.7078	0.0096 5.1104
0.00080 1/ 1250	0.0056 0.5363	0.0058 0.6447	0.0061 0.7658	0.0066 1.0483	0.0071 1.3884	0.0073 1.5813	0.0080 2.2577	0.0086 3.0908	0.0088 3.4056	0.0093 4.0935	0.0097 4.8622	0.0099 5.2780
0.00085 1/ 1176	0.0057 0.5528	0.0060 0.6645	0.0063 0.7893	0.0068 1.0806	0.0073 1.4311	0.0075 1.6300	0.0082 2.3272	0.0089 3.1859	0.0091 3.5104	0.0096 4.2194	0.0100 5.0119	0.0102 5.4405
0.00090 1/ 1111	0.0059 0.5689	0.0062 0.6838	0.0065 0.8122	0.0070 1.1119	0.0075 1.4726	0.0077 1.6772	0.0085 2.3946	0.0092 3.2783	0.0094 3.6121	0.0098 4.3418	0.0103 5.1572	0.0105 5.5982
0.00095 1/ 1053	0.0061 0.5845	0.0064 0.7025	0.0066 0.8345	0.0072 1.1424	0.0077 1.5130	0.0080 1.7232	0.0087 2.4603	0.0094 3.3681	0.0096 3.7111	0.0101 4.4607	0.0105 5.2985	0.0108 5.7516
0.00100 1/ 1000	0.0062 0.5996	0.0065 0.7208	0.0068 0.8561	0.0074 1.1721	0.0079 1.5523	0.0082 1.7680	0.0089 2.5242	0.0097 3.4556	0.0099 3.8075	0.0104 4.5766	0.0108 5.4361	0.0110 5.9010
0.00110 1/ 909	0.0065 0.6289	0.0068 0.7560	0.0071 0.8979	0.0077 1.2293	0.0083 1.6280	0.0086 1.8543	0.0094 2.6474	0.0101 3.6243	0.0104 3.9934	0.0109 4.8000	0.0113 5.7015	0.0116 6.1890
0.00120 1/ 833	0.0068 0.6569	0.0071 0.7896	0.0075 0.9378	0.0081 1.2839	0.0087 1.7004	0.0089 1.9367	0.0098 2.7651	0.0106 3.7854	0.0108 4.1709	0.0113 5.0134	0.0118 5.9550	0.0121 6.4642
0.00130 1/ 769	0.0071 0.6837	0.0074 0.8218	0.0078 0.9761	0.0084 1.3364	0.0090 1.7699	0.0093 2.0158	0.0102 2.8780	0.0110 3.9400	0.0113 4.3413	0.0118 5.2182	0.0123 6.1981	0.0126 6.7282
0.00140 1/ 714	0.0074 0.7095	0.0077 0.8528	0.0081 1.0130	0.0087 1.3868	0.0094 1.8367	0.0097 2.0919	0.0106 2.9866	0.0114 4.0887	0.0117 4.5051	0.0123 5.4151	0.0128 6.4321	0.0131 6.9822
0.00150 1/ 667	0.0076 0.7344	0.0080 0.8828	0.0083 1.0485	0.0090 1.4355	0.0097 1.9011	0.0100 2.1653	0.0109 3.0915	0.0118 4.2323	0.0121 4.6633	0.0127 5.6052	0.0132 6.6579	0.0135 7.2272
0.00160 1/ 625	0.0079 0.7585	0.0083 0.9117	0.0086 1.0829	0.0093 1.4825	0.0100 1.9635	0.0103 2.2363	0.0113 3.1929	0.0122 4.3711	0.0125 4.8162	0.0131 5.7890	0.0137 6.8762	0.0140 7.4643
0.00170 1/ 588	0.0081 0.7818	0.0085 0.9398	0.0089 1.1163	0.0096 1.5282	0.0103 2.0239	0.0106 2.3051	0.0116 3.2911	0.0126 4.5056	0.0129 4.9644	0.0135 5.9672	0.0141 7.0878	0.0144 7.6940
0.00180 1/ 556	0.0084 0.8045	0.0088 0.9670	0.0091 1.1486	0.0099 1.5725	0.0106 2.0826	0.0110 2.3720	0.0120 3.3865	0.0130 4.6362	0.0133 5.1083	0.0139 6.1402	0.0145 7.2933	0.0148 7.9170
0.00190 1/ 526	0.0086 0.8266	0.0090 0.9935	0.0094 1.1801	0.0102 1.6156	0.0109 2.1397	0.0113 2.4370	0.0123 3.4793	0.0133 4.7632	0.0136 5.2483	0.0143 6.3085	0.0149 7.4932	0.0152 8.1340
0.00200 1/ 500	0.0088 0.8480	0.0092 1.0193	0.0096 1.2108	0.0104 1.6575	0.0112 2.1953	0.0115 2.5003	0.0126 3.5697	0.0137 4.8870	0.0140 5.3847	0.0147 6.4723	0.0153 7.6878	0.0156 8.3453
0.00220 1/ 455	0.0092 0.8894	0.0097 1.0691	0.0101 1.2699	0.0109 1.7384	0.0117 2.3024	0.0121 2.6223	0.0132 3.7440	0.0143 5.1255	0.0147 5.6475	0.0154 6.7882	0.0160 8.0631	0.0164 8.7526
0.00240 1/ 417	0.0097 0.9290	0.0101 1.1166	0.0106 1.3263	0.0114 1.8157	0.0122 2.4048	0.0127 2.7389	0.0138 3.9104	0.0150 5.3534	0.0153 5.8986	0.0160 7.0901	0.0168 8.4216	0.0171 9.1418
0.00260 1/ 385	0.0100 0.9669	0.0105 1.1622	0.0110 1.3805	0.0119 1.8899	0.0127 2.5030	0.0132 2.8508	0.0144 4.0701	0.0156 5.5720	0.0160 6.1395	0.0167 7.3796	0.0174 8.7655	0.0178 9.5151
0.00280 1/ 357	0.0104 1.0034	0.0109 1.2061	0.0114 1.4326	0.0123 1.9612	0.0132 2.5975	0.0137 2.9584	0.0149 4.2238	0.0162 5.7824	0.0166 6.3712	0.0173 7.6582	0.0181 9.0964	0.0185 9.8743
0.00300 1/ 333	0.0108 1.0386	0.0113 1.2484	0.0118 1.4829	0.0128 2.0301	0.0137 2.6886	0.0141 3.0622	0.0155 4.3720	0.0167 5.9853	0.0171 6.5948	0.0179 7.9270	0.0187 9.4156	0.0191 10.221
0.00320 1/ 313	0.0111 1.0727	0.0117 1.2894	0.0122 1.5315	0.0132 2.0966	0.0141 2.7768	0.0146 3.1626	0.0160 4.5154	0.0173 6.1816	0.0177 6.8111	0.0185 8.1869	0.0193 9.7244	0.0197 10.556
0.00340 1/ 294	0.0115 1.1057	0.0120 1.3290	0.0126 1.5786	0.0136 2.1612	0.0146 2.8623	0.0151 3.2600	0.0165 4.6543	0.0178 6.3718	0.0182 7.0207	0.0191 8.4389	0.0199 10.024	0.0204 10.881
0.00360 1/ 278	0.0118 1.1378	0.0124 1.3676	0.0129 1.6244	0.0140 2.2238	0.0150 2.9452	0.0155 3.3545	0.0169 4.7893	0.0183 6.5566	0.0188 7.2243	0.0197 8.6835	0.0205 10.314	0.0209 11.196
0.00380 1/ 263	0.0121 1.1689	0.0127 1.4050	0.0133 1.6689	0.0144 2.2848	0.0154 3.0259	0.0159 3.4464	0.0174 4.9205	0.0188 6.7362	0.0193 7.4223	0.0202 8.9215	0.0211 10.597	0.0215 11.503

S = 0.00075 to 0.0038

n is Manning coefficient
S = 0.0040 to 0.0190

Water at normal temperature; full bore conditions.

ie hydraulic gradient = 1 in 250 to 1 in 53

nV in ms^{-1}
$nQ \times 1000$ in m^{3}s^{-1}
ie nQ in litres/sec.

Gradient	(Equivalent) Pipe diameters in m											
	0.350	0.375	0.400	0.450	0.500	0.525	0.600	0.675	0.700	0.750	0.800	0.825
0.00400	0.0125	0.0131	0.0136	0.0147	0.0158	0.0163	0.0179	0.0193	0.0198	0.0207	0.0216	0.0221
1/ 250	1.1993	1.4415	1.7123	2.3441	3.1046	3.5359	5.0483	6.9112	7.6151	9.1533	10.872	11.802
0.00420	0.0128	0.0134	0.0140	0.0151	0.0162	0.0167	0.0183	0.0198	0.0203	0.0212	0.0222	0.0226
1/ 238	1.2289	1.4771	1.7546	2.4020	3.1812	3.6233	5.1730	7.0819	7.8031	9.3793	11.141	12.093
0.00440	0.0131	0.0137	0.0143	0.0155	0.0166	0.0171	0.0187	0.0203	0.0208	0.0217	0.0227	0.0232
1/ 227	1.2578	1.5119	1.7958	2.4585	3.2561	3.7085	5.2948	7.2486	7.9868	9.6000	11.403	12.378
0.00460	0.0134	0.0140	0.0146	0.0158	0.0170	0.0175	0.0191	0.0207	0.0212	0.0222	0.0232	0.0237
1/ 217	1.2861	1.5459	1.8362	2.5138	3.3293	3.7919	5.4137	7.4115	8.1663	9.8158	11.659	12.656
0.00480	0.0137	0.0143	0.0149	0.0161	0.0173	0.0179	0.0196	0.0212	0.0217	0.0227	0.0237	0.0242
1/ 208	1.3138	1.5791	1.8757	2.5679	3.4009	3.8734	5.5302	7.5709	8.3419	10.027	11.910	12.928
0.00500	0.0139	0.0146	0.0152	0.0165	0.0177	0.0183	0.0200	0.0216	0.0221	0.0232	0.0242	0.0247
1/ 200	1.3409	1.6117	1.9144	2.6208	3.4710	3.9533	5.6442	7.7270	8.5139	10.234	12.156	13.195
0.00550	0.0146	0.0153	0.0160	0.0173	0.0185	0.0192	0.0209	0.0226	0.0232	0.0243	0.0254	0.0259
1/ 182	1.4063	1.6904	2.0078	2.7487	3.6404	4.1462	5.9197	8.1041	8.9295	10.733	12.749	13.839
0.00600	0.0153	0.0160	0.0167	0.0181	0.0194	0.0200	0.0219	0.0237	0.0242	0.0254	0.0265	0.0270
1/ 167	1.4688	1.7655	2.0971	2.8709	3.8023	4.3306	6.1829	8.4645	9.3265	11.210	13.316	14.454
0.00650	0.0159	0.0166	0.0174	0.0188	0.0202	0.0208	0.0228	0.0246	0.0252	0.0264	0.0276	0.0281
1/ 154	1.5288	1.8376	2.1827	2.9882	3.9575	4.5074	6.4354	8.8101	9.7073	11.668	13.859	15.045
0.00700	0.0165	0.0173	0.0180	0.0195	0.0209	0.0216	0.0236	0.0255	0.0262	0.0274	0.0286	0.0292
1/ 143	1.5865	1.9070	2.2651	3.1010	4.1069	4.6776	6.6783	9.1427	10.074	12.109	14.383	15.613
0.00750	0.0171	0.0179	0.0187	0.0202	0.0217	0.0224	0.0244	0.0264	0.0271	0.0284	0.0296	0.0302
1/ 133	1.6422	1.9739	2.3446	3.2098	4.2511	4.8418	6.9127	9.4636	10.427	12.534	14.887	16.161
0.00800	0.0176	0.0185	0.0193	0.0208	0.0224	0.0231	0.0253	0.0273	0.0280	0.0293	0.0306	0.0312
1/ 125	1.6961	2.0387	2.4215	3.3151	4.3905	5.0006	7.1394	9.7740	10.769	12.945	15.376	16.691
0.00850	0.0182	0.0190	0.0199	0.0215	0.0230	0.0238	0.0260	0.0282	0.0288	0.0302	0.0315	0.0322
1/ 118	1.7483	2.1014	2.4960	3.4171	4.5256	5.1545	7.3592	10.075	11.101	13.343	15.849	17.204
0.00900	0.0187	0.0196	0.0204	0.0221	0.0237	0.0245	0.0268	0.0290	0.0297	0.0311	0.0324	0.0331
1/ 111	1.7989	2.1623	2.5684	3.5162	4.6568	5.3039	7.5725	10.367	11.423	13.730	16.308	17.703
0.00950	0.0192	0.0201	0.0210	0.0227	0.0244	0.0252	0.0275	0.0298	0.0305	0.0319	0.0333	0.0340
1/ 105	1.8482	2.2216	2.6388	3.6125	4.7844	5.4492	7.7800	10.651	11.736	14.106	16.755	18.188
0.01000	0.0197	0.0206	0.0215	0.0233	0.0250	0.0258	0.0282	0.0305	0.0313	0.0328	0.0342	0.0349
1/ 100	1.8963	2.2793	2.7073	3.7064	4.9087	5.5908	7.9821	10.928	12.040	14.473	17.191	18.661
0.01100	0.0207	0.0216	0.0226	0.0244	0.0262	0.0271	0.0296	0.0320	0.0328	0.0344	0.0359	0.0366
1/ 91	1.9888	2.3905	2.8395	3.8873	5.1483	5.8637	8.3717	11.461	12.628	15.179	18.030	19.571
0.01200	0.0216	0.0226	0.0236	0.0255	0.0274	0.0283	0.0309	0.0335	0.0343	0.0359	0.0375	0.0382
1/ 83	2.0772	2.4968	2.9657	4.0601	5.3772	6.1244	8.7440	11.971	13.190	15.854	18.831	20.442
0.01300	0.0225	0.0235	0.0246	0.0266	0.0285	0.0294	0.0322	0.0348	0.0357	0.0374	0.0390	0.0398
1/ 77	2.1621	2.5988	3.0868	4.2259	5.5968	6.3745	9.1010	12.459	13.728	16.501	19.600	21.276
0.01400	0.0233	0.0244	0.0255	0.0276	0.0296	0.0306	0.0334	0.0361	0.0370	0.0388	0.0405	0.0413
1/ 71	2.2437	2.6969	3.2034	4.3854	5.8081	6.6151	9.4446	12.930	14.246	17.124	20.340	22.080
0.01500	0.0241	0.0253	0.0264	0.0285	0.0306	0.0316	0.0346	0.0374	0.0383	0.0401	0.0419	0.0428
1/ 67	2.3224	2.7915	3.3158	4.5394	6.0119	6.8473	9.7761	13.384	14.747	17.725	21.054	22.855
0.01600	0.0249	0.0261	0.0273	0.0295	0.0316	0.0327	0.0357	0.0386	0.0396	0.0414	0.0433	0.0442
1/ 62	2.3986	2.8831	3.4245	4.6882	6.2091	7.0719	10.097	13.822	15.230	18.307	21.744	23.604
0.01700	0.0257	0.0269	0.0281	0.0304	0.0326	0.0337	0.0368	0.0398	0.0408	0.0427	0.0446	0.0455
1/ 59	2.4724	2.9718	3.5299	4.8325	6.4002	7.2895	10.407	14.248	15.699	18.870	22.414	24.330
0.01800	0.0264	0.0277	0.0289	0.0313	0.0335	0.0346	0.0379	0.0410	0.0420	0.0440	0.0459	0.0468
1/ 56	2.5441	3.0580	3.6323	4.9726	6.5858	7.5008	10.709	14.661	16.154	19.417	23.064	25.036
0.01900	0.0272	0.0284	0.0297	0.0321	0.0345	0.0356	0.0389	0.0421	0.0431	0.0452	0.0471	0.0481
1/ 53	2.6138	3.1418	3.7318	5.1089	6.7662	7.7064	11.003	15.063	16.597	19.949	23.696	25.722

S = 0.0040 to 0.0190

n is Manning coefficient
S = 0.020 to 0.100

Water at normal temperature;
full bore conditions.

ie hydraulic gradient =
1 in 50 to 1 in 10

nV in ms^{-1}
$nQ \times 1000$ in m^{3}s^{-1}
ie nQ in litres/sec.

Gradient (Equivalent) Pipe diameters in m

Gradient	0.350	0.375	0.400	0.450	0.500	0.525	0.600	0.675	0.700	0.750	0.800	0.825
0.02000	0.0279	0.0292	0.0305	0.0330	0.0354	0.0365	0.0399	0.0432	0.0442	0.0463	0.0484	0.0494
1/ 50	2.6817	3.2234	3.8288	5.2416	6.9420	7.9066	11.288	15.454	17.028	20.467	24.311	26.390
0.02200	0.0292	0.0306	0.0320	0.0346	0.0371	0.0383	0.0419	0.0453	0.0464	0.0486	0.0507	0.0518
1/ 45	2.8126	3.3807	4.0156	5.4974	7.2808	8.2925	11.839	16.208	17.859	21.466	25.498	27.678
0.02400	0.0305	0.0320	0.0334	0.0361	0.0387	0.0400	0.0437	0.0473	0.0485	0.0508	0.0530	0.0541
1/ 42	2.9377	3.5311	4.1942	5.7419	7.6046	8.6612	12.366	16.929	18.653	22.421	26.631	28.909
0.02600	0.0318	0.0333	0.0347	0.0376	0.0403	0.0416	0.0455	0.0492	0.0504	0.0528	0.0551	0.0563
1/ 38	3.0576	3.6752	4.3655	5.9763	7.9151	9.0149	12.871	17.620	19.415	23.336	27.719	30.089
0.02800	0.0330	0.0345	0.0361	0.0390	0.0418	0.0432	0.0472	0.0511	0.0524	0.0548	0.0572	0.0584
1/ 36	3.1730	3.8140	4.5302	6.2019	8.2139	9.3552	13.357	18.285	20.148	24.217	28.765	31.225
0.03000	0.0341	0.0357	0.0373	0.0404	0.0433	0.0447	0.0489	0.0529	0.0542	0.0567	0.0592	0.0605
1/ 33	3.2844	3.9478	4.6892	6.4196	8.5022	9.6835	13.825	18.927	20.855	25.067	29.775	32.321
0.03200	0.0353	0.0369	0.0385	0.0417	0.0447	0.0462	0.0505	0.0546	0.0560	0.0586	0.0612	0.0624
1/ 31	3.3921	4.0773	4.8430	6.6302	8.7810	10.001	14.279	19.548	21.539	25.889	30.751	33.381
0.03400	0.0363	0.0381	0.0397	0.0430	0.0461	0.0476	0.0521	0.0563	0.0577	0.0604	0.0631	0.0644
1/ 29	3.4965	4.2028	4.9921	6.8342	9.0513	10.309	14.718	20.150	22.202	26.686	31.698	34.409
0.03600	0.0374	0.0392	0.0409	0.0442	0.0474	0.0490	0.0536	0.0579	0.0594	0.0622	0.0649	0.0662
1/ 28	3.5979	4.3246	5.1368	7.0323	9.3137	10.608	15.145	20.734	22.845	27.460	32.617	35.406
0.03800	0.0384	0.0402	0.0420	0.0454	0.0487	0.0503	0.0550	0.0595	0.0610	0.0639	0.0667	0.0680
1/ 26	3.6965	4.4432	5.2776	7.2251	9.5689	10.898	15.560	21.302	23.471	28.212	33.511	36.376
0.04000	0.0394	0.0413	0.0431	0.0466	0.0500	0.0517	0.0565	0.0611	0.0626	0.0655	0.0684	0.0698
1/ 25	3.7925	4.5586	5.4147	7.4127	9.8175	11.182	15.964	21.855	24.081	28.945	34.381	37.321
0.04200	0.0404	0.0423	0.0442	0.0478	0.0512	0.0529	0.0579	0.0626	0.0641	0.0671	0.0701	0.0715
1/ 24	3.8862	4.6712	5.5484	7.5958	10.060	11.458	16.359	22.395	24.676	29.660	35.230	38.243
0.04400	0.0413	0.0433	0.0452	0.0489	0.0524	0.0542	0.0592	0.0641	0.0656	0.0687	0.0717	0.0732
1/ 23	3.9776	4.7811	5.6790	7.7746	10.297	11.727	16.743	22.922	25.256	30.358	36.059	39.143
0.04600	0.0423	0.0443	0.0462	0.0500	0.0536	0.0554	0.0605	0.0655	0.0671	0.0703	0.0733	0.0749
1/ 22	4.0670	4.8885	5.8066	7.9493	10.528	11.991	17.120	23.437	25.824	31.040	36.870	40.023
0.04800	0.0432	0.0452	0.0472	0.0511	0.0548	0.0566	0.0619	0.0669	0.0685	0.0718	0.0749	0.0765
1/ 21	4.1545	4.9937	5.9315	8.1203	10.754	12.249	17.488	23.941	26.379	31.708	37.663	40.883
0.05000	0.0441	0.0461	0.0482	0.0521	0.0559	0.0577	0.0631	0.0683	0.0700	0.0733	0.0765	0.0781
1/ 20	4.2402	5.0966	6.0538	8.2877	10.976	12.501	17.849	24.435	26.923	32.362	38.439	41.726
0.05500	0.0462	0.0484	0.0505	0.0547	0.0586	0.0606	0.0662	0.0716	0.0734	0.0768	0.0802	0.0819
1/ 18	4.4471	5.3454	6.3493	8.6922	11.512	13.112	18.720	25.628	28.237	33.941	40.315	43.763
0.06000	0.0483	0.0506	0.0528	0.0571	0.0612	0.0633	0.0692	0.0748	0.0766	0.0802	0.0838	0.0855
1/ 17	4.6449	5.5831	6.6316	9.0787	12.024	13.695	19.552	26.767	29.493	35.450	42.108	45.709
0.06500	0.0502	0.0526	0.0549	0.0594	0.0637	0.0658	0.0720	0.0779	0.0798	0.0835	0.0872	0.0890
1/ 15	4.8345	5.8111	6.9024	9.4494	12.515	14.254	20.351	27.860	30.697	36.898	43.827	47.575
0.07000	0.0521	0.0546	0.0570	0.0617	0.0661	0.0683	0.0747	0.0808	0.0828	0.0867	0.0905	0.0924
1/ 14	5.0170	6.0304	7.1629	9.8061	12.987	14.792	21.119	28.912	31.856	38.291	45.482	49.371
0.07500	0.0540	0.0565	0.0590	0.0638	0.0685	0.0707	0.0773	0.0836	0.0857	0.0897	0.0937	0.0956
1/ 13	5.1931	6.2421	7.4144	10.150	13.443	15.311	21.860	29.927	32.974	39.635	47.078	51.104
0.08000	0.0557	0.0584	0.0609	0.0659	0.0707	0.0730	0.0798	0.0864	0.0885	0.0927	0.0967	0.0987
1/ 13	5.3634	6.4468	7.6575	10.483	13.884	15.813	22.577	30.908	34.056	40.935	48.622	52.780
0.08500	0.0575	0.0602	0.0628	0.0679	0.0729	0.0753	0.0823	0.0890	0.0912	0.0955	0.0997	0.1018
1/ 12	5.5285	6.6452	7.8932	10.806	14.311	16.300	23.272	31.859	35.104	42.194	50.119	54.405
0.09000	0.0591	0.0619	0.0646	0.0699	0.0750	0.0775	0.0847	0.0916	0.0939	0.0983	0.1026	0.1047
1/ 11	5.6888	6.8379	8.1220	11.119	14.726	16.772	23.946	32.783	36.121	43.418	51.572	55.982
0.09500	0.0607	0.0636	0.0664	0.0718	0.0771	0.0796	0.0870	0.0941	0.0964	0.1010	0.1054	0.1076
1/ 11	5.8447	7.0252	8.3446	11.424	15.130	17.232	24.603	33.681	37.111	44.607	52.985	57.516
0.10000	0.0623	0.0653	0.0681	0.0737	0.0791	0.0817	0.0893	0.0966	0.0989	0.1036	0.1081	0.1104
1/ 10	5.9965	7.2077	8.5614	11.721	15.523	17.680	25.242	34.556	38.075	45.766	54.361	59.010

S = 0.020 to 0.100

D2

n is Manning coefficient
S = 0.00010 to 0.00048

ie hydraulic gradient =
1 in 10000 to 1 in 2083

Water at normal temperature;
full bore conditions.

nV in ms^{-1}
$nQ \times 1000$ in m^3s^{-1}
ie nQ in litres/sec.

| Gradient | (Equivalent) Pipe diameters in m | | | | | | | | | | | |
	0.900	0.975	1.000	1.050	1.100	1.200	1.350	1.500	1.650	1.800	1.950	2.100
0.00010	0.0037	0.0039	0.0040	0.0041	0.0042	0.0045	0.0048	0.0052	0.0055	0.0059	0.0062	0.0065
1/10000	2.3534	2.9134	3.1168	3.5499	4.0188	5.0683	6.9386	9.1895	11.849	14.943	18.499	22.541
0.00011	0.0039	0.0041	0.0042	0.0043	0.0044	0.0047	0.0051	0.0055	0.0058	0.0062	0.0065	0.0068
1/ 9091	2.4683	3.0556	3.2690	3.7232	4.2150	5.3157	7.2773	9.6380	12.427	15.673	19.402	23.641
0.00012	0.0041	0.0043	0.0043	0.0045	0.0046	0.0049	0.0053	0.0057	0.0061	0.0064	0.0068	0.0071
1/ 8333	2.5780	3.1914	3.4143	3.8888	4.4024	5.5521	7.6009	10.067	12.980	16.369	20.264	24.692
0.00013	0.0042	0.0044	0.0045	0.0047	0.0048	0.0051	0.0055	0.0059	0.0063	0.0067	0.0071	0.0074
1/ 7692	2.6833	3.3217	3.5538	4.0476	4.5821	5.7788	7.9112	10.478	13.510	17.038	21.092	25.700
0.00014	0.0044	0.0046	0.0047	0.0049	0.0050	0.0053	0.0057	0.0062	0.0066	0.0069	0.0073	0.0077
1/ 7143	2.7846	3.4471	3.6879	4.2003	4.7551	5.9969	8.2099	10.873	14.020	17.681	21.888	26.671
0.00015	0.0045	0.0048	0.0049	0.0050	0.0052	0.0055	0.0059	0.0064	0.0068	0.0072	0.0076	0.0080
1/ 6667	2.8823	3.5681	3.8173	4.3478	4.9220	6.2074	8.4980	11.255	14.512	18.302	22.656	27.607
0.00016	0.0047	0.0049	0.0050	0.0052	0.0053	0.0057	0.0061	0.0066	0.0070	0.0074	0.0078	0.0082
1/ 6250	2.9768	3.6851	3.9425	4.4904	5.0834	6.4110	8.7767	11.624	14.988	18.902	23.399	28.512
0.00017	0.0048	0.0051	0.0052	0.0053	0.0055	0.0058	0.0063	0.0068	0.0072	0.0077	0.0081	0.0085
1/ 5882	3.0685	3.7986	4.0639	4.6286	5.2399	6.6083	9.0468	11.982	15.449	19.484	24.119	29.389
0.00018	0.0050	0.0052	0.0053	0.0055	0.0057	0.0060	0.0065	0.0070	0.0074	0.0079	0.0083	0.0087
1/ 5556	3.1574	3.9087	4.1817	4.7627	5.3918	6.7999	9.3091	12.329	15.897	20.048	24.819	30.242
0.00019	0.0051	0.0054	0.0055	0.0057	0.0058	0.0062	0.0067	0.0072	0.0076	0.0081	0.0085	0.0090
1/ 5263	3.2439	4.0158	4.2963	4.8933	5.5395	6.9862	9.5642	12.667	16.332	20.598	25.499	31.070
0.00020	0.0052	0.0055	0.0056	0.0058	0.0060	0.0063	0.0069	0.0074	0.0078	0.0083	0.0088	0.0092
1/ 5000	3.3282	4.1201	4.4079	5.0204	5.6834	7.1677	9.8127	12.996	16.757	21.133	26.161	31.877
0.00022	0.0055	0.0058	0.0059	0.0061	0.0063	0.0066	0.0072	0.0077	0.0082	0.0087	0.0092	0.0097
1/ 4545	3.4907	4.3212	4.6230	5.2654	5.9608	7.5176	10.292	13.630	17.575	22.164	27.438	33.433
0.00024	0.0057	0.0060	0.0061	0.0064	0.0066	0.0069	0.0075	0.0081	0.0086	0.0091	0.0096	0.0101
1/ 4167	3.6459	4.5134	4.8286	5.4995	6.2259	7.8518	10.749	14.236	18.356	23.150	28.658	34.920
0.00026	0.0060	0.0063	0.0064	0.0066	0.0068	0.0072	0.0078	0.0084	0.0089	0.0095	0.0100	0.0105
1/ 3846	3.7947	4.6977	5.0258	5.7241	6.4801	8.1725	11.188	14.818	19.106	24.095	29.828	36.346
0.00028	0.0062	0.0065	0.0066	0.0069	0.0071	0.0075	0.0081	0.0087	0.0093	0.0098	0.0104	0.0109
1/ 3571	3.9380	4.8750	5.2155	5.9402	6.7247	8.4810	11.611	15.377	19.827	25.005	30.954	37.718
0.00030	0.0064	0.0068	0.0069	0.0071	0.0073	0.0078	0.0084	0.0090	0.0096	0.0102	0.0107	0.0113
1/ 3333	4.0762	5.0461	5.3985	6.1487	6.9608	8.7786	12.018	15.917	20.523	25.882	32.041	39.042
0.00032	0.0066	0.0070	0.0071	0.0073	0.0076	0.0080	0.0087	0.0093	0.0099	0.0105	0.0111	0.0116
1/ 3125	4.2099	5.2116	5.5756	6.3503	7.1890	9.0665	12.412	16.439	21.196	26.731	33.092	40.322
0.00034	0.0068	0.0072	0.0073	0.0076	0.0078	0.0083	0.0089	0.0096	0.0102	0.0108	0.0114	0.0120
1/ 2941	4.3395	5.3720	5.7472	6.5458	7.4103	9.3456	12.794	16.945	21.848	27.554	34.110	41.563
0.00036	0.0070	0.0074	0.0075	0.0078	0.0080	0.0085	0.0092	0.0099	0.0105	0.0111	0.0118	0.0123
1/ 2778	4.4653	5.5277	5.9138	6.7355	7.6251	9.6165	13.165	17.436	22.481	28.353	35.099	42.768
0.00038	0.0072	0.0076	0.0077	0.0080	0.0082	0.0087	0.0094	0.0101	0.0108	0.0114	0.0121	0.0127
1/ 2632	4.5876	5.6792	6.0759	6.9201	7.8341	9.8800	13.526	17.914	23.097	29.130	36.061	43.940
0.00040	0.0074	0.0078	0.0079	0.0082	0.0085	0.0090	0.0097	0.0104	0.0111	0.0117	0.0124	0.0130
1/ 2500	4.7068	5.8267	6.2337	7.0999	8.0376	10.137	13.877	18.379	23.698	29.886	36.997	45.081
0.00042	0.0076	0.0080	0.0081	0.0084	0.0087	0.0092	0.0099	0.0107	0.0114	0.0120	0.0127	0.0133
1/ 2381	4.8230	5.9706	6.3876	7.2752	8.2361	10.387	14.220	18.833	24.283	30.624	37.911	46.195
0.00044	0.0078	0.0082	0.0083	0.0086	0.0089	0.0094	0.0102	0.0109	0.0116	0.0123	0.0130	0.0137
1/ 2273	4.9365	6.1111	6.5380	7.4464	8.4299	10.631	14.555	19.276	24.854	31.345	38.803	47.282
0.00046	0.0079	0.0084	0.0085	0.0088	0.0091	0.0096	0.0104	0.0112	0.0119	0.0126	0.0133	0.0140
1/ 2174	5.0475	6.2485	6.6849	7.6138	8.6194	10.870	14.882	19.709	25.413	32.050	39.675	48.344
0.00048	0.0081	0.0085	0.0087	0.0090	0.0093	0.0098	0.0106	0.0114	0.0121	0.0129	0.0136	0.0143
1/ 2083	5.1560	6.3829	6.8287	7.7775	8.8048	11.104	15.202	20.133	25.959	32.739	40.529	49.384

S = 0.00010 to 0.00048

Gradient	(Equivalent) Pipe diameters in m											
	0.900	0.975	1.000	1.050	1.100	1.200	1.350	1.500	1.650	1.800	1.950	2.100
0.00050 1/ 2000	0.0083 5.2624	0.0087 6.5145	0.0089 6.9695	0.0092 7.9379	0.0095 8.9863	0.0100 11.333	0.0108 15.515	0.0116 20.548	0.0124 26.495	0.0131 33.414	0.0139 41.364	0.0146 50.403
0.00055 1/ 1818	0.0087 5.5192	0.0092 6.8324	0.0093 7.3097	0.0096 8.3253	0.0099 9.4249	0.0105 11.886	0.0114 16.272	0.0122 21.551	0.0130 27.788	0.0138 35.045	0.0145 43.383	0.0153 52.863
0.00060 1/ 1667	0.0091 5.7646	0.0096 7.1363	0.0097 7.6347	0.0100 8.6955	0.0104 9.8440	0.0110 12.415	0.0119 16.996	0.0127 22.510	0.0136 29.023	0.0144 36.603	0.0152 45.312	0.0159 55.213
0.00065 1/ 1538	0.0094 6.0000	0.0099 7.4277	0.0101 7.9464	0.0105 9.0506	0.0108 10.246	0.0114 12.922	0.0124 17.690	0.0133 23.429	0.0141 30.209	0.0150 38.098	0.0158 47.163	0.0166 57.468
0.00070 1/ 1429	0.0098 6.2265	0.0103 7.7080	0.0105 8.2464	0.0108 9.3923	0.0112 10.633	0.0119 13.410	0.0128 18.358	0.0138 24.313	0.0147 31.349	0.0155 39.536	0.0164 48.943	0.0172 59.637
0.00075 1/ 1333	0.0101 6.4451	0.0107 7.9786	0.0109 8.5358	0.0112 9.7219	0.0116 11.006	0.0123 13.880	0.0133 19.002	0.0142 25.167	0.0152 32.449	0.0161 40.924	0.0170 50.661	0.0178 61.730
0.00080 1/ 1250	0.0105 6.6564	0.0110 8.2402	0.0112 8.8158	0.0116 10.041	0.0120 11.367	0.0127 14.335	0.0137 19.625	0.0147 25.992	0.0157 33.513	0.0166 42.266	0.0175 52.322	0.0184 63.755
0.00085 1/ 1176	0.0108 6.8613	0.0114 8.4938	0.0116 9.0871	0.0120 10.350	0.0123 11.717	0.0131 14.777	0.0141 20.229	0.0152 26.792	0.0162 34.545	0.0171 43.566	0.0181 53.933	0.0190 65.717
0.00090 1/ 1111	0.0111 7.0602	0.0117 8.7401	0.0119 9.3505	0.0123 10.650	0.0127 12.056	0.0134 15.205	0.0145 20.816	0.0156 27.569	0.0166 35.546	0.0176 44.829	0.0186 55.496	0.0195 67.622
0.00095 1/ 1053	0.0114 7.2537	0.0120 8.9796	0.0122 9.6068	0.0126 10.942	0.0130 12.387	0.0138 15.622	0.0149 21.386	0.0160 28.324	0.0171 36.520	0.0181 46.058	0.0191 57.017	0.0201 69.475
0.00100 1/ 1000	0.0117 7.4421	0.0123 9.2129	0.0125 9.8563	0.0130 11.226	0.0134 12.709	0.0142 16.028	0.0153 21.942	0.0164 29.060	0.0175 37.469	0.0186 47.254	0.0196 58.498	0.0206 71.280
0.00110 1/ 909	0.0123 7.8053	0.0129 9.6625	0.0132 10.337	0.0136 11.774	0.0140 13.329	0.0149 16.810	0.0161 23.013	0.0172 30.478	0.0184 39.298	0.0195 49.561	0.0205 61.353	0.0216 74.759
0.00120 1/ 833	0.0128 8.1524	0.0135 10.092	0.0137 10.797	0.0142 12.297	0.0146 13.922	0.0155 17.557	0.0168 24.036	0.0180 31.833	0.0192 41.045	0.0203 51.765	0.0215 64.081	0.0225 78.083
0.00130 1/ 769	0.0133 8.4853	0.0141 10.504	0.0143 11.238	0.0148 12.799	0.0152 14.490	0.0162 18.274	0.0175 25.018	0.0187 33.133	0.0200 42.721	0.0212 53.878	0.0223 66.698	0.0235 81.272
0.00140 1/ 714	0.0138 8.8056	0.0146 10.901	0.0148 11.662	0.0153 13.283	0.0158 15.037	0.0168 18.964	0.0181 25.962	0.0195 34.384	0.0207 44.334	0.0220 55.912	0.0232 69.216	0.0244 84.340
0.00150 1/ 667	0.0143 9.1147	0.0151 11.283	0.0154 12.072	0.0159 13.749	0.0164 15.565	0.0174 19.630	0.0188 26.873	0.0201 35.591	0.0215 45.890	0.0227 57.875	0.0240 71.645	0.0252 87.300
0.00160 1/ 625	0.0148 9.4136	0.0156 11.653	0.0159 12.467	0.0164 14.200	0.0169 16.075	0.0179 20.273	0.0194 27.754	0.0208 36.758	0.0222 47.395	0.0235 59.773	0.0248 73.995	0.0260 90.163
0.00170 1/ 588	0.0153 9.7033	0.0161 12.012	0.0164 12.851	0.0169 14.637	0.0174 16.570	0.0185 20.897	0.0200 28.609	0.0214 37.889	0.0228 48.854	0.0242 61.612	0.0255 76.272	0.0268 92.938
0.00180 1/ 556	0.0157 9.9846	0.0166 12.360	0.0168 13.224	0.0174 15.061	0.0179 17.050	0.0190 21.503	0.0206 29.438	0.0221 38.988	0.0235 50.270	0.0249 63.398	0.0263 78.483	0.0276 95.632
0.00190 1/ 526	0.0161 10.258	0.0170 12.699	0.0173 13.586	0.0179 15.474	0.0184 17.518	0.0195 22.092	0.0211 30.245	0.0227 40.056	0.0242 51.648	0.0256 65.136	0.0270 80.634	0.0284 98.253
0.00200 1/ 500	0.0165 10.525	0.0175 13.029	0.0177 13.939	0.0183 15.876	0.0189 17.973	0.0200 22.666	0.0217 31.030	0.0233 41.097	0.0248 52.989	0.0263 66.828	0.0277 82.729	0.0291 100.81
0.00220 1/ 455	0.0174 11.038	0.0183 13.665	0.0186 14.619	0.0192 16.651	0.0198 18.850	0.0210 23.773	0.0227 32.545	0.0244 43.103	0.0260 55.576	0.0275 70.090	0.0291 86.767	0.0305 105.73
0.00240 1/ 417	0.0181 11.529	0.0191 14.273	0.0194 15.269	0.0201 17.391	0.0207 19.688	0.0220 24.830	0.0237 33.992	0.0255 45.019	0.0271 58.047	0.0288 73.206	0.0303 90.625	0.0319 110.43
0.00260 1/ 385	0.0189 12.000	0.0199 14.855	0.0202 15.893	0.0209 18.101	0.0216 20.492	0.0229 25.844	0.0247 35.380	0.0265 46.858	0.0283 60.417	0.0299 76.195	0.0316 94.325	0.0332 114.94
0.00280 1/ 357	0.0196 12.453	0.0206 15.416	0.0210 16.493	0.0217 18.785	0.0224 21.265	0.0237 26.819	0.0257 36.716	0.0275 48.626	0.0293 62.698	0.0311 79.072	0.0328 97.886	0.0344 119.27

S = 0.00050 to 0.00280

D2
continued

n is Manning coefficient
S = 0.0030 to 0.014

ie hydraulic gradient =
1 in 333 to 1 in 71

Water at normal temperature;
full bore conditions.

nV in ms^{-1}
nQ × 1000 in m^{3}s^{-1}
ie nQ in litres/sec.

Gradient	(Equivalent) Pipe diameters in m											
	0.900	0.975	1.000	1.050	1.100	1.200	1.350	1.500	1.650	1.800	1.950	2.100
0.00300 1/ 333	0.0203 12.890	0.0214 15.957	0.0217 17.072	0.0225 19.444	0.0232 22.012	0.0245 27.760	0.0266 38.004	0.0285 50.333	0.0304 64.898	0.0322 81.847	0.0339 101.32	0.0356 123.46
0.00320 1/ 313	0.0209 13.313	0.0221 16.480	0.0224 17.632	0.0232 20.081	0.0239 22.734	0.0254 28.671	0.0274 39.251	0.0294 51.984	0.0313 67.027	0.0332 84.531	0.0350 104.64	0.0368 127.51
0.00340 1/ 294	0.0216 13.723	0.0228 16.988	0.0231 18.174	0.0239 20.700	0.0247 23.433	0.0261 29.553	0.0283 40.459	0.0303 53.584	0.0323 69.090	0.0342 87.133	0.0361 107.87	0.0379 131.43
0.00360 1/ 278	0.0222 14.120	0.0234 17.480	0.0238 18.701	0.0246 21.300	0.0254 24.113	0.0269 30.410	0.0291 41.632	0.0312 55.137	0.0332 71.093	0.0352 89.659	0.0372 110.99	0.0390 135.24
0.00380 1/ 263	0.0228 14.507	0.0241 17.959	0.0245 19.214	0.0253 21.883	0.0261 24.774	0.0276 31.243	0.0299 42.772	0.0321 56.648	0.0342 73.041	0.0362 92.116	0.0382 114.03	0.0401 138.95
0.00400 1/ 250	0.0234 14.884	0.0247 18.426	0.0251 19.713	0.0259 22.452	0.0267 25.417	0.0283 32.055	0.0307 43.884	0.0329 58.120	0.0350 74.938	0.0371 94.509	0.0392 117.00	0.0412 142.56
0.00420 1/ 238	0.0240 15.252	0.0253 18.881	0.0257 20.199	0.0266 23.006	0.0274 26.045	0.0290 32.847	0.0314 44.967	0.0337 59.555	0.0359 76.789	0.0381 96.843	0.0401 119.89	0.0422 146.08
0.00440 1/ 227	0.0245 15.611	0.0259 19.325	0.0263 20.675	0.0272 23.548	0.0281 26.658	0.0297 33.620	0.0322 46.026	0.0345 60.956	0.0368 78.596	0.0390 99.122	0.0411 122.71	0.0432 149.52
0.00460 1/ 217	0.0251 15.962	0.0265 19.759	0.0269 21.140	0.0278 24.077	0.0287 27.257	0.0304 34.375	0.0329 47.060	0.0353 62.326	0.0376 80.362	0.0398 101.35	0.0420 125.46	0.0441 152.88
0.00480 1/ 208	0.0256 16.305	0.0270 20.184	0.0275 21.594	0.0284 24.595	0.0293 27.843	0.0310 35.115	0.0336 48.072	0.0360 63.667	0.0384 82.091	0.0407 103.53	0.0429 128.16	0.0451 156.17
0.00500 1/ 200	0.0262 16.641	0.0276 20.601	0.0281 22.039	0.0290 25.102	0.0299 28.417	0.0317 35.839	0.0343 49.063	0.0368 64.980	0.0392 83.783	0.0415 105.66	0.0438 130.81	0.0460 159.39
0.00550 1/ 182	0.0274 17.453	0.0289 21.606	0.0294 23.115	0.0304 26.327	0.0314 29.804	0.0332 37.588	0.0359 51.458	0.0386 68.151	0.0411 87.873	0.0436 110.82	0.0459 137.19	0.0483 167.17
0.00600 1/ 167	0.0287 18.229	0.0302 22.567	0.0307 24.143	0.0318 27.498	0.0328 31.129	0.0347 39.259	0.0375 53.746	0.0403 71.182	0.0429 91.780	0.0455 115.75	0.0480 143.29	0.0504 174.60
0.00650 1/ 154	0.0298 18.974	0.0315 23.488	0.0320 25.129	0.0331 28.621	0.0341 32.401	0.0361 40.862	0.0391 55.941	0.0419 74.088	0.0447 95.528	0.0473 120.48	0.0499 149.14	0.0525 181.73
0.00700 1/ 143	0.0310 19.690	0.0326 24.375	0.0332 26.077	0.0343 29.701	0.0354 33.624	0.0375 42.405	0.0406 58.053	0.0435 76.885	0.0464 99.134	0.0491 125.02	0.0518 154.77	0.0544 188.59
0.00750 1/ 133	0.0320 20.381	0.0338 25.230	0.0344 26.993	0.0355 30.743	0.0366 34.804	0.0388 43.893	0.0420 60.090	0.0450 79.584	0.0480 102.61	0.0509 129.41	0.0536 160.20	0.0564 195.21
0.00800 1/ 125	0.0331 21.049	0.0349 26.058	0.0355 27.878	0.0367 31.752	0.0378 35.945	0.0401 45.333	0.0434 62.061	0.0465 82.194	0.0496 105.98	0.0525 133.66	0.0554 165.46	0.0582 201.61
0.00850 1/ 118	0.0341 21.697	0.0360 26.860	0.0366 28.736	0.0378 32.729	0.0390 37.052	0.0413 46.728	0.0447 63.971	0.0479 84.723	0.0511 109.24	0.0541 137.77	0.0571 170.55	0.0600 207.81
0.00900 1/ 111	0.0351 22.326	0.0370 27.639	0.0376 29.569	0.0389 33.678	0.0401 38.126	0.0425 48.083	0.0460 65.825	0.0493 87.179	0.0526 112.41	0.0557 141.76	0.0588 175.49	0.0617 213.84
0.00950 1/ 105	0.0361 22.938	0.0380 28.396	0.0387 30.379	0.0400 34.601	0.0412 39.170	0.0437 49.400	0.0472 67.629	0.0507 89.568	0.0540 115.49	0.0572 145.65	0.0604 180.30	0.0634 219.70
0.01000 1/ 100	0.0370 23.534	0.0390 29.134	0.0397 31.168	0.0410 35.499	0.0423 40.188	0.0448 50.683	0.0485 69.386	0.0520 91.895	0.0554 118.49	0.0587 149.43	0.0619 184.99	0.0651 225.41
0.01100 1/ 91	0.0388 24.683	0.0409 30.556	0.0416 32.690	0.0430 37.232	0.0444 42.150	0.0470 53.157	0.0508 72.773	0.0545 96.380	0.0581 124.27	0.0616 156.73	0.0650 194.02	0.0683 236.41
0.01200 1/ 83	0.0405 25.780	0.0427 31.914	0.0435 34.143	0.0449 38.888	0.0463 44.024	0.0491 55.521	0.0531 76.009	0.0570 100.67	0.0607 129.80	0.0643 163.69	0.0679 202.64	0.0713 246.92
0.01300 1/ 77	0.0422 26.833	0.0445 33.217	0.0452 35.538	0.0467 40.476	0.0482 45.821	0.0511 57.788	0.0553 79.112	0.0593 104.78	0.0632 135.10	0.0670 170.38	0.0706 210.92	0.0742 257.00
0.01400 1/ 71	0.0438 27.846	0.0462 34.471	0.0470 36.879	0.0485 42.003	0.0500 47.551	0.0530 59.969	0.0574 82.099	0.0615 108.73	0.0656 140.20	0.0695 176.81	0.0733 218.88	0.0770 266.71

S = 0.0030 to 0.014

n is Manning coefficient
S = 0.0150 to 0.0750

ie hydraulic gradient =
1 in 67 to 1 in 13

Water at normal temperature;
full bore conditions.

nV in ms^{-1}
nQ × 1000 in m^{3}s^{-1}
ie nQ in litres/sec.

Gradient	(Equivalent) Pipe diameters in m											
	0.900	0.975	1.000	1.050	1.100	1.200	1.350	1.500	1.650	1.800	1.950	2.100
0.01500	0.0453	0.0478	0.0486	0.0502	0.0518	0.0549	0.0594	0.0637	0.0679	0.0719	0.0759	0.0797
1/ 67	28.823	35.681	38.173	43.478	49.220	62.074	84.980	112.55	145.12	183.02	226.56	276.07
0.01600	0.0468	0.0494	0.0502	0.0519	0.0535	0.0567	0.0613	0.0658	0.0701	0.0743	0.0784	0.0823
1/ 62	29.768	36.851	39.425	44.904	50.834	64.110	87.767	116.24	149.88	189.02	233.99	285.12
0.01700	0.0482	0.0509	0.0517	0.0535	0.0551	0.0584	0.0632	0.0678	0.0723	0.0766	0.0808	0.0849
1/ 59	30.685	37.986	40.639	46.286	52.399	66.083	90.468	119.82	154.49	194.84	241.19	293.89
0.01800	0.0496	0.0524	0.0532	0.0550	0.0567	0.0601	0.0650	0.0698	0.0743	0.0788	0.0831	0.0873
1/ 56	31.574	39.087	41.817	47.627	53.918	67.999	93.091	123.29	158.97	200.48	248.19	302.42
0.01900	0.0510	0.0538	0.0547	0.0565	0.0583	0.0618	0.0668	0.0717	0.0764	0.0809	0.0854	0.0897
1/ 53	32.439	40.158	42.963	48.933	55.395	69.862	95.642	126.67	163.32	205.98	254.99	310.70
0.02000	0.0523	0.0552	0.0561	0.0580	0.0598	0.0634	0.0686	0.0735	0.0784	0.0830	0.0876	0.0920
1/ 50	33.282	41.201	44.079	50.204	56.834	71.677	98.127	129.96	167.57	211.33	261.61	318.77
0.02200	0.0549	0.0579	0.0589	0.0608	0.0627	0.0665	0.0719	0.0771	0.0822	0.0871	0.0919	0.0965
1/ 45	34.907	43.212	46.230	52.654	59.608	75.176	102.92	136.30	175.75	221.64	274.38	334.33
0.02400	0.0573	0.0605	0.0615	0.0635	0.0655	0.0694	0.0751	0.0806	0.0858	0.0910	0.0960	0.1008
1/ 42	36.459	45.134	48.286	54.995	62.259	78.518	107.49	142.36	183.56	231.50	286.58	349.20
0.02600	0.0596	0.0629	0.0640	0.0661	0.0682	0.0723	0.0782	0.0839	0.0894	0.0947	0.0999	0.1049
1/ 38	37.947	46.977	50.258	57.241	64.801	81.725	111.88	148.18	191.06	240.95	298.28	363.46
0.02800	0.0619	0.0653	0.0664	0.0686	0.0708	0.0750	0.0811	0.0870	0.0927	0.0983	0.1036	0.1089
1/ 36	39.380	48.750	52.155	59.402	67.247	84.810	116.11	153.77	198.27	250.05	309.54	377.18
0.03000	0.0641	0.0676	0.0687	0.0710	0.0732	0.0776	0.0840	0.0901	0.0960	0.1017	0.1073	0.1127
1/ 33	40.762	50.461	53.985	61.487	69.608	87.786	120.18	159.17	205.23	258.82	320.41	390.42
0.03200	0.0662	0.0698	0.0710	0.0733	0.0756	0.0802	0.0867	0.0930	0.0991	0.1050	0.1108	0.1164
1/ 31	42.099	52.116	55.756	63.503	71.890	90.665	124.12	164.39	211.96	267.31	330.92	403.22
0.03400	0.0682	0.0720	0.0732	0.0756	0.0780	0.0826	0.0894	0.0959	0.1022	0.1083	0.1142	0.1200
1/ 29	43.395	53.720	57.472	65.458	74.103	93.456	127.94	169.45	218.48	275.54	341.10	415.63
0.03600	0.0702	0.0740	0.0753	0.0778	0.0802	0.0850	0.0920	0.0987	0.1051	0.1114	0.1175	0.1235
1/ 28	44.653	55.277	59.138	67.355	76.251	96.165	131.65	174.36	224.81	283.53	350.99	427.68
0.03800	0.0721	0.0761	0.0774	0.0799	0.0824	0.0874	0.0945	0.1014	0.1080	0.1145	0.1207	0.1269
1/ 26	45.876	56.792	60.759	69.201	78.341	98.800	135.26	179.14	230.97	291.30	360.61	439.40
0.04000	0.0740	0.0780	0.0794	0.0820	0.0846	0.0896	0.0969	0.1040	0.1108	0.1174	0.1239	0.1302
1/ 25	47.068	58.267	62.337	70.999	80.376	101.37	138.77	183.79	236.98	298.86	369.97	450.81
0.04200	0.0758	0.0800	0.0813	0.0840	0.0867	0.0918	0.0993	0.1066	0.1136	0.1203	0.1269	0.1334
1/ 24	48.230	59.706	63.876	72.752	82.361	103.87	142.20	188.33	242.83	306.24	379.11	461.95
0.04400	0.0776	0.0819	0.0832	0.0860	0.0887	0.0940	0.1017	0.1091	0.1162	0.1232	0.1299	0.1365
1/ 23	49.365	61.111	65.380	74.464	84.299	106.31	145.55	192.76	248.54	313.45	388.03	472.82
0.04600	0.0793	0.0837	0.0851	0.0879	0.0907	0.0961	0.1040	0.1115	0.1188	0.1259	0.1329	0.1396
1/ 22	50.475	62.485	66.849	76.138	86.194	108.70	148.82	197.09	254.13	320.50	396.75	483.44
0.04800	0.0810	0.0855	0.0869	0.0898	0.0926	0.0982	0.1062	0.1139	0.1214	0.1287	0.1357	0.1426
1/ 21	51.560	63.829	68.287	77.775	88.048	111.04	152.02	201.33	259.59	327.39	405.29	493.84
0.05000	0.0827	0.0873	0.0887	0.0917	0.0946	0.1002	0.1084	0.1163	0.1239	0.1313	0.1385	0.1455
1/ 20	52.624	65.145	69.695	79.379	89.863	113.33	155.15	205.48	264.95	334.14	413.64	504.03
0.05500	0.0868	0.0915	0.0931	0.0961	0.0992	0.1051	0.1137	0.1220	0.1300	0.1377	0.1453	0.1526
1/ 18	55.192	68.324	73.097	83.253	94.249	118.86	162.72	215.51	277.88	350.45	433.83	528.63
0.06000	0.0906	0.0956	0.0972	0.1004	0.1036	0.1098	0.1187	0.1274	0.1357	0.1438	0.1517	0.1594
1/ 17	57.646	71.363	76.347	86.955	98.440	124.15	169.96	225.10	290.23	366.03	453.12	552.13
0.06500	0.0943	0.0995	0.1012	0.1045	0.1078	0.1143	0.1236	0.1326	0.1413	0.1497	0.1579	0.1659
1/ 15	60.000	74.277	79.464	90.506	102.46	129.22	176.90	234.29	302.09	380.98	471.63	574.68
0.07000	0.0979	0.1032	0.1050	0.1085	0.1119	0.1186	0.1283	0.1376	0.1466	0.1554	0.1639	0.1722
1/ 14	62.265	77.080	82.464	93.923	106.33	134.10	183.58	243.13	313.49	395.36	489.43	596.37
0.07500	0.1013	0.1069	0.1087	0.1123	0.1158	0.1227	0.1328	0.1424	0.1518	0.1608	0.1696	0.1782
1/ 13	64.451	79.786	85.358	97.219	110.06	138.80	190.02	251.67	324.49	409.24	506.61	617.30

D3

n is Manning coefficient
S = 0.00010 to 0.00048

Water at normal temperature;
full bore conditions.

ie hydraulic gradient =
1 in 10000 to 1 in 2083

nV in ms^{-1}
nQ in m^{3}s^{-1}

(Equivalent) Pipe diameters in m

Gradient	2.400	2.700	3.000	3.400	4.000	5.000	6.400	8.000	10.00	12.60	16.00	20.00
0.00010	0.0071	0.0077	0.0083	0.0090	0.0100	0.0116	0.0137	0.0159	0.0184	0.0215	0.0252	0.0292
1/10000	0.0322	0.0441	0.0583	0.0815	0.1257	0.2278	0.4401	0.7979	1.4467	2.6794	5.0664	9.1861
0.00011	0.0075	0.0081	0.0087	0.0094	0.0105	0.0122	0.0143	0.0166	0.0193	0.0225	0.0264	0.0307
1/ 9091	0.0338	0.0462	0.0612	0.0854	0.1318	0.2390	0.4616	0.8369	1.5173	2.8102	5.3137	9.6344
0.00012	0.0078	0.0084	0.0090	0.0098	0.0110	0.0127	0.0150	0.0174	0.0202	0.0235	0.0276	0.0320
1/ 8333	0.0353	0.0483	0.0639	0.0892	0.1377	0.2496	0.4821	0.8741	1.5848	2.9351	5.5500	10.063
0.00013	0.0081	0.0088	0.0094	0.0102	0.0114	0.0132	0.0156	0.0181	0.0210	0.0245	0.0287	0.0333
1/ 7692	0.0367	0.0502	0.0665	0.0929	0.1433	0.2598	0.5018	0.9098	1.6495	3.0550	5.7766	10.474
0.00014	0.0084	0.0091	0.0098	0.0106	0.0118	0.0137	0.0162	0.0188	0.0218	0.0254	0.0298	0.0346
1/ 7143	0.0381	0.0521	0.0690	0.0964	0.1487	0.2696	0.5207	0.9441	1.7118	3.1703	5.9947	10.869
0.00015	0.0087	0.0094	0.0101	0.0110	0.0122	0.0142	0.0168	0.0194	0.0226	0.0263	0.0309	0.0358
1/ 6667	0.0394	0.0540	0.0715	0.0998	0.1539	0.2790	0.5390	0.9772	1.7719	3.2816	6.2051	11.251
0.00016	0.0090	0.0097	0.0104	0.0114	0.0126	0.0147	0.0173	0.0201	0.0233	0.0272	0.0319	0.0370
1/ 6250	0.0407	0.0557	0.0738	0.1031	0.1590	0.2882	0.5567	1.0093	1.8300	3.3892	6.4086	11.620
0.00017	0.0093	0.0100	0.0108	0.0117	0.0130	0.0151	0.0178	0.0207	0.0240	0.0280	0.0329	0.0381
1/ 5882	0.0420	0.0574	0.0761	0.1062	0.1638	0.2971	0.5738	1.0404	1.8863	3.4935	6.6058	11.977
0.00018	0.0095	0.0103	0.0111	0.0120	0.0134	0.0156	0.0184	0.0213	0.0247	0.0288	0.0338	0.0392
1/ 5556	0.0432	0.0591	0.0783	0.1093	0.1686	0.3057	0.5904	1.0705	1.9410	3.5948	6.7973	12.324
0.00019	0.0098	0.0106	0.0114	0.0124	0.0138	0.0160	0.0189	0.0219	0.0254	0.0296	0.0347	0.0403
1/ 5263	0.0444	0.0607	0.0804	0.1123	0.1732	0.3141	0.6066	1.0998	1.9942	3.6933	6.9836	12.662
0.00020	0.0101	0.0109	0.0117	0.0127	0.0141	0.0164	0.0193	0.0224	0.0261	0.0304	0.0356	0.0414
1/ 5000	0.0455	0.0623	0.0825	0.1152	0.1777	0.3222	0.6224	1.1284	2.0460	3.7892	7.1650	12.991
0.00022	0.0106	0.0114	0.0122	0.0133	0.0148	0.0172	0.0203	0.0235	0.0273	0.0319	0.0374	0.0434
1/ 4545	0.0477	0.0653	0.0865	0.1208	0.1864	0.3379	0.6527	1.1835	2.1458	3.9742	7.5147	13.625
0.00024	0.0110	0.0119	0.0128	0.0139	0.0155	0.0180	0.0212	0.0246	0.0285	0.0333	0.0390	0.0453
1/ 4167	0.0499	0.0683	0.0904	0.1262	0.1947	0.3530	0.6818	1.2361	2.2412	4.1509	7.8489	14.231
0.00026	0.0115	0.0124	0.0133	0.0145	0.0161	0.0187	0.0221	0.0256	0.0297	0.0346	0.0406	0.0471
1/ 3846	0.0519	0.0710	0.0941	0.1314	0.2026	0.3674	0.7096	1.2866	2.3328	4.3204	8.1694	14.812
0.00028	0.0119	0.0129	0.0138	0.0150	0.0167	0.0194	0.0229	0.0266	0.0308	0.0360	0.0422	0.0489
1/ 3571	0.0539	0.0737	0.0976	0.1363	0.2103	0.3813	0.7364	1.3352	2.4208	4.4835	8.4778	15.371
0.00030	0.0123	0.0133	0.0143	0.0155	0.0173	0.0201	0.0237	0.0275	0.0319	0.0372	0.0436	0.0506
1/ 3333	0.0557	0.0763	0.1011	0.1411	0.2177	0.3946	0.7622	1.3820	2.5058	4.6409	8.7753	15.911
0.00032	0.0127	0.0138	0.0148	0.0161	0.0179	0.0208	0.0245	0.0284	0.0330	0.0384	0.0451	0.0523
1/ 3125	0.0576	0.0788	0.1044	0.1457	0.2248	0.4076	0.7872	1.4274	2.5880	4.7931	9.0631	16.433
0.00034	0.0131	0.0142	0.0152	0.0165	0.0184	0.0214	0.0252	0.0293	0.0340	0.0396	0.0465	0.0539
1/ 2941	0.0593	0.0812	0.1076	0.1502	0.2317	0.4201	0.8115	1.4713	2.6676	4.9406	9.3420	16.938
0.00036	0.0135	0.0146	0.0157	0.0170	0.0190	0.0220	0.0260	0.0301	0.0349	0.0408	0.0478	0.0555
1/ 2778	0.0611	0.0836	0.1107	0.1546	0.2384	0.4323	0.8350	1.5139	2.7449	5.0838	9.6129	17.429
0.00038	0.0139	0.0150	0.0161	0.0175	0.0195	0.0226	0.0267	0.0309	0.0359	0.0419	0.0491	0.0570
1/ 2632	0.0627	0.0859	0.1137	0.1588	0.2450	0.4441	0.8579	1.5554	2.8202	5.2231	9.8763	17.907
0.00040	0.0142	0.0154	0.0165	0.0179	0.0200	0.0232	0.0274	0.0317	0.0368	0.0430	0.0504	0.0585
1/ 2500	0.0644	0.0881	0.1167	0.1629	0.2513	0.4557	0.8802	1.5958	2.8934	5.3588	10.133	18.372
0.00042	0.0146	0.0158	0.0169	0.0184	0.0205	0.0238	0.0280	0.0325	0.0378	0.0440	0.0516	0.0599
1/ 2381	0.0660	0.0903	0.1196	0.1670	0.2575	0.4669	0.9019	1.6352	2.9649	5.4911	10.383	18.826
0.00044	0.0149	0.0161	0.0173	0.0188	0.0210	0.0243	0.0287	0.0333	0.0386	0.0451	0.0529	0.0613
1/ 2273	0.0675	0.0924	0.1224	0.1709	0.2636	0.4779	0.9231	1.6737	3.0347	5.6204	10.627	19.269
0.00046	0.0153	0.0165	0.0177	0.0192	0.0214	0.0249	0.0293	0.0340	0.0395	0.0461	0.0540	0.0627
1/ 2174	0.0690	0.0945	0.1251	0.1747	0.2695	0.4887	0.9439	1.7113	3.1029	5.7467	10.866	19.702
0.00048	0.0156	0.0169	0.0181	0.0197	0.0219	0.0254	0.0300	0.0348	0.0404	0.0471	0.0552	0.0641
1/ 2083	0.0705	0.0965	0.1278	0.1785	0.2753	0.4992	0.9642	1.7481	3.1696	5.8703	11.100	20.126

S = 0.00010 to 0.00048

n is Manning coefficient S = 0.00050 to 0.00280

Water at normal temperature; full bore conditions.

ie hydraulic gradient = 1 in 2000 to 1 in 357

nV in ms^{-1}
nQ in m^{3}s^{-1}

Gradient	(Equivalent) Pipe diameters in m 2.400	2.700	3.000	3.400	4.000	5.000	6.400	8.000	10.00	12.60	16.00	20.00
0.00050	0.0159	0.0172	0.0185	0.0201	0.0224	0.0259	0.0306	0.0355	0.0412	0.0480	0.0563	0.0654
1/ 2000	0.0720	0.0985	0.1305	0.1822	0.2810	0.5095	0.9840	1.7842	3.2350	5.9913	11.329	20.541
0.00055	0.0167	0.0180	0.0194	0.0210	0.0235	0.0272	0.0321	0.0372	0.0432	0.0504	0.0591	0.0686
1/ 1818	0.0755	0.1033	0.1368	0.1911	0.2947	0.5343	1.0321	1.8713	3.3928	6.2837	11.882	21.543
0.00060	0.0174	0.0188	0.0202	0.0220	0.0245	0.0284	0.0335	0.0389	0.0451	0.0526	0.0617	0.0716
1/ 1667	0.0788	0.1079	0.1429	0.1996	0.3078	0.5581	1.0780	1.9545	3.5437	6.5632	12.410	22.501
0.00065	0.0181	0.0196	0.0210	0.0229	0.0255	0.0296	0.0349	0.0405	0.0470	0.0548	0.0642	0.0745
1/ 1538	0.0820	0.1123	0.1488	0.2077	0.3204	0.5809	1.1220	2.0343	3.6884	6.8312	12.917	23.420
0.00070	0.0188	0.0204	0.0218	0.0237	0.0265	0.0307	0.0362	0.0420	0.0487	0.0569	0.0667	0.0774
1/ 1429	0.0851	0.1166	0.1544	0.2155	0.3325	0.6028	1.1643	2.1111	3.8276	7.0890	13.405	24.304
0.00075	0.0195	0.0211	0.0226	0.0246	0.0274	0.0318	0.0375	0.0435	0.0504	0.0588	0.0690	0.0801
1/ 1333	0.0881	0.1207	0.1598	0.2231	0.3441	0.6240	1.2052	2.1852	3.9620	7.3378	13.875	25.157
0.00080	0.0201	0.0218	0.0233	0.0254	0.0283	0.0328	0.0387	0.0449	0.0521	0.0608	0.0713	0.0827
1/ 1250	0.0910	0.1246	0.1650	0.2304	0.3554	0.6444	1.2447	2.2568	4.0919	7.5785	14.330	25.982
0.00085	0.0207	0.0224	0.0241	0.0262	0.0292	0.0338	0.0399	0.0463	0.0537	0.0626	0.0735	0.0852
1/ 1176	0.0938	0.1284	0.1701	0.2375	0.3664	0.6643	1.2830	2.3263	4.2179	7.8117	14.771	26.782
0.00090	0.0213	0.0231	0.0248	0.0269	0.0300	0.0348	0.0410	0.0476	0.0553	0.0645	0.0756	0.0877
1/ 1111	0.0965	0.1322	0.1750	0.2444	0.3770	0.6835	1.3202	2.3937	4.3401	8.0382	15.199	27.558
0.00095	0.0219	0.0237	0.0254	0.0277	0.0308	0.0358	0.0422	0.0489	0.0568	0.0662	0.0777	0.0901
1/ 1053	0.0992	0.1358	0.1798	0.2511	0.3873	0.7023	1.3564	2.4593	4.4591	8.2585	15.616	28.313
0.00100	0.0225	0.0243	0.0261	0.0284	0.0316	0.0367	0.0433	0.0502	0.0582	0.0680	0.0797	0.0925
1/ 1000	0.1018	0.1393	0.1845	0.2576	0.3974	0.7205	1.3916	2.5232	4.5749	8.4730	16.021	29.049
0.00110	0.0236	0.0255	0.0274	0.0298	0.0332	0.0385	0.0454	0.0526	0.0611	0.0713	0.0836	0.0970
1/ 909	0.1067	0.1461	0.1935	0.2702	0.4168	0.7557	1.4596	2.6464	4.7982	8.8866	16.803	30.467
0.00120	0.0246	0.0267	0.0286	0.0311	0.0346	0.0402	0.0474	0.0550	0.0638	0.0744	0.0873	0.1013
1/ 833	0.1115	0.1526	0.2021	0.2822	0.4353	0.7893	1.5245	2.7641	5.0116	9.2817	17.551	31.821
0.00130	0.0256	0.0277	0.0298	0.0324	0.0361	0.0418	0.0493	0.0572	0.0664	0.0775	0.0909	0.1054
1/ 769	0.1160	0.1589	0.2104	0.2937	0.4531	0.8215	1.5867	2.8769	5.2162	9.6607	18.267	33.121
0.00140	0.0266	0.0288	0.0309	0.0336	0.0374	0.0434	0.0512	0.0594	0.0689	0.0804	0.0943	0.1094
1/ 714	0.1204	0.1648	0.2183	0.3048	0.4702	0.8525	1.6466	2.9855	5.4131	10.025	18.957	34.371
0.00150	0.0276	0.0298	0.0320	0.0348	0.0387	0.0449	0.0530	0.0615	0.0713	0.0832	0.0976	0.1132
1/ 667	0.1246	0.1706	0.2260	0.3155	0.4867	0.8824	1.7044	3.0903	5.6031	10.377	19.622	35.577
0.00160	0.0285	0.0308	0.0330	0.0359	0.0400	0.0464	0.0547	0.0635	0.0737	0.0860	0.1008	0.1170
1/ 625	0.1287	0.1762	0.2334	0.3259	0.5027	0.9114	1.7603	3.1917	5.7869	10.718	20.266	36.744
0.00170	0.0293	0.0317	0.0340	0.0370	0.0412	0.0478	0.0564	0.0655	0.0759	0.0886	0.1039	0.1206
1/ 588	0.1327	0.1817	0.2406	0.3359	0.5181	0.9394	1.8145	3.2899	5.9650	11.047	20.889	37.875
0.00180	0.0302	0.0326	0.0350	0.0381	0.0424	0.0492	0.0580	0.0673	0.0782	0.0912	0.1069	0.1241
1/ 556	0.1365	0.1869	0.2476	0.3456	0.5331	0.9667	1.8671	3.3853	6.1379	11.368	21.495	38.973
0.00190	0.0310	0.0335	0.0360	0.0391	0.0436	0.0506	0.0596	0.0692	0.0803	0.0937	0.1098	0.1275
1/ 526	0.1403	0.1920	0.2543	0.3551	0.5478	0.9931	1.9183	3.4780	6.3061	11.679	22.084	40.041
0.00200	0.0318	0.0344	0.0369	0.0401	0.0447	0.0519	0.0612	0.0710	0.0824	0.0961	0.1127	0.1308
1/ 500	0.1439	0.1970	0.2609	0.3643	0.5620	1.0189	1.9681	3.5684	6.4699	11.983	22.658	41.081
0.00220	0.0334	0.0361	0.0387	0.0421	0.0469	0.0544	0.0642	0.0745	0.0864	0.1008	0.1182	0.1371
1/ 455	0.1509	0.2066	0.2737	0.3821	0.5894	1.0687	2.0641	3.7425	6.7857	12.567	23.764	43.086
0.00240	0.0349	0.0377	0.0404	0.0440	0.0490	0.0568	0.0670	0.0778	0.0902	0.1053	0.1234	0.1432
1/ 417	0.1577	0.2158	0.2859	0.3991	0.6156	1.1162	2.1559	3.9090	7.0874	13.126	24.820	45.002
0.00260	0.0363	0.0392	0.0421	0.0458	0.0510	0.0592	0.0698	0.0809	0.0939	0.1096	0.1285	0.1491
1/ 385	0.1641	0.2246	0.2975	0.4154	0.6408	1.1618	2.2440	4.0686	7.3768	13.662	25.834	46.840
0.00280	0.0376	0.0407	0.0437	0.0475	0.0529	0.0614	0.0724	0.0840	0.0975	0.1137	0.1333	0.1547
1/ 357	0.1703	0.2331	0.3088	0.4311	0.6649	1.2056	2.3287	4.2222	7.6553	14.178	26.809	48.608

S = 0.00050 to 0.00280

n is Manning coefficient
S = 0.0030 to 0.014

ie hydraulic gradient =
1 in 333 to 1 in 71

Water at normal temperature;
full bore conditions.

nV in ms^{-1}
nQ in m^3s^{-1}

(Equivalent) Pipe diameters in m

Gradient	2.400	2.700	3.000	3.400	4.000	5.000	6.400	8.000	10.00	12.60	16.00	20.00
0.00300 1/ 333	0.0390 0.1763	0.0421 0.2413	0.0452 0.3196	0.0491 0.4462	0.0548 0.6883	0.0636 1.2479	0.0749 2.4104	0.0869 4.3704	0.1009 7.9240	0.1177 14.676	0.1380 27.750	0.1602 50.314
0.00320 1/ 313	0.0402 0.1820	0.0435 0.2492	0.0467 0.3301	0.0508 0.4609	0.0566 0.7109	0.0656 1.2889	0.0774 2.4895	0.0898 4.5137	0.1042 8.1838	0.1216 15.157	0.1425 28.660	0.1654 51.964
0.00340 1/ 294	0.0415 0.1877	0.0449 0.2569	0.0481 0.3402	0.0523 0.4750	0.0583 0.7327	0.0677 1.3285	0.0798 2.5661	0.0926 4.6526	0.1074 8.4357	0.1253 15.623	0.1469 29.542	0.1705 53.563
0.00360 1/ 278	0.0427 0.1931	0.0462 0.2643	0.0495 0.3501	0.0538 0.4888	0.0600 0.7540	0.0696 1.3671	0.0821 2.6405	0.0952 4.7875	0.1105 8.6803	0.1289 16.076	0.1512 30.399	0.1754 55.116
0.00380 1/ 263	0.0439 0.1984	0.0474 0.2716	0.0509 0.3597	0.0553 0.5022	0.0616 0.7746	0.0715 1.4045	0.0843 2.7128	0.0979 4.9187	0.1135 8.9181	0.1325 16.517	0.1553 31.232	0.1802 56.627
0.00400 1/ 250	0.0450 0.2035	0.0487 0.2786	0.0522 0.3690	0.0568 0.5153	0.0632 0.7948	0.0734 1.4410	0.0865 2.7833	0.1004 5.0464	0.1165 9.1498	0.1359 16.946	0.1594 32.043	0.1849 58.098
0.00420 1/ 238	0.0461 0.2086	0.0499 0.2855	0.0535 0.3781	0.0582 0.5280	0.0648 0.8144	0.0752 1.4766	0.0887 2.8520	0.1029 5.1711	0.1194 9.3758	0.1393 17.364	0.1633 32.834	0.1895 59.532
0.00440 1/ 227	0.0472 0.2135	0.0510 0.2922	0.0548 0.3870	0.0595 0.5404	0.0663 0.8336	0.0770 1.5113	0.0907 2.9191	0.1053 5.2928	0.1222 9.5964	0.1425 17.773	0.1671 33.607	0.1940 60.933
0.00460 1/ 217	0.0482 0.2183	0.0522 0.2988	0.0560 0.3957	0.0609 0.5526	0.0678 0.8523	0.0787 1.5453	0.0928 2.9848	0.1077 5.4117	0.1249 9.8121	0.1457 18.173	0.1709 34.362	0.1983 62.303
0.00480 1/ 208	0.0493 0.2230	0.0533 0.3052	0.0572 0.4043	0.0622 0.5644	0.0693 0.8706	0.0804 1.5785	0.0948 3.0489	0.1100 5.5281	0.1276 10.023	0.1489 18.563	0.1746 35.101	0.2026 63.643
0.00500 1/ 200	0.0503 0.2276	0.0544 0.3115	0.0584 0.4126	0.0634 0.5761	0.0707 0.8886	0.0821 1.6111	0.0967 3.1118	0.1122 5.6421	0.1303 10.230	0.1519 18.946	0.1782 35.825	0.2068 64.955
0.00550 1/ 182	0.0528 0.2387	0.0571 0.3267	0.0612 0.4327	0.0665 0.6042	0.0742 0.9319	0.0861 1.6897	0.1015 3.2637	0.1177 5.9175	0.1366 10.729	0.1594 19.871	0.1869 37.574	0.2169 68.126
0.00600 1/ 167	0.0551 0.2493	0.0596 0.3413	0.0639 0.4520	0.0695 0.6311	0.0775 0.9734	0.0899 1.7649	0.1060 3.4088	0.1230 6.1806	0.1427 11.206	0.1664 20.755	0.1952 39.244	0.2265 71.155
0.00650 1/ 154	0.0574 0.2595	0.0620 0.3552	0.0666 0.4704	0.0723 0.6568	0.0806 1.0131	0.0936 1.8369	0.1103 3.5480	0.1280 6.4330	0.1485 11.664	0.1732 21.602	0.2032 40.847	0.2357 74.060
0.00700 1/ 143	0.0595 0.2693	0.0644 0.3686	0.0691 0.4882	0.0751 0.6816	0.0837 1.0514	0.0971 1.9063	0.1145 3.6820	0.1328 6.6758	0.1541 12.104	0.1798 22.417	0.2108 42.389	0.2446 76.856
0.00750 1/ 133	0.0616 0.2787	0.0666 0.3815	0.0715 0.5053	0.0777 0.7055	0.0866 1.0883	0.1005 1.9732	0.1185 3.8112	0.1375 6.9101	0.1595 12.529	0.1861 23.204	0.2182 43.877	0.2532 79.554
0.00800 1/ 125	0.0636 0.2878	0.0688 0.3941	0.0738 0.5219	0.0803 0.7287	0.0894 1.1240	0.1038 2.0379	0.1224 3.9362	0.1420 7.1368	0.1648 12.940	0.1922 23.965	0.2254 45.316	0.2615 82.163
0.00850 1/ 118	0.0656 0.2967	0.0709 0.4062	0.0761 0.5380	0.0827 0.7511	0.0922 1.1586	0.1070 2.1006	0.1261 4.0573	0.1464 7.3564	0.1698 13.338	0.1981 24.703	0.2323 46.710	0.2696 84.691
0.00900 1/ 111	0.0675 0.3053	0.0730 0.4180	0.0783 0.5536	0.0851 0.7729	0.0949 1.1921	0.1101 2.1615	0.1298 4.1749	0.1506 7.5697	0.1747 13.725	0.2039 25.419	0.2391 48.064	0.2774 87.147
0.00950 1/ 105	0.0693 0.3137	0.0750 0.4294	0.0805 0.5687	0.0875 0.7941	0.0975 1.2248	0.1131 2.2207	0.1333 4.2893	0.1547 7.7771	0.1795 14.101	0.2094 26.116	0.2456 49.382	0.2850 89.535
0.01000 1/ 100	0.0711 0.3218	0.0769 0.4406	0.0825 0.5835	0.0897 0.8147	0.1000 1.2566	0.1160 2.2784	0.1368 4.4008	0.1587 7.9791	0.1842 14.467	0.2149 26.794	0.2520 50.664	0.2924 91.861
0.01100 1/ 91	0.0746 0.3375	0.0807 0.4621	0.0866 0.6120	0.0941 0.8545	0.1049 1.3180	0.1217 2.3896	0.1435 4.6156	0.1665 8.3686	0.1932 15.173	0.2254 28.102	0.2643 53.137	0.3067 96.344
0.01200 1/ 83	0.0779 0.3525	0.0843 0.4826	0.0904 0.6392	0.0983 0.8924	0.1095 1.3766	0.1271 2.4959	0.1499 4.8208	0.1739 8.7407	0.2018 15.848	0.2354 29.351	0.2760 55.500	0.3203 100.63
0.01300 1/ 77	0.0811 0.3669	0.0877 0.5023	0.0941 0.6653	0.1023 0.9289	0.1140 1.4328	0.1323 2.5978	0.1560 5.0177	0.1810 9.0976	0.2100 16.495	0.2450 30.550	0.2873 57.766	0.3334 104.74
0.01400 1/ 71	0.0842 0.3808	0.0910 0.5213	0.0977 0.6904	0.1062 0.9640	0.1183 1.4869	0.1373 2.6959	0.1619 5.2071	0.1878 9.4410	0.2180 17.118	0.2543 31.703	0.2982 59.947	0.3460 108.69

S = 0.0030 to 0.014

n is Manning coefficient
S = 0.0150 to 0.0750

Water at normal temperature;
full bore conditions.

ie hydraulic gradient =
1 in 67 to 1 in 13

nV in ms^{-1}
nQ in m^3s^{-1}

| Gradient | (Equivalent) Pipe diameters in m | | | | | | | | | | | |
	2.400	2.700	3.000	3.400	4.000	5.000	6.400	8.000	10.00	12.60	16.00	20.00
0.01500 1/ 67	0.0871 0.3941	0.0942 0.5396	0.1011 0.7146	0.1099 0.9978	0.1225 1.5391	0.1421 2.7905	0.1675 5.3898	0.1944 9.7724	0.2256 17.719	0.2632 32.816	0.3086 62.051	0.3581 112.51
0.01600 1/ 62	0.0900 0.4071	0.0973 0.5573	0.1044 0.7381	0.1135 1.0305	0.1265 1.5895	0.1468 2.8820	0.1730 5.5666	0.2008 10.093	0.2330 18.300	0.2718 33.892	0.3187 64.086	0.3699 116.20
0.01700 1/ 59	0.0928 0.4196	0.1003 0.5744	0.1076 0.7608	0.1170 1.0622	0.1304 1.6385	0.1513 2.9707	0.1784 5.7379	0.2070 10.404	0.2402 18.863	0.2802 34.935	0.3285 66.058	0.3812 119.77
0.01800 1/ 56	0.0954 0.4318	0.1032 0.5911	0.1107 0.7828	0.1204 1.0930	0.1342 1.6860	0.1557 3.0568	0.1835 5.9043	0.2130 10.705	0.2471 19.410	0.2883 35.948	0.3381 67.973	0.3923 123.24
0.01900 1/ 53	0.0981 0.4436	0.1061 0.6073	0.1138 0.8043	0.1237 1.1230	0.1378 1.7322	0.1599 3.1406	0.1886 6.0661	0.2188 10.998	0.2539 19.942	0.2962 36.933	0.3473 69.836	0.4030 126.62
0.02000 1/ 50	0.1006 0.4551	0.1088 0.6231	0.1167 0.8252	0.1269 1.1521	0.1414 1.7772	0.1641 3.2222	0.1935 6.2236	0.2245 11.284	0.2605 20.460	0.3039 37.892	0.3564 71.650	0.4135 129.91
0.02200 1/ 45	0.1055 0.4773	0.1141 0.6535	0.1224 0.8655	0.1331 1.2084	0.1483 1.8639	0.1721 3.3795	0.2029 6.5274	0.2354 11.835	0.2732 21.458	0.3187 39.742	0.3738 75.147	0.4337 136.25
0.02400 1/ 42	0.1102 0.4986	0.1192 0.6825	0.1279 0.9040	0.1390 1.2621	0.1549 1.9468	0.1798 3.5297	0.2119 6.8177	0.2459 12.361	0.2854 22.412	0.3329 41.509	0.3904 78.489	0.4530 142.31
0.02600 1/ 38	0.1147 0.5189	0.1241 0.7104	0.1331 0.9409	0.1447 1.3137	0.1612 2.0263	0.1871 3.6739	0.2206 7.0960	0.2560 12.866	0.2970 23.328	0.3465 43.204	0.4063 81.694	0.4715 148.12
0.02800 1/ 36	0.1190 0.5385	0.1288 0.7372	0.1381 0.9764	0.1501 1.3632	0.1673 2.1028	0.1942 3.8125	0.2289 7.3639	0.2656 13.352	0.3082 24.208	0.3596 44.835	0.4216 84.778	0.4893 153.71
0.03000 1/ 33	0.1232 0.5574	0.1333 0.7631	0.1430 1.0106	0.1554 1.4111	0.1732 2.1766	0.2010 3.9464	0.2369 7.6224	0.2749 13.820	0.3190 25.058	0.3722 46.409	0.4364 87.753	0.5065 159.11
0.03200 1/ 31	0.1273 0.5757	0.1377 0.7881	0.1477 1.0438	0.1605 1.4574	0.1789 2.2479	0.2076 4.0758	0.2447 7.8723	0.2840 14.274	0.3295 25.880	0.3844 47.931	0.4508 90.631	0.5231 164.33
0.03400 1/ 29	0.1312 0.5934	0.1419 0.8124	0.1522 1.0759	0.1655 1.5022	0.1844 2.3171	0.2140 4.2012	0.2522 8.1146	0.2927 14.713	0.3397 26.676	0.3962 49.406	0.4646 93.420	0.5392 169.38
0.03600 1/ 28	0.1350 0.6106	0.1460 0.8359	0.1566 1.1071	0.1703 1.5458	0.1897 2.3843	0.2202 4.3230	0.2596 8.3499	0.3012 15.139	0.3495 27.449	0.4077 50.838	0.4781 96.129	0.5548 174.29
0.03800 1/ 26	0.1387 0.6273	0.1500 0.8588	0.1609 1.1374	0.1749 1.5881	0.1949 2.4496	0.2262 4.4415	0.2667 8.5787	0.3094 15.554	0.3591 28.202	0.4189 52.231	0.4912 98.763	0.5700 179.07
0.04000 1/ 25	0.1423 0.6436	0.1539 0.8811	0.1651 1.1670	0.1795 1.6294	0.2000 2.5133	0.2321 4.5569	0.2736 8.8016	0.3175 15.958	0.3684 28.934	0.4298 53.588	0.5040 101.33	0.5848 183.72
0.04200 1/ 24	0.1458 0.6595	0.1577 0.9029	0.1692 1.1958	0.1839 1.6696	0.2049 2.5753	0.2378 4.6694	0.2804 9.0189	0.3253 16.352	0.3775 29.649	0.4404 54.911	0.5164 103.83	0.5992 188.26
0.04400 1/ 23	0.1492 0.6751	0.1614 0.9242	0.1732 1.2240	0.1882 1.7089	0.2098 2.6359	0.2434 4.7793	0.2869 9.2311	0.3330 16.737	0.3864 30.347	0.4507 56.204	0.5286 106.27	0.6133 192.69
0.04600 1/ 22	0.1526 0.6902	0.1650 0.9449	0.1770 1.2515	0.1925 1.7473	0.2145 2.6952	0.2489 4.8867	0.2934 9.4386	0.3405 17.113	0.3951 31.029	0.4609 57.467	0.5404 108.66	0.6271 197.02
0.04800 1/ 21	0.1559 0.7051	0.1686 0.9653	0.1809 1.2784	0.1966 1.7849	0.2191 2.7532	0.2542 4.9918	0.2997 9.6416	0.3478 17.481	0.4036 31.696	0.4708 58.703	0.5521 111.00	0.6406 201.26
0.05000 1/ 20	0.1591 0.7196	0.1721 0.9852	0.1846 1.3047	0.2006 1.8217	0.2236 2.8099	0.2595 5.0947	0.3059 9.8404	0.3550 17.842	0.4119 32.350	0.4805 59.913	0.5635 113.29	0.6538 205.41
0.05500 1/ 18	0.1668 0.7547	0.1805 1.0332	0.1936 1.3684	0.2104 1.9106	0.2345 2.9471	0.2721 5.3434	0.3208 10.321	0.3723 18.713	0.4320 33.928	0.5040 62.837	0.5910 118.82	0.6857 215.43
0.06000 1/ 17	0.1743 0.7883	0.1885 1.0792	0.2022 1.4293	0.2198 1.9956	0.2449 3.0781	0.2842 5.5810	0.3351 10.780	0.3888 19.545	0.4512 35.437	0.5264 65.632	0.6172 124.10	0.7162 225.01
0.06500 1/ 15	0.1814 0.8205	0.1962 1.1232	0.2105 1.4876	0.2288 2.0771	0.2550 3.2038	0.2958 5.8089	0.3488 11.220	0.4047 20.343	0.4696 36.884	0.5479 68.312	0.6424 129.17	0.7455 234.20
0.07000 1/ 14	0.1882 0.8515	0.2036 1.1657	0.2184 1.5438	0.2374 2.1555	0.2646 3.3247	0.3070 6.0282	0.3619 11.643	0.4200 21.111	0.4874 38.276	0.5685 70.890	0.6667 134.05	0.7736 243.04
0.07500 1/ 13	0.1948 0.8813	0.2107 1.2066	0.2261 1.5980	0.2457 2.2311	0.2739 3.4414	0.3178 6.2397	0.3746 12.052	0.4347 21.852	0.5045 39.620	0.5885 73.378	0.6901 138.75	0.8008 251.57

S = 0.0150 to 0.0750